T0213729

Communications
in Computer and Information Science   640

Commenced Publication in 2007
Founding and Former Series Editors:
Alfredo Cuzzocrea, Dominik Ślęzak, and Xiaokang Yang

More information about this series at http://www.springer.com/series/7899

Jerzy Mikulski (Ed.)

Challenge of Transport Telematics

16th International Conference
on Transport Systems Telematics, TST 2016
Katowice-Ustroń, Poland, March 16–19, 2016
Selected Papers

 Springer

Editor
Jerzy Mikulski
Polish Association of Transport Telematics
Katowice
Poland

ISSN 1865-0929   ISSN 1865-0937   (electronic)
Communications in Computer and Information Science
ISBN 978-3-319-49645-0   ISBN 978-3-319-49646-7   (eBook)
DOI 10.1007/978-3-319-49646-7

Library of Congress Control Number: 2016957482

Printed on acid-free paper

This Springer imprint is published by Springer Nature
The registered company is Springer International Publishing AG
The registered company address is: Gewerbestrasse 11, 6330 Cham, Switzerland

Preface

I am pleased to present another volume of proceedings related to the Transport System Telematics (TST) Conference published in the *Communications in Computer and Information Science* series. It will help familiarize the reader with practical aspects of the development of technologies, services, and applications required for the appropriate functioning of modern transportation systems.

For many years, we have been gathering at the TST conference as a group of people interested in the development of transport telematics. The conference is the oldest event of its kind in Poland, and this year's conference was the 16th edition. The conference was inaugurated in 2001 by the Polish Association of Transport Telematics and it has been organized with the association's participation ever since. The conference is organized under the patronage of Polish self-governmental and governmental authorities, the Transport Committee of the Polish Academy of Sciences, as well as many Polish and foreign universities and companies within the field of transport telematics. The TST Conference allows its participants to exchange their experiences and familiarize themselves with the latest intelligent transportation systems (ITS) technologies and directions of development. The conference refers to the ITS development program within the scope of infrastructure of all kinds of transportation. In order to take further actions related to this kind of infrastructure, it is necessary to coordinate activities implemented by the whole community in the field in order to create a new quality of transport services.

The book presents and the knowledge gathered from the field of ITS, the specific solutions applied in the field, and their influence on improving the efficiency of transport systems. The publication is a collection of the scientific achievements presented by authors during the conference sessions.

This year the papers were accepted after detailed reviews made by three independent reviewers in a double-blind way. In addition, a Scientific Editorial Special Committee was appointed to work on the final qualifying items. The acceptance rate was circa 38%; the book consists of 37 articles and five short scientific reports.

The book presents a wide range of current issues related to the development of transport telematics. Readers interested in the current issues related to ITS will find here a vast range of material presenting the different views and research of the book's authors.

I hope that by reading this book you will gain knowledge of the various examples of telematics technologies and services that have been successfully implemented in the existing transport infrastructure. As the same time, I am sure that the series dedicated to the best ITS practices will encourage you to take part in future editions of the conference.

March 2016   Jerzy Mikulski

Organization

Organizers

Polish Association of Transport Telematics

Co-organizers

Transport Committee of the Polish Academy of Sciences
University of Economics in Katowice
Katowice School of Technology

Co-operating Universities

Gdynia Maritime University
Maritime University of Szczecin
University of Bielsko-Biała
Silesian University of Technology
Warsaw University of Technology
Wrocław University of Technology
University of Technology and Humanities in Radom
WSB Schools of Banking in Wrocław
Silesian School of Management in Katowice

Scientific Program Committee

J. Mikulski	Polish Association of Transport Telematics, Poland
A. Bujak	WSB Schools of Banking in Wroclaw, Poland
M. Bukljaš-Skočibušić	University of Zagreb, Croatia
W. Choromański	Warsaw University of Technology, Poland
T. Čorejová	University of Zilina, Republic of Slovakia
M. Dado	University of Zilina, Republic of Slovakia
A. Dewalska-Opitek	University of Economics in Katowice, Poland
J. Dyduch	Transport Committee, Polish Academy of Sciences, Poland
M. Franeková	University of Zilina, Republic of Slovakia
J. Gnap	University of Zilina, Republic of Slovakia
S. Iwan	Maritime University of Szczecin, Poland
M. Jacyna	Warsaw University of Technology, Poland
A. Janota	University of Zilina, Republic of Slovakia
J. Januszewski	Gdynia Maritime University, Poland
Z. Jóźwiak	Maritime University of Szczecin, Poland

A. Kalašová	University of Zilina, Republic of Slovakia
J. Klamka	Polish Academy of Sciences, Poland
B. Kos	University of Economics in Katowice, Poland
A. Križanová	University of Zilina, Republic of Slovakia
A. Lewiński	University of Technology and Humanities in Radom, Poland
M. Luft	University of Technology and Humanities in Radom, Poland
Z. Łukasik	University of Technology and Humanities in Radom, Poland
A. Maczyński	University of Bielsko-Biala, Poland
R. Madleňák	University of Zilina, Republic of Slovakia
M. Michałowska	University of Economics in Katowice, Poland
G. Nowacki	Military University of Technology, Poland
T. Nowakowski	Wrocław University of Technology, Poland
D. Peraković	University of Zagreb, Croatia
Z. Pietrzykowski	Maritime University of Szczecin, Poland
A. Prokopowicz	Center for Analyses in Transport and Infrastructure, Poland
K. Rástočný	University of Zilina, Republic of Slovakia
M. Siergiejczyk	Warsaw University of Technology, Poland
J. Skorupski	Warsaw University of Technology, Poland
L. Sladkeviciene	Vilnius College of Technologies and Design, Lithuania
J. Spalek	University of Zilina, Republic of Slovakia
M. Svitek	Czech Technical University in Prague, Czech Republic
J. Szpytko	AGH University of Science and Technology, Poland
E. Szychta	University of Technology and Humanities in Radom, Poland
R. Tomanek	University of Economics in Katowice, Poland
R. Wawruch	Gdynia Maritime University, Poland
W. Wawrzyński	Warsaw University of Technology, Poland
A. Weintrit	Gdynia Maritime University, Poland
M. Wierzbik-Strońska	Katowice School of Technology, Poland
E. Załoga	University of Szczecin, Poland
J. Żurek	Air Force Institute of Technology, Poland
J. Ždánsky	University of Zilina, Republic of Slovakia

Contents

Knowledge-Based Approach to Selection of Weight-in-Motion Equipment

Aleš Janota[(✉)], Dušan Nemec, Marián Hruboš, and Rastislav Pirník

Faculty of Electrical Engineering, Department of Control
and Information Systems, University of Žilina, Univerzitná 8215/1,
Žilina 010 26, Slovak Republic
{ales.janota,dusan.nemec,marian.hrubos,
rastislav.pirnik}@fel.uniza.sk

Abstract. The paper deals with the knowledge-based approach applied to selection of a proper equipment for measurement of road vehicles weights, preferably based on a weight-in-motion principle. After brief introduction the authors identify potentially usable attributes whose values create a data set characterizing particularly collected measurement equipment. The data mining software is used to create the initial version of the decision tree that is further manually modified to change "a classification model" to "a selection model" providing more information needed for user's decision making. The knowledge is then transformed into the rule-base form, implemented and validate by a rule-based expert system shell.

Keywords: Weight-in-motion · Decision making · Data mining · Knowledge-based selection tree

1 Introduction

High quality of road networks is a basic condition of safety and fluency of traffic operation. To keep required quality standards prevention of overloaded vehicles excessively damaging the road surface is needed together with regular maintenance. Some overloaded vehicles whose total weight or axle loads are higher than those the road is dimensioned for can cause unexpectedly severe and premature road degradation or induce emergency conditions of the road structure significantly increasing maintenance costs. To avoid such situations weight limits have been defined and countries are attempting to protect their road networks by installing stations equipped with technical systems for weighing vehicles and axles. Generally, there are two kinds of such systems: static and dynamic ones. The former require a standing vehicle, the latter (called weigh-in-motion systems, WiM) are able to measure vehicles that are moving, not exceeding a certain speed limit. Vehicles identified as overloaded then may be stopped at a safe place for enforcement purpose where precise measurements may be done with all legal consequences and penalization. At the moment there are four existing international sets of specifications on WiM of road vehicles [1]:

© Springer International Publishing AG 2016
J. Mikulski (Ed.): TST 2016, CCIS 640, pp. 1–12, 2016.
DOI: 10.1007/978-3-319-49646-7_1

- The COST323 European WiM Specification;
- The ASTM E-1318 "Standard Specification for Highway Weigh-in-Motion Systems with User Requirements and Test Methods" from the American Society for Testing Materials;
- The OIML R-134 as a recommendation for "Automatic Instruments for weighing road vehicles in motion" from the International Organization for Legal Metrology; and
- The Measuring Instrument Directive (MID) as a set of uniform European specifications and a European framework for type and product approval.

Since overload is an international problem, there is a need to create a common standard. However, there is still no consensus within the industry on the physical requirements for a WiM system, the calibration of data, and data quality checks. Different types of WiM technology are available at the market; of which bending plates, strip sensors (e.g. piezo electric sensors) and capacitive sensors are the most common. It is not easy at all for a user to make good and qualified choice of a proper measurement method and/or instrument.

This paper has been written with the motivation to show how a knowledge-based approach could be applied in the process of choosing a user-best measurement method and/or measurement instrument. To do that a set of common attributes must be defined first which makes possible to compare available technologies. Then a data mining tool is used to discover knowledge hidden in the data set, in this case in the form of a decision tree. However, various applied effectiveness criteria ensure minimum size of the decision tree which seems to be contra-productive since a minimum set of chosen attributes may be insufficient for a user to make a qualified decision. Therefore the expert-based manual modifications are further performed to extend the tree size, intentionally implementing alternative subtrees. The last step covers transformation of the obtained knowledge into the IF-THEN rules and testing the rule base in an expert system shell.

Generally, the problem of selection of the appropriate sensor for a particular measurement problem can be found e.g. in [2]; however, the particular selection is usually very closely related to and depending on features given by application domain. The use of a knowledge-based approach to selection of a sensor, detector or measurement equipment is not completely new: [3] presents a knowledge-based system developed to allow users, who may not be knowledgeable about sensors, to select sensors suitable for their specific needs; [4] uses ontologies to capture the crucial domain knowledge and semantic matchmaking to perform sensor-task matching; [5] summarizes a knowledge-based assistant system which is able to automatically select edge detectors and their scales to extract a given edge, etc.

2 WiM Measurement

2.1 WiM Systems Purpose and Components

The primary purposes of WiM systems operation are these [6]:

- To record weight data for road analysis and pavement research projects;
- To pre-screen trucks as a part of commercial vehicle weight enforcement operation;
- To use weight information to calculate tolls on toll roads, bridges or tunnels.

Having collected the required data, they may be used plan future road networks, to reduce cost for road repairs and maintenance, to reduce number of accidents and congestions, to reduce accident consequences, etc. A typical WiM system basically consists of four typical components (Fig. 1).

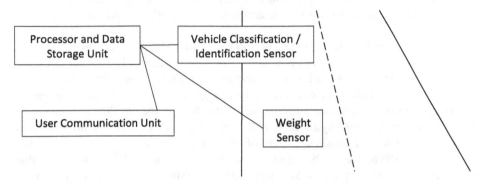

Fig. 1. Components of a typical WiM system [own study]

2.2 WiM Systems Classification

Several different classifications of the WiM systems are available. For example, the source [7] provides the following three categories:

- High Speed Weigh-in-Motion (HS-WiM): measurements are performed under normal speed conditions, usually without affecting the traffic flow and commanding driver's attention. Since achievable accuracy of such measurements is only about 15%, obtained values may not serve as basis of penalization (i.e. for enforcement purpose). Vehicles (trucks) identified as overloaded are to be diverted to another check point where more precise measurement methods may be applied;
- Low Speed Weigh-in-Motion (LS-WiM): vehicles (trucks) are measured under certain speed restrictions (usually from 1 up to 10–16 $km \cdot h^{-1}$) to minimize the dynamic effects but static measurements are usually also possible. The same technological systems may be found in logistic centres, industrial areas, etc.;
- Bridge Weigh-in-Motion (B-WiM): these systems use an existing bridge to weigh vehicles (trucks) via measurement of the structural response of the bridge while a vehicle crosses and usage of an algorithm to back-calculate the axle weights of the vehicle.

In our study we make do with above first 2 categories only, the B-WiM has not been considered.

Quantification of accuracy of dynamic (WiM) systems is more complicated than accuracy of static systems. Operation of any WiM system depends on conditions of road surface, vehicle conditions and road geometry as well. For static measurements an axle, group of axles or the whole vehicle are situated on the weighing plate. The acting force is constant and the only factor influencing accuracy results from properties of the measurement system. If the vehicle is in movement, in addition to its mass there are other influences such as acceleration, deceleration or changes of the load (e.g. liquid moving in a tank). Combination of all these factors is what really the WiM system can measure. Acting dynamic forces also depend on the vehicle speed. Thus the WiM system measures the presence of a moving vehicle and the related dynamic tyre forces at specified locations with respect to time; estimated tyre loads; calculated speed, axle spacing, vehicle class according to axle arrangement, and other parameters concerning the vehicle [8].

To eliminate or reduce dynamic effects in the WiM systems various techniques are applied. One of important factors is a number of measurements. The higher sampling frequency is used, the more information (and higher accuracy) can be achieved. Other factors are size and shape of measurement plane (e.g. several centimetres for piezo-electric sensors or several meters for one load cell). Neither the absolute accuracy may be specified for the WiM systems. For that reason accuracy of any WiM system is given as a certain % Confidence Level of error (typically set to 68% or 95%). For example, the ASTM E1318-09 Standard uses 95% Confidence Level. It means that 95% values measured by the WiM system appear within the given accuracy interval. WiM systems are evaluated to quantify system efficiency for the given application.

3   Selection of Attributes

3.1   Evaluation of WiM Equipment

Criteria used for initial or going evaluation may serve as inspiration for attributes selection. However, many times they are ambiguous, interconnected or interdependent. The most typical criteria are these:

- *Application:* reflects different needs, requirements for network connection, low or high speed of weighed vehicles, etc.;
- *Installation requirements:* requirements for location, type of weighed vehicles, vehicle velocity, installation time, etc.;
- *Simplicity of use:* power supply requirements, automation, temporality of operation (temporal, semi-permanent, permanent systems), communication abilities, need for staff training, availability of user guidelines and manuals;– reflect different needs, requirements for network connection, low or high speed of weighed vehicles, etc.;
- *Data quality:* data storage and reading, output data requirements, special application reports, kinds of required data such as serial count, date and time of exact pass of a vehicle, total length of a vehicle, movement velocity, vehicle class, distance of particular vehicle axles, distance between individual vehicles, weight of individual axles, weight of axle groups, total weight, measurement validity, etc.;
- *Calibration:* initial and/or continuous;
- *Accuracy:* reflects conformance ratio between measured and real weigh of the measured vehicle. The measured value may express weight of individual axles,

a group of axles or the total weight of the whole vehicle. There are three types of errors having effects on accuracy:

- Real error (resulting from the process of determining the real vehicle weight, i.e. error of static weighing);
- Random error (caused by various properties of the vehicle); and
- Systematic error (resulting from offset or caused by zero drift in calibration setting);
- *Lifetime:* depends on lifetime of the WiM system's individual components which is a typical property for all modular systems. The most important and sensitive part of the WiM system is the weight sensor. For that reason a cost of the sensor is an essential part of the total WiM system cost. Lifetime of weight sensors is usually between 3 and 12 years.

3.2 Case Study Attributes

To consider what measurement instrument suits the individual needs of the user best we must choose comparable attributes first. The main and general problem is to find comparable data characterizing individual measurement systems of different manufacturers. Some of them may not be publicly available or specified at all (for marketing reasons or for being irrelevant, e.g. because of different physical principle of operation). That makes comparison of existing solutions a non-trivial task.

Obviously, one of the basic parameters worth of user's considerations is the installation cost covering cost price and price for installation and calibration. Another "cost-related" attribute sometimes available is the cost for the whole life-cycle. To make comparison of different cost values possible we have established and worked with the following 3 cost intervals defined as:

- *Low* - cost may not exceed the value €10.000;
- *Medium* - cost is within the interval €10.000 up to €20.000;
- *High* - cost value is higher than €20.000.

The levels have been defined empirically, based on real equipment cost considerations, and anytime may be changed. Analysing other available catalogue data we have also defined the following attributes:

- *Identification of the considered measurement equipment or system:* in our case study a set of 22 products has been processed; their data were selected in 2014 (mainly from Internet sources – web pages of manufactures and vendors, catalogues sheets, WiM station operators); particularly we consider the following products: IRD Bending Plate, Bending Plate Scale - ZCS-30B, PAT DAW 100, PAT DAW 300, PAT DAW 300 Bluetooth, Trevor Deakin Consultants BIMS, Golden River Capacitive pad, Mikros HSWIM, Golden River Marksman 660, ARRB TR HSEMU, IRD Single Load Cell, IRD Piezoelectric VIBRACOAX, Piezoelectric RoadTrax BL, Trevor Deakin Consultants PIMS Series, KISTLER, ARRB TR CULWAY, ARRB TR Multi-lane CULWAY, IRD Model 6700 WIM mat, Mikros VLM, ARRB TR PCEMU, PAT DAW 50, and Transcale AS1 Axle Scale. The dataset is depicted in Tables 1 and 2.

Table 1. Measurement equipment features to compare (part 1) [own study]

N	WiM system	Sensor type	Category	Installa-tion cost	Life-cycle cost	Accuracy [%] CL 95%	Sensiti-vity	v_{min} [km/h]	v_{max} [km/h]
01	IRD Bending Plate	Bending plate	HS-WiM	Medium	Medium	10	Medium	5	200
02	Bending Plate Scale - ZCS-30B	Bending plate	HS-WiM	Medium	Medium	10	Medium	0	120
03	PAT DAW 100	Bending plate	HS-WiM	Medium	Medium	10	Medium	5	200
04	PAT DAW 300	Bending plate	LS-WiM	Medium	Medium	3	Medium	0	10
05	PAT DAW 300 Bluetooth	Bending plate	LS-WiM	Medium	Medium	3	Medium	0	10
06	Trevor Deakin Consultants BIMS	Bending plate	HS-WiM	Medium	Medium	5	Medium		
07	Golden River Capacitive pad	Capacitive pad	HS-WiM	High		10	Low	50	180
08	Mikros HSWIM	Capacitive pad	HS-WiM	High		9	Low		
09	Golden River Marksman 660	Capacitive pad	HS-WiM	High		10	Low	50	180
10	ARRB TR HSEMU	Load cell	HS-WiM	High	High	5	Low		
11	IRD Single Load Cell	Load cell	HS-WiM	High	High	3	Low	3	130
12	IRD Piezoelectric VIBRACOAX	Piezo electric cable	HS-WiM	Low	Low	8	High		
13	Piezoelectric RoadTrax BL	Piezo electric cable	HS-WiM	Low	Low	7	High		
14	Trevor Deakin Consultants PIMS Series	Piezo electric cable	HS-WiM	Low	Low	8	High		
15	KISTLER	Quartz piezo	HS-WiM	Medium	High	10	High	15	150
16	ARRB TR CULWAY	Strain Gauge	HS-WiM	Medium	Medium	10	Medium		
17	ARRB TR Multi-lane CULWAY	Strain Gauge	HS-WiM	Medium	Medium	10	Medium		
18	IRD Model 6700 WIM mat	Capacitive pad	LS-WiM	High		3	Low		
19	Mikros VLM	Capacitive pad	LS-WiM	High		3	Low		
20	ARRB TR PCEMU	Load cell	LS-WiM	High	High	3	Low		
21	PAT DAW 50	Load cell	LS-WiM	High	High	1	Low	0	10
22	Transcale AS1 Axle Scale	Load cell	LS-WiM	High	High	0.5	Low		

- *Type of the sensor seen from a technological point of view:* considered types are Bending plate, Capacitive pad, Load cell, Piezo electric cable, Quartz piezo, and Strain gauge;
- *Category of the measurement system:* represented by the discrete values HS-WiM or LS-WiM;
- *Accuracy of the measurement system:* expressed in % of the Confidence Level (CL);
- *Sensitivity:* represented by discrete values Low, Medium and High;

Table 2. Measurement equipment features to compare (part 2) [own study]

N	Temporality	Invasiveness	Inst. time [days]	Max No of lanes	Wheels to weigh	Max load per axle [t]	ϑ_{min} [°C]	ϑ_{max} [°C]	Life [yrs]	Link
01	Permanent	Invasive	3	8	All	20	-40	50	10	Cable
02	Permanent	Invasive	3		All	30	-40	100		Cable
03	Semi-permanent	Invasive	3	6	All	20	-40	70	6	Cable
04	Temporal	Non-invasive	0.5	1	All	20	-20	70	6	Cable
05	Temporal	Non-invasive	0.5	1	All	20	-20	70	6	Blue tooth
06	Semi-permanent	Invasive	3	8	All	20	-40	85	6	Cable
07	Temporal	Non-invasive	1	1	Left side		0	80	5	Cable
08	Semi-permanent	Invasive	1	4	All	20	-30	65	5	Cable
09	Permanent	Invasive	1	8	All		-40	80	10	Cable
10	Semi-permanent	Invasive	3		All	80	-20	70	15	Cable
11	Permanent	Invasive	3	8	All	25	-40	50	12	Cable
12	Permanent	Non-invasive	1	8	All	20	-40	50	3	Cable
13	Permanent	Non-invasive	1	1	All		-40	70		Cable
14	Permanent	Invasive	1	16	All		-40	75	5	Cable
15	Permanent	Invasive	0.5	1	All		-40	80	15	Cable
16	Semi-permanent	Invasive	3	1	All	50	-10	70	10	Cable
17	Permanent	Invasive	3	4	All	50	-10	70	10	Cable
18	Temporal	Non-invasive	0.5	1	All	20	-25	55	3	Cable
19	Temporal	Non-invasive	0.5	1	All	20	-30	65		Cable
20	Semi-permanent	Invasive		1	All	30	-20	70	20	Cable
21	Semi-permanent	Invasive		1	All	25	-40	75		Cable
22	Semi-permanent	Invasive		1	All	40	0	60		Cable

- *Max velocity of the weighed vehicles:* v_{max} measured in $km \cdot h^{-1}$;
- *Min velocity of the weighed vehicles:* v_{min} measured in $km \cdot h^{-1}$;
- *Type of system installation:* seen from the viewpoint of temporality - temporal, semi-permanent and permanent;
- *Type of sensor installation:* seen from the view of intentional effect on road surface integrity - invasive and non-invasive;
- *Installation time:* expressed in a number of days needed to complete installation;
- *Max number of traffic lanes covered by the measurement equipment:* experimentally set from 1 up to 16;
- *Number of wheels (of transport means) to be weighed:* all or one side;
- *Max load per axle:* expressed in tons;
- *Min operating temperature* ϑ_{min}: expressed in °C;
- *Max operating temperature* ϑ_{max}: expressed in °C;
- *Life of the weighing equipment:* expressed in years;
- *The way how the sensor is linked* (communication medium) to the control unit: two possibilities considered – cable line and wireless line (Bluetooth).

4   Knowledge Discovery and Data Mining

The process of knowledge discovery in databases is well-known and consists of a set of standard steps to be performed: data preparation and selection, pre-processing, transformation, data mining and interpreting accurate solutions from the observed results, including their evaluation.

In our case data collected on particular WiM equipment may be inconsistent (hardly comparable) since some data may be unavailable, e.g. for marketing reasons, for different physical principles applied, for unapplicability, etc. In our case the data set available in the form of the Excel sheet is provided to the data mining software (Rapid Miner 6.4), applying the procedure consisting of 5 steps: selection of the file to be imported; selection of a sheet and marking the range of values; annotation of attributes; definition of data types (Fig. 2) and saving the result.

Then we are able to define the data mining process (Fig. 3).

After making necessary configuration settings (e.g. setting the label to the classification attribute WiM system and setting the pruning choices settings) we could build decision trees using various effectiveness criteria (information gain, gain-ratio, etc.). One of possible (and most branched) solutions is illustrated in Fig. 4. Different settings bring different structure and sizes of decision trees.

The applied criterion caused the decision tree was built in a streamlined way that fits perfectly for building an effective classifier model. However, for the selection task this approach seems a little bit contra-productive since many of collected attributes (potentially helpful in the selection process) are omitted. Therefore the next step consists in expert-based modification of the tree to re-build it and include as many as attributes as possible. The result decision tree contains the set of 22 IF-THEN rules, each representing a path from the root node to the particular equipment (Fig. 5).

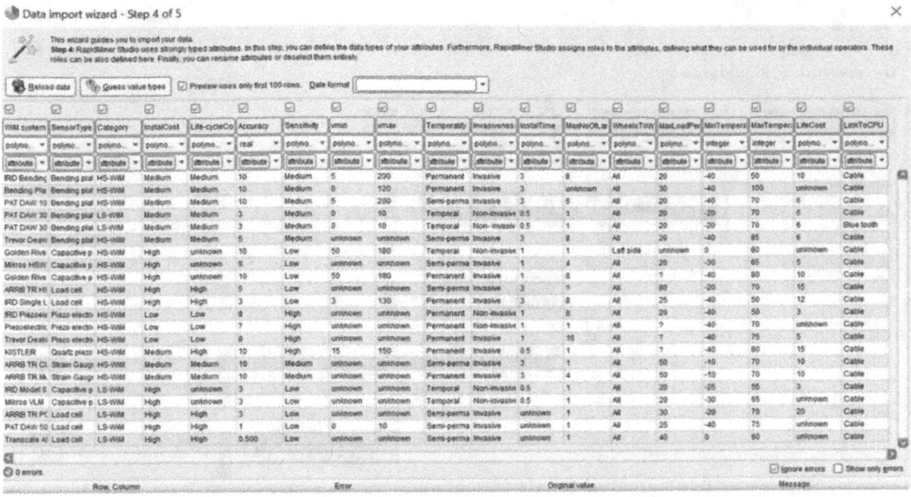

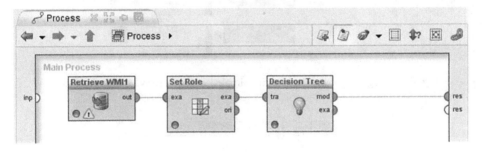

Fig. 2. Illustration of data types definition in Rapid Miner 6.4 [own study]

Fig. 3. Process model definition in Rapid Miner 6.4 [own study]

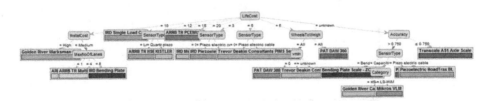

Fig. 4. The result decision tree (criterion: gain-ratio, neither pruning nor pre-pruning applied) [own study]

Completeness and consistency of the created rule-base was tested using the built-in validation engine (Fig. 6). The meaning of symbols used in the validation process is explained using the following examples:

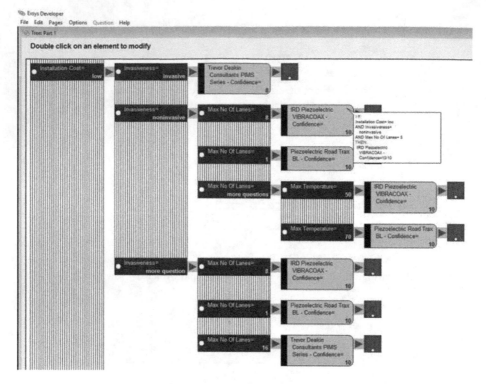

Fig. 5. A fragment of the decision tree implemented in Exsys 8.0 [own study]

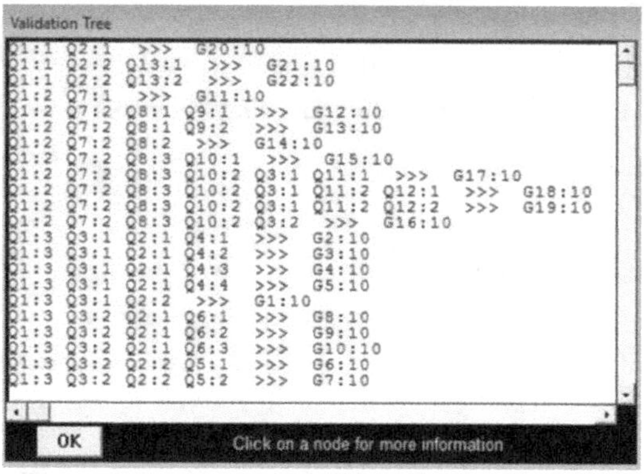

Fig. 6. The validation tree [own study]

Q1:2   Question 1 with assigned the value 2
G3:10   Goal 3, with certainty 10 (scale 0–10)

Analysis of the collected data set indicates that in some cases there are more variants of distinguishing between the goals. If the user is not capable of moving further in the decision process using the proposed attribute at a certain level of the decision tree (e.g. for lack of information needed to make decision), variant subtrees may be proposed by adding an extra value to the attribute (here named "more info needed") as illustrated in Fig. 7 (bold lines and italic text). Using this approach the considered rule base was experimentally extendable up to 95 rules utilizing maximum of collected data. In our case precise discrete values were used in some attributes, for bigger data sets it would be more practicable to substitute them with several continuous ranges cumulating groups of choices.

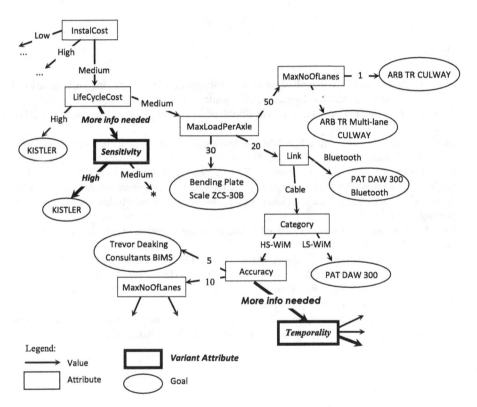

Fig. 7. Examples of the variant subtrees in the se-lection tree fragment [own study]

5   Conclusion

To demonstrate applicability of the presented approach the authors considered a set of 22 typical and sufficiently representative samples of various technological solutions available at the market at the given time (2014). However, this representative set could further be extended using data of other products coming from other manufacturers and properly updated. The more products are available, the more problems the user may have comparing them and making competent decision on selection. In that case a knowledge-based system may serve as a good support tools. The main disadvantage is that the presented approach requires an expert knowledge when modifying the final "selection tree" and thus a complicated repeatability of the whole process (i.e. how many experts, so many solutions – given by variant subtrees). The role of a data mining support tool in the process remains important; however, decision trees are not the only representation formalism and many more data mining approaches could be experimented.

Acknowledgments. This work is supported by the project ITMS: 26210120021, co-funded from EU sources and the European Regional Development Fund.

References

1. Jacob, B., van Loo, H.: Standardization of weigh-in-motion in Europe. In: 1st International Seminar of Weight in Motion, Florianópolis – Santa Catarina, Brazil (2011)
2. Regtien, P.P.: Selection of sensors. In: Handbook of Measuring System Design, Part 7:2:116 (2005)
3. Ong, J.B., Eyada, O.K., Masud, A.S.M.: A knowledge-based system for sensor selection. Comput. Ind. Eng. **17**(1–4), 85–89 (1989)
4. De Mel, G.R., Vasconcelos, W., Norman, T.J.: Intelligent resource selection for sensor-task assignment: a knowledge based approach. In: International Conference on Advanced Topics in Artificial Intelligence. Global Science and Technology Forum, Phuket, Thailand, 29–30 November 2010 (2010)
5. Ziou, D., Koukam, A.: Knowledge-based assistant for the selection of edge detectors. Pattern Recogn. **31**(5), 587–596 (1998)
6. Klebe, F.: Meeting the revised requirements of the ASTM 1318-02 standard specification for highway weigh-in-motion (WIM) systems. White Paper, Metler Toledo (2002)
7. Hussain, I.H.M., Grűnner, K.: Using the results of weighing in motion in pavement management systems. Slovak J. Civil Eng. **3**, 12–22 (2004)
8. Technical Specification: Transport and Main Roads Specifications. MRTS203 Provision of Weigh-in-Motion System. Department of Transport and Main Roads (2010)

Geoinformatics in Shipping and Marine Transport

Adam Weintrit[✉]

The Faculty of Navigation, Gdynia Maritime University,
Al. Jana Pawla II 3, 81-345 Gdynia, Poland
weintrit@am.gdynia.pl

Abstract. This paper discusses the role of Geoinformatics in Shipping and Marine Transport as a new independent, autonomous self-contained, scientific discipline created for dealing with geospatial information. Depending on the scientific background of those involved in shaping the emerging discipline, little emphasis may be placed on various aspects of Geoinformatics. Maritime applications and small developments may address spatial planning, surveying, or computer science related matters. The scientific field of Geoinformatics spans the acquisition and storage of spatial data, the modelling and presentation of spatial information, geoscientific analyses and spatial planning, and the development of algorithms and geospatial database systems, including Geographical Information System (GIS). The current paper emphasizes international and national trends of the discipline and provides a set of geospatial initiatives in the field of marine navigation, safety at sea, maritime shipping, sea transportation and logistics, transport telematics and geomatics.

Keywords: Geoinformatics · GIS · Shipping · e-Navigation · ITS · Marine transport · Transport telematics

1 Introduction

Increasingly used to analyze and manage marine, offshore and coastal areas, Geographical Information Systems (GIS) assure a powerful set of tools for integrating and processing spatial information. These technologies are increasingly used in the management and analysis of the shipping and marine transport processes. Supplying the guidance necessary to use these tools, marine GIS created for ocean and coastal zone management explores essential technical, theoretical, practical, and applications issues. Drawing on the practical experience of expert users in the maritime field, the paper presents samples of recent developments and specific applications.

The sections of paper present groundbreaking marine, offshore and coastal applications of GIS based decision support tools, spatial data infrastructures, remote sensing technology including GPS/GNSS, ECDIS/ECS, Radar/ARPA, VTS/RIS, AIS, LRIT, LiDAR, CASI, and more. A comprehensive, reliable, and up-to-date overview of the state-of-the-art in ocean and coastal areas GIS applications makes the information not only easily accessible but ready for immediate use [18].

© Springer International Publishing AG 2016
J. Mikulski (Ed.): TST 2016, CCIS 640, pp. 13–25, 2016.
DOI: 10.1007/978-3-319-49646-7_2

The use of GIS has seen unprecedented growth in the last twenty years. With more and more sophisticated, powerful technology becoming cheaper and system memories expanding, which means that we can handle much bigger volumes of data. We can say that GIS is in a golden age. It was once the preserve of the cartographer or surveyor - recently GIS has become a core part of modern sciences and technologies.

Many fields benefit from geoinformatics, including smart city planning and land use management, in-car navigation systems, virtual globes, geology, geography, agriculture, meteorology, climatology, archaeology, oceanography, environmental modelling and analysis, public health and epidemiology, biodiversity conservation, local and national press management, military service, transportation network planning and management, business site planning, architecture and archeological reconstruction, telecommunications, criminology and crime simulation, aviation, and maritime transport, logistics and shipping as well. The significance of the spatial dimension in assessing, monitoring and modelling various issues and problems related to sustainable management of natural resources is recognized around the world, the same as merchant fleet management and fright transport (the physical process of transporting commodities and merchandise goods and cargo). Geoinformatics is very important technology to decision-makers across a wide range of disciplines, industries, international organisations and agencies, local and national government, commercial sector, environmental management agencies, research and academia, emergency services, crime mapping, transportation and infrastructure, information technology industries, tourist industry, utility companies, national survey and mapping organisations, GIS consulting firms, market analysis and e-commerce, shipping, exploration of resources, etc. A lot of government (including maritime administration) and non-government agencies started to use spatial data for managing their day-to-day activities ashore and at sea [5, 16].

2   Geoinformation, GIS, Geomatics and Geoinformatics

Ordinary people often confuse with each other the following terms: Geoinformatics, Geomatics, Geographic(al) Information System (GIS), GIScience, Geoinformatics, Geomatics, and Geoinformation, Geospatial and GIS Technologies [2]. Geoinformatics has been described as "the science and technology dealing with the structure and character of spatial information, its capture, its classification and qualification, its storage, processing, portrayal and dissemination, including the infrastructure necessary to secure optimal use of this information" [16] or "the art, science or technology dealing with the acquisition, storage, processing production, presentation and dissemination of geoinformation" [6].

Geomatics is a similarly used term which contains geoinformatics, but geomatics focuses more so on surveying. Geoinformatics has at its core the technologies supporting the processes of acquiring, analysing and visualizing spatial data [2]. Both geomatics and geoinformatics include and rely heavily upon the theory and practical implications of geodesy. But geoinformatics is a subset of geomatics. GIScience is highly related with the term Geoinformatics that is a shorter name for Geographic Information Technology.

Geomatics is a similarly used term, geoinformatics, but geomatics focuses more so on survey covers. Geoinformatics has to support the technology in the core of data, the processes of detection, analysis and visualization. Both geomatics and geoinformatics are and rely heavily on the theory and practical implications of Geodesy. But Geoinformatics is a subset of geomatics. GIScience is highly in the concept Geoinformatics related, which is a shorter name for Geographic Information Technology [2].

Geography and Earth science increasingly rely on digital spatial data acquired from remotely sensed images analysed by Geographical Information Systems and presented on paper or visualized on the computer screen.

Geoinformatics combines geospatial analysis and modelling, information systems design, development of geospatial databases, human-computer interaction and both wired and wireless networking technologies. Geoinformatics uses geocomputation and geovisualization for analyzing geoinformation [2].

All terms, the meaning of which is explained above, they are now very trendy words. In the last ten years, most universities have offered masters degrees specifically in GIS, or post-graduate certificates and diplomas with undergraduate degree in Geoinformatics.

3   Geoinformatics in Transport

In a broad sense a Geographic Information System (GIS) is an information system specializing in the input, management, analysis and reporting of geographical (spatially related) information. Among the wide range of potential applications GIS can be used for, transportation issues have received a lot of attention, including maritime branch. A specific branch of GIS applied to transportation issues, commonly labelled as GIS-T, is one of the first, pioneer application areas of GIS [17]. GIS for Transportation is a large application area of GIS.

Geographic Information Systems for Transportation (GIS-T) refers to the principles, rules and applications of applying geographic information technologies in the field of transportation, and the transport-related problems. Technologically, GIS-T, like GIS as a whole, benefited from developments in management information systems and database techniques in general, and relational databases in particular.

GIS-T is well represented in GIS journals such as: *Computers, Environment and Urban Systems, the International Journal of Geographical Information Science, Geographical Systems, Transactions in GIS, and Geographic Information Sciences, the Journal of Advanced Transportation, TransNav, the International Journal on Marine Navigation and Safety of Sea Transportation, Archives of Transport System Telematics,* and the *Journal of Transportation Planning and Technology.* In addition, in the last decade there have been numerous government and research reports, books and other materials written about GIS-T and closely-related topics.

GIS-T research can be approached from two different, but complementary, directions. While some GIS-T research focuses on issues of how GIS can be further developed and enhanced in order to meet the needs of transportation applications, other GIS-T research investigates the questions of how GIS can be used to facilitate and improve transportation studies. In general, topics related to GIS-T studies, according to Shaw and Rodrigue [17], can be grouped into three main categories:

– data representations (How can different components of transport systems be represented in a GIS-T?),
– analysis and modelling (How can transport methodologies and procedures be used in a GIS-T?),
– applications (What types of applications are particularly proper for GIS-T?).

3.1 GIS-T Data Representations

Data representation is a core research topic of GIS. Before a GIS can be used to tackle real world problems, data must be properly represented in a digital computing environment. One unique characteristic of GIS is the capability of integrating spatial and non-spatial data in order to support both display and analysis needs. There have been various data models developed for GIS. The two basic approaches are object-based data models and field-based data models [17]:

– An object-based data model treats geographic space as populated by discrete and identifiable objects. Features are often represented as points, lines, and/or polygons;
– On the other hand, a field-based data model treats geographic space as populated by real-world features that vary continuously over space. Features can be represented as regular tessellations (a raster grid, e.g. RNC) or irregular tessellations (e.g., triangulated irregular network - TIN).

GIS-T studies have employed both object-based and field-based data models to represent the relevant geographic data. Some transportation problems tend to fit better with one type of GIS data model than the other. For example, network analysis based on the graph theory typically represents a network as a set of nodes interconnected with a set of links. The object-based GIS data model therefore is a better candidate for such transportation applications. Other types of transportation data exist which require extensions to the general GIS data models. One well-known example is linear referencing data (e.g. canal, waterway, mileposts). Transportation agencies often measure locations of features or events along transportation network links. Such a one-dimensional linear referencing system (i.e. linear measurements along a waterway segment with respect to a pre-specified starting point of the waterway segment) cannot be properly handled by the two-dimensional Cartesian coordinate system used in most GIS data models. Consequently, the dynamic segmentation data model was developed to address the specific need of the GIS-T community. Origin-destination (O-D) flow data are another type of data that are frequently used in transportation studies. Such data have been traditionally represented in matrix forms (i.e. as a two-dimensional array in a digital computer) for analysis. Unfortunately, the relational data model widely adopted in most commercial GIS software does not provide adequate support for handling matrix data. Some GIS-T software vendors therefore have developed additional file formats and functions for users to work with matrix data in a GIS environment. The above examples illustrate how the conventional GIS approaches can be further extended and enhanced to meet the needs of transportation applications. Modern information and communication technologies (ICT) such as the Internet and cellular phones have changed the ways that people and businesses conduct their activities. These changing activity and interaction patterns in turn lead to changing spatio-temporal traffic patterns. Our world has become more mobile and dynamic than ever before due to modern ICT. With the advancements of location-aware technologies (e.g., GPS/GNSS, ECDIS/ECS, AIS, LRIT, cellular phone tracking system, and Wi-Fi positioning system), it is now feasible and affordable to collect large volumes of tracking data at the individual level. Consequently, how to best represent and manage dynamic data of moving objects (vessels, or shipments) in a GIS environment presents new research challenges to GIS-T, especially when we have to deal with the Big Data issues. In short, one critical component of GIS-T is how transportation-related data in a GIS environment can be best represented in order to facilitate and integrate the needs of various transportation applications. Existing GIS data models provide a good foundation of supporting many GIS-T applications. However, due to some unique characteristics of transportation data and application needs, many challenges still exist to develop better GIS data models that will improve rather than limit what we can do with different types of transportation studies.

3.2 GIS-T Analysis and Modelling

GIS-T applications have benefited from many of the standard GIS functions (query, geocoding, buffer, overlay, etc.) to support data management, analysis, and visualization needs. Like many other fields, transportation has developed its own unique analysis methods and models. Examples include shortest path and routing algorithms (e.g. route planning, voyage planning, optimisation of the trajectory), spatial interaction models (e.g. gravity model), network flow problems (e.g. traffic flow, container/cargo flow problem, minimum cost flow problem, maximum flow problem, network flow equilibrium models), vessel traffic problems, facility location problems (e.g. set covering problem, maximal covering problem), travel demand models (e.g. the four-step trip generation, trip distribution, modal split, traffic assignment models, and more recent activity-based travel demand models), and land use-transportation interaction models. While the basic transportation analysis procedures (e.g. route planning, shortest path finding, RL/GC sailing) can be found in most commercial GIS software, other transportation analysis procedures and models (e.g. travel demand models) are available only selectively in some commercial software packages. Fortunately, the component GIS design approach adopted by GIS software companies provides a better environment for experienced GIS-T users to develop their own custom analysis procedures and models. It is essential for both GIS-T practitioners and researchers to have a thorough understanding of transportation analysis methods and models. For GIS-T practitioners, such knowledge can help them evaluate different GIS software products and choose the one that best meets their needs. It also can help them select appropriate analysis functions available in a GIS package and properly interpret the analysis results. GIS-T researchers, on the other hand, can apply their knowledge to help improve the design and analysis capabilities of GIS-T. Due to the increasing availability of tracking data that include both spatial and temporal elements, development of spatio-temporal GIS analysis functions to help better understand the dynamic movement patterns in today's mobile world has attracted significant research attention in recent years [17].

3.3 GIS-T Maritime Applications

GIS-T is one of the leading GIS application fields. Many GIS-T maritime applications have been implemented at various shipping companies, maritime administrations, maritime research, education and training institutions, transportation agencies and private firms. They cover much of the broad scope of transportation and logistics, such as infrastructure planning and management, transportation safety analysis, travel demand analysis, route planning, traffic monitoring and control, public transit planning and operations, environmental impacts assessment, intelligent transportation systems (ITS), routing and scheduling, vessel tracking and dispatching (LRIT, AIS), fleet management, site selection and service area analysis, and supply chain management. Each of these applications tends to have its specific data and analysis requirements. A maritime traffic engineering application, on the other hand, may require a detailed representation of individual traffic lanes. Turn movements at intersections also could be critical to a traffic engineering study, but not to a area-wide travel demand study. These different application needs are directly relevant to the GIS-T data representation and the GIS-T analysis and modelling issues discussed above. When a need arises to represent transportation networks of a study area at different scales, what would be an appropriate GIS-T design that could support the analysis and modeling needs of various applications? In this case, it is desirable to have a GIS-T data model that allows multiple geometric representations of the same transportation network. Research on enterprise and multidimensional GIS-T data models discussed above aims at addressing these important issues of better data representations in support of various transportation applications. With the rapid growth of the Internet and wireless communications in recent years, a growing number of Internet-based and wireless GIS-T applications can be found. ECDIS/ECS and Global Positioning System (GPS) navigation systems also are available as a built-in device in watercrafts/vehicles (vessels or boats) or as a portable device. Coupled with wireless communications, these devices can offer real-time traffic information and provide helpful location-based services (LBS). The concept of e-Navigation gives a huge field of new possibilities of applications in this field. Another trend observed in recent years is the growing number of GIS-T applications in the private sector, particularly for logistics applications. Since many businesses involve operations at geographically dispersed locations (e.g., broker/ship chandler/supplier sites, distribution centres, warehouses, terminals, retail stores, and customer location), GIS-T can be a useful tool for a variety of logistics applications. Many of these logistics applications are based on the GIS-T analysis and modelling procedures such as the routing and facility location problems. Transportation GIS (GIS-T) is interdisciplinary in nature and has many possible applications. Transportation geographers, who have appropriate backgrounds in both geography and transportation (and of course in disciplines related to them), are well positioned to continue GIS-T studies [14].

3.4   More Samples of Maritime Applications of GIS Technology

Geospatial Technology (GST or GIS) has become pervasive nowadays in a wide variety of applications. The maritime industry has increasingly applied geospatial technologies such as GNSS/GPS, remote sensing (RS), hydrographic surveying and coastal mapping [20], ports planning and management, and charting, as well as development of a marine spatial data infrastructure (MSDI). The GIS technology in marine transportation is certainly useful in varied areas namely: routing of vessels and the type of vessel, knowing the positions of vessels in real time, mapping and analyzing incidence, selecting new sites and analyzing marine aids to navigation such as AtoNs, signal lights, and other manmade coastal and offshore objects and structures, hydrographic and bathymetric mapping of harbours, approaches, and channels, delineating shipping channels, maritime zones, and marine protected areas, producing, managing and upgrading IMO compliant navigation charts, designing and analyzing transportation networks, and monitoring and analyzing climate patterns and ocean currents [1]. Maritime mapping can also best be accomplished by GIS software [20]. To address the challenges in maritime mapping and charting, the International Hydrographic Office (IHO) has proposed competencies in several spatial related skills such as cartography, hydrography, geodesy, GPS, IMO compliant electronic navigational chart (ENC) and digital nautical chart (DNC) production based on IHO former S-57 and new IHO S-100 and S-101 standards, spatial database management system (SDBMS), GIS software such as ArcGIS and SevenCs, and electronic chart display and information system (ECDIS) [11, 19]. Some examples of software tools used in maritime GIS are ESRI suite of software, CARIS, dKart, and SevenCs.

Amongst several technological solutions that might contribute to the emergence of maritime-based decision-aid systems, integration of Geographical Information Systems (GIS) with maritime navigation systems appears as one of the promising directions to explore. There are several contributions to such a field of maritime GIS: from the real-time monitoring of navigations for a local authority and maritime clients, to the diffusion of maritime data to mobile interfaces, and the development of a relative-based model and visualisation system for maritime trajectories [4].

4   GIS Solutions for Ports and Maritime Transport

Port operators today face increased demands for operational efficiency, effective facility management, comprehensive safety and security, and sensitive environmental management. These diverse challenges require access to detailed, up-to-date information and careful analysis to achieve optimum results. Geographic Information System (GIS) technology provides management solutions that incorporate the position of operator assets to gain a decisive competitive advantage.

There are the following areas of interest (potential fields) for GIS application where GIS can improve port and maritime efficiency [7]: infrastructure and expansion planning, port design, environmental management (storm water management, environmental compliance), facility and utility management, asset and inventory management, maintenance/work order management, utility operations and control, property and lease management, security operations, emergency response and management (spill response and management, and incident tracking), port operations (real-time vehicle and asset location, vessel routing and tracking, berth occupancy and assignment, cargo and berth time calculations, and dangerous cargo display), intermodal management, meteorological monitoring, water depth assessment and visualization, marine navigation, nautical charting, public information (shipping channels location, and restricted area awareness).

5   GIS for Offshore Zone Management

Except of port and maritime applications Geographic Information Systems technology offer also the following applications for coastal and offshore areas, especially for Offshore Zone Management [18, 21].

5.1   Coastal Spatial Data Infrastructure (SDI)

The term Spatial Data Infrastructure (SDI) is now in common use in many countries around the world, although definitions for the term differ quite considerably. The stated objectives of SDI initiatives vary as much as do the definitions, legal mandates, types of organisation responsible for specifying and implementing SDI and actual progress achieved in creating national and regional SDIs. One complication in specifying any SDI is the nature of spatial information, i.e. information with an important location attribute. The visionaries and designers of SDI must accommodate the widely varying information needs of highly diverse disciplines and sectors of society, business and government. Then knowledge of the coastal zone fauna and flora, hydrography, tides and tidal currents, nearby land use practices of industry and agriculture and transport routes, roads, anchorages, fishing areas, wind farms and zones all become intertwined. The complex relationships between various types of spatial information are one reason that countries take different ways to specify their SDI, ranging from visions to strategies to goals to detailed content and implementation plans. We all recognize that the coastal area is a difficult geographical zone to manage due to temporal issues (tides and seasons) and the overlapping of physical geography and hydrography (offshore, near shore, shoreline, inshore), of jurisdictions, legal mandates and remits of government agencies and the often competing needs of stakeholders. Typically, many different local, national and regional government agencies are responsible for different aspects of the same physical areas and uses of the coastal zone, e.g. fisheries, environment, agriculture, transport (inland, coastal, offshore and marine), urban planning and cadastre, national mapping agency and the hydrographic service [12].

5.2   Bridging the Land-Sea Divide Through Digital Technologies

There are many different types of users of coastal zone information, from the casual user who may only want to browse, to the sophisticated user who makes frequent use of mapping and demands continuous improvement. These user communities are diverse in the topics they address, covering such areas as local and regional government, environmental and economic analysis, and also increasingly leisure use. A common mapping framework that bridges the land-sea divide allows users to build applications and decision-making tools necessary to promote the shared use of such data throughout all levels of Government, the private and non-profit sectors and academia. A consistent framework also serves to stimulate growth, potentially resulting in significant savings in data collection, enhanced use of data and assist better decision-making. As well as a physical division, the land-sea divide has also, for many spatial data suppliers, acted as a limit to their area of responsibility, or formed a data product boundary. As a result users wanting to model the diverse aspect of the coastal zone across this divide have had to identify, obtain and combine separate datasets to provide the data coverage they require. The combination process must resolve integration problems resulting from the differing projections, scale of capture and other specification issues of the source data-sets. This process can be time consuming, result in inconsistent data and can cause a hindrance to the management of a particularly sensitive environmental zone [10].

5.3   Spatial Uncertainty in Marine, Offshore and Coastal GIS

The dynamic nature of coastal landscapes and the inherent complexity of the biophysical processes operating in these environments challenge the application of GIS methods. It is well recognised that spatial data models representing static objects are rife with uncertainty [8]. However, the mobility of many coastal and marine phenomena and the nebulous nature of boundaries in these environments provide an additional dimension to the problems associated with spatial data uncertainty. In abstracting the infinite complexity of reality into a finite computer based storage structure, multiple levels of uncertainty are introduced. The more encompassing or inclusive a data set, often the more complex the process of abstraction. Users of coastal and marine GIS are faced with both uncertainty in the information derived from spatial data, and uncertainty that inher-ently exists in the models. The ubiquitous nature of uncertainty in spatial analysis high-lights the need to examine the implications for coastal and marine decision-making. This paper examines the sources of uncertainty, methods for assessing reliability, model uncertainty and the cognitive and practical implications associated with the communi-cation and incorporation of uncertainty in coastal and marine [3].

5.4   Visualisation of Offshore Areas

3D landscape visualisation is increasingly used in spatial sciences and planning, including marine, offshore and coastal areas applications. Currently available visual-isation tools permit creating highly realistic representations of landscapes based on geodata, such as digital elevation models, aerial photographs, or remote-sensing data. However, in contrast to traditional 2D representations, photorealistic 3D landscape visualisations offer a higher degree of visual clarity, thus contributing to a better understanding of spatial structures and processes and promoting visual thinking. Photorealistic landscape visualisations can be generated with the aid of either pure landscape-rendering systems, which usually do not support interactivity, or real-time visualisation environments.

Methods to visualise landscapes and landscape processes in three dimensions are increasingly used in regional planning. Based on a steadily growing volume of geodata with a high degree of geometrical resolution, such as digital elevation models, topographical data, and aerial or satellite images, today's visualisation environments are capable of rendering landscapes in three dimensions with a photorealistic effect. Compared to conventional two-dimensional maps, these visualisations are more illustrative, enabling the information contained in maps and plans, which is generally abstract and difficult to interpret for nonexperts, to be communicated in a format that is more easily understood. 3D landscape visualisations are of particular interest in the context of integrated Coastal Zone Management and Offshore Zone Management which aims for general public participation to gain acceptance for future protection, preservation, and development measures in coastal areas at an early time [9].

5.5   Application of a GIS-Based Decision Support System (DSS) in the Development of a Hydrodynamic Model for a Coastal Area

Recent advances in numerical modelling of physical processes and field survey technology nowadays allow the development of numerical models with extensive data sets. As a consequence, model developers are facing new challenges to handle the increasing amount of data and its analysis. Furthermore, model development and application concerning coastal areas are heavy time demanding tasks that need tools to assist the researcher. Most of the time they involve analysis of field measured data and its comparison with numerical model outputs. An application of a Decision Support System (DSS) in the development of a hydrodynamic model for a coastal area, as well as description of the DSS components, their interaction, and its GIS capabilities to handle spatial data are presented in details in items [13, 15].

5.6   Developing an Environmental Oil Spill Sensitivity Maps for Offshore

Marine oil spill sensitivity mapping has become widespread. The purpose is to provide oil spill response planners and responders with tools to identify resources at risk, establish protection priorities and identify appropriate response and clean-up strategies. GIS is an important tool in the development of oil spill sensitivity maps and can also be used for presentation [18].

To improve the performance of satellite-based Synthetic Aperture Radar (SAR) oil spill detection and monitoring, in coastal area, a combination of model data and SAR data has been developed. Such concepts may include oil drift components and SAR image models. If a possible oil spill is detected in an SAR image, the GIS-based models are used to try to reconstruct the spill given wind, current and wave height history of the area.

5.7   GIS Applications in Integrated Coastal Zone Management (ICZM)

The emergence of Integrated Coastal Zone Management (ICZM) represents a paradigm shift for a range of practitioners who work in the complex, dynamic area where land meets sea. The structure and implementation of geomatics technologies has been strongly affected by this shift, and GIS/RS practitioners have stepped up to meet the information needs of ICZM by creating coastal information systems featuring increased rigour, openness, and usability. As a result of this forcing, coastal GIS/RS is increasingly differentiating itself from the marine sciences and emerging as a unique discipline. In 2000, Wright wrote: "it may be fair to say that marine applications of GIS have been more in the realm of basic science whereas coastal applications, due in part to the intensity of human activities, have encompassed both basic and applied science, as well as policy and management" [21]. This move towards an integrated approach is being realized in parallel with similar shifts in the approach of other actors in the coastal zone, including scientists, managers and planners. The principles of Integrated Coastal Zone Management make explicit what coastal GIS/RS practitioners have known for years: that the coastal zone is a uniquely complex system that requires new and innovative management approaches. Here we outline the nature of this paradigm shift and the ways in which the authors in this volume have tackled the challenges of applying GIS to the integrated management of the coastal zone. The outcome of these principles is a planning methodology that employs an ecosystem approach to management (considering interconnected elements of the ecosystem) and incorporates adaptive management to deal with uncertainty, variability, and change [18].

6 Conclusion

The development of integrated maritime and GIS systems still requires the integration of different geographical information sources to be combined, adapted and shared in real-time between different levels of users acting in the maritime environment. The development of information and telecommunication technologies brings new and often unexpected possibilities for integrating, analyzing and delivering maritime traffic data within GIS. Integrating GIS information architectures and services with maritime information systems should improve the economic and technological benefits of transportation information by allowing the diffusion of traffic information to a larger community of decision-makers, engineers and final end-users (GIS-T, ECDIS, e-Navigation, e-Maritime, and maritime ITS concepts).

Research challenges are varied: development of cross-domain protocols and exchange standards for the transmission and interoperability of traffic data. Conventional statistical, geographical data analysis and visualization methods should also be adapted to the specific nature of traffic information often associated with large volumes of data. At the implementation level, there is a need for the development of GIS-based distributed computing environment, computational and processing capabilities as traffic data and applications are usually physically allocated in different geographical locations and computationally expensive in terms of the data volumes generated. The diversity of concepts and ideas presented in this paper illustrates the range of opportunities of the integration of GIS and Intelligent Transportation Systems (ITS) for marine transport (GIS-T) and navigation (ECDIS). The Author believes that all these application domains in shipping and marine transport should benefit for this information integration and those methodological findings should be shared and cross-fertilized amongst the research communities active in these fields.

References

1. Baylon, A.M., Santos, E.M.R.: Introducing GIS to TransNav and its extensive maritime application: an innovative tool for intelligent decision making? TransNav Int. J. Mar. Navig. Safety Sea Transp. **7**(4), 557–566 (2013)
2. Bhatta, B.: Remote Sensing and GIS. Oxford University Press, New York (2008)
3. Bruce, E.: Spatial uncertainty in marine and coastal GIS, Chap. 5. In: Bartlett, D., Smith, J. (eds.) GIS for Coastal Zone Management. CRC Press, Boca Raton (2004)
4. Claramunt, C., Devogele, T., Fournier, S., Noyon, V., Petit, M., Ray, C.: Maritime GIS: from monitoring to simulation systems. In: Popovich, V.V., Schrenk, M., Korolenko, K.V. (eds.) Information Fusion and Geographic Information Systems. Springer, Heidelberg (2007)
5. Clark, J.R.: Coastal Zone Management Handbook. Lewis Publishers, CRC Press, Taylor and Francis Group, New York (1996)
6. Ehlers, M.: Geoinformatics and digital earth initiatives: a German perspective. Int. J. Digit. Earth **1**(1), 17–30 (2008)
7. ESRI: Port operators today face increased demands for operational efficiency (2007). www.esri.com. Accessed 12 Dec 2015
8. Fisher, P.F.: Models of uncertainty in spatial data. Geograph. Inf. Syst. **1**, 191–205 (1999)
9. Gabler-Mieck, R., Duttman, R.: Application of geovisualisation techniques in coastal-zone management. In: 10th AGILE International Conference on Geographic Information Science 2007, Aalborg University, Denmark (2007)
10. Gomm, S.: Bridging the land-sea divide through digital technologies, Chap. 2. In: Bartlett, D., Smith, J. (eds.) GIS for Coastal Zone Management. CRC Press, Boca Raton (2004)
11. Hecht, H., et al.: The Electronic Chart: Fundamentals, Functions and other Essentials. A Textbook for ECDIS Use and Training, 3rd edn. Geomares Publishing, Lemmer (2011)
12. Longhorn, R.A.: Coastal spatial data infrastructure technologies, Chap. 1. In: Bartlett, D., Smith, J. (eds.) GIS for Coastal Zone Management. CRC Press, Boca Raton (2004)
13. Mayerle, R., Toro, F.: Application of a decision support system in the development of a hydrodynamic model for a coastal area, Chap. 9. In: Bartlett, D., Smith, J. (eds.) GIS for Coastal Zone Management. CRC Press, Boca Raton (2004)
14. Murray, A.T.: GIS and transportation, Chap. 29. In: Teodorovic, D. (ed.) Routledge Handbook of Transportation. Routledge, Taylor and Francis Group (2016)
15. Populus, J., et al.: Decision-making in the coastal zone using hydrodynamic modelling with a GIS interface, Chap. 10. In: Bartlett, D., Smith, J. (eds.) GIS for Coastal Zone Management. CRC Press, Boca Raton (2004)
16. Raju, P.L.N.: Fundamentals of geographic information systems. In: Sivakumar, M.V.K. (ed.) Satellite Remote Sensing and GIS Applications in Agricultural Meteorology. World Meteorological Organization, Geneva (2003)
17. Shaw, S.L., Rodrigue, J.P.: Geographic information systems for transportation (GIS-T). In: Rodrigue, J.P., Comtois, C., Slack, B. (eds.) Geography in Transport Systems. Routledge, New York (2013)
18. Smith, J.L., Bartlett, D.J.: Meeting the needs of integrated coastal zone management. In: Bartlett, D., Smith, J. (eds.) GIS for Coastal Zone Management. CRC Press, Boca Raton (2004)

19. Weintrit, A.: Handbook on Operational Use of ECDIS, A Balkema Book. CRS Press Taylor & Francis Group, Boca Raton, London, New York, Leiden (2009)
20. Weintrit, A.: Six in one or one in six variants. electronic navigational charts for open sea, coastal, off-shore, harbour, sea-river and inland navigation. TransNav Int. J. Mar. Navig. Safety Sea Transp. 4(2), 165–177 (2010)
21. Wright, D., Bartlett, D.J.: Working on the frontiers of science: applying GIS to the coastal zone. In: Wright, D.J., Bartlett, D.J. (eds.) Marine and Coastal Geographical Information Systems. Taylor & Francis, London (2000)

Tasks of Independent Assessment Bodies in Risk Management in Rail Transport – the Polish Experience

Adam Jabłoński[✉]

OTTIMA Plus Sp. z o.o., Południowy Klaster Kolejowy,
Ul. Gallusa 12, 40-594 Katowice, Poland
adam.jablonski@ottima-plus.com.pl

Abstract. In recent years, the issue of safety management in rail transport has been widely developed, which has consequences for the participants in this market, namely infrastructure managers, railway undertakings, ECM, as well as manufacturers and distributors. This results from the need to verify the conformity of products or services with the safety criteria. A special role in this process is played by assessment bodies (AsBo). Due to the requirements of the European Commission Regulation, the place and role of these bodies in safety management in rail transport has been clearly specified. The tasks of AsBo are to verify the adequacy of the process of risk evaluation and assessment in the event of technical, operational and organizational changes in the railway system. Previous experience in Poland related to the activity of AsBo indicates that the railway market does not fully understand their essence, importance and key impact on safety in rail transport. The purpose of this paper is to identify the key tasks and challenges of assessment bodies related to improving rail transport safety in Poland.

Keywords: Assessment body · Risk management · Safety management · Significant change

1 Introduction

The issue of safety management in rail transport, especially in recent years, is a priority throughout the European Union. Based on the defined requirements, Regulation 402/2013 [1] defines the involvement of independent assessment bodies in the process of the adequacy of risk evaluation and assessment in rail transport. The use of a common safety method for risk evaluation and assessment shows the key role of independent bodies in the process of safety management in rail transport.

The main changes introduced by the new regulation concern, inter alia:

– Formalizing the process of validating the competence of Assessment Bodies (AsBo) - according to new rules, an assessment body can only be a body, whose competence was recognized by the authorized body (in Poland, the process of accreditation by the PCA - the Polish Centre for Accreditation).

© Springer International Publishing AG 2016
J. Mikulski (Ed.): TST 2016, CCIS 640, pp. 26–34, 2016.
DOI: 10.1007/978-3-319-49646-7_3

– Obliging the entities in charge of maintenance to observe the Regulation - according to new rules, a structured risk management process in accordance with the requirements of the common safety methods included the entities in change of maintenance (ECM). Thus, it has become important to use assessment bodies in the process of risk and safety management in rail transport. The mechanisms defined for the purposes of their operation are now becoming one of the key factors of their effective management. The aim of this paper is to present and discuss some aspects of the functioning of independent assessment bodies operating in the rail transport sector in Poland.

2   Risk and Change Management in Rail Transport

The modern principles of rail transport operation are based on the mechanisms of risk and safety management in rail transport. Every decision is burdened with some kind of risk, hence the concept of risk begins to set a new dimension of railway undertakings management. It is important to approach risk in rail business analytics in a systemic way, so as to make it an indispensable factor influencing the organizational behaviour of managers in railway undertakings. Then business analytics enables management supported by forecasting and planning processes, based on the mutual synthesis of cause and effect relationships. The cause and effect approach strengthens the place and role of effective risk management in rail transport. Risk management is based on both the utility theory and probability theory [2]. The approach to a risk process should be standardized. Standards should be changed only in the context of the specific characteristics of the functioning of various modes of transport. Hazard identification should be very similar, and the differences should result only from specific technical conditions [3]. Safety understood as a subjective state in which there is no hazard determines the activity of the railway sector and opens up new perspectives. The new requirements of the European Union create a proactive dimension of safety management having its particular dynamics. Safety can also be regarded from the perspective of the following:

Safety as a feature of the quality of services provided to railway, safety as an element of achieving customer satisfaction and loyalty, safety as an element of creating company brand in the market, safety as an element of pursuing environmental objectives, safety as an extension of requirements related to health and safety at work, safety as an element of introducing innovative products to the railway market, safety as an element of the life cycle of a railway vehicle.

Examining the operation of rail transport in the legal dimension, three most important directives related to rail transport can be distinguished: railway safety directives [4, 5] and the railway interoperability directive [6]. They determine the direction of operation and improvement of railway systems. It should be highlighted that a safe system is a system in which all components are certified in accordance with the technical specifications for interoperability (TSI). TSI are covered by the risk analysis in the scope of all interfaces between these components. This approach combines the principles of interoperability with safety management in rail transport. Factors strengthening these processes include investment and innovation, showing new areas to improve the rail transport sector. Innovation and railway technology can be transferred by cluster initiatives [7].

3 The Place and Role of Independent Assessment Bodies in the Polish Legislation

Current management principles with the use of assessment bodies define three types of bodies: Notified Body NoBo, Designed Body DeBo, and Assessment Body AsBo. The Notified Body NoBo assesses the compatibility of interoperability constituents and interoperable sub-systems with Community law. The Designed Body DeBo assesses the conformity of buildings used for railway traffic operation and the types of devices for railway traffic with national law.

An Assessment Body AsBo means the independent and competent external or internal individual, organisation or entity which undertakes investigation to provide a judgement, based on evidence, of the suitability of a system to fulfil its safety requirements [1].

Among these three bodies, independent assessment bodies, AsBo play an important role. The role of the assessment body is:

- to support the entity that introduces a change in the risk management process by verifying and confirming that the risk management process was carried out correctly and comprehensively (with regard to form and factual content).
- to take further analytical action or implement additional solutions as regards control measures to ensure the appropriate level of safety after introducing the planned change.

These tasks are described in detail in Art. 6 of the Regulation 402/2013, which includes the independent assessment of the adequacy of the risk management process described in Annex I and its outcomes.

There are three types of assessment bodies:

- Type A- legally and functionally autonomous entities, conducting inspections only as
- a third party.
- Type B- bodies which are functionally part of a larger organization (parent), conducting inspections solely for the larger organization.
- Type C- bodies that are a branch or department of a larger organization. They can provide services to external organizations, but they are not bodies acting as an independent third party.

In Poland, bodies of types A and B are likely to be more common.

4 The Tasks of Independent Assessment Bodies in Risk Management in Rail Transport

An Assessment Body AsBo deals with the independent assessment of procedures and practices in the analysis of risk evaluation and safety measures, applied by infrastructure managers and railway undertakings, as well as of their assessment of risk acceptability in the case of significant changes.

These changes include, in particular, the sub-systems development and introducing into operation, including the verification of their technical compatibility with railway systems which they are introduced into and the safe integration of these sub-systems. The safe integration of the system means linking two devices or railway systems, where the interface is assessed in terms of safety criteria. Then not a system component or the system is analysed, but the system with the interface in real conditions.

The tasks of the assessment body related to granting permission to operate railway systems result, inter alia, form the rail system interoperability regulation [8]. The regulation says that the technical documentation attached to the EC declaration of sub-system verification includes a report on the safety assessment issued by the assessment body.

Accreditation requirements and the conditions for granting and maintaining accreditation for assessment bodies are contained in a document describing the accreditation rules [9].

In order for an assessment body to be recognized as reliable and competent, it should fulfill:

- General accreditation requirements specified in the PN-EN ISO/IEC 17020 Conformity assessment. Requirements for the operation of various types of bodies performing inspection
- Specific accreditation requirements set out in the Commission Implementing Regulation (EU) No 402/2013 of 30 April 2013., on a common safety method for risk evaluation and assessment, and repealing Regulation (EC) No 352/2009 (OJ EC L 121 of 05.03.2013)
- DA-05 Policy for Participation in Proficiency Testing (if the assessment body uses studies conducted by research laboratories);
- DA-06 Policy for Ensuring Traceability of Measurement Results.
- DAK 08- the accreditation of assessment bodies for the activities covered by the Implementing Regulation (EU) No 402/2013, Issue 1 Warsaw, 13.03.2015

The scope of assessment body accreditation, according to DAK 08, includes the inspections of the adequacy of applying a common safety method for risk evaluation and assessment of the following sub-systems.

Structural subsystems:

- Infrastructure
- Energy
- Control - onboard devices
- Control - track-side devices
- Rolling stock

Functional subsytems:

- Rail traffic
- Maintenance
- Telematics applications

Management systems:

- SMS - Safety Management System
- MMS - Maintenance Management System
- QMS - Quality Management System

It should be highlighted that the inspection of a significant change in rail transport is of different nature than other inspections. This type of inspection is primarily based on examining documents, verifying their completeness, verifying the process of documenting the required mechanisms of evaluation and assessment carried out by the customer. This specific character determines the different range of interpretation, both in the process and selection of experts for its implementation. An important document supporting the activities of assessment bodies and a risk management process is the EU guide on a common safety method for evaluation and assessment [10]. To facilitate an inspection process, the National Safety Authority (NSA) has developed, in the form of a guide, an expert opinion on the practical application of a common safety method for risk evaluation and assessment (CSM RA) by railway sector entities [11].

5   The Documentation of Independent Assessment Bodies in Risk Management in Rail Transport

An assessment body determines the processes needed for the Quality Management System and their application throughout the organization. The sequence of these processes and their interaction has been determined in the form of process maps. Processes include management, main and support processes. The activities of the inspection department included in the documentation of the Quality Management System proceed as planned and have a systemic character. They are characterized by a clear distribution of competences, responsibilities and division of tasks allowing for quality planning.

An assessment body cooperates with a group of consultants, eminent specialists in the rail industry. When choosing a subcontractor, the assessment body uses only the database of experienced and tested experts. The assessment body enters into civil law contracts with subcontractors, which contain relevant provisions for the provision of services and supervision of the service provision.

The following is the map of processes carried out by an independent assessment body (Fig. 1).

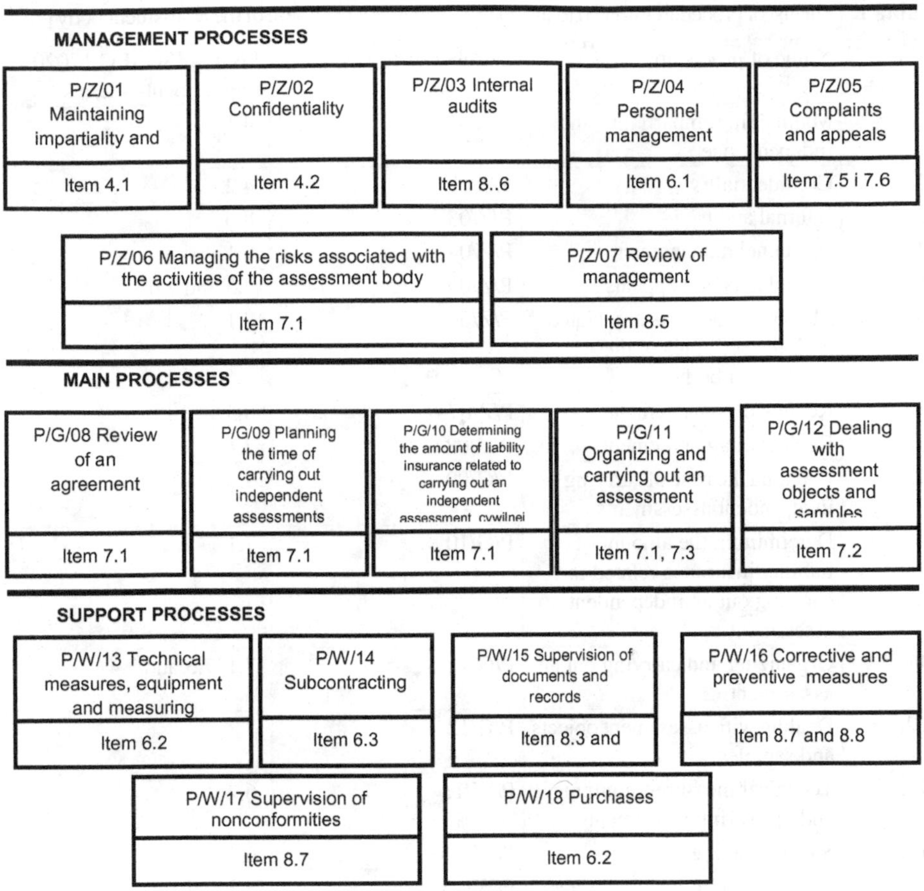

MANAGEMENT PROCESSES

P/Z/01 Maintaining impartiality and	P/Z/02 Confidentiality	P/Z/03 Internal audits	P/Z/04 Personnel management	P/Z/05 Complaints and appeals
Item 4.1	Item 4.2	Item 8..6	Item 6.1	Item 7.5 i 7.6

P/Z/06 Managing the risks associated with the activities of the assessment body	P/Z/07 Review of management
Item 7.1	Item 8.5

MAIN PROCESSES

P/G/08 Review of an agreement	P/G/09 Planning the time of carrying out independent assessments	P/G/10 Determining the amount of liability insurance related to carrying out an independent assessment cywilnei	P/G/11 Organizing and carrying out an assessment	P/G/12 Dealing with assessment objects and samples
Item 7.1	Item 7.1	Item 7.1	Item 7.1, 7.3	Item 7.2

SUPPORT PROCESSES

P/W/13 Technical measures, equipment and measuring	P/W/14 Subcontracting	P/W/15 Supervision of documents and records	P/W/16 Corrective and preventive measures
Item 6.2	Item 6.3	Item 8.3 and	Item 8.7 and 8.8

P/W/17 Supervision of nonconformities	P/W/18 Purchases
Item 8.7	Item 6.2

Fig. 1. The map of processes of the OTTIMA plus assessment body – the department of railway inspection [12].

Legend: P/G/....- The procedure describing main processes
P/Z/...- The procedure describing management processes
P/W/...- The procedure describing support processes
Item.... – Item in PN-EN ISO/IEC 17020

Clear procedures determine the correct model of the assessment body functioning. Table 1 presents a list of procedures related to the quality management system of the assessment body.

Table 1. The list of procedures related to the quality management system of the assessment body [12].

No.	Name of procedure	Number of procedure	PN-EN ISO/IEC 17020 requirements - items
1.	Maintaining impartiality and independence	P/Z/01	4.1
2.	Confidentiality	P/Z/02	4.2
3.	Internal audits	P/Z/03	8.6
4.	Personnel management	P/Z/04	6.1
5.	Complaints and appeals	P/Z/05	7.5 and 7.6
6.	Managing the risks associated with the activities of the assessment body	P/Z/06	7.1
7.	Review of management	P/Z/07	8.5
8.	Review of an agreement	P/G/08	7.1
9.	Planning the time of carrying out independent assessments	P/G/09	7.1
10.	Determining the amount of liability insurance related to carrying out an independent assessment	P/G/10	7.1
11.	Organizing and carrying out an assessment	P/G/11	7.1, 7.3 and 7.4
12.	Dealing with assessment objects and samples	P/G/12	7.2
13.	Technical measures, equipment and measuring instruments	P/W/13	6.2
14.	Subcontracting	P/W/14	6.3
15.	Supervision of documents and records	P/W/15	8.3 and 8.4
16.	Corrective and preventive measures	P/W/16	8.7 and 8.8
17.	Supervision of nonconformities	P/W/17	8.7
18.	Purchases	P/W/18	6.2

The result of the work of the assessment body is a safety assessment report. The safety assessment report means the document containing the conclusions of the assessment performed by an assessment body on the system under assessment [1]; The report is prepared on the basis of full documentation provided by the customer and visits and audit in real conditions.

A safety assessment report should include the following elements:

1. The identification of the assessment body
2. The identification of the proposer
3. The plan of an independent assessment
4. The limitations of an independent assessment
5. The scope of independent assessment

6. The results of the various stages of assessment
   6.1. The main principles of risk management
   6.2. Interface and common risk management
   6.3. The description of the process of risk evaluation
   6.4. The identification of hazards
   6.5. Using the code of practice during risk evaluation
   6.6. Using a reference system during risk evaluation
   6.7. The application and evaluation of explicit risk
   6.8. Management of risks- a risk management process
   6.9. The evidence resulting from applying the risk management process
7. Final conclusions of the independent assessment - recommendations
8. Conclusions
9. Reference documents related to the assessment
10. Annexes

In order for the assumptions contained in the report to be fully implemented, a checklist with questions is used to collect the necessary data, information and knowledge necessary to prepare a required safety assessment report.

6   Recommendations Related to the Operation of Independent Assessment Bodies in Polish Conditions

Carrying out a multi-dimensional assessment of the operation of assessment bodies in Polish conditions, the following strategic problems can be identified:

1. The improper classification and interpretation of the accreditation held in relation to the type of change in the railway system.
2. Not having full, competent personnel capable of carrying out the inspection process correctly.
3. The excessive formalization of activities related to the possible, improper functioning of the quality management system of the assessment body.
4. Errors in the process of reviewing the agreement before the inspection process begins.
5. Errors in classifying a technical, operational and organizational change.
6. Inadequate cooperation with notified bodies, or its lack, an inability to find oneself in the environment of certification and cooperation with the NoBo and DeBo.
7. The ambiguous scope of inspection compared to DeBo and NoBo.
8. Failure to find the place and role of the assessment body in cooperation with the NSA (National Safety Authority).
9. Failure to compare activities with foreign assessment bodies, which have a different classification of certification ranges.

7   Conclusion

The proper supervision of rail traffic safety should be supported through the effective use of the place and role of independent assessment bodies in the process of managing changes in the railway system. These changes, determining the development of rail transport through their impact on safety, are a key factor of supervision. This supervision should have a stable logic that ensures the effective management of rail transport safety. The author has shown such factors and strategic problems that determine the effective use of independent assessment bodies in the process of creating safe conditions for railway traffic. This represents a significant challenge for the development of the railway sector in Poland. Thus, it is worth highlighting in the summary that in addition to other entities, assessment bodies play an important role in railway safety.

References

1. Commission Implementing Regulation (EU) No 402/2013 of 30 April 2013 on the common safety method for risk evaluation and assessment and repealing Regulation (EC) No 352/2009
2. Jabłoński A., Jabłoński M.: Ryzyko techniczne i zawodowe w transporcie kolejowym – kluczowe aspekty integracji, Technika Transportu Szynowego, September 2014
3. Jabłoński, A., Jabłoński, M.: Key challenges and problems in conducting independent evaluations of the adequacy of the risk management process in rail transport. In: Mikulski, J. (ed.) Tools of Transport Telematics, pp. 1–11. Springer, Heidelberg (2015)
4. Directive 2004/49/EC of the European Parliament and of the Council of 29 April 2004 on safety on the Community's railways and amending Council Directive 95/18/EC on the licensing of railway undertakings and Directive 2001/14/EC on the allocation of railway infrastructure capacity and the levying of charges for the use of railway infrastructure and safety certification (Directive on rail safety)
5. Directive of the European Parliament and Council Directive 2008/110/EC of 16 December 2008 amending Directive 2004/49/EC on safety on the Community's railways (Directive on rail safety)
6. Directive of the European Parliament and Council Directive 2008/57/EC of 17 June 2008 on the interoperability of the rail system within the Community (Recast)
7. Jabłoński, A., Jabłoński, M.: Transfer of technology in the field of rail transport through cluster initiatives management, communications in computer and information science. In: Mikulski, J. (ed.) Activities of Transport Telematics, pp. 58–66. Springer, Heidelberg (2013)
8. Regulation of the Minister of Transport, Construction and Maritime Economy of 6 November 2013 on the interoperability of the rail system, Item.1297
9. Accreditation of Inspection Bodies, specific requirements DAK-07 Issue 7 Warszawa, 13 February, 2014
10. Common Safety Method for risk evaluation and assessment Guidance on the application of Commission Regulation (EU) 402/2013, March 2015
11. An expert opinion on the practical application of requirements of a common safety method for risk assessment (CSM RA) - a guide, UTK, Warszawa (2015)
12. The documentation of the quality management system of OTTIMA plus assessment body

Transaction Costs in the IT Project Implementation Illustrated with the Example of Silesian Card of Public Services

Anna Urbanek[✉]

Department of Transport, University of Economics, Katowice, Poland
anna.urbanek@ue.katowice.pl

Abstract. Transaction costs theory explains the existence of the costs of economic exchange, which are very often not shown in the costs analysis. Transactions costs are the consequence of the relationships between entities and occur both before and after signing the contract. In practice, transaction costs are not identified, although proper assessment of the total investment costs requires knowledge of the size of transaction costs. The aim of the paper is to identify and estimate the size of transaction costs in the IT project implementation on the example of Silesian Card of Public Services in Poland (ŚKUP project).

Keywords: Transaction costs · IT project · ICT · Electronic payment systems · Electronic ticket · Public urban transport · Contracts

1 Introduction

IT projects rank among very difficult investment undertakings. The projects differ much, they often are of unique type. Also the structure of IT projects - from the point of view of project management - is very complex, as it requires taking into account not only the issues of software and hardware, but – first of all – relating those elements with the already existing infrastructure and the structure of the organization, is which the project is implemented. Moreover, one should also note that IT projects are of evolutionary type, they develop in the course of their implementation, as well as after the implementation in a given entity. Thus, the implementation and maintenance of the IT system in a given entity requires co-operation with a specialized external entity, active on the market. Implementation of an IT system by means of outsourcing is the most frequently used method of execution of IT projects, even in case of enterprises that have their own, relatively well developed, IT departments.

Complex, unique and evolutionary nature of IT projects makes the employer ordering such a system undertake many activities related to execution of such project, as well as risks related to it. This leads to generation of costs, which in economic theory are referred to as transaction costs. Those costs are not always known, as they are incurred in various places, and in economic practice the employer does not keep a register of those costs. Most often, at the stage of assessing the costs of project implementation those costs are narrowed down solely to the per-unit costs in the entity responsible for project management, as well as supervision of contract execution.

J. Mikulski (Ed.): TST 2016, CCIS 640, pp. 35–46, 2016.
DOI: 10.1007/978-3-319-49646-7_4

Less frequently they are increased by adding the financial services or organization and legal services. Commonly, cost accounting does not take into consideration the costs of project preparation, risk-related costs, or involvement of other department of the employer, that are indirectly engaged in project execution, which often has the consequences in underestimating the service costs of IT project.

In the paper, using the example of the Silesian Card of Public Services (ŚKUP – Śląska Karta Usług Publicznych) project, implemented in the central part of the Province of Silesia, certain theoretical assumptions for transaction costs have been presented, and attempt has been made to assess the amount of those costs, both for the investment period and for project operation.

2 Outsourcing of IT Projects in the Perspective of Transaction Costs Theory

The notion of transaction costs was introduced to economic analysis by Ronald Coase in the 1930s, and then developed by Oliver Williamson as theory of transaction cost, developed from new institutional economics. In the pioneering paper entitled The Nature of the Firm, published in 1937, R. Coase stated that the reason why firms emerged was the cost of using the price mechanism, which was connected, among other things, with: discovering suitable prices offered on the markets products and services, searching for trade partners, and negotiating contracts [2]. The theory of transaction costs makes people aware of the existence of costs of trade, which are neglected in traditional accounting. According to Coase, conducting market transactions requires the parties to find each other, to inform the parties involved about transaction conditions, conducting the negotiation process, drawing a contract, and monitoring its execution, which requires bearing certain costs [3].

Transaction costs of using the price mechanism are, first of all, the result of limited rationality of market participants. It causes the concluded contracts to be incomplete, while all changes and amendments to the signed contracts are costly and inevitably connected with occurrence of conflicts between parties, which are intensified by the possibility that opportunistic behaviour may occur [12]. Williamson defines opportunism as self-interest seeking with guile. This comprises its more extreme forms, such as lying, stealing or cheating, still opportunism more often uses subtle forms of trickery. Limited rationality of a firm's activity entails that in practice firm do not maximize profits, but merely strive to achieve satisfactory profits. Firms apply only routine, not optimizing principles of financial accounting [11].

Transaction costs appear in various areas of the enterprises activities, and at various stages of contract execution. Those costs may be divided into three basic groups [11]:

– Costs of information asymmetry – decisions concerning conclusion of contracts are always made in information asymmetry conditions, the knowledge about transaction subject is not distributed evenly, parties to the contract do not have access to the same information resources, they have to pay the costs of searching information, that is costs of market research, obtaining the knowledge concerning potential suppliers, availability of goods and services, that is knowledge concerning prices.

- Costs of contract conclusion and management – each contract demands proper preparation: from unambiguous formulation of its subject, through selection of the employer, e.g. through tender procedure, preparation of contract content, negotiations, and establishment of provisions for due execution of the contract [8].
- The costs of control and verification of contract execution – costs of supervision of due execution of the contract, incurred in connection with monitoring, control of processes and results, quality of foods and services delivered, costs of insurance to cover losses caused by improper behaviour of parties and other risks, e.g. those related to the necessary changes in contract content, and adjusting it to the changing conditions, or related to the risk of contract breach.

Among the factors that influence the level of transaction costs, the following are listed [13]:

- Specificity of assets indispensable for the transaction;
- Uncertainty and complexity of transactions;
- As well as their frequency and regularity.

Generally speaking, the assets of an enterprise are specific if they cannot be used easily in other enterprises, due to physical properties, specificity of location or of labour force. Nonspecific transactions are regulated most efficiently by the market. A certain repeatability of transactions is connected with a standardized product or service, usually with many possible substitutes or suppliers, thanks to which transaction costs turn out to be lower than in case of coordination of unique and rare transactions. Transaction costs increase if the enterprises functions in changing and unpredictable environment, in which the development of specific mechanisms and procedures of conduct is difficult or even impossible [13].

The basic unit used in the analysis of new institutional economics is transaction, while the enterprise is understood as a nexus of contracts. The key element, decisive for the efficiency of the enterprise is the way of its organization and internal regulation. The entrepreneur is a contractor, who deals with organizing the firm, conclusion of contracts, establishment of unique connection of production factors. Decisive for the efficiency of firm activities will be the knowledge about useful forms of contracts, technology of contract conclusion, as well as the knowledge concerning securing and monitoring of contracts [1].

The problems of transaction costs are thus strictly related to the necessity of making a choice between generation of goods on one's own, and purchasing of goods and services from independent specialized entities, present on the market. Cooperation with other entities, agreements with suppliers, client, or even competitors – e.g. under strategic alliances – is the expression of rational striving for reduction of transaction costs. In accordance with the transaction costs theory, specialized enterprises, which use external sources of foods and services as a rule function more efficiently than the enterprises that perform a wide scope of activities on their own. Purchasing goods and services from external sources is connected, however, with higher costs resulting from the necessity of planning, supervising, and coordinating the functioning of enterprise, as well as the network of organizations that cooperate with it [3, 11].

In the light of transaction costs theory, the implementation and maintenance of IT systems is an example of a complex transaction, with a high degree of specificity and uncertainty, generating – as a rule – high transaction costs. Additionally, legal references are decisive for high degree of specificity of IT projects, in particular the regulations concerning copyright, as well as highly specialized producers of software. The complexity of IT projects and their evolutionary nature are the reasons why such contracts are always concluded in information asymmetry conditions, while the risk of occurrence of opportunistic behaviour, the so-called moral hazard is very substantial, for both the supplier and the client. This can be confirmed by the studies conducted in Poland in the years 2010–2014, which show that as many as 39% of IT projects implemented in outsourcing mode, that concerned the implementation of IT systems assisting management in enterprises, end in complete or partial failure [11].

Special attention in that respect should be devoted to IT projects executed by the public sector. Striving for increased efficiency of public sector functioning and good quality of public services provided makes the public sector executes the vast majority of big IT projects, having substantial budgets [6, 7]. Those investment projects are executed in cooperation with private entities specialized in that respect, selected in public tender procedures. The forms of cooperation between the public party and private partner may differ much, and may range from public procurement to close cooperation under public-private partnership. Whatever form of cooperation has been selected, due to the contract concluded, legal requirements concerning the public procurement related to transparency of the transaction, issues of ownership, organization structure, and project financing structure, as well as distribution of risks between parties – the investment projects undertaken by the public sector with a private partner will generate much higher transaction costs, when compared with other contracts concluded on the market.

3 Transaction Costs in the Process of Implementation of Silesian Card of Public Services Project

Silesian Card of Public Services is a project implemented by the Municipal Transport Union of the Upper Silesian Industrial District (KZK GOP), which is the biggest organizer of public transport services in Poland, and one of the biggest in Europe. The Union is presently made up of 29 municipalities from the central part of the province of Silesia. Also two municipalities that are not members of KZK GOP participate in the project (Tychy and Jaworzno). The aim of the project is to develop a trans-local IT system, which will increase the range and availability of services provided electronically by public institutions, and at the same time will become a tool assisting the process of management in public administration. Contactless ŚKUP smart card is a carrier of electronic money (e-purse), which enables making payments not only for the use of public transport, but also for municipal administrative services, culture-related services, mass recreation, library services, and parking. The card is thus not only a carrier of season electronic tickets in urban mass transport, but also allows for dynamic settlements of journey costs and time, using the pay as you go systems (where registration of the time one gets on and off is required) in various tariff systems [10].

In the scope of the project implemented so far, the ŚKUP project required developing IT infrastructure which consisted, among other things, of [5, 9]:

- A portal for customers, that is an internet platform to provide card-related services and for ticket purchase;
- Establishing and equipping 40 customer service points, and 800 agents selling public transport tickets;
- Provision of suitable equipment to the 7 Customer Service Points that already existed;
- 109 ticket vending machines with the function of ŚKUP cards topping-up function;
- 223 parking meters adjusted to the use of ŚKUP cards;
- 410 modules for collecting fees and charges for municipal services;
- Provision of 1300 public transport vehicles with devices for collecting fares, on board computers, data transmission devices, and other indispensable technical elements;
- Provision of 20 bus and tram depos with suitable infrastructure for data collection and transmission;
- 320 devices for ticket inspection;
- Development of two modern Data Processing Centres;
- Provision of suitable software for the system, which enables management of devices and business analysis of the data received.

Within the framework of ŚKUP project also the maintenance and issuing of contactless electronic cards is necessary, which is provided by the bank, that in consortium with the IT company is the contractor for the system. During the execution of the project and operation of the system, it is planned to issue 700 thousand cards (385 thousand during implementation, and 325 thousand within 5 years from the implementation completion, which is connected with card wear out and necessity of issuing them for other reasons) [5].

The implementation of ŚKUP project was carried out in two stages. Stage one involved the implementation of the system which, according to the contract, was to take place within 21 months of its signing. The expenditures at that stage were co-financed from the European Regional Fund, under the Regional Operational Program for the Silesian Voivodship, for the years 2007–2013. The second stage of project, on the other hand, comprised the maintenance of the ŚKUP system by the contractor, for the period of 65 after completion of its implementation. The system maintenance concerned – first of all – servicing and operations of all elements of system infrastructure, as well as provision of authorization and settlement services. The total value of the project, comprising implementation and maintenance, amounts to some 190 million PLN net [4].

The contract for delivery, implementation and maintenance of the ŚKUP system was concluded between Municipal Transport Union of the Upper Silesian Industrial District (KZK GOP) and the contractor on January 9, 2012. If we try to identify the transaction costs of such an IT project, however, it is worth noting that work upon the system commenced 5 years earlier, that is in 2007. Over the period of 5 years preceding the start of the investment project, it was necessary to bear certain costs for its preparation and analysis of its viability, among others the costs related to the preparation of functional and technical design by an external entity, cost estimates, timetable, and feasibility study. Moreover, which often takes place when executing such a big and complex project, the public party also used the services of external experts, whose role was to assess the designed solutions. These are examples of typical transaction costs, related to information asymmetry and contract preparation. It is also worth adding that such a lengthy order preparation procedure was also related to the fact that the project applied for co-financing from European Union funds, which required meeting the specific requirements and formal-legal procedures. Table 1 presents the detailed time schedule concerning preparations for implementation of the ŚKUP project.

Table 1 Time schedule concerning preparations for implementation of the ŚKUP project [own study]

1	KZK GOP submitted the first version of PWP (Project Work Plan) for ŚKUP project	April 2007
2	Submission of the final version of PWP for ŚKUP project and preliminary feasibility study	April 2008
3	Conclusion of contract with the Contractor, selected via the procedure of negotiations with announcement, for preparation of functional – technical design, together with the full set of ŚKUP documents	July 2008
4	Conclusion of contract for execution of joint ŚKUP project by KZK GOP and 21 municipalities – project partners	October 2009
5	Confirmation of the positive substantial and financial assessment of ŚKUP project by the Managing Authority of Regional Operational Programme WSL 2007 – 2013 and referral of the project for co-financing	April 2010
6	Signing the contract for co-financing of the project Silesian Card of Public Services (ŚKUP) by the Government of the Silesian Voivodship, performing the role of Managing Authority of ROP WSL for the years 2007 – 2013 and KZK GOP	October 2010
7	Publication of procurement notice for delivery, implementation, and maintenance of the Silesian Card of Public Services (ŚKUP) system	November 2010
8	Cancellation of the procedure of awarding the contract in the form of unlimited tender	June 2011
9	Repeated initiation of the procedure and Publication of procurement notice for delivery, implementation, and maintenance of the Silesian Card of Public Services (ŚKUP) system	July 2011
10	Announcing the result of unlimited tender and awarding the contract to the winning bidder	November 2011
11	Signing the contract for delivery, implementation and maintenance of the system Silesian Card of Public Services (ŚKUP) with the Contractor	January 2012

As results from the table, the very procedure of contract preparation required conclusion of many contracts, and in connection with that bearing transaction costs, which are not included in the cost accounting for projects. Identification and valuation of costs related to the necessity of hiring external experts or preparation of the functional-technical design and feasibility study is not difficult, as those costs can be obtained from the contracts signed. In case of the ŚKUP project, those costs amounted to some 0.2 % of the entire project costs. Nevertheless, the contract preparation also required transaction costs to be paid, related to selection of contractor for the project in public tender procedure, which was initiated twice. It is worth noting that nearly 14 months elapsed from the first publication of the procurement notice to signing the contract with contractor. It is very difficult to assess those costs, as it requires assessment of involvement of the firm in the activities performed. In the process of contract preparation they will be negligible and will be skipped in the analysis. However, those costs will have very significant importance during project implementation and maintenance.

Implementation of an IT project such as ŚKUP requires, besides bearing the costs related to development of suitable IT infrastructure, also requires paying the costs of project management. They are the costs of establishing, functioning, as well as the costs of settlement of the managing unit, responsible for organizational, technical, and formal execution of the project. Those costs comprise of the costs of remuneration (project implementation required the employment of minimum 25 people), other costs of administration, costs related to information and promotion activities, as well as costs connected with training of system users, purchase of fixed assets as well as office equipment and fittings required for project management purposes, also costs related to consultations, opinions, analyses as part of the tasks related with project needs. They are typical transaction costs of management and coordination of the undertaking, contractor supervision, and monitoring the execution of the undertaking. Costs of project management are kept on a separate sub-account in project cost accounting, they are estimated as early as at the stage of project feasibility study. Project management costs reflect the costs paid in connection with establishing the so-called project management team, the establishment of which is necessary in case of executing big investment projects. In case of the ŚKUP Project, in KZK GOP the Silesian Card of Public Services Department was established, with 27 people of staff, which consists of 5 sections:

- Section for Electronic Card Payment Systems;
- Section for IT support of ŚKUP;
- Section for Transport Organizers and Operators;
- Section for Municipal Services;
- Section for Organization and Legal issues concerning ŚKUP.

It is worth noting, however, that also other units in the organization are involved in the execution of such a project, they very frequently are directly involved in the process of system implementation, and assessment of results achieved. For example, as many as 66 employees from other departments of the organization (29% of all employees of KZK GOP from departments other than the ŚKUP Department) have been involved in commissioning individual modules of the system, separated from functional and business perspective, with as many as 76% of managers (heads of departments and sections) responsible for functioning of units that have been involved directly.

Table 2 presents financial data from KZK GOP, concerning the costs of ŚKUP project implementation, including also the costs of project management, and administrative costs of the KZK GOP office. On the basis of the analysis of project management costs, as well as the analysis of involvement level of specific KZK GOP departments in the work concerning implementation of ŚKUP project, overall transaction costs of project implementation have been assessed. The time-frame of the analysis results from the fact that stage one of system implementation has been extended. Due to the many problems, which occurred during project execution, the implementation stage was only formally completed in October 2015, and 2016 may be considered the first year of system maintenance.

Table 2. Transaction costs assessment in the implementation of ŚKUP project in the investment period (in Polish zloty - PLN, net) [own study]

No.	Investment period						Total costs in the investment period
	Year 1	Year 2	Year 3	Year 4	Year 5	Year 6	
	2010	2011	2012	2013	2014	2015*	
1.	Costs of execution of the ŚKUP project						
	348 540	786 096	921 012	14 992 645	30 585 665	75 299 391	122 933 349
2.	of which: Project management costs						
	348 540	786 096	710 398	1 102 752	1 795 423	3 008 200	7 751 409
3.	Share of Project management costs in the total costs of Project execution						
	100.0%	100.0%	77.1%	7.4%	5.9%	4.0%	6.3%
4.	Administrative costs of the KZK GOP Office (excluding expenditures for EU projects)						
	14 343 064	14 657 228	14 874 155	16 346 580	17 257 811	23 070 920	100 549 757
5.	Estimated costs of involvement of other department of the Office in Project execution						
	-	-	743 708	1 144 261	2 243 515	2 999 220	7 130 703
6.	Total transaction costs related to implementation of ŚKUP (items 2+5)						
	348 540	786 096	1 454 106	2 247 012	4 038 938	6 007 420	14 882 112
7.	Share of transaction costs related to project implementation in the total costs of Project execution						
	100.0%	100.0%	157.9%	15.0%	13.2%	8.0%	12.1%

* Forecasted execution of financial plan.

The costs of involvement of the remaining part of KZK GOP Office have been assessed on the basis of analysis concerning the number of employees not from the ŚKUP Department, who have been directly involved in work on ŚKUP project, and average time devoted to that work. It can be concluded from the studies conducted in KZK GOP that an average employee who participated directly in various work teams and commissioning activities devoted half of her/his work time to those activities.

As can be concluded from Table 2, in the course of the 6 years of investment, transaction costs paid in connection with project implementation amounted, in case of ŚKUP, to 12.1% of total costs of project. In the investment period, asset-related expenditures dominate, linked with the purchase of required elements of system infrastructure. It can be noted, however, that in 2012 the transaction costs de facto exceeded the costs of project implementation, aggregated on cost account (they amounted to 157.9% of the aggregated costs of project execution).

Table 3 presents the estimated transaction costs of ŚKUP project over the 6 years of system maintenance. The costs of execution of the ŚKUP project, and the costs of administration related to KZK GOP office are forecasts derived from long-term financial forecasts prepared by KZK GOP as a unit of public finance sector.

Table 3. Assessment of transaction costs for ŚKUP project, over the 6 years of system maintenance (in Polish zloty – PLN, net) [own study]

No.	Maintenance period						Total costs during the maintenance period
	Year 1	**Year 2**	**Year 3**	**Year 4**	**Year 5**	**Year 6**	
	2016	**2017**	**2018**	**2019**	**2020**	**2021**	
1	**Costs of execution of the ŚKUP project**						
	15 466 207	10 071 263	9 753 263	11 287 030	13 239 981	7 252 257	67 070 001
2	of which: Project management costs						
	2 220 800	2 282 696	2 442 351	2 533 286	2 628 768	1 344 548	13 452 449
3	Share of Project management costs in the total costs of Project execution						
	14.4%	22.7%	25.0%	22.4%	19.9%	18.5%	20.1%
4	**Administrative costs of the KZK GOP Office (excluding expenditures for EU projects)**						
	23 801 920	24 325 561	25 128 300	25 932 380	26 762 200	27 591 820	153 542 181
5	Estimated costs of involvement of other department of the Office in Project execution						
	3 094 250	1 459 534	1 507 698	1 555 943	1 605 732	1 655 509	10 878 665
6	**Total transaction costs related to implementation of ŚKUP (items 2+5)**						
	5 315 050	3 742 230	3 950 049	4 089 229	4 234 500	3 000 057	24 331 114
7	Share of transaction costs related to project implementation in the total costs of Project execution						
	34.4%	37.2%	40.5%	36.2%	32.0%	41.4%	36.3%

The estimates of involvement costs of the remaining parts of the institution in the implementation of ŚKUP project have been made on the basis of the same assumptions as those in Table 3, yet taking into account the fact that due to changes in organization, and development of new management procedures – resulting from the fact that suitable units of organization take over new tasks – the average labour intensity of those activities will be reduced by half in the years 2017–2021.

It can be concluded from the analysis of data presented in Table 3 that the share of transaction costs related to the project implementation in the total costs of its implementation is significantly higher in the system maintenance period than in the investment period. It assumed that altogether in the maintenance period the transaction costs will amount to 36.3% of the total maintenance costs for ŚKUP system, over the period of 6 years. This is, first of all, due to the fact that in the maintenance period the expenditures related to tangible assets are much lower.

Fig. 1 presents the estimated total transaction costs of ŚKUP project as a part of total project execution costs over the 12 years of preparation, investment and maintenance.

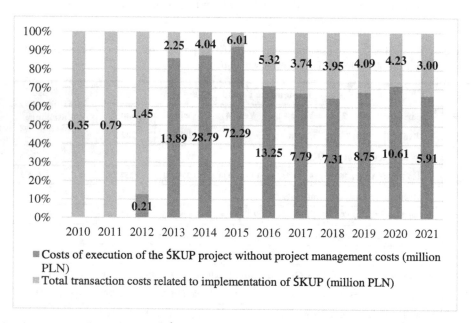

Fig. 1. Total transaction costs of ŚKUP project as a part of total project execution costs, over the 12 years of preparation, investment and maintenance [own study]

It can be concluded from the analysis of data presented in Fig. 1 that considering transaction costs during the whole analyzed period, total costs of ŚKUP project seem to be a considerable underestimate.

4 Conclusion

The economics of transaction costs is one of the most dynamically developing research problems in contemporary economy. It is a very crucial and not completely investigated area of economy. The emergence of the transaction costs theory is the sign of search for new economic models, which more accurately describe the actual conditions for functioning and development of enterprises, as well as co-operation between various business units.

Each firm, which begins the execution of an investment project, which concerns the implementation of an IT project must take into account the necessity of bearing substantial transaction costs, related to the preparation of the project and its management. Although it does not result directly from the analysis performed, one can risk a statement that probably there is an inverse relationship between the level of transactions costs paid during the contract preparation phase and selection of the most advantageous offer, and the investment costs made during its execution. Using more funds for better preparation and management of project may result, on the part of the employer, in reduction of execution costs and limitation of the risk of conflict between parties during project execution.

The Silesian Card of Public Services project is a big and complex one, mainly because of the substantial number of municipalities involved, and the use of electronic money for payment for various type of public services comprised in the project. Transaction costs apply to both sides of the transaction, that is both the employer/client and contractor/supplier of the system. However, the analysis applied only to transaction costs concerning the employer, in the case of the ŚKUP project it was a public entity.

On the basis of calculations made one can state that in the investment period the share of transaction costs related to project implementation in the total costs of project execution amounted to 12.1%, whereas over the 6 years of system maintenance that share will amount to 36.3%. Of course, those are amounts calculated for that specific project, as it should be pointed out that the level of transaction costs depends on many factors, first of all on the degree of project complexity, project size, and its unique character. It cannot remain unnoticed, however, that those costs - in particular those reflecting labour costs – will be significantly lower in case of public administration entities, than in case of private entities, due to the substantial regulations concerning remuneration level in the public sector.

Identification of transaction cost level should not be a big problem, yet such a record is not always kept in economic practice. It is not customary, nor is it required by law, also those costs are not part of financial statements. At the same time, analysis of transaction costs would allow to property analyze the costs of project execution/implementation, and to determine the actual degree of the party's involvement in the investment, which would undoubtedly be a strong and rational foundation in the decision- making process concerning execution of such projects.

References

1. Alchian, A., Demsetz, H.: Production, Information Costs, and Economic Organization, The American Economic Review, no. 62, pp. 777–795 (1972). https://www.aeaweb.org/aer/top20/62.5.777-795.pdf. Accessed 4 December 2015
2. Coase, R.H.: The nature of the firm. Economica **4**(16), 386–405 (1937)
3. Coase, R.H.: The problem of social cost. J. Law Econ. **3**(1), 1–44 (1960)
4. Dydkowski, G.: Effectiveness of the urban services electronic payment systems on the example of Silesian card of public services. Arch. Transp. Syst. Telematics **7**(4), 3–8 (2014)

5. Dydkowski, G.: Transformations in the ticket distribution network for public urban transport in the processes of implementation of electronic fare collection systems. In: Mikulski, J. (ed.) Tools of Transport Telematics, pp. 198–209. Springer, Heidelberg (2015)
6. Kos-Łabęddowicz, J.: Integrated e – ticketing system – possibilities of introduction in EU. In: Mikulski, J. (ed.) Telematics - Support for Transport. CCIS, vol. 471, pp. 376–385. Springer, Heidelberg (2014)
7. Kos, B.: Development of electronics payment in Poland using the example of local and regional collective transport. In: Mikulski, J. (ed.) Tools of Transport Telematics. CCIS, vol. 531, pp. 352–361. Springer, Heidelberg (2015)
8. Miranda, S.M., Kim, Y.M.: Professional versus political contexts: institutional mitigation and the transaction cost heuristic in information systems outsourcing. MIS Q. 30(3), 725–753 (2006). http://faculty-staff.ou.edu/M/Shaila.M.Miranda-1/M%26K06.pdf. Accessed 12 Apr 2015
9. Silesian Card of Public Services. http://www.karta.skup.pl. Accessed 12 Apr 2015
10. Urbanek, A.: Pricing policy after the implementation of electronic ticketing technology in public urban transport: an exploratory study in Poland. In: Mikulski, J. (ed.) Tools of Transport Telematics. CCIS, vol. 531, pp. 322–332. Springer, Heidelberg (2015)
11. Wachnik, B.: The analysis of transaction costs in IT implementation projects with the use of outsourcing, business informatics, 1(35): 70–83 (2015)
12. Williamson, O.E.: The Economic Institutions of Capitalism, Markets, Relational Contracting. The Free Press, New York (1998)
13. Williamson, O.E.: Markets and hierarchies: some elementary considerations. Am. Econ. Rev. 63(2), 316–325 (1973)
14. Williamson, O.E.: Transaction cost economics: how it works, where it is headed. Economist 146(1), 23–58 (1998)

The Impact of Intelligent Transport Systems on an Accident Rate of the Chosen Part of Road Communication Network in the Slovak Republic

Alica Kalašová[1], Jerzy Mikulski[2(✉)], and Simona Kubíková[1]

[1] Faculty of Operation and Economics of Transport and Communications, University of Žilina, Univerzitná 1, 01026 Žilina, Slovakia
{alica.kalasova, simona.kubikova}@fpedas.uniza.sk
[2] University of Economics in Katowice, ul. 1 Maja 50, 40-287 Katowice, Poland
jerzy.mikulski@ue.katowice.pl

Abstract. Nowadays we record a large increase of individual automobile transport on roads. This situation results in still increasing intensity of traffic in cities and also outside cities. It leads to congestions and finally to traffic accidents. One of the possibilities of the number of traffic accidents reduce on the chosen part of communication network is implementation of ITS. On first class roads and highways where higher speed is permitted as on roads in cities is important to alert drivers of dangerous situations in advance to provide more time to decision making of drivers and adapt their driving to road conditions. In the paper we will discuss a model that compares two variants of the transport organisation. As a modelling tool we use software Aimsun which allows a microscopic simulation of real traffic conditions on any network.

Keywords: Model · Telematics applications · Traffic safety · Accident rate

1 Introduction

Nowadays, we record a large increase of individual automobile transport on roads in the Slovak Republic. This situation results in still increasing intensity of traffic in cities and also outside cities. It leads to congestions and finally to traffic accidents. European Commission promotes to reduce a number of road transport fatalities by half compare with 2010 for the years 2011 to 2020 as a priority. This way conceived intent is the key to improving overall performance of the transport system and meet the needs and expectations of citizens and society. The Slovak republic as a member of European Union has committed to meet this goal and reduce the number of road fatalities by half until 2020. Statistics shows that this aim is realistic. The Fig. 1 shows the number of traffic accidents for the years 2005 till 2014 in Slovakia [1].

The decrease of a number of traffic accidents is caused by a development and an evolution of automotive technologies, better and more quality roads, better markings and traffic management, as well as organising measures. The application of intelligent transport systems can be a next step how to decrease a number of traffic accidents.

© Springer International Publishing AG 2016
J. Mikulski (Ed.): TST 2016, CCIS 640, pp. 47–58, 2016.
DOI: 10.1007/978-3-319-49646-7_5

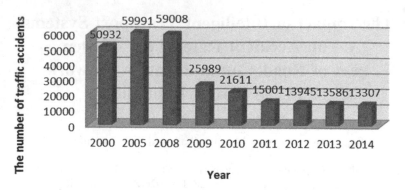

Fig. 1. ·The number of traffic accidents in the Slovak Republic [own study]

ITS are also very important to hold a sustainable development and to increase traffic safety, which is based on collecting, processing, evaluating and distributing the information. The information technologies create the base of transport telematics systems which include information about transport chain and participants of transport. To be sufficiently informed about traffic situation is the key to make decisions which route between origin and destination is the most appropriate for a driver [2].

2   Characteristic of the Chosen Part of Road Network

It is necessary to focus on sections of road network, where the highest intensity of traffic is concentrated. A communication network between Bytča and Ivachnová is the one of these sections which need to be solved.

Chosen part of network includes highway D1 connecting city of Bytča and Žilina, roads of first class I/11, I/18, I/59, I/61 and I/70 and roads of second class II/507, II/583. First class roads and highway are a part of the main transport corridor connecting the east and the west Slovakia. According to The National Traffic Census average intensity on road I/18 between Žilina and Martin is over 27 000 vehicles per 24 h. It is a lot of vehicles and congestions and traffic incidents occur very often.

One of the possibilities of the number of traffic accidents reduction on the chosen part of communication network is implementation of ITS. On first class roads and highways where higher speed is permitted it is important to alert drivers of dangerous situations in advance to provide more time to decision making and adapt driving to road and traffic conditions. Therefore, it is important to alert drivers on congestions, traffic accidents and also obstacles, communication status and visibility [1, 3].

On the chosen part of communication network there are a few critical accidents locations. These locations are chosen according to traffic accidents consequences and these sections must have length no more than 0,5 km. On Fig. 2. you can see the critical accident locations on our chosen network. Triangles highlight the locations. In 2014 in this area 216 traffic accidents were occurred according to statistics of Slovak Road Administration.

Fig. 2. Critical accident locations on the chosen part of communication network [own study]

As you can see traffic situation on this network need to be solved and safety and fluency of traffic flow has to be provided.

2.1 Questionnaire Survey Focused on Drivers' Perception of ITS

We made a questionnaire survey to find out the perception of ITS by drivers. The results of this survey are shown in the next chapter.

The drivers' voluntary acceptance of information is needed to traffic management using information systems and because of this fact the credibility of provided information is very important. This chapter is focused on drivers' perception of ITS. We performed a questionnaire survey which provides information how drivers obtain traffic information, which criteria are the most important for choosing a route, etc. [4].

Every driver when choosing his route, prefers route which best suits his needs and criteria. The basic criteria for selecting the most appropriate route are time availability of destination, fuel consumption, distance between origin and destination, risk of traffic accidents, etc. Figure 3. shows the weighted arithmetic averages of the importance of individual criteria. The most important criterion for men is quality of road infrastructure. For women the most important criterion is time availability of destination.

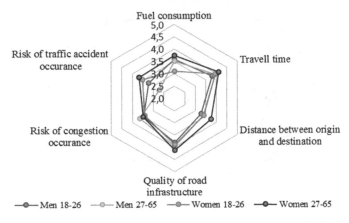

Fig. 3. The weighted arithmetic averages of the importance of individual criteria [own study]

We asked whether the respondents would accept an alternative way which is offered by information system, when traffic accident or other obstacle on their route occurs. The most common response was that respondents would be willing to adopt an alternative route, even if the distance to their destination would be longer more than 2 km and time availability of more than 16 min (Fig. 4).

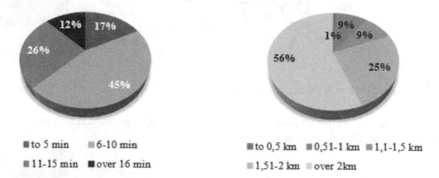

Fig. 4. Use of an alternative route according to time availability and length of detour [own study]

The next question was whether drivers appreciate intelligent transport systems to alert them to a possible danger and which of dangers they assign as the most important. We analysed the importance of warning to various danger traffic situations from the women's point of view and also from the men's point of view. The survey shows that women consider as the most important to be aware of a black ice and the least of a blind road. Men consider as the most important to be aware of a traffic accident and a black ice and the least of a blind road and a one-way road (see Fig. 5).

The overall analysis of the questionnaire shows that drivers are opening to the new possibilities of ITS use, especially to increase their own safety during travel [1].

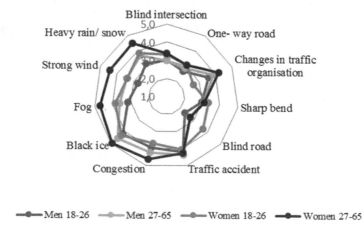

Fig. 5. The evaluation of warnings' importance of dangerous traffic situations [own study]

3 Modelling of Transport at the Chosen Part of Road Network

We have made a simulation of traffic flow in two variants. The first one represents a current state of transport at the chosen part of road network and the second one includes changes in network as construction a missing part of highway D1 between Žilina and Ivachnová and implementation of ITS elements. As a simulation tool we have used software Aimsun.

AIMSUN (Advanced Interactive Microscopic Simulator for Urban and Non- Urban Networks) is software of Spanish company TSS (Transport Simulation Systems) which is able to reproduce on computer real traffic conditions in any transport network. By simulation in this program it is possible to get a lot of outputs such as the average intensity, density, mean speed, section speed, travel time, delay time, stop time, number of stops, total travelled distance, total travel time, fuel consumption and quantity of produced emissions.

The base of microscopic simulation is to find optimal relation to every single situation and to verify it empirically, which is nearly always complicated. Compared with the conventional calculations, it allows much better approximation to reality and also much easier and better verifiable entering of the input data [5].

3.1 Creation of Model

First step was to import a background of communication network. We used free open street map, specifically the file OpenStreetMap data for Slovakia. On this background we create a road network by hand including nodes, sections, lanes, junctions and defined possible turnings, maximal speed and width of an each lane in every direction and transport management (see Fig. 6).

Fig. 6. Road network created in Aimsun [own study]

When the basic network was complete we made a description of travel demand. So on every section we defined the intensity of traffic flow and made origin/destination matrix. For transport generating the exponential function was used. Next step was to create signal plans for each junction (see Fig. 7). When the network was completely defined we ran microscopic simulation. the network.

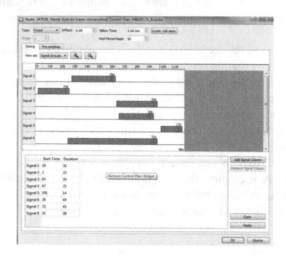

Fig. 7. Description of signal plan for junction in Žilina [own study]

3.2   Outputs of Microscopic Simulation (Variant A)

The variant A represents current state of traffic situation on the chosen part of network. We simulated traffic for 4 h since 7:00 a.m. to 11:00 a.m. The input data were obtained from road traffic surveys which were made by University of Žilina. For modelling we used data of traffic peak. The characteristics we were interested in were delay time, density, flow, fuel consumption, speed and travel time. Table 1 shows the outputs of variant A. On Fig. 8 you can see running of traffic flow in microscopic simulation [6].

Table 1. Outputs of microscopic simulation for variant A [own study]

Time series	Value	Standard deviation	Units
Delay time - Car	6,15	3,29	sec/km
Density - Car	1,46	N/A	veh/km
Flow - Car	2544,5	N/A	veh/h
Speed - Car	71,04	6,07	km/h
Travel time - Car	51,05	4,47	sec/km

In Table 1 there are outputs of simulation. Delay time for whole network is 6,15 s/km which is not so high number but on the most loaded section between Žilina and Martin delay time is 15,3 s/km and the density is 29 veh/km. It is a section of the

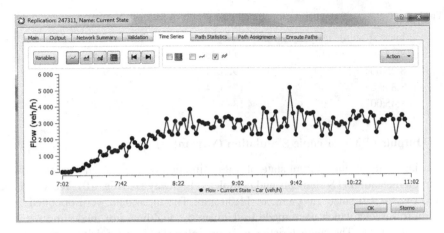

Fig. 8. Traffic flow running during simulation [own study]

highest number of traffic incidents because of high intensity. It is necessary to implement ITS elements and build missing highway connections.

The outputs for the chosen group of sections were recorded too. This chosen group is shown in the Fig. 9. During the simulation the traffic accident has occurred on the section StreČno. Data were recorded in intervals of ten minutes during the whole simulation. The traffic accident was simulated from 8:15 a.m. to 8:45 a.m. Within traffic accident there were blocked two lanes of section in direction Žilina – Martin. Maximum delay time of vehicles was 31 min and travel time increased triple time. Vehicles which passed through accident section did not have information about traffic accident in time so they did not have either an option to use an alternative way. After removal of an accident, transport has restored in 35 min. Simulation outputs for this group of sections during traffic accident are shown in Table 2.

Fig. 9. The chosen group of sections [own study]

Table 2. Outputs of simulation for the chosen group of setions– Variant A [own study]

Interval [min]	Intensity [veh/h]	Delay time [s]	[km/h]	Travel time [s]
8:15:00	714	118,26	75,19	867,16
8:25:00	0	925,54	0	1792,7
8:35:00	0	1756,34	0	2523,5
8:45:00	96	1883,42	25,7	2634,98

3.3   Outputs of Microscopic Simulation (Variant B)

The variant B represents current state of traffic situation on the chosen part of network with implementation ITS elements and completed highway D1 (Fig. 10). We simulated traffic flow for 4 h since 7:00 a.m. to 11:00 a.m. The input data were obtained from road traffic surveys which were made by University of Žilina. For modelling we used data of traffic peak. The characteristics we were interested in were delay time, density, flow, fuel consumption, speed and travel time. In this variant the alternative ways were defined and ITS elements as variable message signs were installed. The traffic accident has occurred during the simulation in the same time and on the same place as in the variant A. The alternative ways are shown in the Fig. 11. Drivers were attended on traffic accidents using VMS situated on the network. Drivers who passed through incident section did not have an option of alternative way, but they were attended on this situation and it has been recommended to reduce their speed on 20 km/h. This speed was recommended because of the occurrence of secondary accidents caused by not adapt the driving speed to traffic situation. Table 3 shows the outputs of variant B for whole network.

Fig. 10. Network with completed highway D1 [own study]

As you can see the delay time for whole network is lower than in variant A. When highway D1 will be completed and ITS implemented also the density on section between Žilina and Martin is reduced on 16 veh/km and delay time is 9, 36 s/km.

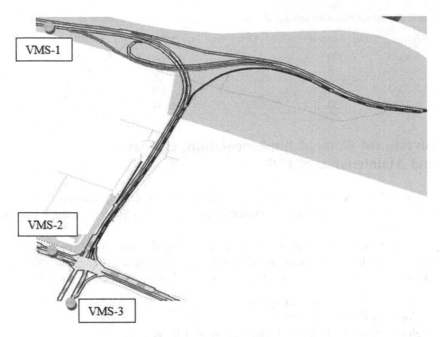

Fig. 11. VMS and alternative ways [own study]

Table 3. Outputs of microscopic simulation for Variant B [own study]

Time series	Value	Standard deviation	Units
Delay time - Car	3,11	1,09	sec/km
Density - Car	0,88	N/A	veh/km
Flow - Car	2544,5	N/A	veh/h
Speed - Car	93,11	4,21	km/h
Travel time - Car	41,05	2,98	sec/km

Vehicles can use highway D1 when travel from Bytča to Martin, Ružomberok or Liptovský Mikuláš. We suggest that every fifth vehicle of traffic flow use highway and the rest of vehicles use roads of first class. When the intensity of traffic flow is divided on more routes, it results to reduction of risk of traffic incidents and congestions. So we can say that build missing part of highway D1 and implementation of ITS is crucial for transport in region of Žilina.

In variant B were also recorded data for the chosen group of sections. Data were recorded in intervals of ten minutes during the whole simulation. The traffic accident was simulated from 8:15 a.m. to 8:45 a.m. Within traffic accident there were blocked two lanes of section in direction Žilina – Martin. Maximum delay time of vehicles was 25 min and travel time decreased to 38.5 min. After removal of an accident, transport has restored in 15 min. Simulation outputs for this group of sections during traffic accident are shown in Table 4.

Table 4. Simulation Outputs for the chosen group of sections–Variant B [own study]

Interval [min]	Intensity [veh/h]	Delay time [s]	Speed [km/h]	Travel time [s]
8:15:00	336	72,02	80,3	813,7
8:25:00	0	897,65	0	1711,35
8:35:00	0	1501,35	0	2315,05
8:45:00	132	1524,35	29,34	2241,52

4 Investment Costs of Implementation, Operation and Maintenance of ITS

For Variant B the economic analysis were made. We took into account investment costs and costs of operation and maintenance and compared them to external costs of traffic accidents.

Investment costs to build telematics systems are very different in case of technology, communications transport mean and climate area. Costs in this paper are calculated only as approximate average of costs. [9] Calculated investment costs include:

– development of management system,
– development of infrastructure,
– development of single technological equipment in terrain.

To calculate investment costs it is necessary to take into account the length of sections, and tunnels. Total investment costs are shown in Table 5.

Table 5. Total investment costs to build ITS [own study]

Section	Length of section	Costs on 1 km [€]	Total costs [€]
Free	62,98	280 000	17 634 400
Tunnel	17,22	420 000	7 232 400
All	80,2	700 000	24 866 800

Costs of operation and maintenance is difficult to settle down. We can only assume how many devices would have a malfunction during their service life. In Table 6 there are average approximate costs of operation and maintenance.

Total approximate average costs of operation and maintenance are 1 989 344 €.

Costs related on traffic accidents calculation included traffic accidents, which were recorded on highway D1 and on roads of first and second class between Bytča and Ivachnová. Total costs are shown in Table 7. Costs on person are calculated in specific document [10].

Table 6. Operation and maintenance costs [own study]

Section	Length of section	Costs on 1 km [€]	Total costs [€]
Free	62,98	22 400	1 410 752
Tunnel	17,22	33 600	578 592
All	80,2	56 000	1 989 344

Table 7. Total costs related on traffic accidents [own study]

	Costs on person [€/person]	Number of persons/TA	Total costs [€]
Fatal injured	426 008,74	58	24 708 506,80
Seriously injured	92 423,53	134	12 384 753,02
Slightly injured	13 564,32	660	8 952 451,20
Material damage	8 098,49	1101	8 916 437,49
		Total	54 962 148,51

Measures to increase the safety on the chosen part of network is economical effective according to analysis of costs. The main contribution of ITS applications is that due to their effects is possible to decrease number of traffic accidents and their consequences, which has a positive effect on the whole society.

5 Conclusion

Traffic modelling doesn't include only traffic simulation. It is wide range of tools ranging from simple application for only one purpose to complex tools which enable to perform complicated analysis of transport networks. Traffic modelling can be used to many operations and there is assumption that the importance of the traffic modelling will be increasing with the increase of transport. In our study we focused on possibilities to reduce the number of traffic accidents at the chosen part of road network. Using simulation we proved that the implementation of ITS elements and building missing part of highway D1 has large impact on decreasing of intensity and delay times on the chosen network, which leads to reduction of probability of traffic accidents. Using ITS in transport is crucial for transport policy of the Slovak Republic [7, 8].

References

1. Kalašová, A., Kubíková, S.: The interaction of safety and intelligent transport systems in road transport. In: Young Researches Seminar 2015, 17–19 June 2015. Sapienza - Universita di Roma - [S.l.: S.n.] - CD-ROM, s. 3–13 (2015)
2. Černický, L'., Kalašová, A., Kubíková, S.: Possibilities of using simulation software in a traffic-capacity assessment of uncontrolled intersections. In: MOSATT 2015: Modern Safety Technologies in Transportation: Proceedings of the International Scientific Conference, Zlata Idka, Slovakia, 16–18 September 2015 - Košice: Perpetis - S. 29–34 (2015). ISBN 978-80-971432-2-0
3. Madleňák, R., et al.: Analysis of website traffic dependence on use of selected internet marketing tools. In: Procedia - Economics and Finance, vol. 23, pp. 123–128 (2015). ISSN 2212-5671
4. Kapusta J., Černický, L'.: Effects of autonomous vehicles on road safety. In: 11-th European Conference of Young Researchers and Scientists TRANSCOM 2015, Žilina, 22–24 June 2015, Slovak Republic - Žilina: University of Žilina - CD-ROM, s. 40–44 (2015). ISBN 978-80-554-1043-2

5. TSS – Transport Simulation Systems. Aimsun 7 Dynamic Simulators User's Manual May 2012
6. Černický, Ľ.: Tvorba modelu pre posúdenie uplatnenia inteligentných dopravných systémov v riadení mestskej dopravy. In: Ph.D. progress: vedecký Časopis študentov doktorandského štúdia Fakulty prevádzky a ekonomiky dopravy a spojov Žilinskej univerzity v Žiline . - RoČ. 2, Č. 1, s. 32–43 (2014). ISSN 1339-1712
7. Madleňák, R., Madleňáková, L.: The differences in online advertising acceptance in China and Slovakia In: ICEMI 2015 = International Conference on Management Engineering and Innovation, Changsha, China, 10–11 January 2015, pp. 45–49. Atlantis Press, Paris (2015). ISBN 978-94-62520-45-5
8. Gogola, M.: Modelové riešenie preferencie MHD na vybranej Časti mesta Žilina. Doprava a spoje 2012. číslo. 2012-1. str. 94–103. http://fpedas.uniza.sk/dopravaaspoje/2012/1/gogola.pdf
9. Koncepcia budovania informačného a riadiaceho systému. NDS (2015). intern document
10. Spoločenské náklady z dopravnej nehodovosti, Ministerstvo dorpavy, výstavby a regionálneho rozvoja. intern document of Ministry of transport

GPRS Network as a Cloud's Tool in EU-Wide Real Time Traffic Information Services

Elzbieta Grzejszczyk[(✉)]

Electrical Department, Warsaw University of Technology, 00-661 Warsaw, Poland
elzbieta.grzejszczyk@ee.pw.edu.pl

Abstract. Modern communication technologies such as Smartphone's and PND (*Personal Navigation Devices*) platforms provide new services for users IT'S as well. A base of above solutions is GPRS network described in this article. This paper shortly presents types of current real time traffic services also an example of the remotely controlled step motor by GPRS network and analyzed Microsoft Azure Notification Hub facilitating dynamic communication with end user RTTI system.

Keywords: Tools of cloud computing technology in ITS · wireless communication in GPRS network

1 Introduction

On 16 December 2008 the European Commission adopted the ITS Action Plan (next 2012, 2014). The ITS (*Intelligent Transport System*) Action Plan defines the necessary requirements (*and elements*) to make EU-wide real-time traffic information (*RTTI*) services accurate and available across borders to ITS users [22]. The final report of the ITS Directive underlines technological innovations which are fundamentally changing the RTTI services landscape. New technologies have created new ways of communication, collecting road and traffic data and have introduced new services platforms such as Smartphone's and personal navigation devices. Recently Cloud Computing technology seems to be one of the key priorities in RTTI services area. The article concerns to issues of implementing GPRS network as a cloud's tool in RTTI services.

2 RTTI Services Available Today

For the last 10 years, the market of RTTI (*Real Time Traffic Information*) services was mainly served by small and medium-sized companies (*road and service*) that focused on local national markets (*one such example is the Polish company T-Traco* [18]). Currently the leading roles in the market of traffic data acquisition, aggregation and adaptation, as well as providing RTTI service, are played by big companies such as Google, TomTom, NOKIAHERE or INRIX. Services provided by these companies are meant for the dynamically developing Portable Navigation Device (PND) market, as

© Springer International Publishing AG 2016
J. Mikulski (Ed.): TST 2016, CCIS 640, pp. 59–71, 2016.
DOI: 10.1007/978-3-319-49646-7_6

well as the automotive industry. The latest RTTI applications communicate directly with a connected vehicle using on-board SIM card. [5, 6, 23]. These services are a strong and promising foundation for a wider introduction of emerging communication technologies such as V2 V (*Vehicle to Vehicle*) data mining, as well as support services based on packet transmission, both terrestrial and satellite.

One of the first European systems to standardize the exchange of traffic data was and still is DATEX. System DATEX is a multi-part standard, maintained by CEN Technical Committee 278 (*Road Transport and Traffic Telematics*)[1] [1].

CEN/TC 278 has a number of Working Groups (*WG*), each responsible for a specific ITS area. Working group 8 is responsible for the multi part DATEX II standard[2].

In its first implementations (*over the last 10 years*), the system provided the exchange of information between traffic management centers, traffic information centers and consumer services. The new generation of DATEXII (*v. 2.3 - 12.01.2014*) provides the user with new additional functionalities based, among others, on the huge data collections and newly available technologies [2].

For example, the Norwegian implementation of DATEXII [17] lists the following services:

I.   Free access Safety Related Traffic Information (SRTI) data. These data describe, among other things: (1) the estimated travel time along the planned route[3], (2) weather information (*including forecasts obtained in real time from approx. 300 weather stations located along the main national roads*), (3) data on road works, accidents, road closures and road conditions received from 5 regional Traffic Information Centers, (4) CCTV (*Closed Circuit Television*) recordings from cameras located along national roads, available in real time, with on-line playback on a PND.

II. Parking information - for a selected area. These include data concerning the location of parking's, their accessibility, hours, prices or safety.

III. OLR (*Open Local Referencing*)[4] - providing methods, tools and standards for integration of local information resources with other local resources. One example of OLR service may be the connection between the Swedish Transport Administration and the Dutch National Data warehouse for traffic information [13]. Data acquired in this manner are transmitted to users' Smartphone's and other portable devices, such as PND.

The implementation of these services became possible thanks to new technologies, such as DVB (*Digital Vehicle Broadcasting*) or high bandwidth transmissions using

---

[1] CEN, the European Committee for Standardization, is an association that brings together the National Standardization Bodies of 33 European countries.

[2] Another ITS application areas: WG1 - Electronic Fee Collection, WG2 Freight, WG3 Public Transport, WG7 ITS Spatial Data, WG8 Road Traffic Data, WG10 Human-Machine Interfacing, WG12 Vehicle Identification, WG13 ITS Architecture, WG15 eCall, WG16 Cooperative ITS.

[3] In ver. 2.3 travel times were calculated for selected major roads (e.g. Oslo, Bergen, Trondheim and Stavanger) [17].

[4] OLR proposed as an open standard in an Open Source framework. The OpenLR specification is defined in the Open LR whitepaper version 1.5 [13].

mobile internet (LTE). These technologies are described below. Internet protocol (IP) provides the basic requirements for transporting disparate data such as websites, databases, MP3 sound files, or even MPEG4-compressed video sequences and films. (*e.g. obtained from road safety cameras*).

Another one of the RTTI services, offered especially in Eastern European countries, is the RDS/TMC (*Radio Data System/Traffic Message Channel*) [24] which was recently included in the latest version of TomTom application.

TMC provides precise information concerning planned route, with estimated time and any additional information on traffic jams or incidents in the user's language. In case of TomTom's RDS-TMC receiver, the user receives traffic information automatically using TMC connection and the optimal route that avoids any obstacles is chosen. The user can display additional information that shows the reasons for any delays in traffic, such as the location of traffic congestion, incident or road closure. RDS-TMC is available in such countries as Norway, Hungary, Austria, Belgium, Czech Republic, Denmark, Finland, France, Spain, Netherlands, Germany, Italy, Switzerland, Sweden, United Kingdom and selected regions of the United States and Canada. Information in RDS/TMC is broadcast using packet data transmission over encrypted FM waves in the RDS FM system.

The American system INRIX is currently the leading global system for monitoring traffic and its conditions[5], extending to 40 states (*as of Sept. 2014*), i.e. The United States, Canada, Brazil, China and most of Europe [11].

INRIX is based on cloud computing technology, and is provided as SaaS (*Software as a Service*) and DaaS (*Data as a Service*) [4, 9, 15] (Fig. 1).

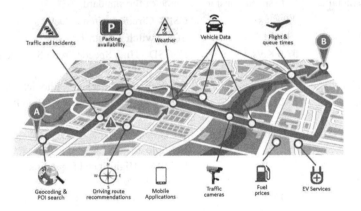

Fig. 1. INRIX Driver Services [11].

INRIX services and systems - are typical for such systems and provide versatile data, such as: traffic congestion, road construction, accidents, events, police activity, road

---

[5] INRIX was founded by former Microsoft employees (Bryan Mistele and Craig Chapman) in July 2004.

weather, parking information, fuel point, traffic cameras, routing and inter-modal routing, drive times & drive time polygons and historical traffic patterns.

The system is currently the largest network serving more than "175 million real-time vehicles and devices from 100's of distinct sources around the world to then analyze through our traffic intelligence platform"(*a quote from* [11]).

It should be noted that INRIX provides both turnkey cloud solutions known as SaaS (*Software as a Service*), as well as a platform and tools for creating one's own, local cloud-based solutions, i.e. Mobile SDK (*Software Development Kit*) platform. The use (*an example*) of these tools is described later in this article.

Ensuring uniform worldwide access[6] to the tools to create traffic mobile applications may contribute significantly in the future (*in the opinion of the author*) to the unification of the design of services concerning traffic management, as well as the dissemination of the format/standard of data used in INRIX.

3 Information Transmission in RTTI Systems

Communication networks that support RTTI exchange of information include GSM/ GPRS networks that use electromagnetic waves of certain frequency. Above networks have undergone rapid development in recent years, and the individual stages of development are dubbed generations. A list of these is given in Table 1.

Table 1. A list of GSM communication standards [3]

Generation	Data speed transfer	Name of the standard
1G	9,6 Kb/s	CSD - Circuit Switched Data
2G	14,4 Kb/s	SDI -Switched Data Transfer
2G	57,6 Kb/s	HSCSD - High Speed Circuit Switched Data
2,5G	115 Kb/s	GPRS -General Packed Radio Service
3G	384 Kb/s	EDGE - Enhanced Data Rates for GSM Evolution
3G	7,2 Mb/s (1,9 Mb/s)	WCDMA - Wideband Code Division Multiple Access
3G	14,4 Mb/s	HSDPA - High-Speed Downlink Packet Access
4G	150 Mb/s	LTE - Long Time Evolution

A special role in the development of the network was played by 2.5G, which provided/created GPRS (*General Packed Radio Service*) packet data transmission, which allowed for transfer with speeds of 144 kb/s of such data as graphic files (*including animations*) or measurement data.

From that moment on, the process of introducing advanced algorithmisation of medium coding started, using electromagnetic waves (*to achieve higher and higher data*

---

[6] Based on cloud computing technology.

transfers). This process still continues, with one of the results being the transmission standard dubbed LTE (*Long Time Evolution*). Generation 3G (*with data transfers from 7.2* Mb/s *to 14.4* Mb/s) marks the beginning of UMTS 3G/4G networks and brought about the introduction of advanced videophone, videoconferencing, interactive TV and interactive maps displayed on mobile devices, as well as teleservice communication with vehicles.

3.1   GSM (Global System of Mobile Communication) Network

An analysis of any teleservice services in wireless networks should be embedded in the concept/logic of GSM networks shown in Fig. 2. Adding a controller dedicated to handling packet transmission (*2.5G*) to a GSM network allows for the coupling of both networks into a GPRS network Fig. 3.

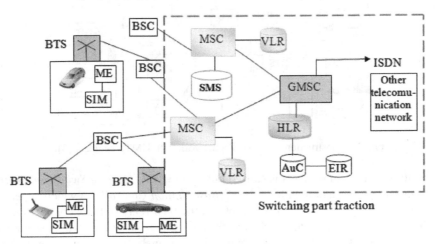

BTS - Base Transiver Station, **BSC** - Base Station Controller, **MSC** - Mobile Switching Centre, **VLR**-Visitor Location Register, **HLR**- Home Location Register, **SMC** - Short Message Controller, **AuC** - Authentication Centre, **EIR** - Equipment Identity Register, **GMSC** - Gateway Mobile Switching Centre, **ME** - Mobile Equipment, **SIM**- Subscriber Identity Module, **PSTN** - Public Switched Telephone Network, **ISDN** - Integrated Services Digital Network

Fig. 2.   Connection topology of system controllers in GSM network [3].

In simplified terms, a GSM network can be defined as a network of connections between specialized controllers executing strictly defined functions and exchanging information using a wireless transmission protocol.

This protocol works in dynamically assigned communication channels (*radio channels in the case of GSM*) and time slots[7] (*in the case of GPRS*). The identification and

---

[7] GPRS networks also use dynamically assigned temporary identifier TMI (*Temporary Flow Identity*).

authentication of the sender and receiver[8] of information is a typical action in analyzing network traffic. Therefore, local databases shown in the diagram store information concerning both users logged into the GSM/GPRS system (*HLR and VLR*)[9], as well as data describing EIR (*Equipment Identity Register*) devices defined in the system and purchased rights/access to AuC (*Authentication Centre*) system functionalities.

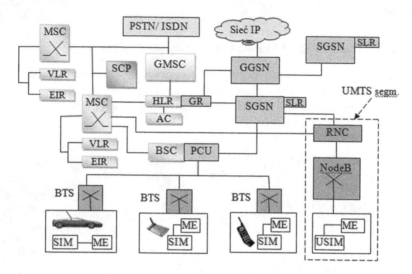

Fig. 3. Construction of the GPRS network with UMTS segment [3].

Identification of subscribers is carried out by the MSC (*Mobile Switching Centre*) by exchanging information with the BSC (*Base Station Controllers*) which supervise BTS (*Base Transceiver Station*) (see footnote 9). BTS relay stations connect mobile terminals (*mobile phones, PCMCIA cards*[10], *SIM card devices/cars*) with a stable part of the network. BSCs allocate communication channels to individual subscribers/stations radio transceivers with which they are equipped. BSCs also route connections between BTSs and the switching-network part of the system (Figs. 2 and 3) (*up-link/down-link transmissions*). End/mobile users are identified using data read from their SIM/USIM cards and the ME (*Mobile Equipment*) modules. Communication between networks with different structures and different communication protocols is implemented by the GMSC (*Gateway Mobile Switching Centre*).

---

[8] With the exception of broadcast type - i.e. to all users logged into the network at a given moment. (*e.g. with LAI data*).

[9] HLR - Home Location Register, VLR - Visitor Location Register.

[10] PCMCIA - Personal Computer Memory Card International Association.

3.2 GPRS (General Packed Radio Service) Network

The construction of a GPRS network is shown in Fig. 3. In addition to the classical GSM network[11] and the UMTS segment (*dashed line*). This structure contains additional elements, such as:

- PCU (*Packet Control Unit*) that manages packet traffic in a network by allocating both radio channel and time slot to GPRS terminals, along with a TFI identifier[12] (*Temporary Flow Identity*). On the other hand, the PCU is responsible for the correct distribution of data received from SGSN to appropriate BTS stations (*so-called data forwarding*).
- SGSN (*Serving GPRS Support Node*) is a node that manages terminals in a controlled area with assigned RA identifiers (*Routing Area*). A change to the location of the terminal that also changes its area (*different RA*) modifies the RAI parameter (*Routing Area Identifier*) saved in the memory of the SGSN. The SGSN node also authorizes the terminal joining the network.
- GGSN (*Gateway GPRS Support Node*) connects the GPRS mobile network with the external Internet or other LANs. It is a GPRS gateway which integrates the two networks by assigning an IP addresses to a mobile user, along with the temporary parameters to identify them (*IMSI*)[13] and their location (*IP & RAI SGSN*)[14]. All these parameters define the so-called PDP Context, which is a data set describing a GPRS user for the duration of a communication session with an external network, e.g. WWW (Fig. 4)
- SCP (*Service Control Point*) is a controller which verifies access to purchased services in the network (*both GSM and GPRS*), and manages billing for GPRS transmissions.

Exchange of information on GPRS networks is primarily geared to communicate with the Internet and other external networks. Figure 4 shows the method of connection between the mobile user MS (*Mobile Station*) and the Internet.

The activation of PDP Context means the assignment of an IP address to a mobile user, whose location is constantly monitored by the system from the moment of activation and saved in the VLR (*SLR*) and HLR (*GR*) registers in real time. Activation request is submitted by the user who wants to connect to the Internet or any other network. The figure below shows the flow of information in a GPRS network used for activating a service.

---

[11] Figure 3 the gray parts of the model.

[12] TFI (*Temporary Flow Identity*) means a temporary identifier of assignment of packets of a given user to a given time slot. PCU compares the TFI stored in the packet with the system identifier of the time slot and, if they are compatible, starts transmission with a given TFI.

[13] IMSI (*International Mobile Subscriber Identity*) is a personal 15-digit identification number of the subscriber, which creates a data structure containing the country code, network code, user id in a given network [3].

[14] RAI SGSN (*Routing Area Identifier SGSN*) is an id of a mobile user's location in a network area with a given RA. Any change to the RA, tracked by SGSN, results in immediate calculation of the RAI stored in the SGSN controller's memory.

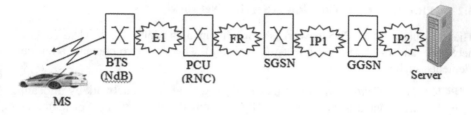

Fig. 4. PDP Context activation procedure (own study based on [3, 14]).

E1:   Communication between the BTS controller and the PCU or NodeB with RNC controller (*for UMTS network*)

FR: Frame Relay network used to send packets to SGSN

IP1: (*Internet Protocol*) SGSN queries GGSN for the IP1 address

IP2: (*Internet Protocol*) GGSN sends a request to an external server about the IP2 number

The order of connection setup includes the activation and joining the GPRS network by the mobiles user and the consecutive definition of PDP Context, along with the assignment of an IP number. Following these activities bilateral packet exchange may happen, with every packet including the sender's and receiver's addresses both in IP and GPRS networks.

In GPRS networks, the frequency band and coverage is the same as in GSM networks due to the fact that both types of networks use a common radio network. The networks differ mainly in the structure of transmitted information.

Data exchange over RTTI networks happens according to the protocol (*according to the definition of transmission frames*) in TPEG standard (*Transport Protocol Experts Group*)[15] [19, 20, 24]. Needless to say, this standard is continually being improved and is the source of searching for the best solutions, numerous patents, and thus more and more areas related to RTTI systems are subject to TPEG standardization. This model is an example of a public network which serves as the basis for dynamically setting up various sub networks, so-called virtual ones, according to users' needs.

The virtual connections are based "on dynamic virtual connections - virtual circuits (a connection in a network between the provider and client, in which data transmission routes and bandwidth for the connection are set up dynamically) or tunnels – non-phys-ical entities that only exist when traffic is present. Such connections may be set up between two separate devices, a device and a network and between two networks" [16].

---

[15] Protocol TPEG validated by TISA (*Traveler Information Services Association*) having issued 31 Service IDs. For example some of them: INRIX (038.051.034; 077.076.069; 080.084.068), Nokia (052.224.003; 065.015.247; 065.100.001), Media Mobile (054.163.061) whether Intelematics (027.015.157).

Virtual Private Network (*VPN*), now referred to as private cloud[16] is a data transmission network that uses communications infrastructure which uses tunneling and security procedures to maintain data confidentiality. The infrastructure may consist of the telecom's backbone network (*e.g. Frame Relay or ATM*), or the global Internet. [16].

The main feature of the private cloud (*similarly to the VPN*) is its security, construed as confidentiality, authenticity of data transmission and its integrity. As defined by the NIST (*National Institute of Standards and Technology, 2011*), a private cloud is a network which is made available for the sole use of a single organization, regardless of whether the infrastructure is owned by the same organization (*company*) or an external provider (*e.g. a data centre*). Its exclusive purpose is vital. The main feature of the cloud is the scalability of the solutions, their automation and adaptation of resources to current demand.

Public clouds provide free access to network infrastructure for a large number of users. Public clouds also provide scalability, automation and adaptability of solutions. One such cloud is Microsoft Azure, in which the application described below has been implemented.

4   Example of On-board Device Control Based on GPRS Network

Remote control of on-board device based on TCP/IP and GPRS networks[17] is shown in Fig. 5, while the control interface is shown in Fig. 6 and the implemented system control features are shown in Fig. 7. The remote control is implemented in a dozen of ways, while the control parameters are input using the remote device's keyboard.

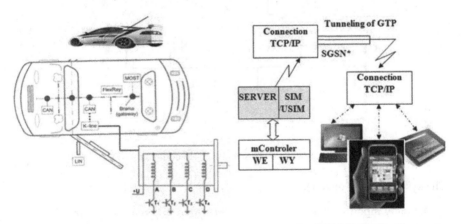

* SGSN - controller Serving GPRS Support Node, GTP - GPRS Tunneling Protocol

Fig. 5.   The TCP/IP remote control model (own study based on [7, 8, 10, 21]).

---

[16] It should be added that the cloud covers not only network connections but also IT hardware and/or software (*IaaS - Infrastructure as a Service, SaaS - Software as a Service*).

[17] The system described here has been developed for the teaching lab, as part of the course in Wireless data transmission systems taught by the author in the faculty of computer studies [10, 21].

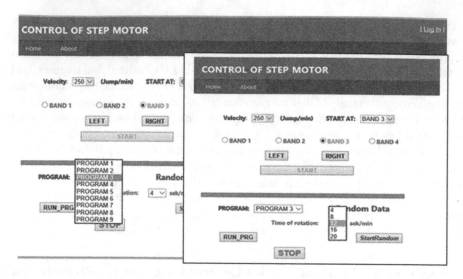

Fig. 6. Remote Control Interface for EDS10 engine - implemented in Windows 10 and Visual Studio 2015 [own study]

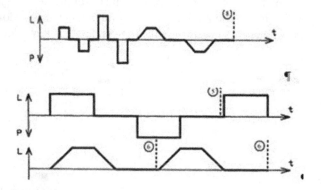

Fig. 7. Selected characteristics of remote control [own study]

The system can be operated in 3 ways:

- by manually inputting control parameters (*engine rotation direction, the duration of rotation, rotational speed and the started band of the engine*).
- selection of a preset control programs (*one of nine*)
- using a random number generator to submit control variants, i.e. rotation direction (*0,1*), rotation speed (*from 1 to 5*) × 10 revs per minute and run time (*from 4 to 200*) [s].

Individual engine bands Fig. 5 (*transistors T1, T2, T3, T4*) are controlled by four bits of port B of the controller, which are provided with control codes[18].

All data are written in the micro controller's memory. After the data are saved (*from the random number generator or the keyboard*), the micro controller port is setup and connected to the executive system. Thanks to the sequential power supply of consecutive bands - the engine's rotor operates with preset frequency and rotates[19] left or right.

5 Microsoft Azure Notification Hubs Developer Center as a Tool of the RTTI Systems

The concept of publicly available cloud solutions that result from the rapid development of wireless networks (*mainly GPRS*) was/is the basis for equally dynamic development of cloud computing applications. One of the leading solutions, Microsoft Azure, allows for building/developing applications for any platform and device that works with Azure. Services provided by the cloud offer:

– development of one's own cloud applications (*Web and Mobile apps - IaaS*) based on such IDEs as Java, .NET Or PHP. (*The example presented in this paper uses .NET on Windows 10*)
– joining corporate networks with the option to share information, as well as social networks such as Facebook or Twitter (*SaaS*).
– building and integrating inter-corporate solutions via the cloud (*B2B - Business to Business*)

as well as ensuring the security, integrity and scalability of the proposed solutions.

The Mobile Applications service dedicated for mobile environments was chosen in the context of analysis of RTTI systems (*for both developers and end users*). The most important tool in the service is the Notification Hubs Developer Center [12]. Notification Hub allows you to send dedicated information to large groups of users at the same time. (*One example of such notifications is the information about accidents in a given area, described in detail in* [4]). The segment of recipients and the information sent may change dynamically and at any rate, while end users may use different mobile platforms such as Android, Windows, Kindle or iOS. The necessary condition is, of course, that the device (*user*) has joined the cloud.

6 Conclusion

GPRS networks that a huge number of mobile services on all types of Portable Assisted Devices are based on are entering or have already entered the automotive market. Car manufacturers have already made us used to mobile Internet based on an in-car SIM/

---

[18] **ZERO** EQU 0000 0000B **STOP**;ONE EQU 0000 0010B **BAND1**;TWO EQU 0000 1000B **BAND2**; THREE EQU 0010 0000B **BAND3**; FOUR EQU 1000 0000B **BAND4**.

[19] 4 red LEDs located in the motor housing flash off and on sequentially to show the powered bands of the engine, while the arrow connected to the rotor shows the rotation.

USIM card. This card provides access to typical Internet services. However, users of cars that are driving fast in changing road conditions expect more technological support. This is undoubtedly provided by the development of real-time ITC systems. The quick and reliable access to a dynamically changing reality plays a key role in them.

Such access may be provided by the technology and software presented in the paper.

References

1. CEN DATEX II European Committee for Standardization/Technical Committee 278, (Road Traffic and Traffic Telematics), CEN/TC 278. www.itsstandards.eu
2. DATEX II version 2.3 available now. Easyway Project. www.datex2.eu/news/2014/12/01/datex-ii-version-23-available-now
3. Fryskowski, B., Grzejszczyk, E.: Data Transmission Systems. Transport and Communication Publishers, Warsaw (2010)
4. Grzejszczyk, E.: Control of coordinated systems traffic lights. Cloud computing technology. In: Mikulski, J. (ed.) TST 2015. CCIS, vol. 531, pp. 45–56. Springer, Heidelberg (2015). doi: 10.1007/978-3-319-24577-5_5
5. Grzejszczyk, E.: Teleservice communication with motor vehicle. Electr. Rev. 94–98 (2011). ISSN 0033-2097, R. 87 NR 12a/2011
6. Grzejszczyk, E.: Selected issues of wireless communication over GSM with cars network. GSTF J. Eng. Technol. 2(2), 106–109 (2013). ISSN 2251-3701
7. Grzejszczyk, E.: Vehicle Stability Control (VSC) and supervision based on CAN network. II GSTF J. Eng. Technol. 2(4), 38–44 (2013). ISSN 2251-3701
8. Grzejszczyk, E.: The control on-line over TCP/IP exemplified by communication with automotive network. In: Federated Conference on Computer Science and Information Systems (FedCSIS), vol. 1, pp. 807–810. IEEE Xplore, Cracow (2013). ISBN 978-1-4673-4471-5
9. Grzejszczyk, E.: Cloud computing as a tool in smart communication with a motor vehicle. Arch. Transp. Syst. Telematic 7(4), 9–16 (2014). ISSN 1899-8208
10. Grzejszczyk, E.: Introduction into Visual Web Developer Express 2008 and ASP.NET 2.0 Technology. Publishing House of Warsaw University of Technology, Warsaw (2010). ISBN 978-83-7207-853-7
11. NRIX. Driving intelligence. http://inrix.com
12. Microsoft Azure Platform, Notification Hubs Developer Centre. https://azure.microsoft.com/en-us/documentation/. Accessed 01 Dec 2015
13. Open LR, white paper, ver. 1.5 revision 2, An open standard for encoding, transmitting and decoding location references in digital maps, Copyright 2009–2012 Tom Tom International B.V. http://openlr.org/data/docs/OpenLR-Whitepaper_v1.5.pdf. Accessed 01 Dec 2015
14. Sanders, G., et al.: GPRS Networks. Wiley, Chichester (2003)
15. Sergiejczyk, M.: Analiza wykorzystania chmur obliczeniowych w zarządzaniu zasobami IT firm transportowych, Logistyka, February 2014
16. Sergiejczyk, M.: Możliwości wykorzystania wirtualnych sieci prywatnych w zarządzaniu firmami transportowymi, Prace Naukowe Politechniki Warszawskiej, Transport z 64 (2008)
17. The Norwegian Public Roads Administration's DATEX II information site. http://www.vegvesen.no/en/Traffic/Ontheroad/Datex2/Dataavailable. Accessed 01 Dec 2015
18. T-Traco - globally system track and trace. http://www.t-traco.com/. Accessed 01 Dec 2015
19. Traveller Information Services Association (TISA), TISA Technical and Standardization Committee, TISA14001/2014-01-27. www.tisa.org

20. TPEG-TEC-Specification-V1_0-20060309.docV1.0.   www.mobile-info.org/prom/
    mobileinfo.nsf/DocID/7FA3503CF5AAB9E7C125723E004B2276/$file/
    TPEG_TEC_Specification_V1_0_20060309.pdf. Accessed 01 Dec 2015
21. Gryszpanowicz, B.: The Implementation of Wide Area Network for Remote Communication
    with Motor Vehicle. Electrical Department, Warsaw University of Technology, Warsaw
    (2004)
22. Van de Ven, T., Wedlock, M.: ITS Action Plan. D5 - Final Report. Action B - EU-wide real-
    time traffic information service. European Commission. Directorate-General Mobility and
    Transport. B-1040 Brussels, Belgium, July 2014
23. V-Traffic: First German traffic information service on DAB. http://www.v-traffic.com and
    http://www.mediamobile.com/en. Accessed 01 Dec 2015
24. We design the future -with you and for you, Institute fur Rundfunktechnik, Activities, TPEG
    Traffic and Travel Information. www.irt.dr/en/activities. Accessed 01 Dec 2015

Public-Private Partnerships for the Implementation of IT Projects

Grzegorz Dydkowski[✉] and Barbara Kos

Department of Transport, University of Economics in Katowice, Katowice, Poland
{grzegorz.dydkowski,barbara.kos}@ue.katowice.pl

Abstract. The execution of investment projects in the public-private partnership formula is a relatively frequent solution in the developer countries of the Western Europe. That solution has numerous virtues and advantages, especially in case of costly and complex investment projects. The public-private partnership formula may be successfully applied in case of projects concerning the supply and implementation of IT systems, it allows to manage the very undertaking better, as well as to manage better the risks that occur in the course of execution of such projects.

Keywords: Public-private partnerships · IT systems · IT projects · Implementation of IT systems

1 Introduction

The public sector is ever more present in everyday life, which is manifested by multiplicity of public institutions, substantial amount of people they employ, as well as ever greater amount of services they commission and deliver. The public sector has also been executing numerous investment projects, not only those focusing on construction works, but also those dealing with development of IT infrastructure, as well as provision of services using IT systems and technologies.

The growth of public sector demands searching for solutions in which the provision of public services will not be connected with involving increasing financial means by the sector. The actions aiming at enhanced efficiency and effectiveness of public sector functioning should take place in parallel in many areas. For sure, in most general terms, this will concern improved management of entities and public services, implementation of innovations, provision of services with the use of modern technologies, and competent as well as better and better personnel.

Improvement of public sector efficiency has been referred to in publications, also the directions of changes needed have been indicated, for example the concept of the so-called new public management [1, 4, 14]. One should not also leave out the solution consisting of involvement of private sector for the execution of public tasks, starting with privatization, through public-private partnership to public procurement. Provision of services in the public-private partnership (PPP) formula is worth particular attention. The paper presents public-private partnership as a formula that is possible to be used during implementation of IT projects. It deserves attention that many publications

J. Mikulski (Ed.): TST 2016, CCIS 640, pp. 72–83, 2016.
DOI: 10.1007/978-3-319-49646-7_7

devoted to public-private partnership are focused on construction oriented projects, whereas those concerning delivery and implementation of IT systems are almost completely disregarded.

2 Traditional Ways of Public Procurement and the Public-Private Partnership Formula

For the execution of public tasks, or activities for the benefit of public sector entities, private sector entities may be involved – as providers of services under public procurement in traditional formula. A public entity is then the employer and pays for a given product, service, investment activity, such as – for example – the construction of a particular motorway section, car park, aquapark or swimming pool, or delivery and implementation of an IT system. The contractor is not interested in the future fate of the goods delivered, whether they bring any benefits – this is the problem and risk of the employer, who also has the future expenditures related to the costs of maintenance, operation, and modernization of a given structure or IT system, developed under the project implemented. In such a model, the public entity practically bears the entire risk that the project will be successful, and will generally sufficient income in the so-called service life, that is the normal operation/use of the object/structure [8]. This risk is particularly high in case of IT systems, they are undertakings, the usefulness of which often results from sufficient functionality, reliability, and easy operation. They are also undertakings that require quite frequent corrections, modifications, and updates, the reasons for which are quite difficult to identify – whether they derive from insufficient development of the delivered module or system, or whether the problem stems from changing expectations of users.

In the traditional formula of public procurement, in which the public entity is held responsible for such system, commissioning the execution of selected activities to external parties; it is difficult to highly motivate such a party to work in order to achieve such parameters of functionality and efficiency of the system, that will be assessed positively by the users, and which thus translate into revenues generated by means of a given IT system. It is often so that systems manager by public entities work correctly, yet they lack innovativeness, easy use, or broader integration with other systems, and in line with expectations of users. The problems with modifications are also partly the result of difficulties with describing them, and the lengthy procedures, in which the performance of specific service actions or development actions is ordered or commissioned.

The situation may be different in case of a solution in which a private entity manages the functioning and development of IT system, and benefits, at least in a certain part, from its efficient functioning. For example, a private entity that manages a given system may receive part of the income from electronic tolling system, or the electronic payments for services of public urban transport or for parking [7, 13]. In such a case, the entity is interested in implementing, on ongoing basis, such correction or modifications of the system, which would meet the expectations of users or even anticipate them. Of course, this is then translated into increased interest in given services, and – as a result – also influences the income generated. Considering the possibility of applying public-private partnership, the very fact of financial participation of a private entity in the execution of the project also is of importance, as this reduces the expenditures of the public party, related to the implementation of a given project (Fig. 1).

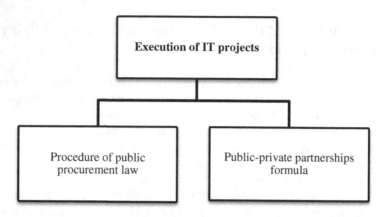

Fig. 1. Two basic procedures of IT projects execution [own study]

Besides the use of know-how of private partners, reduction of risks for the public party, the involvement of private capital in the execution of public projects, public-private partnership also has other advantages. One can mention here the thorough assessment of feasibility of projects to be undertaken, including determination of conditions upon which the execution will be possible, and the level of involvement of public funds. Concluding a contract with a private entity, in the PPP formula, a realistic forecast concerning involvement of public funds is made. In case of a part of the projects, financial involvement of the public party will be visible throughout the life cycle of the project.

The issues concerning public-private partnership were treated differently in legal regulations in the past. After 1990 no special act of law was introduced, which would refer in its content to public-private partnership, what is more, no act of law was introduced that concerned public procurement – awarding of contracts or co-operation in public-private partnership formula followed general rules, of course in such a way as to provide thrifty use of public funds. In June 1994 the public procurement act was passed, which gradually, as of January 1, 2005 or January 1, 2006 came into force, comprising subsequent public entities. After the public procurement law came into force, tender procedures were decided, still there were no detailed regulations concerning public-private partnerships. As can be assumed, in order to simplify the procedures in the public-private partnership formula, on July 28, 2005 the act of law on public-private partnerships was passed, and a few years later – due to lack of such a type of undertakings – the act of law was replaced by another one, dated December 19, 2008, having the same title. Unfortunately, it is still difficult to talk about this form of execution of public tasks as a widespread one. One can thus claim that legal regulations do not matter that much, in this case. When there was no act of law on public-private partnerships, as well as when such an act was binding, there were few projects executed in the public-private partnership formula [8].

3   Prerequisites for Implementation of IT Projects in the Public-Private Partnership Formula

In many publications, public-private partnership is indicated as a solution which allows to involve the private sector in execution of investment projects, most often those demanding substantial funds to be invested. Private sector entities may involve their own funds for the execution of the project. It is also a good solution for implementation of projects serving public objectives, which require efficient use of resources as well as experience, that meet the characteristics of private entities. The private sector can be characterized as being capable of conducting business and having the experience acquired in the course of activity in various domestic and international markets. In addition, the private sector is free from many limitations imposed upon the public sector. The public sector, in turn, has the right to be active in areas often reserved for that sector only, thus unavailable directly for private entities without suitable agreements or concessions. An advantage of public sector entities is also the substantial income stability, including also a relatively high certainty of income generated from taxes and various charges, which is import and particularly as regards the possibility of getting bank loans. All in all, this results in effects that may be achieved from co-operation, in which the different resources of partners are used, such as competences, abilities, and know-how. Most often, during joint execution of projects the public partner brings in the right to deliver a given service, the area, partly the financing of a given project, any possible sureties and guarantees, which it extends to banks or the private partner. The private partner, in turn, involves first of all the investment means, technologies, experience, and know-how. The private partner also takes part of the risk related to the demand for services. It is expected that the co-operation of the public and private sector will foster increased efficiency of the project executed, also additional benefits will be obtained - the so-called synergy effect – which will allow for more effective execution and later on also operations of a given undertaking or project [1].

Also another aspect of public-private partnership should be brought to attention. Readiness of the private partner to enter such a project, to involve own capital in it, possibly also to receive suitable support of the bank – which means that the project is well prepared, stands chances for execution, and that it highly probable it will provide the expected benefits.

In Poland, in comparison with other economically well developed countries of Europe or the world, the public sector has been co-operating but slightly with the private one, in provision of public services [9]. The solutions that dominate are those in which the private sector provides selected products or renders specific work procured by the public sector, yet the role of the former essentially ends with this. The private sector does not participate directly in the provision of services, does not take the risk related to their quality, continuity of provision, as well as efficiency of the entire solution. In various types of guidebooks concerning public-private partnerships, the examples given comprise construction projects, most often toll motorways, car parks, aquaparks, swimming pools, or other types of structures for culture and sport [13]. All these are classical examples of public-private partnership projects, in which the private entity takes the risk concerning demand and financing of the undertaking from sales of services. The prerequisites in favour of implementation of IT projects in public-private partnership formula contain the existence of numerous areas of risk, both during the implementation of such projects, and later, during their operation. The type of risk, and most generally the fact that a given risk should be managed by the entity which can have influence upon it support the use of public-private partnerships in such cases.

The public party, undertaking the delivery and implementation of an IT system, expects the system to be functional and reliable. In practice, however, the problems with functioning of such systems are common. Implementation of many systems takes place with delay, there are also many problems, especially in the early stages of their operations, the assumed functionality is not always achieved, either. In case of the public-private partnership formula those problems, which have the source in the delivery and implementation of systems, still remain the responsibility of the party responsible for implementation of the system, i.e. the private partner. Besides that, problems related to public procurement in connection with development of systems are avoided. Development, enhancement of parameters of the IT system, as well as extension of its functionalities may be carried out without more lengthy public procurement procedures, while the motivation is the proper provision of services, so that it is reflected in income from sales.

4   Procedures of Selection of Private Partners in Public-Private Partnership Systems

In Poland, in accordance with the Act of law of December 19, 2008, on public-private partnership [10], the selection of the private partner under a public-private partnership arrangement, may be executed on the basis of that Act of law, or the Act of law of January 19, 2009, on concessions for works or services [11], or the Act of law of January 29, 2004 – Public procurement law [12]. In accordance with the provisions of art. 4 of the Act of law on public-private partnership, if the remuneration of the private partner is the right to proceeds from the object of the public-private partnership, or first of all it is the right to – together with obtaining payment – also to select the private partner and conclude a public-private partnership agreement, the provisions of the Act of law of January 9, 2009 – on concession for works or services – are applicable. In other cases, the provisions of the Act of law of January 29, 2004 – Public procurement law [12] are applicable for the selection of private partner and conclusion of a public-private partnership agreement. In cases in which neither the Act of law of January 9, 2009 on concessions for works or services nor the Act of law of January 29, 2004 – Public procurement law apply, the private partner selection is made in a way that would guarantee fair and free competition, as well as equal treatment, transparency, and proportionality, with taking properly into consideration the provisions of the act of law on public-private partnership.

It can be derived from the public-private partnership law that the choice of the partner is made on the basis of the Act of law on concessions for works or services, or deliveries – public procurement law. The provisions of art. 6 of the Act of law on concessions for works or services impost the obligation to provide – when selecting the private partner – equal and non-discriminative treatment of the interested entities, as well as to act transparently, observing the principles of fair competition. At the same time, in accordance with art. 7 of the Act of law referred to, the entity carrying out the procedure is obliged to make a description of the subject of concession in a way that would enable the interested entities equal access to carry out the concession and the way that would not create limitations for competition to carry out the concession.

The description of the subject of contract is provided by making reference to the technical specification or functional requirements, provided that such a description enables the interested parties to determine the subject of concession, or refer partly to the technical specification or functional requirements. Technical specification should contain references to Polish standards, being applications of European standards and norms, European technical approvals, joint technical specifications, international norms, other systems of technical references established by European normalization bodies or – in case of their absence – to national norms/standards, national technical approvals, referring to designing, calculation, execution of works, as well as use of products, and should include the procedure for assessment of compliance of the offer with requirements of the concessionaire. Moreover, in case of services – it should concern the required features of the service, particularly the quality level, influence on the environment, adjustment to the needs of all users. The description of concession subject may not include trademarks, patents, type designation, designation of origin or place of manufacturing, which could lead to giving privileges to a given entity, or to eliminate that entity from the procedure.

On the basis of the provisions of the Acts of law referred to above, one can state that there are no regulatory obstacles concerning the execution of IT projects under the formula of public-private partnership procedure. Also other factors which would cause additional difficulties for that type and form of IT projects execution do not occur. The reasons why that formula is scarcely used for the execution of IT projects may lie in the lack of motivation in public entities to apply project-oriented approach to the undertakings. Public-private partnerships require excellent preparation of the undertaking, its realistic assessment from the point of view of expenditures, revenues, as well as cash flow generated, the size of real involvement of the public party, areas of risk and risk management. Those assessments are made not only by public entities preparing those undertakings, but also carried out and verified by private partners that consider participation in the undertaking, as well as banks or other financial institutions, which provide loans. The result of such a verification is the interest, making a bid or offer, and conclusion of the contract by the private partner, or lack of interest and resignation from making an offer. Resignation and not making the final offer – in most cases – implies that the undertaking, within the assumed financial framework and in accordance with the assessment made by the private partner, does not generate sufficient positive cash flow (even putting together the public funds and funds from users of services).

5   Solutions Applied – Partnership Models for Implementation of IT Projects

Public tasks may be carried out in various ways. They may be implemented directly, through public administration agencies, or units they have established, or the so-called internal branches. They are separate entities, being legal persons, established by public agencies in order to carry out a given public task. Such solutions are commonly used in the course of providing various public services; it is not excluded, however, that separate public units may manage big IT systems. However, these will not be common solutions, and we also do not have here the solutions concerning public-private partnership. Although the difference between an undertaking executed as traditional public procurement and public-private partnership is quite fine and not clear, it should be assumed that public-private partnership entails joint execution of an undertaking, in which the private entity involves its capital, experience, and skills, and its remuneration depends first of all on the revenues received from users (persons using a given service – not the public entity). The private entity takes or should at least party take the risk of reduced demand for services.

One can thus enumerate the most popular scopes of tasks comprised in a given undertaking, which are transferred to the private partner during execution and subsequent management of a given undertaking (undertaking execution models) [2, 3]:

– BOT (Build – Operate – Transfer),
– BOO (Build – Operate – Own),
– BOOT (Build – Operate – Own – Transfer),
– DBFO (Design – Build – Finance – Operate).

Those solutions differ, first of all, in the scope of tasks to be performed by the private partner, as well as blurred data concerning the ownership of resources purchased within the framework of the undertaking. Figure 2 presents in darker shades the tasks the private partners are entrusted with in specific models of solutions.

In the BOT model, Build – Operate – Transfer, following a contract concluded between a public sector entity and a private entity, the latter obtains the right (concession) to build, finance, and operate a given undertaking/project, e.g. an IT system, which serves the execution of tasks related to public services, for a time stipulated in the contract. An undertaking in the BOT model is finance partly with public funds – the most often encountered contributions include land or real estate needed for a given undertaking, as well as private funds, yet it is the public party that owns the investment project. During concession validity, the private entity has the right to establish prices and collect fees and charges from system users. After expiry of the contract, which is most often concluded for the duration of a given undertaking, all the assets with the rights of their use are transferred to the public authority [6]. Of course, various detailed solutions are also possible there, among them – particularly in case of IT systems – those related to the dilemma whether to join designing and construction, or to separate the two. Joining designing and delivery, as well as implementation of a given system entails transferring the responsibility for proper designing to the contractor. Additionally, it also entails avoidance of the problem concerning restriction of competition and giving

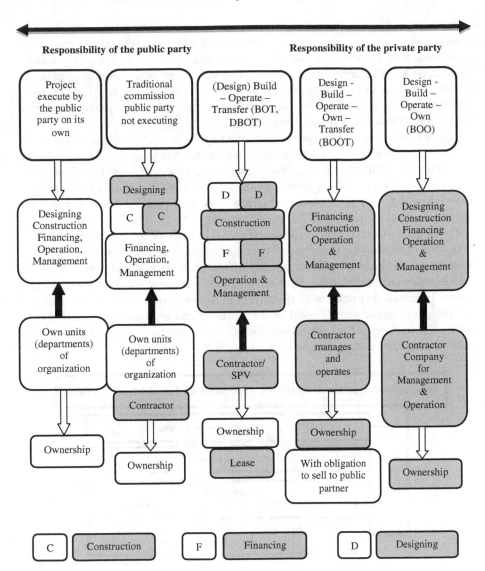

Fig. 2. Models of implementation of investment meant for provision of public services (from traditional forms to contracts in PPP formula) [5]

privileged position to a certain contractor, selecting certain solutions which are convenient for that contractor in the project. Joining designing and building, as well as transferring those tasks to a private entity, also entails depriving the public entity of influence on the solution design, which is not always a good solution.

The contractor may design the system in such a way, that it is easy to implement, not necessarily executing it the way the public entity wanted. In turn, separation of designing and building allows to design without paying attention to whether the implementation is easy or difficult, focusing mainly on the requirements for a given system.

In the "build and operate" solution, the number of tender procedures is also limited, which simplifies the processes related to preparation and implementation of the undertaking. Moreover, knowledge about the complete costs – not only of designing, but also of operation and maintenance of a given IT system – are obtained. Of importance, particularly in case of IT systems, are also the problems concerning copyright. Separation of designing as well as subsequent implementation of such a system may create numerous problems in the area of copyright.

The BOT solution, from the perspective of a public entity, is not beneficial in the IT systems area, as such systems – for example – after several years of use may not have much value left, while transferring them to the private entity may, due to their physical as well as moral wear out, just mean a problem for that entity. On top of that, there is the problem of further maintenance of such a system, in practice this may imply the dependence upon the entity that has been operating is so far. The BOO model, that is Build – Operate – Own, is a better solution (Fig. 3). The private entity remains the owner of the assets related to the execution of the undertaking. The funds for that purpose, in the initial period, are obtained from own capital and from external financial institutions, those funds are recovered in the form of fees paid by users of the investment. Such a solution is convenient in case of IT systems, as in fact the private partner takes over also the problem of liquidation/disposal of the obsolete and used components of the IT system. Those systems have often been and still are a substantial problem, they often inhibit the implementation of more modern solutions.

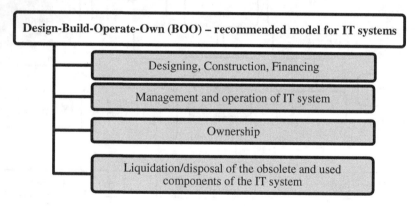

Fig. 3. Design – Build – Operate – Own as a recommended model for IT systems implementation [own study]

In case of concessions of the BOOT type, Build – Own – Operate – Transfer, the private partner designs using only own funds or external funds (not from the public partner), finance the undertaking – purchases and implements the IT system, and subsequently maintains it for a certain time to be operational, as well as introduces development modifications. When the contract expires, the private investor transfers the system ownership to the public entity. In this solution, problems similar to those encountered in the BOT system, concerning the reception from the private partner – after the assumed period of operations – of a used up and obsolete system, as well as the pretty difficult issue of copyright concerning such a system.

Also other models of co-operation based on public-private partnership are known, yet to a large extent they are modifications of those referred to earlier, they may differ regarding the time of transfer, the scope of undertaking – to what extent it is a new undertaking, and how much it is a modernization of the existing system, for example – or the form of contract, that allows the use of assets and resources of the other partner.

Selection of a suitable model of co-operation is an extremely important thing in the development of public-private partnership, it also has a significant influence upon the effectiveness of the undertaking, division of risks, and subsequent management of risks. In assessing every possible variant, one should take into account the potential barriers in partnership execution, the interest of private partners in the given undertaking; one should also determine which of the elements should remain under public party responsibility, and which should be executed in co-operation with the private sector.

Of much importance for finishing a project on time is the ongoing control over work progress, comparison of the real work progress with the schedule. Quite often there are situations, in which the project gets delayed even at the early stages of project execution, due to lack of knowledge about it no actions are taken, which would – for example – consist of involving more resources, or changing time schedules. Later on, this generates additional labour intensive efforts in various areas of the project execution, thus additionally extending the time of its implementation. Also of vital importance for completing the project on time are efficient procedures concerning risk management and changes in the project. Good knowledge about risks enables to reduce the occurrence of delays during various actions taken, while efficient procedures and thorough knowledge about consequences of changes minimizes possible further complications in project execution.

During the execution of the project, there may be situations which have not been foreseen in the contract, for which responsibilities of parties and rules of procedure have not been stipulated. This is partly due to the large variety of events, which may occur during project execution, as well as the fact that some events and situations have been considered hardly probable, thus they have been omitted in the contract. In case of some of the events, e.g. those concerning force majeure, it is often difficult to determine their scale and consequences, for example such events as destruction caused by earthquake, acts of war, riots, destruction of a structure due to fire caused by storm. Such events do not necessarily happen in the place to which the IT system is delivered, yet they are not so rare in the countries where the hardware or other components for a given project are manufactured. It is necessary, then, to assess whether the events having negative influence on the project execution may be classified as force majeure and whether they really had influence on delays. Employers, first of all those that units of the public finance sector, then have the possibility of introducing suitable amendments to the contract, including the order execution time, thus reducing stipulated penalties or cancelling them.

It is quite common during implementations of IT systems that delays occur, in relation to the deadlines stipulated in the contract. Delays pose a threat for both the employer and the contractor. For the employer, they cause various types of complications, which may consist of hindering or even completely precluding the provision of selected services, extension of the scope of activities, obtaining quality standards, thus achieving such effects that were the basis for taking up the execution of that project. In case of the contractor, the delays most often involve additional costs related to project execution, stipulated penalties, the problem concerning obtaining confirmation of correct execution of the contract, or image problems.

As a rule, the contracts contain definite stipulated penalties for retardation or delay. The institution of stipulated penalty does not require the contractor to provide documentation or estimate the losses of the employer, caused by retardation or delay. This simplifies the penalty amount issue, and reduces it to the assessment of the retardation or delay time. In the contract, the amount of stipulated penalty is usually defined as a percentage of remuneration for the contractor, for each day of retardation or delay. In a similar way, reductions of remuneration due to retardation or delay would work – the difference being in onomatology and accounting principles. Stipulated penalties may be proportional to the time that elapsed, but can also be progressive. In the latter case, as the retardation/delay increases, the stipulated penalty per day increases, too.

6   Conclusion

The ever greater scope of tasks executed by the public sector should, in parallel, enforce transformations in that sector, which would consist first of all of more common applications of solutions and tools that increase the efficiency and effectiveness of undertakings implemented. One of such activities is the use of public-private partnership, as a solution in which knowledge, experience, and capital possessed by the private sector are involved. The formula of public-private partnership may be applied in various type of undertakings and projects – which vary in terms of capital expenditure amount, as well as the types of public tasks executed. The undertakings related to delivery, implementation, and use of IT systems may and should be – in the extent wider than so far – executed with the use of public-private partnership formula. This will contribute to better utilization of public funds, as well as better management of the IT systems after implementation.

References

1. Austen, A., Dydkowski, G.: Zarządzanie usługami użyteczności publicznej, pp. 21–64. Wydawnictwo Uniwersytetu Ekonomicznego w Katowicach, Katowice (2011)
2. Cenkier, A.: Partnerstwo publiczno – prywatne jako metoda wykonywania zadań publicznych, pp. 131–144. Monografie i Opracowania 566, Szkoła Główna Handlowa w Warszawie, Warszawa (2009)
3. Dydkowski, G., Urbanek, A.: Partnerstwo publiczno – prywatne. Wydawnictwo Uniwersytetu Ekonomicznego w Katowicach, Katowice (2011)
4. Frączkiewicz–Wronka, A. (ed.): Zarządzanie publiczne – elementy teorii i praktyki. Wydawnictwo Akademii Ekonomicznej w Katowicach, Katowice (2009)
5. Herbst, I.: Finansowanie rozwoju. Partnerstwo Publiczno – Prywatne. http://www.twigger.pl/pliki/PPP.ppt. Accessed 17 Feb 2010

6. Jaśniewicz, A.: Partnerstwo Publiczno – Prywatne, Materiały szkoleniowe Regionalnego Ośrodka Europejskiego Funduszu Społecznego w Pile. www.pila.roefs.pl. Accessed 01 Dec 2015
7. Kos-Łabędowicz, J.: Integrated E - ticketing system – possibilities of introduction in EU. In: Mikulski, J. (ed.) TST 2014. CCIS, vol. 471, pp. 376–385. Springer, Heidelberg (2014). doi: 10.1007/978-3-662-45317-9_40
8. Okoń, J.: Partnerstwo publiczno – prywatne jako narzędzie korzystania z kapitałów prywatnych i efektywnej realizacji zadań publicznych. Unpublished manuscript
9. Partnerstwo publiczno-prywatne w Polsce w latach 2009–2011. Report. Platforma Partnerstwa Publiczno-Prywatnego, Pomoc Techniczna Narodowa Strategia Spójności, Ministerstwo Rozwoju Regionalnego, Unia Europejska Europejski Fundusz Rozwoju Regionalnego, Warszawa, maj 2012. www.ppp.gov.pl
10. The Act of 19 December 2008 on public-private partnership, Dz. U. 2009 Nr 19, poz. 100
11. The Act of 19 January 2009 on concessions for works or services, Dz. U. z 2009 Nr 19, poz. 101
12. The Act of 29 January 2004 on public procurement law, Dz. U. z 2013 r. poz. 907
13. Urbanek, A.: Pricing policy after the implementation of electronic ticketing technology in public urban transport: an exploratory study in Poland. In: Mikulski, J. (ed.) TST 2015. CCIS, vol. 531, pp. 322–332. Springer, Heidelberg (2015). doi:10.1007/978-3-319-24577-5_32
14. Zawicki, M.: Nowe zarządzanie publiczne. PWE, Warszawa (2011)

Assessment of the Method Effectiveness for Choosing the Location of Warehouses in the Supply Network

Ilona Jacyna-Gołda[1]([⊠]), Mariusz Izdebski[2],
and Emilian Szczepański[2]

[1] Faculty of Industrial Engineering,
Warsaw University of Technology, Warsaw, Poland
i.jacyna-golda@wip.pw.edu.pl
[2] Faculty of Transport, Warsaw University of Technology, Warsaw, Poland
{mizdeb,eszczepanski}@wt.pw.edu.pl

Abstract. This paper describes the problem of determining the location of warehouses in the logistics network supplying different type of entities with raw materials. Was presented a mathematical model, i.e. were defined decision variables, were specified constraints and objective function that minimizes the total cost of transporting raw materials from suppliers to the recipients. Were included costs of flow of raw materials between network facilities and costs associated with salaries for drivers engaged in the realization of this flow. Was developed a method of determining the location of storage facilities based on genetic algorithm. In order to check the efficiency of the method was proposed algorithm for evaluating its effectiveness.

1 Introduction

Location problem of warehouses in the logistics and transport networks is a known and widely discussed issue in the literature [1, 2, 7, 8, 11]. In general, it refers to the choice of place to build a new warehouse or a reorganization of the existing network configuration. The article indicates computer applications suitable for solving specific problems in the company.

Among the criteria for selection of warehouse facilities usually are [9, 16]: ease of recruitment of manpower, convenient connections with the market, the possibility of the acquisition of buildings or other real estate, labor costs, ease of acquisition of raw materials for construction, proximity of the motorway, local taxes, the distance from the supply markets and customers etc.

In the case of the reorganization of the network configuration, i.e. the choice of location among the existing warehouse facilities [7], the problem is to plan the deployment of warehouse facilities in the transport and logistics network, so that the total costs of transport were as small as possible.

The article analyzes the problem of the location of warehouses in the supply networks of different types of entities, e.g. production companies, large shopping centers, etc. The selection of warehouse facilities shall be made among the existing

© Springer International Publishing AG 2016
J. Mikulski (Ed.): TST 2016, CCIS 640, pp. 84–97, 2016.
DOI: 10.1007/978-3-319-49646-7_8

warehouses. In this type of issues decidedly important are: distance between all objects network and the size of flow of stream.

Decision model of selection can be assigned to a group of transport and production models [10], in which it is assumed that the location of both supported entities and suppliers is determined. Location problem is solved under the assumption that to meet the demand of consumers, you need to determine the appropriate location of warehouses and other logistics facilities [7].

Due to the many aspects of the problem and the high computational complexity of the issue as an optimization tool to solve was applied genetic algorithm [5]. The calculation procedure is divided into two stages. The first stage concerns the determination of the size of flows of raw materials and the second concerns the routing for riding the vehicles carrying given flows.

In order to check the operational efficiency of proposed method was developed an algorithm that determines the degree of its effectiveness. The concept of efficiency is a concept variously defined in the literature [3, 6, 12, 13, 15, 17] and refers to the different thematic areas, e.g. the system's efficiency, supply chain efficiency. To the need of efficiency evaluation of the method it is assumed that the method is effective if the generated results are better than the results obtained in a random way.

2 Warehouses Location Model in the Recipients Supply Network

Supply network for customers with complex functional structure such as production companies or big commercial centers consists of: suppliers of materials, intermediate warehouses for the storage of materials and recipients: production facilities or shopping centers, etc. Any supplier can deliver the material directly to customers or supply warehouses. With an extensive supply network may exist central, regional or local warehouses. Recipients report the need for a certain amount of material. This means that could also be the flow of materials between warehouse facilities.

Given the above choice of warehouses location depends on the assumed optimization criteria (e.g. delivery time, cost of transportation), whereby it is assumed that all customers' needs should be executed.

In the present model, it is assumed that the amount of material offered by suppliers is at least equal to the size of the demand of consumers (companies, shopping centers, etc.) and the choice of warehouses location depends on the adopted criterion function, which is the cost of delivery.

To the need of building a model of localization was defined:

- $V = \{v : v = 1, 2, \ldots, v', \ldots, V\}$ - a set number of transport and logistics network elements: suppliers, warehouses of multi assortment materials (raw materials for companies, components, subassemblies, finished goods, etc.), recipients.
- $DS = \{v : \alpha(v) = 0 \text{ for } v \in V\}$ - a set of numbers of suppliers, $\alpha(w)$ – the mapping of carrying elements of the set V into the set $\{0,1,2\}$,
- $MS = \{v : \alpha(v) = 1 \text{ for } v \in V\}$ - a set of numbers of warehouses,

- $P = \{v : \alpha(v) = 2 \text{ for } v \in V\}$ - a set of numbers of entities with complex functional structure, e.g. production companies, large shopping centers, etc..,
- $H = \{h : h = 1, 2, \ldots, H\}$ - a set of supply materials (e.g. raw materials for companies, products for shopping centers, etc.),
- $D = \{d : d = 1, 2, \ldots, D\}$ - a set of working days,
- $K = \{k : k = 1, 2, \ldots, K\}$ - a set of drivers,
- $KR(v, v') = \{(kr, v, v') : kr = 1, \ldots, KR\}$ - a set of transport courses without load between v-th object of a network and v'-th object,
- $\mathbf{D1} = [d1(v, v')]$, $\mathbf{D2} = [d2(v, v')]$, $\mathbf{D3} = [d3(v, v')]$, $\mathbf{D4} = [d4(v, v')]$ - distance matrices on the relationship: supplier-warehouse, warehouse-warehouse, supplier-company, warehouse-company,
- $\mathbf{Q1} = [q1(h, v)]$ - matrix of delivery volumes of h-th type of supply materials from v-th supplier,
- $\mathbf{Q2} = [q2(h, v)]$ - matrix of demand volumes of h-th type of materials from v-th customer,
- $\mathbf{POJ} = [poj(v)]$ - matrix of capacity of v-th warehouse,
- $\mathbf{T1} = [t1(v, v')]$, $\mathbf{T2} = [t2(v, v')]$, $\mathbf{T3} = [t3(v, v')]$, $\mathbf{T4} = [t4(v, v')]$ - matrices of driving times on the relationship: supplier-warehouse, warehouse-warehouse, supplier-company, warehouse-company,
- c - the unit cost of transport of the unit load per unit of distance,
- g - hourly wage of driver work,
- s - the cost of fuel consumption while driving without a load,
- $Tdop$ - allowable driving time.

Were introduced two types of decision variables, variables defining the size of the flow of materials between individual objects and binary variables defining the connections between individual objects.

The first type that is written in the form of a **X1** matrix:

$$\mathbf{X1} = \left[x1(h, v, v') : \quad x1(h, v, v') \in \mathbf{R}^+ \cup \{0\}, \quad h \in \mathbf{H}, \ v \in \mathbf{DS}, \quad v' \in \mathbf{MS} \right] \quad (1)$$

relates to determining the optimum size of the flow of stream of materials. The matrix elements have an interpretation of the size of transport of h-th type of supply material between v-th supplier and v'-th warehouse object.

Analogously been defined decision variables determining the flow on the relationship between: warehouses for storing materials – written with matrix **X2**, between supplier and recipient (company/shopping center) – written with matrix **X3**, between warehouse of materials and recipient (company/shopping center) – written with matrix **X4**.

The second type of variables are binary variables. Sample binary variable defining the connection between v-th supplier and v'-th warehouse facility carried out by k-th driver in d-th day in kr-th course is defined as:

$$X5 = \begin{bmatrix} x5(v, v', k, d, kr) : x5(v, v', k, d, kr) \in \{0, 1\}, v \in \mathbf{DS}, v' \in \mathbf{MS}, \\ k \in \mathbf{K}, d \in \mathbf{D}, kr \in \mathbf{KR}(v, v') \end{bmatrix} \quad (2)$$

Analogously been defined decision variables determining the connections on the relationship: warehouse – warehouse – written with matrix $\mathbf{X6}$, suppliers – recipient – written with matrix $\mathbf{X7}$, warehouse – recipient – written with matrix $\mathbf{X8}$.

In addition, the model highlights such limitations as to meet this demand the company, not to exceed the supply capabilities of the supplier, not to exceed the capacity of the warehouse, limiting driving time: driving time should not exceed the allowable driving time in a given working day and the balancing the flows in the warehouse.

The criterion function $\mathbf{KSK}$ minimizes the total cost of delivery of materials to the recipient, takes into account the costs arising from the movement of materials between the objects of network and costs associated with driving without a load, i.e. fuel consumption costs and salaries of the driver.

$$\mathbf{KSK} = \mathbf{KPS} + \mathbf{KWK} \rightarrow \mathbf{min} \quad (3)$$

The cost of the flow of supply materials $\mathbf{KPS}$ between suppliers and warehouses with multi-assortment of materials (the first component of the formula), between warehouses (the second component) between the supplier and the recipient (the third component) and the warehouse and the recipient (the fourth component) can be represented as follows:

$$\mathbf{KPS} =$$

$$\sum_{h \in \mathbf{H}} \sum_{v \in \mathbf{DS}} \sum_{v' \in \mathbf{MS}} x1(h, v, v') \cdot d1(v, v') \cdot c + \sum_{h \in \mathbf{H}} \sum_{v \in \mathbf{MS}} \sum_{v' \in \mathbf{MS}} x2(h, v, v') \cdot d2(v, v') \cdot c$$

$$+ \sum_{h \in \mathbf{H}} \sum_{v \in \mathbf{DS}} \sum_{v' \in \mathbf{P}} x3(h, v, v') \cdot d3(v, v') \cdot c + \sum_{h \in \mathbf{H}} \sum_{v \in \mathbf{MS}} \sum_{v' \in \mathbf{P}} x4(h, v, v') \cdot d4(v, v') \cdot c \quad (4)$$

$$\rightarrow min$$

The cost $\mathbf{KWK}$ associated with salary of drivers on the route of driving without load and the cost of fuel consumption on this route can be presented as follows:

$$
\mathbf{KWK} =
$$

$$
s \cdot \sum_{k \in \mathbf{K}} \sum_{d \in \mathbf{D}} \left(\sum_{v \in \mathbf{DS}} \sum_{v' \in \mathbf{MS}} \sum_{kr \in \mathbf{KR}(v,v')} x5(v, v', k, d, kr) \cdot d1(v, v') \right.
$$

$$
+ \sum_{v \in \mathbf{MS}} \sum_{v' \in \mathbf{MS}} \sum_{kr \in \mathbf{KR}(v,v')} x6(v, v', k, d, kr) \cdot d2(v, v')
$$

$$
+ \sum_{v \in \mathbf{DS}} \sum_{v' \in \mathbf{P}} \sum_{kr \in \mathbf{KR}(v,v')} x7(v, v', k, d, kr) \cdot d3(v, v')
$$

$$
\left. + \sum_{v \in \mathbf{MS}} \sum_{v' \in \mathbf{P}} \sum_{kr \in \mathbf{KR}(v,v')} x8(v, v', k, d, kr) \cdot d4(v, v') \right)
$$

$$
+ g \cdot \sum_{k \in \mathbf{K}} \sum_{d \in \mathbf{D}} \left(\sum_{v \in \mathbf{DS}} \sum_{v' \in \mathbf{MS}} \sum_{kr \in \mathbf{KR}(v,v')} x5(v, v', k, d, kr) \cdot t1(v, v') \right.
$$

$$
+ \sum_{v \in \mathbf{MS}} \sum_{v' \in \mathbf{MS}} \sum_{kr \in \mathbf{KR}(v,v')} x6(v, v', k, d, kr) \cdot t2(v, v')
$$

$$
+ \sum_{v \in \mathbf{DS}} \sum_{v' \in \mathbf{P}} \sum_{kr \in \mathbf{KR}(v,v')} x7(v, v', k, d, kr) \cdot t3(v, v')
$$

$$
\left. + \sum_{v \in \mathbf{MS}} \sum_{v' \in \mathbf{P}} \sum_{kr \in \mathbf{KR}(v,v')} x8(v, v', k, d, kr) \cdot t4(v, v') \right) \rightarrow min
$$

(5)

3 The Procedure of the Method Designating the Location of Warehouses

3.1 Assumption

The method solving the warehouses location problem in supply network for recipients with the assumed functional structure such as the manufacturing companies or large shopping centers can be divided on two stages. The first stage is to designate the transportation tasks which are characterized by the minimal cost of their realization. (the cost **KPS** is designated). In the second stage the delivery schedule of multi-assortment materials to individual facilities in the supply network is indicated. (the cost **KWK** is additionally designated). In order to specify this schedule it is required to designate the routes of vehicles which realize this tasks.

3.2 Realization of the Stage I

The stage I is to designate the transportation tasks of minimum costs of their realization. In this case it is essential:

- to indicate the connections between the facilities of the network which generate the minimum cost associated with the movement of material flow;
- to designate the optimum size of the material flow on this connections. The material flow which is decomposed into the network is the flow which fully meets customers' demands.

In this stage the function **KPS** is minimized which determine the cost of the material flow in the network. In order to minimize this function the genetic algorithm was used. To designate this algorithm of the structure, the adaptation function, the mutation and crossover process must be indicated.

The another steps of the algorithm were defined in the following way:

- **Step 1**. Defining the data input structure. The data input structure was presented as the matrix **M**, where rows and columns of the matrix determine the facilities of the supply network. The facilities in the matrix are located in the following order: suppliers, warehouses, economic operators. Graphic interpretation of the structure of the sample sizes of the stream is shown in Fig. 1 (**DS**- suppliers, **MS**- warehouses, **P**- recipients);

	DS1	DS2	MS1	MS2	MS3	P6	P7
DS1	0	0	7	5	10	10	5
DS2	0	0	5	3	15	5	5
MS1	0	0	0	3	3	5	5
MS2	0	0	3	0	2	2	5
MS3	0	0	1	1	0	8	20
P6	0	0	0	0	0	0	0
P7	0	0	0	0	0	0	0

Fig. 1. The data input structure of the genetic algorithm [own study]

- **Step 2**: Defining the adaptation function. In order to search the minimum cost of the material flow the adaptation function Fp_n for n-this structure takes the form:

$$Fp_n = C - KPS_n \qquad (6)$$

where:
C- value much higher than the value of the costs of the material flow in the network,
KPS_n - the cost of the material flow in n-this structure, formula (4).
The tendency of genetic algorithms is to maximize the function of adaptation. Maximization of the function Fp_n consequently is the process of minimization of the function KPS_n, what is the assumed optimization aim.
- **Step 3**. Defining the crossover operation. The crossover operation is adequate to the adopted matrix structure [14]. In the aim of conducting the crossover operation the two matrix are built: DIV which contains rounded average values with both parents, and the matrix REM which contains the information, whether rounding was necessary. Assuming that the values of matrices **M1** and **M2** (parents) in each cells take

the symbol $m^1_{v,v'}$, $m^2_{v,v'}$, the values of the matrices DIV and REM are calculated with the following relations:

$$\dim_{v,v'} = \left\lfloor (m^1_{v,v'} + m^2_{v,v'})/2 \right\rfloor \tag{7}$$

$$rem_{v,v'} = (m^1_{v,v'} + m^2_{v,v'})/\!\mod 2 \tag{8}$$

The example of the crossover process is presented in Fig. 2.

a)

Structure 1

	DS1	DS2	MS1	MS2	MS3	P6	P7
DS1	0	0	7	5	10	10	5
DS2	0	0	5	3	15	5	5
MS1	0	0	0	3	3	5	5
MS2	0	0	3	0	2	2	5
MS3	0	0	1	1	0	8	20
P6	0	0	0	0	0	0	0
P7	0	0	0	0	0	0	0

Structure 2

	DS1	DS2	MS1	MS2	MS3	P6	P7
DS1	0	0	5	7	5	10	10
DS2	0	0	6	5	5	15	2
MS1	0	0	0	0	10	2	8
MS2	0	0	4	0	10	2	10
MS3	0	0	1	0	0	1	10
P6	0	0	0	0	0	0	0
P7	0	0	0	0	0	0	0

b)

DIV

	DS1	DS2	MS1	MS2	MS3	P6	P7
DS1	0	0	6	6	7	10	7
DS2	0	0	5	4	10	10	3
MS1	0	0	0	1	6	3	6
MS2	0	0	3	0	6	2	7
MS3	0	0	1	0	0	4	15
P6	0	0	0	0	0	0	0
P7	0	0	0	0	0	0	0

REM

	DS1	DS2	MS1	MS2	MS3	P6	P7
DS1	0	0	0	0	1	0	1
DS2	0	0	1	0	0	0	1
MS1	0	0	0	1	1	1	1
MS2	0	0	1	0	0	0	1
MS3	0	0	0	1	0	1	0
P6	0	0	0	0	0	0	0
P7	0	0	0	0	0	0	0

c)

New structure 1

	DS1	DS2	MS1	MS2	MS3	P6	P7
DS1	0	0	6	6	8	10	7
DS2	0	0	5	4	10	10	4
MS1	0	0	0	2	6	3	7
MS2	0	0	4	0	6	2	7
MS3	0	0	1	0	0	5	15
P6	0	0	0	0	0	0	0
P7	0	0	0	0	0	0	0

New structure 2

	DS1	DS2	MS1	MS2	MS3	P6	P7
DS1	0	0	6	6	7	10	8
DS2	0	0	6	4	10	10	3
MS1	0	0	0	1	7	4	6
MS2	0	0	3	0	6	2	8
MS3	0	0	1	1	0	4	15
P6	0	0	0	0	0	0	0
P7	0	0	0	0	0	0	0

Fig. 2. Crossover process [own study]

The used crossover operator guarantees the correct individuals after the conducting process. without using the repair algorithms.

- **Step 4**: Defining the mutation operation. The principle of operation of the operator mutation relies on randomizing the two numbers p and q in the range of: $2 \le p \le k$

and $2 \leq q \leq n$, which define the number of rows and columns of the submatrix which has the dimension: $p \times q$ (p - the number of rows in the main matrix which is processed by the algorithm, q - the number of columns). The submatrix is modified is such a way that the sum of the values in rows and columns before and after the modification process has not changed. The example of the crossover process is presented in Fig. 3.

Matrix

	DS1	DS2	MS1	MS2	MS3	P6	P7
DS1	0	0	7	5	10	10	5
DS2	0	0	5	3	15	5	5
MS1	0	0	0	3	3	5	5
MS2	0	0	3	0	2	2	5
MS3	0	0	1	1	0	8	20
P6	0	0	0	0	0	0	0
P7	0	0	0	0	0	0	0

Sub-matrix

MS1	MS2	MS3
7	5	10
5	3	15
0	3	3

Modyfication

MS1	MS2	MS3
8	4	10
4	4	15
0	3	3

Mutation

	DS1	DS2	MS1	MS2	MS3	P6	P7
DS1	0	0	8	4	10	10	5
DS2	0	0	4	4	15	5	5
MS1	0	0	0	3	3	5	5
MS2	0	0	3	0	2	2	5
MS3	0	0	1	1	0	8	20
P6	0	0	0	0	0	0	0
P7	0	0	0	0	0	0	0

Fig. 3. Mutation process [own study]

The steps of genetic algorithm are repeated until the moment when the stop condition is achieved. After determining the optimum flow of the materials between the facilities in the network one can go to designate the transportation tasks. The task is interpreted as e.g. transport the materials $x1(h, v, v')$ between the loading points (suppliers) and unloading points (the warehouse) l-this mode of transportation of a certain capacity $poj(l)$. An example of a formula that specifies the number of tasks on this connection is presented as:

$$n1(v, v') = \left\lceil \frac{x1(h, v, v')}{poj(l)} \right\rceil \tag{9}$$

The total number of tasks is expressed by the formula ($n2(v, v')$ - the connection between suppliers-recipients, $n3(v, v')$ - between the warehouses, $n4(v, v')$- between the warehouses and suppliers):

$$N = \sum_{v \in DS} \sum_{v' \in MS} n1(v, v') + \sum_{v \in DS} \sum_{v' \in P} n2(v, v') + \sum_{v \in MS} \sum_{v' \in MS} n3(v, v') + \sum_{v \in MS} \sum_{v' \in P} n4(v, v')$$

$$\tag{10}$$

The number of the tasks is depended on the capacity of vehicles which realize transport of materials. In order to minimize the number of tasks transport of materials by the use of the vehicles of the maximum capacities is recommended. In case of unrealized tasks due to the limited number of vehicles the given capacity and time constraints resulting from the date of delivery of materials to the warehouses or undertaking, the tasks must be designated for another types of the vehicles.

3.3 Realization of the Stage II

After designating tasks the main aim is to determine the routes of vehicles in such a way to the tasks in this routes were realized in the minimal transportation cost. The process of storing materials and the fact of the existence of the transportation process this materials between warehouses and the warehouses and the recipient imposes developing the delivery schedule of materials to individual facilities in the network. Transport of the materials must be conducted in the following order:

- Step 1. Designating the routes on the connections: suppliers-recipients, suppliers – warehouses.
- Step 2. Designating the routes on the connections between the warehouses.
- Step 3. Designating the routes on the connections: warehouses – recipients.

This order is the result of the transport processes occurring in the supply network. The realization of the step 2 in the first place is the wrong solution because no material was transported to the warehouses, so there is no possibility of its further processing. In the step 3 the final state of the all warehouses is known, so the routes on the connections: warehouses – recipients are possible to designate. The tasks are realized in successive working days, taking into account the presented schedule (steps from 1 to 3) and time constraints resulting from working time of drivers. In the model it was assumed that the single driver can handle each connections between each facility in the network, there is not separate zones for individual drivers.

The stage II is to determine the delivery schedule of multi assortment of materials (raw materials, intermediates, components etc.) to individual facilities in the supply network. All tasks must be assigned to the appropriate routes of the vehicles. The aim of the stage II is:

- designating the routes of the vehicles which realize transportation tasks,
- the optimization of the designated routes in terms of transportations costs,
- the assignment of the designated routes to the drivers.

In the stage II the function **KWK** is minimized which determines the costs associated with the salary of the drivers and the cost of full consumption on the route without the cargo. In order to designate the routes which realize all tasks the genetic algorithm was used again. The stages of the algorithm were defined as: designating the structure processed by algorithm, the adaptation function, and the crossover and mutation operator.

To represent the chromosome in the problem of determining the transportation routes the string of natural number was used. The chromosome in this problem consists of the tasks and the bases points which are located at the recipients e.g. enterprises or shopping centers.

4 Efficiency of Method for Warehouse Location with the Use of Genetic Algorithm

The task of the algorithm in the case of stage II is to look for such a set of genes in the chromosome so the **KWK** cost of all tasks in given relationship, i.e. in steps 1–3 was minimal. The number of tasks in the chromosome depends on the step of supply of supplying materials, while the size of the chromosome is determined on the basis of the following rules. When the network is one base - e.g., production company, number of genes in the chromosome is calculated from the formula: $2 \cdot zad + 1$. The maximum length of chromosome is dictated by the situation in which is implemented a route base-task, there is no relationship task-task and the vehicle leaving to the task realizes it and returns to the base (Fig. 4a).

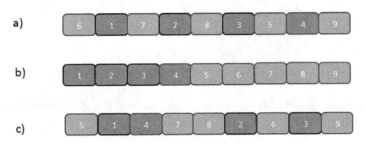

Fig. 4. The structure of chromosome for determining transportation routes [own study]

The use of a chromosome with a smaller number of genes than given relationship does not take into account this situation. The base in the chromosome was designated with numbers of genes from $zad + 1$ to the $2 \cdot zad + 1$ gene (Fig. 4b). This is the same base encoded under different genes. Such polygenic coding of the same base allows to check the situation in which each task is performed in the route base-task (Fig. 4a). Physical sense of the route presented as chromosome from Fig. 4c can be interpreted as a departure of the vehicle from the base to the task 1, directly from the task 1 vehicle goes to task 4, then return to base, departure to task 2, then return to base, departure to task 3, return to the base. From series of digits of the first two genes i.e. 1 and 6 (Fig. 4a) may determine costs between the base (6) and task (1). First and last of chromosome gene is always a base gene. The physical interpretation of the adopted chromosome construction is linked to the fact that the vehicles always start and end the route in the base.

In the case of a larger number of bases, chromosome length is changed. The new length is determined in the same manner as per one base.

The adaptation function takes the form identical as in stage I instead of the costs function associated with supply material flow **KPS**$_n$ in the n-th structure is implemented the $\mathbf{KWK}_n$ function of n-th chromosome. The crossover is carried out by known in the literature PMX operator [4], the mutation involves a random exchange of two genes. Determining the size of the population, the number of iterations, crossover parameters, and mutation is also an experimental stage.

Genetic algorithm should run three times for each of the highlighted steps of supplying materials supply (Step 1–3). It is associated with the respective generating population for given step, an evaluation of individuals and the operations of crossover and mutation.

After determining the chromosome with a minimum transportation route taking into account all the tasks within given step of supply material transportation we can determine individual driving routes for each driver in given relationship. Method of determining individual routes is shown in Fig. 5.

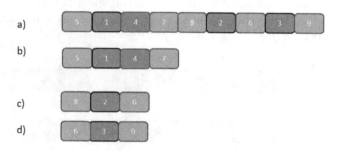

Fig. 5. Individual transport routes (a - minimum chromosome, b, c, d - individual routes) [own study]

Each individual route begins and ends at the base (the company) and consists of tasks. Such approved construction of routes allows you to add them to the total route of individual vehicles while maintaining not exceeding realization of driving time of driver.

The final step of stage II is to determine the schedule of the drivers route in the workday. Knowing the schedule of supplying materials supply i.e. steps 1–3 and with individual routes of driving of individual drivers is constructed work schedule on the driver's working day.

Operation of warehouses location method is based on the action of genetic algorithms. Genetic algorithms belong to a group of heuristic algorithms, which do not guarantee the optimal solution, but only close to the optimal so-called suboptimal. In addition, genetic algorithms take into account the factor of randomness, so for complex optimization problems, each running the algorithm produces a different result. For this purpose, is required an efficiency assessment of the presented method.

To evaluate the efficiency of the method was developed an algorithm, which determines the effectiveness of the method depending on the number of launches. The efficiency of the method is compared to the random result of warehouses locations.

In step 1, was entered input data, which is the minimum cost of location determined with K_i method and the random cost $Klos_i$ and the number of launches of the method $\mathbf{I}$.

In the case of generated the value of -1 by the function $F_i = \text{sgn}(K_i - \text{Klos}_i)$ the variable *sum* is increased by a value of 1. If $K_i - \text{Klos}_i < 0$ then the cost generated from the method is less than in the result of random selection of the location and the same method is effective in a given iteration. The variable *sum* determines the number of launches (iteration), in which method was more effective than random selection.

In the third step is determined efficiency coefficient of the method. It is the ratio of the number of launches in which the method has generated a better solution than a random selection to all launches of the method. The algorithm is shown in Fig. 6.

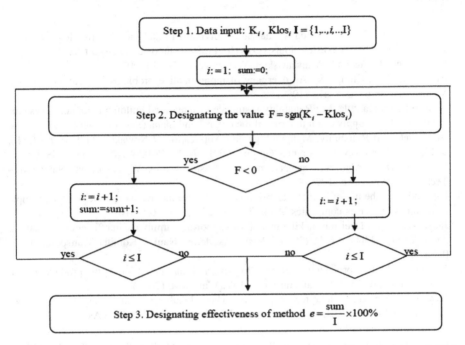

Fig. 6. The algorithm for determining efficiency of methods for locating warehouse facilities [own study]

5 Conclusion

Designing supply networks for customers with complex functional structure such as production companies or big shopping centers consisting of: material suppliers, intermediate warehouses for materials storage and the recipients: production facilities or shopping centers, etc. is a complex decision-making problem.

Important is, in such cases, the selection method in the selection of appropriate network configuration, i.e. the choice of location of warehouse facilities representing storage points and distribution of intermediate materials.

This paper proposes a method for choosing the location of warehouses, the operation of which is based on the action of genetic algorithms. Genetic algorithms belong to a group of heuristic algorithms, which do not guarantee the optimal solution, but only close to the optimal so-called suboptimal. In addition, genetic algorithms take into account the factor of randomness, so for complex optimization problems, each running of the algorithm generates a different result.

Acknowledgments. The scientific work carried out in the frame of PBS 3 project "*System for modeling and 3D visualization of storage facilities*" *(SIMMAG3D)* financed by the NCBR.

References

1. Ambroziak, T., Żak, J.: Metoda wyznaczania optymalnej lokalizacji centrów logistycznych w wybranym obszarze usług logistycznych. Prace Naukowe Transport, z.60, Oficyna Wydawnicza Politechniki Warszawskiej, Warszawa, s.67–75 (2007)
2. Brandeau, M.L., Chiu, S.S.: An overview of representative problems in location research. Manag. Sci. **35**(6), 645–674 (1989)
3. Bukowski, L., Karkula, M.: Reliability assurance of integrated building automation systems reliability of the supply chain in terms of the control and management of logistics processes, [w:] Safety and Reliability: Methodology and Applications / Nowakowski T. [i in.] (red.), 2015, CRC Press Taylor & Francis Group, ISBN 978-1-138-02681-0, ss. 549–558
4. Goldberg, D.E.: Algorytmy genetyczne i ich zastosowanie, Wydawnictwo Naukowo – Techniczne Warszawa (1995)
5. Izdebski, M.: The use of heuristic algorithms to optimize the transport issues on the example of municipal services companies. Arch. Transp. **29**(1), 27–36 (2014)
6. Jacyna-Gołda, I.: Decision-making model for supporting supply chain efficiency evaluation. Arch. Transp. **33**(1), 17–31 (2015). Polska Akademia Nauk - Komitet Transportu, ISSN 0866-9546
7. Jacyna, I.: Metod projektowania sieci logistycznej dla przedsiębiorstw produkcyjnych. Rozprawa doktorska. Wydział Transportu PW, Warszawa (2011)
8. Jacyna, M.: Metoda wielokryterialnej oceny wyboru lokalizacji centrum logistycznego, W.: Transport w logistyce. Łańcuch logistyczny. Wydawnictwo Uczelniane Akademii Morskiej w Gdyni, Gdynia (2003)
9. Pirkul, H., Schilling, D.A.: The maximal covering location problem with capacities on total workload. Manag. Sci. **37**(2), 233–248 (1991)
10. Klincewicz, J.G.: Heuristics for the p-hub location problem. Eur. J. Oper. Res. **53**, 25–37 (1991)
11. Kuehn, A.A., Hamburger, M.J.: A heuristic program for locating warehouses. Manag. Sci. **9** (4), 643–666 (1963)
12. Mishra, R.K.: Measuring supply chain efficiency: a DEA approach. JOSCM **5**(1), 45–68 (2012)
13. Kowalska, K.: Efektywność procesów logistycznych [w]: Kowalska, K., Markusik, S.S.: Sprawność i efektywność zarządzania łańcuchem dostaw. WSzB w Dąbrowie Górniczej, Dąbrowa Górnicza (2011)
14. Michalewicz, Z.: Algorytmy genetyczne + struktury danych = programy ewolucyjne, Wydawnictwo Naukowo – Techniczne, Warszawa (1996)
15. Minnich, D., Maier, F.H.: Supply chain responsiveness and efficiency – complementing or contradicting each other? In: 24th International Conference of the System Dynamics Society, Nijmegen, pp. 1–16 (2006b)

16. Neebe, A.W.: A procedure for locating emergency-service facilities for all possible response distances. J. Oper. Res. Soc. **39**(8), 743–748 (1988)
17. Nowakowski, T.: Reliability model of combined transportation system. In: Spitzer, C., Schmocker, U., Dang, V.N. (eds.) Probabilistic Safety Assessment and Management, pp. 2012–2017. Springer, London (2004)

The Analysis of the Possibility of the Simultaneous Use of Two and More Satellite Navigation Systems in Different Modes of Transport

Jacek Januszewski[✉]

Gdynia Maritime University, al. Jana Pawla II 3, 81-345 Gdynia, Poland
jacekjot@am.gdynia.pl

Abstract. Actually (April 2016) more than 90 operational satellites of navigation (GPS, GLONASS, BeiDou, Galileo) and based augmentation systems (WAAS, EGNOS, MSAS, GAGAN) are in orbit transmitting a variety of signals on multiple frequencies. Several hundred million receivers of all these systems are used in each mode of transportation in the world – land, maritime and air. This paper gives the reply to some important questions as: how many systems can be in given moment and in given point on the Earth taken into account in position fix process, which performance parameters of the multi-constellation receivers designed for given mode of transport are the most important.

Keywords: Satellite navigation system · Mode of transport · Position fix · Integrated receivers · Integrity information

1 Introduction

At the time of this writing (April 2016) more than 90 operational GPS, GLONASS, BeiDou, and Galileo (Satellite Navigation System – SNS), and EGNOS, MSAS, WAAS, GAGAN and SDCM (Satellite Based Augmentation System – SBAS), and QZSS and IRNSS (Regional Satellite Navigation System – RSNS) satellites are in Earth orbit transmitting a variety of signals on multiple frequencies. All these systems are known also as GNSS (Global Navigation Satellite System). Several hundred million receivers of all these systems are used in each mode of transportation in the world – land, maritime and air. Each user of transport can obtain its own position by one or more receivers of one or more SNS with or without GPS in differential mode (DGPS) and with or without one or more SBAS. Fix position by means of each mentioned above system is possible only on condition that the user are in the range of these systems and his receiver is able to utilize the signals from all these systems appropriately [1, 2, 8, 9, 15, 21–23].

J. Mikulski (Ed.): TST 2016, CCIS 640, pp. 98–108, 2016.
DOI: 10.1007/978-3-319-49646-7_9

2 Global and Regional Satellite Navigation and Based Augmentation Systems

Two global SNSs, American GPS (Global Positioning System) and Russian GLONASS (GLObal NAvigation Satellite System), and four SBASs, European EGNOS (European Geostationary Navigation Overlay System), American WAAS (Wide Area Augmentation System), Japanese MSAS (Multi Satellite Augmentation System) and Indian GAGAN (GPS Aided Geo Augmented Navigation) are operational. Two next global SNSs, Galileo in Europe and BeiDou in China, one SBAS, SDCM (System for Differential Corrections and Monitoring) in Russia, and two RNSSs, QZSS (Quasi Zenith Satellite System) in Japan and IRNSS (Indian Regional Navigation Satellite System) are under construction [2, 14, 22, 23].

Nowadays in each mode of transport at least one SNS is used. Both SNS currently fully operational GPS and GLONASS provide the continuous of the current user's position but without information about integrity [2, 12–14]. As the accuracy of this position is sometimes in some regions, for some users insufficient the need of the construction at least one another SNS or SBAS is indispensable. Which SNS or SBAS is recommended depends on the mode of the transport. Selected parameters of all these systems, integrity information, in particular, and the part of each one in different modes in transport in April 2016 or in the future are presented in the Table 1.

Satellite-based positioning systems have been used in a broad array of fields including car navigation and land surveying. The Quasi-Zenith Satellite System (QZSS – Jun-Ten-Cho) uses multiple satellites that have the same orbital period as geostationary satellites with some orbital inclinations. These satellites are placed in multiple orbital planes, so that one satellite always appears near the zenith above the region of Japan. The system makes it possible to provide high accuracy satellite positioning service covering close to 100% of Japan, including urban canyon and mountain terrain. It is very important in land transport in city streets between high buildings where the major accuracy problem is the fact that most observable signals are reflections [22, 23, 26, 27].

IRNSS is an autonomous regional satellite navigation system being developed by ISRO (Indian Space Research Organization). The objective of this project is to implement an independent and indigenous regional spaceborne navigation system for national applications. The IRNSS design requirements call for a position accuracy of <20 m throughout India and within the region of coverage extending about 1500 km beyond. The system is expected to provide accurate real-time position, velocity and time observables for users on a variety of platforms with a 24 h × 7 day service availability under all weather conditions [22, 24].

As we want to use all mentioned above global and regional SNS and SBAS simultaneously with one multi-constellation receiver only and some parameters of these systems important for theirs exploitation differ, we must solve essential problem for all users of transport – problem of compatibility and interoperability of all these systems, carrier frequency, reference datum and time reference, in particular.

The most important problem is certainly different carrier frequency, that's why in all global and regional SNSs both currently operational and under construction one carrier frequency at least is (GPS, Galileo, BeiDou) or will be (GLONASS) the same in

Table 1. Selected parameters of global (SNS), regional (RSNS) and satellite based augmentation systems (SBAS) in April 2016, A – area transport, L – land transport, M – maritime transport, 1 – significant, 2 – noticeable, 3 – marginal, FOC – Full Operational Capability [2, 18–20, 22, 23, 26, 28, 29]

System		Status	Range	Integrity information	The part			Remarks
					A	L	M	
SNS	GPS	Operational	global	In the future satellite block III	1	1	1	The first III in 2017
	GLONASS			In the future satellite block K, frequency L3	2	3	3	Block K in 2019
	Galileo	Under construction		Service SoL	1	1	1	FOC in 2020
	BeiDou			D2 NAV message 5 GEO satellites, selected area only	1	1	1	FOC in 2018
RSNS	QZSS	Under construction	Japan	L1 – SAIF with integrity function	3	1	2	FOC in 2018
	IRNSS		India	Integrity of the network	2	2	2	FOC in 2016
SBAS	EGNOS	Operational	Europe	Service SoL	1	3	3	FOC in 2012
	WAAS		USA	In message	1	3	3	FOC in 2008
	MSAS		Japan		1	3	3	FOC in 2007
	GAGAN		India		1	3	3	FOC in 2015
	SDCM	Under construction	Russia	About GLONASS & GPS	Unknown			FOC in 2018

two or more SNS, both RSNS and SBAS (Table 2). All mentioned in this table frequencies (except for frequency 1227.60 MHz) currently used or planned in three global SNSs, GPS, Galileo and BeiDou, both regional SNSs and in all SBASs are based on the fundamental frequency $f_O = 10.23$ MHz, in the case of 1176.45 MHz, 1207.14 MHz and 1575.42 MHz, the factor (f_O) is 115, 118 and 154, respectively [2, 10, 22, 23].

One of the ways to improve SNSs compatibility and interoperability is to increase the accuracy of mutual synchronization of national UTC (Universal Time Coordinated) scales and their coordination to UTC and the accuracy of SNS time scales synchronization to national UTC scales. BeiDou system is the first SNS which transmits in its

Table 2. Common carrier frequencies in global and regional satellite navigation and based augmentation systems, today and *in the future* [2, 16, 19–21, 24–26]

Frequency [MHz]	System, signals, satellites						
	GPS	GLONASS	Galileo	BeiDou	QZSS	IRNSS	SBAS
1176.45	L5 satellites IIF, *satellites III*	*L5 OC M satellites KM*	E5a signals 1, 2 all satellites	–	L5, *L5S*	L5	*Some systems, some satellites*
1207.14	–	*L3 OC M satellites KM*	E5b signals 3, 4 all satellites	B2	–	–	–
1227.60	L2 all satellites, *satellites III*	–	–	–	L2 C	–	–
1575.42	L1 all satellites, *satellites III*	*L1 OC M satellites KM*	E2–L1–E1 signals 8, 9, 10 all satellites	*B1*	L1 C/A, L1C, L1 SAIF	–	All systems, all satellites

navigation messages the parameters indicating the relationship between own time (BDT) and all three other global systems times [9, 11, 16].

The IRNSS System Time started at 00:00 UT on Sunday August 22nd 1999. At the start epoch, this time was ahead of UTC by 13 leap seconds. IRNSS time is a continuous time without leap second corrections determined by the IRNSS System Precise Timing Facility (IRNPT). IRNSS times offset from GNSS such as GPS, GLONASS are broadcasted in the form of coefficients in one of the IRNSS navigation messages. Time reference of QZSS is QZSST (QZSS Time), the length of one second identical to TAI (Time Atomic International) is the same as in the case of GPS system [17, 24, 26].

Although the international civil coordinate reference standard is the International Terrestrial Reference Frame (ITRF), each GNSS has its own reference frame, which depends on the control stations'coordinates hence guaranteeing independence among systems. As currently all SBASs augment GPS system only, the reference frame for all these systems is WGS84 also. The QZSS geodetic coordinate system is known as the Japan satellite navigation Geodetic System (JGS). This coordinate system is defined as the approach to ITRF. IRNSS datum is WGS84 [24–26].

Two SNSs are said to be interoperable from a reference frame perspective if the difference between frames is below target accuracy. Four reference frames, WGS84 (GPS), GTRF (Galileo), CGCS 2000 (BeiDou) and ITRF (GLONASS), differ by only a few centimetres (e.g. this difference between WGS84 and GTRF is expected to be within 3 cm), so this is only an issue for high-precision users. Therefore we can say that the problem of compatibility of SNSs and SBASs in the case of reference frame (datum) for transport users does not exist [10].

3 Multi-constellation Receivers

As all carrier frequencies, reference datums and times reference of currently two global operational SNSs (GPS and GLONASS) differ the determination of the user's position by means of one integrated receiver of these systems is not possible because this receiver determines two independent positions – one via GPS system, one via GLONASS system.

Meanwhile the integrated receivers of two or more SNS and SBAS are available on the market already but nowadays these receivers are used in the laboratories of the different institutions and universities mainly. In the case of the maritime transport almost all ships are equipped with one GPS/DGPS receiver at least, the most frequently two receivers (both the same models or two units provided by different manufacturers), sometimes with three even four (special ships). The need of the use of second SNS exists mainly in these restricted areas where the position fix from one SNS only is impossible. On the ship the user can determine own position from other sources, e.g. the radar. That's why at sea GNSS receivers (two or more SNS) are still very rare. In the case of the integrated receivers, GPS and SBAS, the problem of compatibility and interoperability doesn't exist because the navigation messages with appropriate corrections are transmitted by all GEO satellites on frequency L1 = 1575.42 MHz, carrier frequency of GPS, in the future L5 = 1176.45 MHz also.

The information about GNSS receivers can be find in GPS World Receiver Survey and GNSS Report Market.

3.1 GPS World Receiver Survey

The most known and certainly most comprehensive receiver survey of database of GPS and GNSS equipment is published in the magazine GPS World the number January each year [5–7, 22]. In this survey we can find detailed information, i.e. 19 performance parameters, about several hundred receivers provided by several dozen manufacturers. The receiver survey via performance parameter user environment and applications distinguishes 15 different types of the users. Four types of them – aviation (A), marine (M), navigation (N) and vehicle/vessel tracking (V) – were taken into account in this paper [7].

In 2015 and 2014 the numbers of manufacturers and receivers were 48 & 47 and 434 & 380 respectively. In the number from January 2016 this information is provided by 45 manufactures on more than 438 receivers. From among these 438 receivers 304 (83.1%) provided by 33 (73.3%) manufacturers available to be used by at least one from four mentioned above types of applications were designed for GPS system and one other SNS, RSNS or/and SBAS at least [5–7].

Additionally 29 receivers of 5 manufacturers listed in survey 2015 but not mentioned in survey 2016 were taken into account also in the analysis in the paper. Finally it was 333 receivers of 38 manufacturers. The manufacturers with the biggest number of the receivers are JAVAD GNSS (40), Trimble (34), Septentrio (21), NovAtel (20) and Leica Geosystems AG (17). Some manufacturers provide one receiver only, e.g. Galileo Satellite Navigation Ltd., John Deere, Microwave Photonic Systems, Topcon [6, 7].

The number and the percentage of each global SNS (GPS, GLONASS, BeiDou, Galileo), SBAS and each regional SNS (QZSS and IRNSS) in multi-constellation receivers were presented in the Table 3. In the case of SBAS it was considered that this notion includes 5 satellite based augmentation systems mentioned in the Table 1 (all four operational and one under construction). As since 2011 GLONASS system is fully operational the percentage (77) in this case is greater than in the case of two others global SNS, Galileo (60) and BeiDou (52), still under construction. It should be emphasized that GLONASS receiver is in the offer of 37 manufacturers, except one (Nottingham) only. In the case of some manufacturers it concerns only the part of the total offer, e.g. Spectrum Instruments and Unicore 2 and 2 from among 7 and 9 receivers, respectively, in the case of all others offered receivers, e.g. NovAtel (20), Data Grid (8). The number of manufacturers without offer concerning receiver of given system is equal 14 in the case of QZSS, 11 – BeiDou, 9 – Galileo and 8 – SBAS [6, 7].

Table 3. The number and the percentage of selected system receivers in multi-constellation GNSS receivers, total number of the receivers – 333 [6, 7]

Parameter	System						
	GPS	GLONASS	Galileo	BeiDou	SBAS	QZSS	IRNSS
Number of the receivers	333	257	201	172	304	175	10
Percentage	100	77.2	60.4	51.7	91.3	52.6	3.0

The smallest percentage (3.0) is for regional IRNSS, under construction also and with one manufacturer only – Septentrio [6, 7]. That's why in subsequent part of the paper this system was not taken into account and all next calculations and analysis were made for 6 systems only.

The number and the percentage of the integrated GNSS receivers depending on the number of the systems except for GPS system were presented in the Table 4. The greatest percentage (30.0) is in the case of 5 systems, it means that on the world market there are at least one hundred receivers capable of receiving and tracking the signal from several dozen satellites of GPS and 5 other systems, the lowest percentage (11.1) in the case of 3 systems.

Table 4. The number and the percentage of the multi-constellation GNSS receivers depending on the number of the systems except for GPS system, total number of the receivers – 333 [6, 7]

Parameter	GPS system + different number of the systems				
	1	2	3	4	5
Number of the receivers	67	65	37	64	100
Percentage	20.1	19.5	11.1	19.2	30.0

The detailed distribution of GNSS receivers with GPS system and one, two, three and four other systems is showed in the Table 5. In each case the greatest percentage is then if one of this system is SBAS. This percentage is equal 76 if it is SBAS only, more than 60 if it is the combination of GLONASS and SBAS (both operational), more than 54 – combination of GLONASS, Galileo and SBAS, more than 34 – the latter and BeiDou.

Table 5. Distribution of multi-constellation GNSS receivers depending on the number of the systems [6]

GPS system + one system			GPS system + two systems		
System	number of the receivers	%	systems	number of the receivers	%
			GLO, SBAS	40	61.5
			GAL, SBAS	8	12.3
SBAS	51	76.1	QZSS, SBAS	6	9.2
GLO	8	11.9	GLO, GAL	5	7.6
BeiDou	5	7.5	GLO, BeiDou	4	6.2
GAL	3	4.5	GAL, QZSS	1	1.6
			BD, SBAS	1	1.6
	67	100		65	100

GPS system + three systems			GPS system + four systems		
systems	number of the receivers	%	systems	number of the receivers	%
GLO, GAL, SBAS	20	54.1	GLO, GAL, BeiDou, SBAS	22	34.3
GLO, SBAS, QZSS	7	18.9	GLO, GAL, SBAS, QZSS	17	26.6
GAL, SBAS, QZSS	5	13.5	GAL, BeiDou, SBAS, QZSS	16	25.0
GLO, SBAS, BeiDou	3	8.1	GLO, GAL, BeiDou, QZSS	5	7.8
GLO, BeiDou, QZSS	1	2.7	GLO, BeiDou, SBAS, QZSS	4	6.3
GLO, GAL, BeiDou	1	2.7			
	37	100		64	100

On the world market there are several hundred GNSS receivers provided by several dozen manufacturers, from cheap standard GPS receiver aided by one other system, e.g. Crescent P102OEM Board (SBAS), Origin Multi SISO Hornet ORG4502 (GLONASS) to professional GNSS receivers designed for all four types of the user, e.g. NovAtel OEM 615 or Septentrio AsteRx–U [12].

The number and the percentage of the GNSS receivers designed for each type of the users (A, M, N and V) are presented in the Table 6. The percentage of navigation (93) and marine (86) user are both greater than in the case of aviation (63) and vehicle/vessel tracking (62). There are 117 receivers (35.1%) designed for all four mentioned about types of the user, 31 among these receivers are at the same time the multi-constellation GNSS receivers of all global and regional SNS and SBAS; e.g. 8 NovAtel models, 6 Septentrio models. The number of manufacturers without offer concerning receiver designed for given type of the user is equal 14 and 13 in the case of aviation and vehicle/vessel tracking user respectively.

Table 6. The number and the percentage of the multi-constellation GNSS receivers designed for different types of the user (A – aviation, M – marine, N – navigation, V – vehicle/vessel tracking), total number of the receivers – 333 [6, 7]

Parameter	Type of the user			
	A	M	N	V
Number of the receivers	208	285	309	206
Percentage	62.5	85.6	92.8	61.9

The comparison of 9 selected performance parameters of 10 multi-constellation GNSS receivers designed for maritime and navigation applications is showed in the Table 7. The number of the systems in one receiver is between 3 (e.g. model Cartesio PLUS) and 6 (e.g. model TRIUMPH–1). In the case of model SXBlue III–L GNSS one of these systems is Omnistar, a satellite based augmentation system service provider (subscription must be bought). In all 10 receivers one of the systems is SBAS. The number of channels is in most cases greater than 100, in mentioned above model TRIUMPH–1 this number is equal 216 because the signals are from several dozen satellites on several frequencies and with different codes.

Table 7. The comparison of 9 selected performance parameters of 10 multi-constellation GNSS receivers designed for maritime and navigation applications [5–7]

No	Manufacturer	Model	Channels/tracking mode	System	Size (W × H × D)
1	Altus Positioning Systems	APS–U	136 Parallel	GPS, GLONASS, Galileo, WAAS, EGNOS	17.7 × 16.7 × 4.8 cm
2	Aschtech	MB 800 Board	120 Parallel	GPS, GLONASS, Galileo, SBAS	3.9 × 3.1 × 0.5 in.
3	Geneq inc.	SXBlue III–L GNSS	117 channel	GPS, GLONASS, SBAS, OmniSTAR	8.0 × 5.6 × 14.1 cm
4	Hemisphere GPS	S 320	117 parallel + 1	GPS, GLONASS, SBAS	4.5 × 7.8 cm
5	JAVAD GNSS	TRIUMPH-1	216	GPS, GLONASS, Galileo, SBAS, QZSS, BeiDou	17.8 × 9.6 × 17.8 cm
6	Leica Geosystems AG	Viva GS25	120	GPS, GLONASS, Galileo, SBAS, BeiDou	20.0 × 9.4 × 22.0 cm
7	NovAtel	SMART6	120	GPS, GLONASS, Galileo, SBAS, BeiDou	15.5 × 15.5 × 8.1 cm
8	STMicroelectronics	Cartesio PLUS	32	GPS, Galileo, SBAS	16 × 16 × 1.2 mm
9	Trimble	BX 982	220 + 2	GPS, GLONASS, SBAS, Galileo, QZSS	26.2 × 14.0 × 5.5 cm
10	u-blox	UBX-G6010-NT	50 parallel	GPS, EGNOS, WAAS, MSAS, GAGAN	5 × 6 × 1.1 mm

3.2   GNSS Report Market

In Europe one of the most comprehensive source of knowledge and information on the dynamic, global GNSS market is report published on average 15 months by European Global Navigation Satellite System Agency (GSA). In the first issue (October 2010) the number of GNSS market segments was equal 4, in the latest (issue 4, March 2015) this number increases to 8 [3, 5, 6]. Among these segments we can list four connected with different modes of transport: road, aviation, maritime and rail. The cumulative core revenue (in percentage) in all 8 segments is as follows: Location Based Service (LBS) – 53.2, Road – 38.0, Surveying – 4.5, Aviation – 1.1, Maritime – 1.1, Agriculture – 1.9, Rail – 0.2, Timing Synchronization – 0.1. Since the beginning the segment with the biggest revenue is LBS, additionally its value increases each year, in mentioned above issue it was more than 50% [6].

Table 7. Cont (additional columns) [5–7]

No	Position autonomous (code) [m]	Time (nanosec)	Start [s] Cold	Warm	No. of ports	Port type
1	1.3	10	<45	<15	8	3 RS-232, 1 USB, 1 Bluetooth, 2 TNC, 1 Ethernet
2	3.0	No response	45	35	4	RS-232, 2 LV-TTL, USB 2.0
3	2.5	Not applicable	60	35	3	Bluetooth, USB, RS-232,
4	1.5	20	60	30	6	2 RS-232, USB, 2 Bluetooth, SD
5	<2	3	<35	<5	8	2 RS-232, Bluetooth, USB, Ethernet, PPS Event Marker, Wi-Fi
6	2 ÷ 3	<20	50	35	8	2 RS-232, 2 Event, USB, TNC, Combined, Bluetooth, Power, PPS
7	1.5	20	50	35	2	3 RS-232, CAN, Bluetooth, Emulated Radar
8	2	10	35	34	22	11 different
9	1 ÷ 5	100	<45	<30	6	3 RS-232, Ethernet, USB, CAN
10	<2.5	50	26	26	4	UART, USB, SPI, I2C

In 2015 the market size of core revenue refers to the value of only GNSS receivers and chipsets in different devices was about 75 billion euro. The capability of GNSS integrated receivers (in percentage) according to GNSS Market Report in maritime segment and GPS World receiver survey (marine user) in 2016 is presented in the Table 8. The total numbers of the manufacturers and the receivers are in both surveys very similar, 31 & 37 and 301 & 308 respectively.

Table 8. Capability of multi-constellation GNSS receivers (in percentage) according to GNSS Market Report and GPS World receiver survey [4, 7]

GPS system and one of the four systems mentioned above	GNSS market report	GPS world
	Maritime segment	Marine user
SBAS	77	91
GLONASS	68	77
Galileo	53	60
BeiDou	34	52
Number of the manufacturers	31	37
Number of GNSS receivers	300	308

4 Conclusion

- Only when all Satellite Navigation and Based Augmentation Systems (SNS and SBAS) will be compatible and interoperable in respect of carrier frequency, reference datum and time reference simultaneous use of these systems will be possible. At this moment the determination of resultant position from integrated receiver will be possible.
- The most effective simultaneous use of SNS and SBAS is and will be in these areas where the satellite visibility is limited. It concerns e.g. land transport in urban canyon.
- As currently operational global SNS (GPS and GLONASS) don't ensure information about integrity and the main goal of SBAS is to provide integrity assurance more than 91% of multi-constellation GNSS receivers contain SBAS receiver. This solutions is very useful in air transportation.
- Two global SNSs (Galileo, BeiDou), two regional SNSs (IRNSS, QZSS) and one SBAS (SDCM) are under construction, but all theirs future users can wait for information about Full Operational Capability (FOC) of these systems with calmness because several hundred GNSS integrated receivers of two, three or even all five mentioned above systems and all SNSs and SBASs currently operational provided by several dozen manufacturers are available on the world market already.
- The percentage of the multi-constellation GNSS receivers (GPS system and/or SBAS or other SNS) depending on the system presented in GPS World receiver survey and GNSS Market Report is almost the same, in maritime segment, in particular.

References

1. Admiralty List of Radio Signals (ALRS): The United Kingdom Hydrographic Office, vol. 2 (2014/2015) (2015/2016)

2. Betz, J.W.: Engineering Satellite-Based Navigation and Timing: Global Navigation Satellite Systems, Signals and Receivers. Wiley-IEEE Press, Hoboken (2016)
3. GNSS Market Report (2010). The European GNSS Agency, Prague, Issue 1, October 2010
4. GNSS Market Report (2015). The European GNSS Agency, Prague, Issue 4, March 2015
5. GPS World, Receiver Survey, vol. 25, no. 1 (2014)
6. GPS World, Receiver Survey, vol. 26, no. 1 (2015)
7. GPS World, Receiver Survey, vol. 27, no. 1 (2016)
8. Januszewski, J.: Satellite navigation systems in the transport, today and in the future. Arch. Transp. **22**(2), 175–187 (2010)
9. Januszewski, J.: The problem of compatibility and interoperability of satellite navigation systems in computation of user's position. Artif. Satell. **46**(3), 93–102 (2011)
10. Januszewski, J.: Choice of the final number of satellite navigation and based augmentation systems in the immediate and not too distant future. In: Mikulski, J. (ed.) TST 2014. CCIS, vol. 471, pp. 146–155. Springer, Heidelberg (2014). doi:10.1007/978-3-662-45317-9_16
11. Januszewski, J.: BeiDou, Chiński globalny nawigacyjny system satelitarny wchodzi na rynek światowy, Przegląd Telekomunikacyjny+Wiadomości Telekomunikacyjne, str. 1379–1386, nr 11 (2014) in polish
12. Januszewski, J.: Shipborne satellite navigation systems receivers, exploitation remarks. Sci. J. Maritime Univ. Szczecin **40**(112), 67–72 (2014)
13. Januszewski, J.: Integralność nawigacyjnych systemów satelitarnych, globalnych i wspomagających, Przegląd Telekomunikacyjny+Wiadomości Telekomunikacyjne, str. 573–579, nr 5 (2015) in polish
14. Januszewski, J.: Visibility and geometry of global satellite navigation systems constellations. Artif. Satell. J. Planet. Geodesy **50**(4), 169–180 (2015)
15. Munich Satellite Navigation Summit, Munich (2015)
16. www.beidou.gov.cn. Accessed 15 Dec 2015
17. www.directory.eoportal.org. Accessed 15 Dec 2015
18. www.egnos-user-support.essp-sas.eu. Accessed 15 Dec 2015
19. www.esa.int. Accessed 15 Dec 2015
20. www.glonass-ianc.rsa.ru. Accessed 15 Dec 2015
21. www.gps.gov. Accessed 15 Dec 2015
22. www.gpsworld.com [date of access: 15.12.2015]
23. www.insidegnss.com. Accessed 15 Dec 2015
24. www.isro.gov.in. Accessed 15 Dec 2015
25. www.navipedia.net. Accessed 15 Dec 2015
26. www.qzss.go.jp. Accessed 15 Dec 2015
27. www.qzss.jaxa.jp. Accessed 15 Dec 2015
28. www.rtca.org. Accessed 15 Dec 2015
29. www.sdcm.ru. Accessed 15 Dec 2015

Availability Protection of IoT Concept Based Telematics System in Transport

Ivan Cvitić[1(✉)], Dragan Peraković[1], Marko Periša[1], and Branimir Jerneić[2]

[1] Faculty of Transport and Traffic Sciences,
Department of Information and Communication Traffic, University of Zagreb, Zagreb, Croatia
{ivan.cvitic,dragan.perakovic,marko.perisa}@fpz.hr
[2] Faculty of Transport and Traffic Sciences, PhD, University of Zagreb, Zagreb, Croatia
bjerneic@gmail.com

Abstract. This paper will analyze the security disadvantages of the IoT concept based telematics system in the transport environment in terms of availability and the possibility of denial of access to information resources. Analysis will be conducted on Electronic Toll Collection (ETC) system implemented in Republic of Croatia. Based on this analysis, protection mechanisms that can be implemented in order to maintain agreed quality of service level within the telematics systems based on the IoT concept such as ETC system will be proposed.

Keywords: Internet of Things · Electronic toll collection · Communication security · Resource availability · DDoS attack

1 Introduction

In a past several years the Internet of Things (IoT) concept is frequently applied in a wide range of environments and very important application is visible in the field of traffic and transport, respectively in transport telematics systems. Telematics system form the foundation of intelligent transport systems (ITS). ITS as a holistic management and information and communication upgrade of classic transport systems, through its services enables increased performance of the traffic system, efficient transportation of goods and passengers and increase traffic safety. The emergence and use of the IoT concept in such an environment has enabled the development and delivery of new and upgrade existing services available to users of the transport system. Because of the requirements of connection with the public communication network (IP/Internet) brings into question the level of security of such systems. One of the primary threats to these systems are Distributed Denial of Service attacks (DDoS), which aims to prevent access to service for legitimate users. The increase in the number and traffic volume of attacks aimed at denial of service indicates the need for further and more intensive research of this issue. The development of new concepts that are quickly accepted and implemented within the information and communication (IC) environments further contributes to the importance of mentioned problematic. An example of a transport telematics system based on IoT concept is an electronic toll collection (ETC) system in Republic of Croatia. This system was used for the research of described problem.

© Springer International Publishing AG 2016
J. Mikulski (Ed.): TST 2016, CCIS 640, pp. 109–121, 2016.
DOI: 10.1007/978-3-319-49646-7_10

The aim of this paper is to research the possibilities for denial of the ETC system's availability in order to maintain the agreed service quality level of ETC system by defining the proposals for the implementation of the protection methods against DDoS attacks.

1.1 Previous Research

The importance of IC services availability, security methods, and attack methods aimed at denial of service are subject of numerous studies for last two decades. A large number of studies was focused on DDoS attacks in the classic IC environments (environment which do not apply concepts such as IoT, M2M, cloud computing, etc.). The paper [1] shows the taxonomy of DDoS attacks as well as methods to prevent, detect and defend against these types of attacks. The paper [2] also shows the taxonomy of DDoS attacks and protection methods in order to better understand the problem of denial of access, and to encourage the development of more efficient protection methods.

Class of DDoS attacks that is using the method of network resources flooding is the most common class of this type of attacks. The importance of developing protection methods against these attacks is shown in the paper [3] trough the classification of the currently available protection methods. The same paper has highlighted the need for collaborative and distributed access in developing efficient protection methods.

Development of IoT concept and its application has resulted in exponential growth of connected devices, development of new services and amount of data appose to known IC environment. Previous implies additional increase in the number of vulnerabilities within such systems, and thus the growth potential surface of DDoS attacks enforcement [4]. Therefore, numerous studies are directed towards the protection of the entire IoT architecture or particular elements of the architecture in order to preserve the availability of resources of such systems.

Taxonomy of DDoS attacks on Wireless Sensor Networks (WSN) is shown in the paper [5]. This paper refers to the identification method of attackers, their ability, attacks goals, used vulnerabilities, and the results of the attack. The paper [6] presents a taxonomy of existing protection methods against DDoS attacks in the cloud computing (CC) environment and it provided proposals to eliminate shortcomings of discussed protection methods. Protection against DDoS attacks in the CC environment is also contemplated in the paper [7]. The paper presents a detailed overview of DDoS attacks in the CC environment and specifies the shortcomings of current protection methods that are not adapted to CC environment.

Although a large number of studies dealing with the issue of service availability in the conventional IC systems and systems based on the IoT concept there is lacks of research of availability protection in telematics systems based on the IoT concept in a transport environment. Some papers such as [8, 9] partly emphasize mentioned issues, but not as the primary research problem.

1.2   Research Methodology

Within this paper architectures of the IoT concept and the ETC system are shown and compared. The purpose of the comparison is to show that the ETC transport telematics system is based on the IoT concept. The importance of the availability of the ETC system is shown in the analysis of statistical data collected by the Croatian motorway concessionaire (HAC Ltd. and ARZ Ltd.) on the number of users of the ETC system since its implementation in 2006 until the end of 2015. In addition, from the same source, data was collected on the total number of vehicles which have passed the motorway and share of vehicles using the ETC system services in 2015. The collected data was used for the projection of the number of users who would be unable to use the ETC system services in case of unavailability caused by DDoS attacks. A systematic analysis of the ETC infrastructure identified the process of information exchange between the individual elements of the system in purpose of defining DDoS attacks zone. Finally, given the ETC system based on the IoT architecture and the possibility of the DDoS attacks a certain methods available to protect the different layers of architecture have been proposed.

2   Attacks Directed at the Service Availability

Congestion in the communication node affects the quality of service (QoS) as an important element of providing any form of service. According to [14], QoS is defined in several ways and according to recommendation of the ITU-T E.800 (International Telecommunication Union) it is "joint effect of the performance service that determines the level of satisfaction of users of the service." From the aspect of service providers, QoS is expected condition of service quality offered to the user defined in the basic parameters such as bandwidth, packet loss, latency/delay and jitter [11].

Intentionally causing traffic congestion in the network by generating large amounts of illegitimate network traffic has had a negative impact on the QoS (a direct impact on one or more parameters that determine it) and can impair Service Level Agreement (SLA) established between the end user and the service provider [12].

Denial of service attacks, implies a general class of attacks aimed at the availability of information and communication (IC) services and resources. As the purpose of each IC service is to provide requested information, DoS can be defined as an attack aimed at preventing access to the data [10]. Availability, as one of three key principles of ICS security, is referred in the availability of the required information to legitimate users within the required time and under given conditions. If that is not fulfilled then its primary function is meaningless, and the system is unable to meet the requirements set by the end-users [13]. According to the method of distribution DoS attacks can be divided into two categories, denial of services with a single source (SDoS) and distributed denial of service (DDoS) [14]. Source of SDoS attack is one computer or device on the network. In DDoS attacks multiple devices are coordinated for the purpose of routing large volumes of illegitimate traffic to attack target.

Reason for the appearance of DDoS attack methods is increase in the speed of processing packets within the router and end devices (e.g. server) causing one device in the network often not been able to generate a sufficient traffic volume to create congestion in the network. Other reason is to camouflage real attack source by applying a large number of mostly geographically dislocated, attack generating devices. An additional reason for the application of DDoS attacks is a high probability of creating congestion in unwanted network segment using the SDoS attack methods.

Denial of service attacks represent a growing problem therefore it is necessary to research and analyze trends of applied protocols and traffic volume and bandwidth of attacks with the aim of timely response to future attacks.

Distribution of DDoS attack methods based on application layer protocols is shown in Fig. 1. The data is presented on a quarterly basis for the period from Q1 2013 to Q1 2015. From the visible data application of the GET method of attack based on the HTTP protocol is the most common. Decrease of 10.63% is visible in transition from Q4 2013 (19.91%) to Q1 2014 (29.9%). From Q4 2013 to Q1 2015 use of application layer protocols for the realization of DDoS attacks have consistently fallen. The highest incidence was recorded in Q2 2013 (25.29%) while in Q1 2015 amounted to (9.32%) as a total decline of 15.97%.

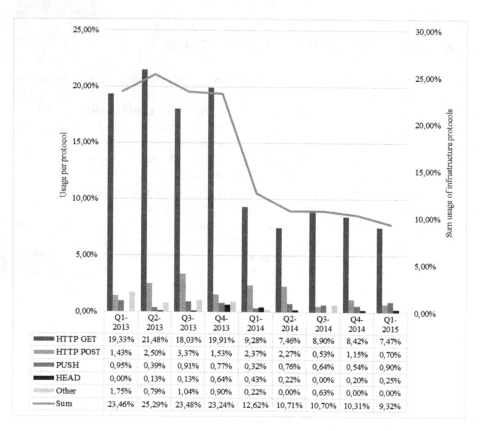

	Q1-2013	Q2-2013	Q3-2013	Q4-2013	Q1-2014	Q2-2014	Q3-2014	Q4-2014	Q1-2015
HTTP GET	19,33%	21,48%	18,03%	19,91%	9,28%	7,46%	8,90%	8,42%	7,47%
HTTP POST	1,43%	2,50%	3,37%	1,53%	2,37%	2,27%	0,53%	1,15%	0,70%
PUSH	0,95%	0,39%	0,91%	0,77%	0,32%	0,76%	0,64%	0,54%	0,90%
HEAD	0,00%	0,13%	0,13%	0,64%	0,43%	0,22%	0,00%	0,20%	0,25%
Other	1,75%	0,79%	1,04%	0,90%	0,22%	0,00%	0,63%	0,00%	0,00%
Sum	23,46%	25,29%	23,48%	23,24%	12,62%	10,71%	10,70%	10,31%	9,32%

Fig. 1. Frequency of application layer protocol used in conducting DDoS attacks period Q1-2013–Q1-2015 [15]

Distribution of infrastructure layer (OSI network and transport layer) for the realization of DDoS attacks is shown in Fig. 2. For the entire analyzed period of time we can see continued growth. The overall increase from Q1 2013 to Q1 2015, a summary of all protocols, is 14.14%. In Q1 2015, 90.68% of all recorded DDoS attacks used infrastructure layer protocols. The primary protocol used for the implementation of DDoS attacks from Q1 2013 to Q4 2014 was the TCP SYN.

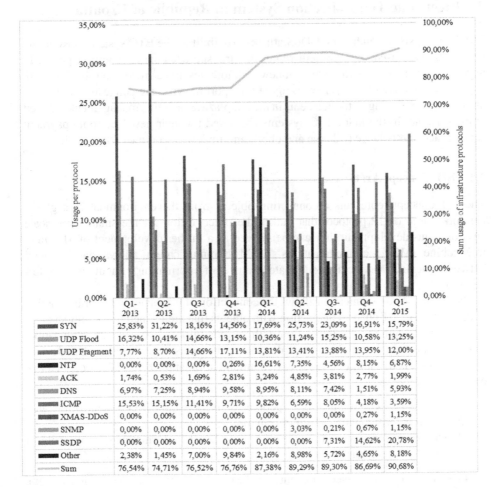

	Q1-2013	Q2-2013	Q3-2013	Q4-2013	Q1-2014	Q2-2014	Q3-2014	Q4-2014	Q1-2015
SYN	25,83%	31,22%	18,16%	14,56%	17,69%	25,73%	23,09%	16,91%	15,79%
UDP Flood	16,32%	10,41%	14,66%	13,15%	10,36%	11,24%	15,25%	10,58%	13,25%
UDP Fragment	7,77%	8,70%	14,66%	17,11%	13,81%	13,41%	13,88%	13,95%	12,00%
NTP	0,00%	0,00%	0,00%	0,26%	16,61%	7,35%	4,56%	8,15%	6,87%
ACK	1,74%	0,53%	1,69%	2,81%	3,24%	4,85%	3,81%	2,77%	1,99%
DNS	6,97%	7,25%	8,94%	9,58%	8,95%	8,11%	7,42%	1,51%	5,93%
ICMP	15,53%	15,15%	11,41%	9,71%	9,82%	6,59%	8,05%	4,18%	3,59%
XMAS-DDoS	0,00%	0,00%	0,00%	0,00%	0,00%	0,00%	0,00%	0,27%	1,15%
SNMP	0,00%	0,00%	0,00%	0,00%	0,00%	3,03%	0,21%	0,67%	1,15%
SSDP	0,00%	0,00%	0,00%	0,00%	0,00%	0,00%	7,31%	14,62%	20,78%
Other	2,38%	1,45%	7,00%	9,84%	2,16%	8,98%	5,72%	4,65%	8,18%
Sum	76,54%	74,71%	76,52%	76,76%	87,38%	89,29%	89,30%	86,69%	90,68%

Fig. 2. Frequency of infrastructure layer protocols used in DDoS attacks in the period Q1-2013–Q1-2015 [15]

From Q3 - 2014 a growth of application SSDP protocol (7.31%) is seen, and in Q1 2015 usage of the same protocol was increased by 13.47% and amounted to 20.78%, which is 4.99% more than the TCP SYN.

Previously conducted analysis shows DDoS as a growing threat to all forms of information and communication environments. This type of attack has the potential to cause the unavailability of telematics systems in transport based on the IoT concept, but IoT concept also plays a role in generating greater volume of attacks. This statement is proved trough increase in the application of the SSDP protocol which is one of the fundamental communication protocols in IoT concept devices.

3   Electronic Toll Collection System in Republic of Croatia

For the analysis of conducting DDoS attacks possibilities the ETC system in Republic of Croatia was used. ETC system represents transport telematics solution primarily applied in a function of introduction of new technologies in the process of toll collection. Additional positive effects achieved in the transport network are congestion reduction at toll plazas, reducing traffic accidents in the toll plaza area, traffic management, reduced emissions, etc. In the field of IC systems are noted for their new electronic payment service, account balance recharge and check and review of completed transactions.

3.1   IoT Concept Architecture

The IoT concept represents a connecting objects from the environment in a global network based on IP protocol that makes the precondition for the smart large scale environments development. IoT can be viewed as a further development of Machine-to-Machine (M2M) communications. M2M communication allows data transfer between machines and includes automated transfer of information without human intervention.

While M2M communication allows the connection of machines and their interaction over a network, IoT enables interaction with objects that are located in the human environment by extending their interaction with various information, such as geolocation, time, etc. The endpoints of the communication can be a person or objects such as devices or machines. As a result, two modes of communication applied in IoT concept can be observed [16]:

- Person - object communication - people as users establish communication with devices in order to obtain certain information
- Object - object communication – the object supplies the information to another object with or without human intervention.

From the IoT aspect, objects are things in the physical environment (physical things), or in a virtual environment (virtual things). Such objects have the capacity of integration within the IC networks and to become active participants in terms of information exchange, events recognition and changes in the environment and autonomous reactions to the same events and changes [17]. Generic layered architecture of IoT concept consists of four basic layers, shown in Fig. 3.

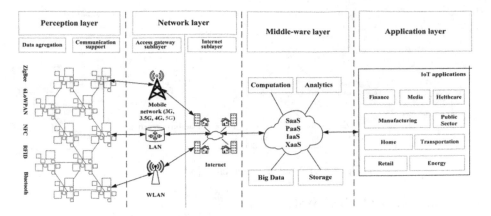

Fig. 3. IoT concept architecture [19]

According to [18] perception layer is consist of two basic functionalities, data collection and collaboration between elements of that layer. The network layer consists of two sublayers, access sublayer with the role of collecting the data from perception layer and sending it to the Internet sublayer. Internet sublayer is the backbone of IoT environment and its main task is the transfer of data to the next layer, middleware, a layer which perform the processes such as intelligent routing and the network address translation. Middleware layer is responsible for data collecting, its filtering, transformation and the intelligent processing most commonly with the use of cloud computing concept. After processing, the data is passed to the application layer, which uses the given data in order to provide and present various services to the end user [19].

3.2 ETC System Architecture

ETC system is based on the IoT concept which can be observed from the system architecture shown in Fig. 4. Within the perception layer of ETC system are ETC devices that unambiguously define the user within the system. The information recorded in the ETC device is detected by the ETC reader that is forwarding the information to middleware layer through network communication layers. ETC devices and readers are based on Dedicated Short-Range Communications (DSRC) technology, which is often used in transport telematics systems. Vulnerability to DDoS attack exist and can be carried out at the level of ETC device and ETC reader trough jamming, black hole and similar DoS attacks methods with the aim of preventing mutual communication [8].

The entire process of transferring information from the perception to middleware layer takes place through dedicated, and not through public communication infrastructure. Middleware layer has the role of processing data collected at the perception layer such as the payment transactions, analytical processing of big data, report generation, etc. The security disadvantage of this layer is a direct connection to the services provided by application layer to end-users which makes the system vulnerable to DDoS attacks. Application layer services are available through public communication network (Internet) that is making this layer directly exposed to potential DDoS attacks.

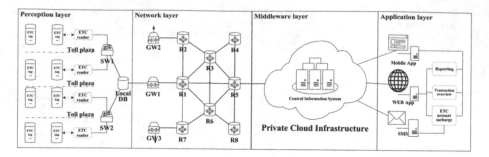

Fig. 4. ETC system architecture [own study]

4   Availability of ETC System

On a total length of 1,289.4 km of motorways there are 105 tolls with 686 toll lanes, of which the 466 toll lanes has implemented ETC system. Linear growth in the number of users can be noticed since the implementation of the ETC system in 2006. Extrapolation of trend predicts around 412,000 ETC users in 2016 and around 450 000 ETC users in 2017 with a determination coefficient $R^2 = 98.46\%$, as shown in Fig. 5.

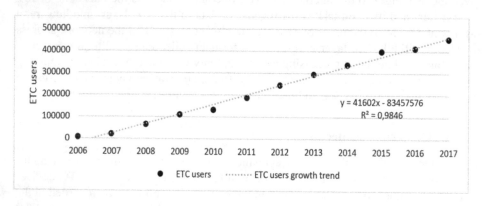

Fig. 5. Growth of ETC users [20]

The information exchange process between elements of the ETC system is shown in UML sequence diagram in Fig. 6. The enforcement of the network or the application layer DDoS attacks directed at web, mobile or SMS server will deny access to services that these servers provide (primarily service of ETC account charge). The scenario in which the DDoS attack prevents operation of the central information system (IS) will cause inability of information exchange between user services and the central IS, and between the toll plazas and the central IS. Doing so, the toll plaza is not able to check the status of incoming ETC user, and users are not able to access services such as charge of ETC account, verification of transactions, etc., which implies a complete unavailability of ETC services.

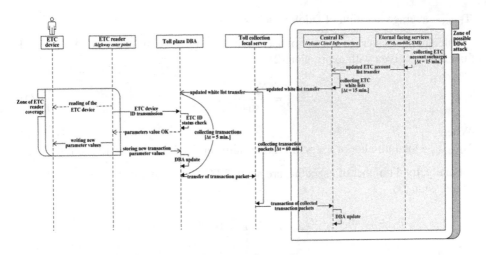

Fig. 6. Information exchange process in ETC system

Figure 7 shows the share of vehicles using the ETC system service in the total number of vehicles in 2015, which varies from 20.1% to 41.1% depending on the observed month. The total number of vehicles that used the ETC system services in 2015 is 17,822,971 of the total 55,457,536 vehicles, or 32.1%.

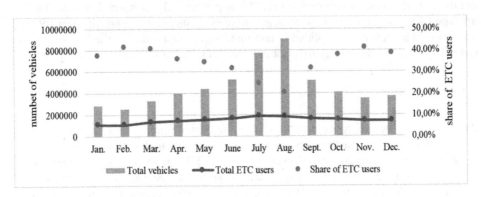

Fig. 7. The share of ETC users in the total number of motorway users in Croatia

The number of users that wouldn't be able to use ETC system services in the event of successfully conducted DDoS attack can be calculated using the mathematical expression (1).

$$K_{DDoS} = 2N_{ETC/h} * T_{DDoS}. \tag{1}$$

Where:

K_{DDoS} – number of users that wouldn't be able to use ETC system services in case of successfully conducted DDoS attacke

$N_{ETC/h}$ – number of ETC services users in period of one hour

T_{DDoS} – average duration of DdoS attack = 24 h.

Calculation of average number of users that are using ETC system services in period of one hour is given with expression (2).

$$N_{ETC/h} = \frac{N_{uk(ETC)}}{8640}. \tag{2}$$

Where:
$N_{uk(ETC)}$ – total number of users of ETC system service in period of 12 month.

Finally, total number of users that are using ETC system services annually is shown in expression (3).

$$N_{uk(ETC)} = \sum_{i=1}^{n} N_{uk_i} * p\left(N_{uk(ETC)_i}\right); \quad i = 1 \ldots 12. \tag{3}$$

Where:
N_{uk} – total number of motorway users
$p(N_{uk(ETC)})$ – user share of ETC system in total number of motorway users
i – months in year.

Given the previously showed data and mathematical expressions (1), (2) i (3) average of 4125 vehicles (users) used ETC system service in time period of one hour. The average duration of DDoS attacks according to [21] is approximately 24 h which means that the average DDoS attack to the ETC system will prevent its use for approximately 99,000 users. For this reason the application and middleware layer, due to the impact on the overall system, are defined as zones of possible DDoS attacks, as shown in Fig. 6.

5   Discussion

Application of the IoT concept in the ETC system, with the possibility of contactless toll payment, resulted in the development of services available to users through the Internet network. Primarily those are the services of ETC account charge and review of transactions via the web or mobile applications and SMS. The offer of such services has caused the exposure of the overall architecture of the ETC system to public communications network, and thus to a large number of malicious users who consequently have the option of DDoS attacks enforcement.

The increase in the number of users of the ETC system results in greater negative impact and consequences in case of successful enforcement of DDoS attacks. This primarily refers to the number of users in the event of an attack that will not be able to use the services of the ETC system. According to extrapolation of users' growth trend in the coming years the negative impact of the unavailability of the ETC system will be even greater, therefore, the protection of system availability is of major importance.

Table 1 shows some of the methods that can be applied in different layers of the system architecture to protect against certain types of DDoS attacks.

Table 1. Protection methods of the ETC system from DDoS attacks [own study]

IoT layers	DDoS attacks	Protection methods
Perception	Jammnig, black hole, spamming, flooding attacks	Channel switching, frequency hopping, APDA [8, 22]
Network	Isolated local network infrastructure, low probability of DDoS attacks	–
Middleware	ICMP Flood, Amplification, Reflector, DNS, UDP, SYN attack	Extended DERM [3], load balancer, ScreenOS, multops, DNSSEC [23], IDS/IPS systems
Application	Middleware based DDoS attacks, OSI Layer 7 DDoS attacks (HTTP Get/Post flooding, Session flooding, Request flooding)	Middleware protection methods, web application firewall, anomaly detection

In the IoT based telematics system such as ETC, protection from DDoS attacks needs to be implemented at all layers of the system architecture. It is especially important to protect from attack the system's application layer that is exposed to public communications network (Internet) and is intended to provide services to end users of the ETC system (SMS surcharge and Web-based services and charge information). Due to the direct connection with the application layer, middleware layer should be protected.

Enforcement of DDoS attacks on previously identified layers of architecture would cause the global unavailability of ETC system services. Although the DDoS attack can be carried out at the perception layer of the ETC architecture its influence is limited to the local area (one toll plaza or set of toll plazas in the same geographical area) due to the need for the physical presence of the attackers at the site of conducting attacks. Such DDoS attack would have an impact on a much smaller number of users than is the case with the attacks carried out on the application and middleware layer.

6   Conclusion

Resource availability of ETC system is essential for normal operation of providing services to end users. One of the biggest threats to the availability of any information and communication systems are DDoS attacks. With implementation of transport telematics systems based on the IoT concept DDoS became a growing threat, which is result of the openness of such systems to the public communications network.

This paper presents and analyzes the ETC system in the Republic of Croatia, which is based on the IoT architecture. The system analysis defined architecture layers where the possibility of DDoS attacks enforcement exists. Based on the statistical data, approximately 99,000 users would not have the ability to access the ETC system services during the execution of the average DDoS attacks.

Finally guidelines to protect the availability of the ETC system and methods applicable to the protection of the individual layers of the system architecture were proposed.

References

1. Patrikakis, C., Masikos, M., Zouraraki, O.: Distributed denial of service attacks. Internet Protoc. J. **7**(4), 13–36 (2004)
2. Mirkovic, J., Reiher, P.: A taxonomy of DDoS attack and DDoS defense mechanisms. SIGCOMM Comput. Commun. Rev. **34**(2), 39–53 (2004)
3. Saurabh, S., Roy, S., Sairam, A.S.: Extended deterministic edge router marking. Int. J. Commun. Netw. Distrib. Syst. **13**(2), 169–183 (2014)
4. Covington, M.J., Carskadden, R.: Threat implications of the Internet of Things. In: 5th International Conference on Cyber Conflict, pp. 1–12. IEEE Press, Tallinn (2013)
5. Wood, D., Stankovic, J.A.: A taxonomy for denial-of-service attacks in wireless sensor networks. In: Handbook Sensor Networks Compact Wireless Wired Sensing System, pp. 739–763 (2004)
6. Shameli-Sendi, A., et al.: Taxonomy of distributed denial of service mitigation approaches for cloud computing. J. Netw. Comput. Appl. **58**, 1–42 (2015)
7. Somani, G., et al.: DDoS attacks in cloud computing: issues, taxonomy, and future directions. ACM Comput. Surv. **1**(1), 1–44 (2015)
8. Hamida, E., Noura, H., Znaidi, W.: Security of cooperative intelligent transport systems: standards, threats analysis and cryptographic countermeasures. Electronics **4**(3), 380–423 (2015)
9. Matharu, S., Upadhyay, P., Chaudhary, L.: The Internet of Things: challenges & security issues. In: 2014 International Conference on Emerging Technologies, pp. 54–59 (2014)
10. Imperva: Hacker Intelligence Initiative Overview. Imperva Inc., Redwood City (2012)
11. Mrvelj, Š.: Dynamic allocation of capacity of the Internet node according the service demands (Dinamičko dodjeljivanje kapaciteta internetskog čvora prema zahtjevima usluge). Ph.D. dissertation, Faculty of Transport and Traffic Sciences, University of Zagreb, Zagreb, Croatia (2008)
12. Fowler, S., Zeadally, S., Chilamkurti, N.: Impact of denial of service solutions on network quality of service impact of denial of service solutions on network quality of service. Secur. Commun. Netw. **4**(10), 1089–1103 (2011)
13. Ciampa, M.: Guide to Network Fundamentals. Course Technology, Boston (2012)
14. Hussain, A., Heidemann, J., Papadopoulos, C.: A framework for classifying denial of service attacks. In: Applications Technologies Architectures and Protocols for Computer Communications (SIGCOMM 2003), pp. 99–110 (2003)
15. Peraković, D., Periša, M., Cvitić, I.: Analysis of the IoT impact on volume of DDoS attacks. In: 33rd Symposium on New Technologies in Postal and Telecommunication Traffic (PosTel 2015), pp. 295–304. University of Belgrade, Belgrade (2015)
16. Peraković, D., Husnjak, S., Cvitić, I.: IoT infrastructure as a basis for new information services in the ITS environment. In: 22nd Telecommunication Forum (TELFOR 2014), Belgrade, pp. 39–42 (2014)
17. Farooq, M., et al.: A critical analysis on the security concerns of Internet of Things (IoT). Int. J. Comput. Appl. **111**(7), 1–6 (2015)
18. Zheng, L., et al.: Technologies, applications, and governance in the Internet of Things. In: Internet of Things - Global Technological and Societal Trends from Smart Environments and Spaces to Green ICT, pp. 141–175. River Publishers (2011)

19. Cvitić, I., Vujić, M., Husnjak, S.: Classification of security risks in the IoT environment. In: 26th International DAAAM Symposium on Intelligent Manufacturing and Automation, pp. 731–740. DAAAM International, Vienna (2016)
20. HUKA: HUKA - Croatian Association of Toll Motorways Concessionaires, National Report 2006–2014. http://www.huka.hr. Accessed 10 Dec 2015
21. Akamai Technologies: Faster Forward to the Latest Global Broadband Trends (Q1-2015). https://www.akamai.com/us/en/multimedia/documents/state-of-the-internet/2015-q1-internet-security-report.pdf. Accessed 10 Dec 2015
22. Malla, A.M.: Security attacks with an effective solution for DOS Attacks in VANET. Int. J. Comput. Appl. **66**(22), 45–49 (2013)
23. Prabadevi, B., Jeyanthi, N.: Distributed denial of service attacks and its effects on cloud environment-a survey. In: The 2014 International Symposium on Networks, Computers and Communications, pp. 1–5 (2014)

ICT and the Future of Urban Transportation. European Perspective

Joanna Kos-Łabędowicz[✉]

University of Economics, 40-287 Katowice, Poland
joanna.kos@ue.katowice.pl

Abstract. The increasing population of urban areas in the European Union creates new challenges in the areas of policing and planning urban transportation and urban mobility. Modern solutions based on ICT change the conditions of the urban transport system and residents' expectations about the scope and quality of available transport solutions. Attention should also be paid to the growing awareness of the problems associated with the negative effects of transport on both the natural environment and society. The goal of this paper is to present possibilities of improving the efficiency of city transport infrastructure with the application of various ICT-based solutions. Solutions for both public and private transportation are considered.

Keywords: ICT · Urban transportation · Mobility

1 Introduction

The increasing population of urban areas in the European Union creates new challenges in the areas of policing and planning urban transportation and urban mobility. Steadily increasing traffic, the growing number of vehicles - including passenger cars, commercial vehicles, and public transportation units - increase congestion and transport costs - both for commuters and for the rest of society.

City policy makers undertake various activities in order to facilitate mobility along with reducing negative transportation externalities. One of the possibilities is the improvement of the efficiency of the existing infrastructure. A goal of this paper is to present the possibilities of improving the efficiency of city transportation infrastructure with the applications of various ICT-based solutions.

Firstly an article briefly presents the city transport problems in the light of the recommendations of the European Union regarding the promotion of urban mobility according to the principles of sustainable development. In the next part ICT solutions, which may increase urban mobility while reducing the external costs of transport, are discussed. The solutions discussed are mainly directed towards the better use of existing public and private transport infrastructure, and towards the facilitating of decision-taking of both policymakers and passengers.

© Springer International Publishing AG 2016
J. Mikulski (Ed.): TST 2016, CCIS 640, pp. 122–134, 2016.
DOI: 10.1007/978-3-319-49646-7_11

2   Urban Transport in the European Union

Almost three quarters of EU citizens live in urban areas (defined as cities, towns and suburbs) (72.4% of the population) [12]. Urbanization in the European Union takes place by both increasing the proportion of the population living in urban areas, by expanding urban areas and by blurring the boundaries between urban areas and rural areas [11]. No city can function without an efficient transport network, which would allow citizens to move freely between different destinations. It is in the cities that all activities: economic, cultural, social and educational are concentrated.

The development of sustainable mobility in urban areas encourages economic development, employment growth and social inclusion, which is in line with the overall objectives of the EU's growth strategy Europe 2020. The vital importance of sustainable mobility entails promotion of environmentally-friendly transport and influencing preferences and behaviour patterns of residents regarding modes of transport. The challenges facing urban transport include [36]: traffic congestion and parking difficulties, longer commuting, public transport inadequacy, difficulties for non-motorised transport, loss of public space, high maintenance costs, environmental impacts and energy consumption, accidents and safety, land consumption and freight distribution.

The establishment and development of urban transport systems are inextricably linked to the characteristics of a given city, its history, geographical location and economic and social development. The main factors identified in literature and relevant for the development of the transport system are [51]:

- technology - technological progress can contribute to the development of a sustainable transport system and more efficient use of available infrastructure and public transport vehicles as well as deciding the most suitable mode of transport through the use of ICT for the journey undertaken,
- economic development - depending on changes in the economy, some means of transport can gain or lose their attraction, for example the price of fuel can encourage greater interest in green technologies and in the use of active or public transportation,
- spatial and land-use patterns - the demand for transport services depends on the size of the urban area, the density of housing development and population and location of places of special interest to residents,
- government policy - the involvement of public administration at local level has a large influence on the shape of the urban transport system,
- social and behavioural trends - behaviour, norms and values of individuals largely influence the decisions concerning the choice of a place of residence and means of transport used every day.

Urban transport is an important element of the common transport network, and in most cases local institutions are responsible for its formation in a given city or agglomeration. The growing requirements for mobility in urban areas have been of interest to the European Union since the 90s, with particular emphasis on the "Action Plan on urban mobility" [6] and the "Urban mobility package" [10] providing support to local institutions for the development of convergent policies relating to mobility and the use of

Intelligent Transport Systems (ITS) in urban transport (i.e. in the fields of multimodal information, smart ticketing and traffic management).

Any decisions related to planning and development of a sustainable urban transport system are extremely complex because they require taking into account a number of factors and criteria, which often contradict each other [43]. The criteria most often used in the decision-making process regarding projects for urban transport are presented in Table 1. A notable problem which might arise at various stages of implementation of the project is the change of priorities of individual criteria, especially if the implementation period is extended - e.g. placing more importance on environmental criteria in the planning phase, only to emphasise the relevance of social criteria while assessing the implementation of the project, and settling the project on this basis [2].

Table 1. Criteria most often employed for decision-making in urban transport - passenger transport projects - example adapted from [43].

Criterion	Description
Economic	(Monetary) resources that have to be accounted for in a given project
Technical and logistics	Technical requirements of a given project
Environmental	Concerning the impact on the natural environment
Safety	Concerning the safety of people involved in the operation of transport systems and their use
Social	Positive and negative impacts on society cased by the decision made
Land use	If land adaptation as a component of the transport system is required

Table 2. Features of three types of urban fabrics, adapted from [23, 28]

Type	Walking city	Transit city	Car city
Description	Network of streets and public spaces suitable for active transport (walking, cycling)	Public transport system and network of urban areas with qualities that enable transit systems	Found in all cities that have car traffic, more so in low density, car-oriented and car-dependent urban areas
Basic elements	Appropriate infrastructure for walking and cycling (e.g. car-free zones)	One or more public transport modes with appropriate infrastructure	Infrastructure of freeways, roads, streets and parking facilities
Location	Most city centres and biggest sub centres	Around public transport systems	Not spatially limited
Accessibility	Based on co-existence with other city fabrics	Based on public transport system	Based on road infrastructure
Examples	Asian cities	European-type cities	Cities in USA, Australia and Canada

Another important issue which should be taken into account when considering various activities related to mobility and sustainable urban development is determining which of the urban fabrics they relate to. Table 2 shows the characteristics of different urban fabrics [28].

Ensuring sustainable urban mobility requires actions which take into account the interaction of the different fabrics of the city, especially the importance of traffic in generating negative externalities. Subsequently the potential of ICT in improving urban transport and better implementation of the demand for ensuring sustainable mobility are discussed.

3   Use of ICT in Urban Transport

Activities related to ensuring sustainable urban mobility are to increase the efficiency of the transport system through better use of existing infrastructure on the one hand, and on the other hand they are to reduce harmful effects on the community and the environment. A major factor contributing to the existence of harmful effects (such as noise, exhaust fumes, congestion) is the extensive use of private cars. Many of the efforts made at local level ("push-and-pull measures") are designed to encourage residents to reduce car use. Push measures are any measures aimed at discouraging the use of cars through appropriate regulations, fees or physical planning of the infrastructure, for example: restricting traffic, charges changing with city-zones, redesigning streets and public spaces. Pull measures on the other hand consist of all actions focusing on the promotion of public and non-motorised, active transport [38]. Studies show that the simultaneous application of both measures is the most effective, taking into account the interconnectedness of individual urban fabrics. If discouraging car use is to succeed, there is a need for a feasible substitute to the whole planned journey or at least the possibility of a seamless alternative to other modes of transport for part of it [42].

There are plenty of solutions that enhance sustainable urban mobility, and improve the efficiency of the urban transport system, which operate based on the use of ICT. The most popular; car sharing, ride sharing, integrated fare management, real-time traffic management, real-time traffic information, p2p car rental, bike sharing, personal travel or assistance applications let us affirm that ICT has become a regular and an essential element of the urban transport system. Unfortunately, a major limitation to the implementation of some ICT solutions in urban transport is the existing infrastructure and quite often the high cost of implementation [29]. Subsequently, some examples of the application of ICT are discussed, divided by type of use in urban transport. Firstly, the application of ICT in order to better use and manage the existing infrastructure and resources (vehicles) are discussed, then systems and applications supporting decision-making and social participation are presented, and finally, current potential scenarios for the development of urban transport in the future are shown.

3.1 Infrastructure and Resources Management Using ICT

ICT in urban transport is used most frequently to develop integration. The integration of urban public transport can be of varied range and apply to the selected elements of the whole system, such as the common tariff – ticket system, coordination of transport schedules, coordination of communication lines, a common information system, unified standards of technical - operational services rendered, or a common transfer and stopover infrastructure. ICT systems supporting the process of integration of transport and supporting transport management most frequently perform the following functions: location of vehicles, charging (using electronic cards, payments over the Internet, or using mobile phones), identification of the size of passenger flows on individual lines and sections of routes, management of production factors in public transport entities [27]. The integration of urban transport is a great convenience for both service providers and travellers, especially if it involves the integration of charges for different services [48]. A manifestation of integration from the point of view of passengers, for example, is an electronic city card, which can serve as a tool for cashless payments in the area of public transport and other public services on the basis of an appropriate information system [50]. In addition to increased convenience, these cards allow for the collection of data on the demand for services and they considerably improve the quality of services; in particular they improve transport coordination. The growing popularity of mobile devices enables the implementation of innovative solutions such as ticket sales based on personal mobile devices, which can significantly reduce the cost of implementation and modification of the system for both the public transport operator and passengers [13].

ICT also facilitates, or even in some cases enables, shared use of vehicles by multiple users. Such solutions involve different vehicles (mainly cars and bicycles) and various entities (public transport operators, businesses, individuals) can deal with their organization.

Bicycles are used quite often to promote active transport and as last-mile transport. Bicycle-sharing programs most frequently form a part of urban transport, under the responsibility of the operator or carried out by or in cooperation with an external company on behalf of the operator [22]. Most systems of bike-sharing, like Warsaw Veturilo, are third generation bike-sharing schemes, based on the use of ICT and smart bikes. Without an efficient system allowing the users to rent, give back and pay for bicycle use and allowing the operators to manage the fleet and relocate the bicycles between particular docking-bays as needed, these programs would not be so popular [5]. Obviously, bike-sharing schemes fit more into the implementation of the urban mobility system when they are accompanied by other promotional activities, such as organization of competitions, creating a community around an active lifestyle or the possibility of free transport of bicycles in other modes of transport [34].

In the case of programs based on the concept of using one car by many users there are more options. Car-sharing programs may be offered within urban transport and smart card by a public transport operator [41], by a company as part of commercial activity or by individuals. In the last case, there are several options: car-sharing (when one vehicle is used by several people at different times), car-pooling or ridesharing (when several people use the same vehicle at the same time to reduce costs), taxi-sharing (when a few people decide to travel by taxi together) [19]. Also, the type of vehicles (according to the engine drive) and the methods of making them available (points or free-floating) change, depending on solutions used in various cities [22]. Table 3 shows some variations in vehicle-sharing schemes.

Table 3. Differences in systems involving the sparing of vehicles.

Criterion	Types
What's being shared?	Vehicle
	A single journey on a given route
	Repeated journeys on a given route
Type of activity	Commercial endavour
	Private endavour
	Collaborative endavour
	Part of public transport system
Vehicle ownership	Company providing service
	Company leasing/providing vehicle
	Person/driver providing service
	Person providing use of vehicle
Who is providing insurance?	Company providing vehicle
	Company providing service
	Company acting as intermediary
	Vehicle owner
Who is providing maintenance?	Company providing vehicle
	Company providing service
	Company acting as intermediary
	Vehicle owner
Fee scheme	Enrolment subscription fee and payment per ride
	Monthly subscription and payment per ride
	Payment per ride based on fixed price-list
	Payment per ride set on ride-by-ride basis
	Payment as a share of the fuel costs
Type of trip [25]	Point-to-point trip – free-floating fleet
	Round-trip – fixed access points
Type of engine	Combustion engines
	Hybrid engines
	Electrical vehicles (evs)
Access to vehicle	Smart card
	Smartphone application
	Vehicle key (requiring or not contact with company/car-owner)
Verification of participants	Done by company providing services in case of car owners or drivers
	Done by company acting as an intermediary between car owner and potential user/passenger
	Based on social networking profiles of users
	Based on recommendation system within particular sharing scheme
	None except contact between interested parties (driver and potential passengers)

Vehicle sharing potentially reduces the number of vehicles on the road and the aggravation for other participants and the environment, but it is not an ideal solution – it still requires the allocation of public space for the infrastructure needed by cars.

Predicting the effects of popularizing individual solutions is very difficult. And so, for example, research on the different solutions in the area of car- sharing indicates a number of benefits for all parties [1, 14, 15] but also pays attention to the various problems and difficulties, such as: the need to attain critical mass so that the scheme is profitable [26], the need for cooperation with public administration at various levels [7], incentives for entrepreneurs and potential users [7, 9], the need to take into account conditions concerning the capacity to implement car- sharing schemes taking into account differences such as population density and demographic data [40, 49] between the areas being considered for future activities.

3.2   Use of Applications and Systems Supporting Decision Making

As already mentioned, the widespread use of ICT solutions in urban transport allows for the collection of large amounts of detailed information about vehicles, passengers and traffic density on individual routes. This information not only supports the more efficient management of infrastructure and the available vehicle (e.g. anticipated demand for bicycles or cars during rush hour), but it can be the basis for creating various types of applications, operating on the grounds of data obtained.

The most common type are applications supporting decision-making, both by passengers and by the organizations responsible for the operation of the urban transport system. From the point of view of operators, firstly ICT applications allow for in-depth analysis of information concerning passenger behaviour and on this basis, decide on the shape of the transport system (e.g. the concepts of Flexible Transport Services and Demand Responsive Transport [31]) or the analysis of the results of the decisions already taken [33]. Applications supporting decisions used by passengers, based on passenger information (in the optimal version the real-time information, for example; taking delays into account), aim to help decide on the choice of modes of transport and facilitate movement between different modes of transport [43].

One of the basic and most common systems facilitating decision-making by the passengers is the Dynamic Passenger Information System (SDIP). The main purpose of this system is to inform passengers about current arrivals and departures of various modes of transport from a given stop and about possible connecting transfers. Information is presented in a visual form, on electronic boards (LED/LCD), located at bus stops or in public transport vehicles [37]. The boards may be of various sizes and have different functions, for example, they may show the waiting time for a mode of transport, help decide about choosing a place to wait for one or change one's travel plans for alternative ones [30]. The information presented includes the most frequently updated times of arrival and departure of vehicles, modified depending on traffic through the use of information provided by compatible onboard equipment installed in public transport vehicles as well as other information such as, among others, stop name, line number, direction, special messages (information on traffic jams, diversions, prices of tickets, etc.) and advertising [4]. Presenting real-time information on arrivals and departures is also possible via the Internet and the necessary equipment allowing for the collection and transmission of information in real time is the same as in the case of information boards. Additionally, adapting information to mobile devices gives travellers permanent access to public transport timetables on any mobile phone that supports this function [32]. A particularly interesting solution is the application supporting travellers with disabilities and helping them make decisions during the trip concerning choice of transport vehicle or providing information on which stop to get off or change the mode of transport [39].

Another interesting use of the application is the attempt to involve passengers in the co-creation of services related to urban transport. Activities aimed at involving passengers in the process of providing services (co-creation) generally go in two directions. Firstly, applications are available allowing passengers to share relevant information from the point of view of both the service provider (e.g. obstructions on the road, accidents) and other passengers (e.g. bus delays, the lack of free parking spaces in the area) [20]. Secondly, they are to encourage city dwellers to share ideas and contribute to innovative projects undertaken - the big problem is that the operators responsible for urban transport quite frequently are not interested enough in customers [43], which unfortunately often means a lack of knowledge about the population groups that do not use public transport, despite being one of the target groups (e.g. the elderly) [39].

One should also mention the possibility of using social media and their impact on various elements of the urban transport system. One of the previously mentioned solutions improving the use of existing infrastructure was able to develop so quickly thanks to applying social media to solving trust issues among participants of the particular sharing schemes (so called "Goldilocks Complex"). People using both p2p car-sharing programmes or carpooling schemes were reluctant to allow strangers use of their private vehicles or undertake a journey with drivers they did not know. Applying the use of social media-like solutions allowed for not only negotiating trust issues, but also promoting this type of behaviour (by sharing information about participation among social networks users) [44]. Also, because social media are so popular, urban public transport operators decide to use them more and more often. In 2012, as part of the Transit Cooperative Research Program, in order to assess the use of social media in urban public transport, a study concerning organizers of urban public transport in the United States and Canada was conducted. The research showed that, in spite of many problems associated with the use of social media, they are an effective channel of communication with the environment, both for transmitting information and establishing a dialogue both with passengers and current and potential employees of the service provider [47]. In addition to the use of social media for information (either by promoting new initiatives or the transmission of real-time information), or activities in the field of CR [18], their potential application as an easy way to obtain information from customers is evident, both through facilities in conducting research and more effective ways of reaching out to the respondents [8] and using data mining techniques [17].

3.3   Urban Transport in the Future

The described methods of using ICT in urban transport allow for forecasting mobility of the urban population in the future, taking into account existing solutions, further directions of development and mutual interaction and deeper integration. Clearly exact predictions concerning future urban transport are not possible, but different institutions create more or less likely scenarios. Table 4 presents three scenarios for the future shape of the urban transport system proposed by Deloitte, prepared on the basis of the key features of a modern transport system: massively networked, user centred, integrated, dynamically priced and reliant on new models of public-private collaboration [16].

Table 4. Three scenarios for digital-age transportation adapted from [16].

Scenario	Description	Requirements
The internet of cars	Autonomous driving, environmental friendly fleet of cars, managed by integrated platform but suited to the individual needs of users	Combine vehicle communications in single platform
		Extensive coverage of its infrastructure
		Unification of connectivity standards
		Need for high standards of data and connection security
		Need to resolve the issues concerning privacy
Dynamic pricing	More direct portion of actual costs of service is paid by users and the prices respond to changes in demand	Common use of wireless payments
		Extensive coverage of its infrastructure
		Introduction of new payment models
		Overcoming users' resistance to different aspects of the system
		Building one common vision of the shape of transportation market
Social transport	All aspects of transportation system built on collaboration of the all stakeholders	Dynamic, up-to-the-minute information source (both public and private) that helps users to make decisions
		Gratification from the experience, competition among users and rewords
		Planning for network effects
		Overcoming issues concerning privacy

In modern urban transport systems, a few elements of each of the scenarios can be seen. Some are more visible (e.g. the increasing use of smart cards within integrated ticketing systems [24]), others are not yet widespread although they could actually function (as there are some legal controversies related to driveless, autonomus vehicles and liability in case of an accident [3]).

4 Conclusion

Taking into consideration the importance which is attributed to urban mobility in the European Union both at local and state levels, the implementation of innovative solutions using ICT is promoted in order to ensure the best services while limiting the negative impact of it on society and the environment.

The realization of these demands is not simple and requires integrated and comprehensive action in several areas. The implementation of modern solutions, as described in the article, (facilitating efficient management, using existing infrastructure and supporting decision-making) should be linked with measures aimed at better understanding the needs of residents of urban areas and providing them with services rendered by both operators of mass public transport and entrepreneurs and predicting future trends in mobility in urban areas (e.g. caused by demographic processes, such as an aging population and associated with increased traffic during the tourist season). It is also important to promote changes in the behaviour of the inhabitants, which becomes especially important in the light of the perceived generation gap (i.e. ownership vs. access in the case of cars or active transport as a lifestyle) [35] and noticeable difficulties in changing habits regarding the choice of the preferred means of transport [21, 42].

Nevertheless, it can be concluded that the actions taken are going in the right direction, which are indicated by the results of the Urban Mobility Ranking. In 2011, the consulting agency Arthur D. Little developed the Urban Mobility Index for 84 cities [45], which was subsequently modified in 2014 - Urban Mobility Index 2.0 [46]. Unfortunately, due to the changes implemented, the two indexes cannot be directly compared, but it should be noted that in both cases among the ten best rated cities in terms of mobility, there were respectively 7 (in 2011) [45] and 8 (in 2014) [46] cities located in the European Union (presented in Table 5).

Table 5. European City top performance Urban Mobility Index and Urban Mobility Index 2.0 adapted from [45, 46].

Urban Mobility Index (2011)		Urban Mobility Index 2.0 (2014)	
Place	City	Place	City
2	Amsterdam	2	Stockholm
3	London	3	Amsterdam
4	Stockholm	4	Copenhagen
5	Goteborg	5	Vienna
7	Vienna	7	Paris
8	Paris	9	London
9	Munich	9	Helsinki
–	–	10	Munich

Such a high position in the ranking suggests that the actions taken contribute to the implementation of the demands for the provision of sustainable urban mobility - in both versions of the index, parameters such as bike-share and car-share schemes and penetration of smart cards are taken into account. Nevertheless, it should be noted that these are the cities belonging to the so-called "old 15 countries", that is before the 2004 enlargement of the European Union and that apart from two cases (Gothenburg and Munich), all of them are the capitals of the countries. A conclusion can be drawn that ensuring sustainable urban mobility across the whole European Union requires more time, especially taking into account differences in development among the member countries.

References

1. Baptista, P., Melo, S., Rolim, C.: Energy, environmental and mobility impacts of car-sharing systems: Empirical results from Lisbon, Portugal. Procedia Soc. Behav. Sci. **111**, 28–37 (2014). doi:10.1016/j.sbspro.2014.01.035
2. Brorström, S.: Implementing innovative ideas in a city: good solutions on paper but not in practice? IJPSM **28**(3), 166–180 (2015). http://dx.doi.org/10.1108/IJPSM-11-2014-0137
3. Cepolina, E.: Farina, A: A methodology for planning a new urban car sharing system with fully automated personal vehicles. Eur. Transp. Res. Rev. **6**, 191–204 (2014). doi:10.1007/s12544-013-0118-9
4. Chandurkar, S., et al.: Implementation of real time bus monitoring and passenger information system. Int. J. Sci. Res. Publ. (IJSRP) **3**(5) (2013)
5. Cichosz, M.: IT solutions in logistics of smart bike-sharing systems in urban transport. Management **17**(2), 272–283 (2013). doi:10.2478/manment-2013-0071
6. Communication from the Commission to the European Parliament, The Council, the European Economic and Social Committee and the Committee of the Regions: Action Plan on Urban Mobility (COM 2009), 490 final, Brussels (2009)
7. Dowling, R., Kent, J.: Practice and public-private partnerships in sustainable transport governance: the case of car sharing in Sydney, Australia. Transport Policy **40**, 58–64 (2015). http://dx.doi.org/10.1016/j.tranpol.2015.02.007
8. Efthymiou, D., Antoniou, C.: Use of social media for transport data collection. Procedia Soc. Behav. Sci. **48**, 775–785 (2012). doi:10.1016/j.sbspro.2012.06.1055
9. Engel-Yan, J., Passmore, D.: Carsharing and car ownership at the building scale: examining the potential for flexible parking requirements. J. Am. Plan. Assoc. **79**(1), 82–91 (2013). doi:10.1080/01944363.2013.790588
10. European Commission: Clean transport, Urban transport (2013). http://ec.europa.eu/transport/themes/urban/urban_mobility/ump_en.htm. Accessed 31 Jan 2016
11. European Environment Agency: Urban environment (2013). http://www.eea.europa.eu/themes/urban/intro. Accessed 31 Jan 2016
12. Eurostat: Eurostat regional yearbook 2015, Luxemburg (2015), http://ec.europa.eu/eurostat/documents/3217494/7018888/KS-HA-15-001-EN-N.pdf/6f0d4095-5e7a-4aab-af28-d255e2bcb395. Accessed 31 Jan 2016
13. Ferreira, M., et al.: A proposal for a public transport ticketing solution based on customers' mobile devices. Procedia Soc. Behav. Sci. **111**(2014), 233–241 (2014). doi:10.1016/j.sbspro.2014.01.056
14. Firnkorn, J., Müller, M.: What will be the environmental effects of new free-floating car-sharing systems? The case of car2go in ULM. Ecol. Econ. **70**, 1519–1528 (2011). doi:10.1016/j.ecolecon.2011.03.014

15. Firnkorn, J., Müller, M.: Free-floating electric carsharing-fleets in smart cities: The dawning of post-private car era in urban environments? Environ. Sci. Policy **45**, 30–40 (2015). doi: 10.1016/j.envsci.2014.09.005
16. Fishman, T.: Digital-Age Transportation: The Future of Urban Mobility. Deloitte University Press (2012). http://dupress.com/articles/digital-age-transportation/. Accessed 2 Feb 2016
17. Gal-Tzur, A., et al.: The potential of social media in delivering transport policy goals. Transp. Policy **32**, 115–123 (2014). doi:10.1016/j.tranpol.2014.01.007
18. Gal-Tzur, A., et al.: The impact of social media usage on transport policy: issues, challenges and recommendations. Procedia Soc. Behav. Sci. **111**, 937–946 (2014). doi:10.1016/j.sbspro. 2014.01.128
19. Gansky, L.: The Mesh: Why the Future of Business is Sharing. Portfolio/Penguin, New York (2012)
20. van der Graaf, S., Veeckman, C.: Designing for participatory governance: assessing capabilities and toolkits in public service delivery. Info **16**(6), 74–88 (2014). doi:10.1108/ info-07-2014-0028
21. Huwer, U.: Public transport and car-sharing - benefits and effects of combined services. Transp. Policy **11**, 77–87 (2004). doi:10.1016/j.tranpol.2003.08.002
22. Kaltenbrunner, A., Meza, R., Grivolla, J., Codina, J., Banchs, R.: Urban cycles and mobility patterns: Exploring and predicting trends in a bicycle-based public transport system. Pervasive Mob. Comput. **6**, 455–466 (2010). doi:10.1016/j.pmcj.2010.07.002
23. Kosenen, L.: Model of three urban fabrics (2014). www.urbanfabrics.fi. Accessed 31 Jan 2016
24. Kos-Łabędowicz, J.: Integrated E-ticketing system – possibilities of introduction in EU. In: Mikulski, J. (ed.) TST 2014. CCIS, vol. 471, pp. 376–385. Springer, Heidelberg (2014). doi: 10.1007/978-3-662-45317-9_40
25. Le Vine, S., et al.: A new approach to predict the market and impacts of round-trip and point-to-point carsharing systems: case study of London. Transp. Res. Part D **32**, 218–229 (2014). http://dx.doi.org/10.1016/j.trd.2014.07.005
26. Lee, A., Savelsbergh, M.: Dynamic ridesharing: is there a role for dedicated drivers? Transp. Res. Part B **81**, 483–497 (2015). http://dx.doi.org/10.1016/j.trb.2015.02.013
27. Lubieniecka - Kocoń, K., Kos, B., Kosobucki, Ł., Urbanek, A.: Modern tools of passenger public transport integration. In: Mikulski, J. (ed.) TST 2013. CCIS, vol. 395, pp. 81–88. Springer, Heidelberg (2013). doi:10.1007/978-3-642-41647-7_11
28. Mäkinen, K., Kivimaa, P., Helminen, V.: Path creation for urban mobility transitions. Manag. Environ. Qual. Int. J. **26**(4), 485–504 (2015). http://dx.doi.org/10.1108/MEQ-07-2014-0115
29. Mikulski, J.: The possibility of using telematics in urban transportation. In: Mikulski, J. (ed.) TST 2011. CCIS, vol. 239, pp. 54–69. Springer, Heidelberg (2011). doi: 10.1007/978-3-642-24660-9_7
30. Molecki, B.: Przystankowe tablice dynamicznej informacji pasażerskiej. In: Transport Miejski i Regionalny Nr 7/8, pp. 38–44. Stowarzyszenie Inżynierów i Techników Komunikacji Rzeczpospolitej Polskiej, Warszawa (2011)
31. Mulley, C., et al.: Barriers to implementing flexible transport services: an international comparison of the experiences in Australia, Europe and USA. Res. Transp. Bus. Manag. **3**, 3–11 (2012). doi:10.1016/j.rtbm.2012.04.001
32. Park, D., Kim, H.: SBISURBAN - secure urban bus information system based on smart devices. Int. J. Secur. Appl. **9**(1), 205–220 (2015). doi:10.14257/ijsia.2015.9.1.21
33. Pensa, S., et al.: Planning local public transport: a visual support to decision-making. Procedia Soc. Beha. Sci. **111**, 596–603 (2014). doi:10.1016/j.sbspro.2014.01.093

34. Pospischil, F., Mailer, M.: The potential of cycling for sustainable mobility in metropolitan regions - the facts behind the success story of innsbruck. Transport Res. Procedia **4**, 80–89 (2014). doi:10.1016/j.trpro.2014.11.007

35. Rifkin, J.: The Zero Marginal Cost Society: The Internet of Things, the Collaborative Commons, and the Eclipse of Capitalism. Palgrave MacMillan, New York (2014)

36. Rodrigue, J.-P.: Urban transport challenges (2013). http://people.hofstra.edu/geotrans/eng/ch6en/conc6en/ch6c4en.html. Accessed 2 Feb 2016

37. Rojowski, R., Gancarz, T.: System dynamicznej informacji pasażerskiej. In: Autobusy: technika, eksploatacja, systemy transportowe, R. 10, Nr 4, pp. 24–31, Instytut Naukowo-Wydawniczy "SPATIUM". sp. z o.o., Radom (2009)

38. Saliara, K.: Public transport integration: the case study of Thessaloniki. Greece. Transport Res. Procedia **4**, 535–552 (2014). doi:10.1016/j.trpro.2014.11.041

39. Schlingensiepen, J., et al.: Empowering people with disabilities using urban public transport. Procedia Manuf. **3**, 2349–2356 (2015). doi:10.1016/j.promfg.2015.07.382

40. Scholler, S., Bogenberger, K.: Analyzing external factors on the spatial and temporal demand of car sharing systems. Procedia Soc. Behav. Sci. **111**, 8–17 (2014). doi:10.1016/j.sbspro.2014.01.033

41. Cruz, I.S., Katz-Gerro, T.: Urban public transport companies and strategies to promote sustainable consumption practices. J. Cleaner Prod. (2016). doi:10.1016/j.jclepro.2015.12.007

42. Şimşekoğlu, Ö., Nordfjærn, T., Rundmo, T.: The role of attitudes, transport priorities, and car use habit for travel mode use and intentions to use public transportation in an urban Norwegian public. Transp. Policy **42**, 113–120 (2015). http://dx.doi.org/10.1016/j.tranpol.2015.05.019

43. Sindakis, S., Depeige, A., Anoyrkati, E.: Customer-centered knowledge management: challenges and implications for knowledge-based innovation in the public transport sector. J. Knowl. Manag. **19**(3), 559–578 (2015). http://dx.doi.org/10.1108/JKM-02-2015-0046

44. Stephany, A.: The Business of Sharing: Making It in the New Sharing Economy. Palgrave MacMillan, Hampshire (2015)

45. Little, A.D.: The Future of Urban Mobility: Towards networked, multimodal cities of 2050 (2011) http://www.adlittle.com/downloads/tx_adlreports/ADL_Future_of_urban_mobility.pdf. Accessed 10 Dec 2015

46. Little, A.D.: The Future of Urban Mobility 2.0: Imperatives to Shape Extended Mobility Ecosystems of Tomorrow (2014). http://www.adlittle.com/downloads/tx_adlreports/2014_ADL_UITP_Future_of_Urban_Mobility_2_0_Full_study.pdf. Accessed 10 Dec 2015

47. Transportation Research Board (TRB): Uses of Social Media in Public Transportation. TRB's Transit Cooperative Research Program (TCRP) Synthesis 99, Washington, DC, Na-tional Academy of Sciences (2012)

48. Urbanek, A.: Pricing policy after the implementation of electronic ticketing technology in public urban transport: an exploratory study in Poland. In: Mikulski, J. (ed.) TST 2015. CCIS, vol. 531, pp. 322–332. Springer, Heidelberg (2015). doi:10.1007/978-3-319-24577-5_32

49. Wappelhorst, S., Sauer, M., Hinkeldein, D., Bocherding, A., Glass, T.: Potential of electric carsharing in urban and rural areas. Transp. Res. Procedia **4**, 374–386 (2014). doi:10.1016/j.trpro.2014.11.028

50. Zakonnik, Ł.: Karty miejskie w Polsce jako etap w rozwoju płatności bezgotówkowych opiewających na niskie kwoty. In: Studia i Materiały Polskiego Stowarzyszenia Zarządzania Wiedzą, Nr 29, pp. 167–178, Polskie Stowarzyszenie Zarządzania Wiedzą (2010)

51. Zito, P., Salvo, G.: Toward an urban transport sustainability index: an European comparison. Eur. Transp. Res. Rev. **3**(4), 179–195 (2011). doi:10.1007/s12544-011-0059-0

Analysis of Data Needs and Having for the Integrated Urban Freight Transport Management System

Kinga Kijewska[1], Krzysztof Małecki[2], and Stanisław Iwan[1]([⊠])

[1] Maritime University of Szczecin, Pobożnego 11, 70-507 Szczecin, Poland
{k.kijewska, s.iwan}@am.szczecin.pl
[2] West Pomeranian University of Technology,
Żołnierska 11, 71-210 Szczecin, Poland
kmalecki@wi.zut.edu.pl

Abstract. One of the key factors necessary for controlling the increasing disorder in the area of urban freight transport and conditioning its sustainable growth is ensuring effective data flows between individual parties engaged, to a smaller or greater extent, in its functioning. Lack of knowledge regarding cargo streams and their direction, structure etc. makes it difficult to control and manage them so as to limit their negative impact on the city organism, particularly on the environment and inhabitants. Additionally, development of information society and continuous digitalisation of different areas of life make it necessary to get control over the data resources regarding urban freight transport, as it is becoming indispensable for its correct functioning. This paper is focused on determining the structures of data relevant for the functioning of sustainable urban freight transport, and establishing the sources of data acquisition.

Keywords: Urban freight transport · Freight transport management · Data flows · Integrated systems

1 Introduction

The city as a system of economic entities, which are its users, operates and develops as a result of a variety of activities of each of these entities. Individual entities implement specific functions, striving to achieve their objectives. However, objectives of individual entities are generally different, which often leads to conflict of interest. This is particularly important in respect to the freight transport subsystem and results from its specificity. Lack of consensus between the expectations of different groups of stakeholders can cause many problems and risks relating to social, economic and environmental issues. Therefore, an appropriate management plays an important role in the functioning of UFT.

One of the key factors necessary for efficient management system in urban freight transport (UFT) are the proper data flows between the various parties involved in it. Lack of knowledge contributes to the difficulty in controlling and guiding the freight flows in a way that minimize the negative impact of the urban freight transport, in particular, on the environment and residents [1].

© Springer International Publishing AG 2016
J. Mikulski (Ed.): TST 2016, CCIS 640, pp. 135–148, 2016.
DOI: 10.1007/978-3-319-49646-7_12

2 The Specificity of UFT Management Processes

In the literature there are different approaches to the issue of UFT management, but the vast majority of its are focused on the management of specific actions and solutions, both those functioning independently [2–4], and having integrated nature, such as Urban Consolidation Centres [5–8] or Freight Quality Partnerships [9, 10]. However, there are few studies, which perceive the UFT management processes through the prism of integration at the entire urban system [11]. The starting point for discussion in this respect is an in-depth analysis of the functioning of UFT in terms of the system. The city can be defined as a dynamically changing open system [12], which includes two key subsystems [13]:

- physical, consisting of buildings connected by streets, roads and infrastructure,
- human, which consists of traffic, interactions and activities of its residents.

Locating UFT in the city logistic system requires to extract it from material goods transport subsystem, while stressing its strong correlation with material goods and persons flow control subsystem. It can be assumed that it is a set of the following components [12]:

- transport operations;
- storage and selling points, processes associated with the goods and the types of transport performed;
- types and equipment related to handling physical flows;
- location and management of fundamental structures necessary for the implementation of physical flows, including warehouse processes of incoming and outgoing goods;
- operation of logistics information systems.

According to its classic definition, management can be characterized as a set of activities, which consists of: planning, organizing, leading and controlling [14]. The appropriate quality of this process in relation to the city depends primarily on the ability to look at the processes taking place in it in a holistic way, resulting from all functions of the city and relationships between elements of its structure as a system. In relation to UFT, of crucial importance in this case becomes the operation of the goods distribution subsystem.

UFT management may be viewed from three major perspectives which, taking into consideration the functioning of the urban structure as a whole, may be referred to the particular levels of the classical management pyramid [15]:

- the perspective of individual freight transport market participants, which focuses on ad-hoc measures that are usually taken over a short-term horizon, and are focused on meeting individual needs of single UFT stakeholders;
- the perspective of integrated measures taken by groups of stakeholders who aim at developing more effective methods of transport and deliveries over a longer time horizon, taking into account diversified needs and many a time divergent priorities (e.g. initiatives connected with organizing consolidated deliveries or commercial vehicles sharing);

– the perspective of a city as a whole, which should refer to the measures that integrate any other initiatives taken on lower levels and indicate far-reaching areas of development and the policy for the city's transport system management.

Different management levels are linked directly with time horizons within which management processes are implemented. At each of these levels management processes may relate to two areas of activity:

– managing the implementation of solutions that should be realized in accordance with the PDCA (Plan, Do, Check, Act) cycle [16];
– managing the current operation of the UFT system, which should focus on solving emerging problems in regard to the organization and delivery, transport monitoring, removing threats etc.

Regardless of the accepted range, the time horizon or area, integrated UFT management should:

– be connected to monitoring systems for technical state of all sites and equipment intended for the provision of infrastructure services and the extent to which demand for services is met,
– allow the quick analysis of the impact and influence of inefficiency of sites and equipment in relation to the efficiency of the distribution of goods in the city,
– be connected to a rapid response systems in terms of site usability restoration e.g. road state improvement, rectification of the aftermath of an accidents,
– include a module allowing to forecast the needs for repairs and infrastructure investment resulting from the operation of the commercial vehicles,
– include a module to forecast development of investment and infrastructure services in relation to the needs of freight transport in accordance with the local development plan,
– include a module of rapid alert system informing about obstacles on the road, e.g. caused by accidents,
– be able to collect information about directions of commercial vehicles.

Accordingly, the effective implementation of tasks within such a complex management system requires the use of appropriate methods and tools. Therefore, a key role is played IT tools, without which it is impossible to ensure proper cooperation between the various sub-processes. For this reason, all of the elements of the integrated UFT management must be merged together through the integration of data, information and knowledge. Basis for integration of management system information structure is primarily knowledge of data needs and sources of their acquisition.

When analysing the issue of UFT management, the first of the factors determining the efficiency of this process that should be emphasized is the need to ensure consensus between the expectations of different groups of stakeholders. In addition, it is necessary to identify the key groups that influence management processes directly. From the perspective of efficient management of freight transport system, there should be noted three main categories of stakeholders, who have a direct influence on decisions made within this system. These are:

- local governments, which make decisions regulating and conditioning transport processes in the city;
- transport system monitoring and security entities, which supervise the organization and functioning of transport within the city and react in emergencies in order to reduce the negative impact of events occurring in the urban space on the functioning of the entire transport system;
- shippers and freight carriers, who are directly responsible for the implementation of services and to make key decisions affecting their implementation and effectiveness.

These entities were subjects for further study presented in the paper.

Nowadays effective transport system depends largely on the proper data and information flow and knowledge acquisition allowing for effective management of processes occurring in this system [17]. The importance of data flows for efficient functioning of UFT systems is underlined in the classic definition of city logistics, proposed by [18]: city logistics is the process for totally optimizing the logistics and transport activities by private companies with the support of advanced information systems in urban areas considering the traffic environment, its congestion, safety and energy savings within the framework of a market economy.

Specifying the information needs within the area of UFT, and also the interdependencies between them and the other elements of the city's transport system will make it possible to take control over the increasing disorder in this area and enable its development in accordance with the principles of sustainable growth [17].

In the area of UFT, data can be classified according to three basic criteria: variability, reliability, sources [18]. Additionally, taking to the account identification of data sources it should be expand by the following criteria: the availability of data and data quality [19]. However, providing these data resources faces considerable difficulties. Next to the diversity of needs and expectations of stakeholders, significant fragmentation of transport and high functional complexity of the system itself, a major obstacle to its effective management and implementation of intelligent transport systems is the difficulty in data acquisition [17].

3   Analysis of Data Needs and Having of Chosen Entities in Poland

The analysis carried out under the project "Analysis of information needs of heterogeneous urban environment in a sustainable freight transport system" aimed to establish existing data needs among selected groups of UFT stakeholders, as well as to identify potential sources of data for integrated UFT management. It was the first stage of work aimed at building a generalized model of data flows. The study area was limited to the Polish territory. As demonstrated by the experience of direct contacts with stakeholders UFT in Polish cities, there are considerable disparities and substantial degree of disorder in terms of needs and data resources having.

The research process involved the development of a survey with the questionnaire prepared on the assumptions introduced in [20]. Finally, the questionnaire includes 50 data categories related to 15 UFT indicators, aggregated in 4 area of impact (Table 1).

The study was attended by 33 entities, which were classified into particular groups (each group consisted of the same number of respondents):

- public administration (in Poland there are 4 levels to extract: municipal, county, provincial and national)
- monitoring, security and emergency response entities (this group included professionals such as police, emergency medical service, fire brigade and engineering-technical service)
- system users, such as carriers, TSL sector company, drivers, business entities (recipients of goods).

The task of respondents was to determine which aspects are necessary from the point of view of effective UFT management (Needs), and which data that they have can be provided to the system (Having). The results show that majority of entities are interested in obtaining data from the system and to a lesser extent in providing data – Table 1.

Table 1. Analysis of data needs and having in relation to the area of influence [own study]

Area of impact	Total		Local governments		Traffic monitoring, security		System users	
	Needs	Having	Needs	Having	Needs	Having	Needs	Having
Economy	34,29%	24,03%	56,20%	11,57%	29,34%	2,89%	23,55%	61,98%
Social area	24,08%	24,69%	25,97%	40,91%	34,42%	14,29%	16,23%	23,38%
Environment	34,69%	7,35%	63,64%	3,90%	24,68%	0,00%	22,08%	19,48%
Mobility	37,96%	9,80%	44,16%	11,69%	35,06%	1,30%	41,56%	18,18%

In the economic area, over 34% of all entities need data while 24% declare to have them. In the social area results were comparable with respect to two categories: 24.08% of entities need data and 24.69% have them. The lowest available data could be observed in the other two areas – environment and mobility. 35% and almost 38% of respondents respectively expressed the need for data collection in these areas, but having them was declared only by about 7% and nearly 10% of all respondents. With respect to particular groups of stakeholders, local governments have the greatest interest in the economic and environmental data. At the same time, they have almost no information in these two areas (about 12% in relation to the economic area and 4% to the environmental area), which demonstrates the serious dysfunctions within the data flow in UFT. A valuable group proved to be system users, who have data in economic area. These data, in turn, are of interest to public administration (56.2% of respondents from this group have declared the need for such data).

The following part of the study involved the analysis of data needs in the context of indicators for evaluating UFT. A particularly important result of the study proved to be the relationship between different groups of respondents, for example: system users have the data needed from the point of view of the indicators "the efficiency of deliveries within the city" and "transport needs", and public administration is interested in these data. On the other hand – traffic monitoring and security entities have data on accidents in UFT that are in the circle of interest to system users (e.g. in the context of bypassing the blocked areas of the city). A special case of data that need to be obtained from sources other than those included in the study are "noise level data". Collected information shows that system users can provide little data on this subject (9%), and 72% of public administration entities need this data, which means that it becomes necessary to either carry out regular research in this area by specialized bodies or mounting specialized measuring stations, which will provide up to date information about the noise in the city. Moreover, it is necessary to consider that the indicator "home deliveries", for which the city does not have the data and the system users may provide only a small percentage (9%) of these data.

Due to the complexity of data collection for integrated UFT management, it was necessary to rank different categories of data. This allows to select those that are the most important for stakeholders and the obtaining of which should be focused on in the first place. In addition, hierarchization allows to highlight these categories of data that different stakeholder groups most frequently have. This, in turn, allows to compare the needs with the scope of the data resource having and highlight the fundamental dysfunction in this area.

Hierarchization of data categories was carried out separately for the needs and for the scope of their having, using a two-dimensional ranking based on the result of evaluation in relation to the internal relationship between the ranking made by the various stakeholder groups, as well as the relationships between the different data categories. This method has been used in [21] to determine the weights in evaluating UFT good practice adaptation using the success rate. The analysis is based on a summary of the results r obtained during the questionnaire study in the results matrix (Table 2).

The last column of the matrix involves calculation of ranks for each data category, indicating their importance in the overall analysis:

$$RD_n = \frac{\sum_{m=1}^{M} r_{n,m}}{\sum_{n=1}^{N} \sum_{m=1}^{M} r_{n,m}}, \ n \in \{1,\ldots,N\} \tag{1}$$

A similar method is used to calculate ranks for each stakeholder group, specifying the total number of ranking made:

$$RI_m = \frac{\sum_{n=1}^{N} r_{n,m}}{\sum_{n=1}^{N} \sum_{m=1}^{M} r_{n,m}}, \ m \in \{1,\ldots,M\} \tag{2}$$

The next step involved determining the resultant for each of ranking (for each category of data N and each stakeholder group M), taking into account the parameters RDN and RIM as the forces acting on the resultant value:

Table 2. Results matrix for hierarchization using two-dimensional ranking method [own study]

	Stakeholder group 1	Stakeholder group 2		Stakeholder group M	Sum of ranking of stakeholder groups	Rank for data categories
Data category 1	$r_{1,1}$	$r_{1,2}$	...	$r_{1,M}$	$\sum_{m=1}^{M} r_{1,m}$	RD_1
Data category 2	$r_{2,1}$	$r_{2,2}$		$r_{2,M}$	$\sum_{m=1}^{M} r_{2,m}$	RD_2
...					...	
Data category N	$r_{N,1}$	$r_{N,2}$		$r_{N,M}$	$\sum_{m=1}^{M} r_{N,m}$	RD_N
Sum for data category	$\sum_{n=1}^{N} r_{n,1}$	$\sum_{n=1}^{N} r_{n,2}$	...	$\sum_{n=1}^{N} r_{n,M}$	$\sum_{n=1}^{N}\sum_{m=1}^{M} r_{n,m}$	$=1$
Rank in relations to stakeholder group	RI_1	RI_2		RI_M	$=1$	

$$RV_{n,m} \approx \sqrt{\left(r_{n,m}RP_n\right)^2 + \left(r_{n,m}RR_m\right)^2}, \quad \begin{array}{l} n \in \{1,\ldots,N\} \\ m \in \{1,\ldots,M\} \end{array} \qquad (3)$$

where:

$RV_{n,m}$ – resultant value of the *n-th* data category and the *m-th* stakeholder group.

Two separate matrices were prepared for the analysed data ranges - for data needs and data having. Hierarchy of data categories for both of them were established as the sum of resultant values achieved in each row (in relation to stakeholders groups). The two results of hierarchization were compared in order to determine the disproportion between data needs reported by the surveyed stakeholder groups and data resources having, taking into account the level of significance of particular data categories in the context of the UFT management. Table 3 shows the final results of hierarchization of the data categories in relation to the needs (columns 1 and 2) and having (columns 3 and 4) based on the method discussed above. Finally, the difference between positions of each data category in both tables was determined (column 5). This was the basis to determine the extent of diversification of the surveyed population.

This hierarchization allowed to capture the disproportion between data needs and data resources having of stakeholders. Chart presented in Fig. 1 illustrates the degree of dispersion of data categories, corresponding to the mismatch between data needs and data resource having.

A variance determined for the surveyed population was a measure of the data dispersion and standard deviation as a measure of the average deviation from the arithmetic mean of the surveyed population. These measures amount to 353.8 and 18.81 points respectively and reflect numerical level of mismatch between data needs and data resource having.

Table 3. Summary of hierarchization using two-dimensional ranking method [own study]

Data needs		Data having		Deviation (2–4)
Data category	Sum of resultant values	Data category	Sum of resultant values	
Traffic/congestion level	7,2162	The places, where the trip starts	6,5786	−41
The noise level generated by the commercial vehicle	6,7125	Length of trips	6,5786	−45
Days and hours of deliveries	6,3267	Regularity of trips	6,2781	−4
Length of trips	6,0079	The distance between network nodes	6,1629	2
Time spent within the city area	5,9386	Travel times	6,1629	−8
Pollutant emissions by vehicle type	5,9386	Travel times to and through the city centre	6,1629	−29
The volume of freight brought into the urban area	5,6316	Days and hours of deliveries	6,0527	−18
Total deliveries	5,6316	Number of stops per trip, day	5,6572	−20
The places, where the trip starts	5,6316	Logistics costs	5,2440	8
Travel times	5,1664	The places, where the trip ends	5,0369	5
Travel times to and through the city centre	5,1664	Vehicles used within the city	4,9004	5
The number of cyclists	5,1465	The number of recipients of the goods	4,6246	−19
The number of pedestrians	5,1465	Time spent within the city area	4,3270	−19
Number of stops per trip, day	5,0882	The number of jobs in urban freight transport	4,2802	6
The number of vehicles entering the city	5,0668	The number of vehicles by DMC and age	4,2372	−12
The distance between network nodes	5,0078	Wages in urban freight transport	4,0055	12
The places, where the trip ends	4,9385	Load capacity utilization	4,0055	7
Share of urban freight transport in exhaust emissions	4,8602	The number of drivers of commercial vehicles and trucks	3,9590	−27

(*continued*)

Table 3. (*continued*)

Data needs		Data having		Deviation (2–4)
Data category	Sum of resultant values	Data category	Sum of resultant values	
The share of trucks in total vehicular traffic	4,8493	The number of accidents	3,8182	−27
The number of recipients of the goods	4,6932	The number of deaths	3,8182	8
The average speed of vehicles	4,6510	Deliveries in enclosed areas (within the enterprises)	3,7083	−22
Loading and unloading time	4,5544	Loading and unloading time	3,7083	0
Regularity of trips	4,5345	Total number of car drivers	3,6377	20
The noise level when loading/unloading commercial vehicles	4,2489	Share of freight vehicles in accidents	3,5125	−24
Load capacity utilization	4,0902	The volume of freight brought into the urban area	3,3868	8
Energy consumption in urban freight transport	4,0827	Population density and distribution of the population within urban areas	3,1093	−14
Consumption of non-renewable fuels	3,9437	The number of vehicles entering the city	3,0411	−17
The number of kilometres per capita	3,8620	Total deliveries	2,7683	−22
The number of kilometres travelled by commercial vehicle	3,7155	The delivery transport cost in the overall cost of the supply chain	2,4717	−12
Deliveries in enclosed areas (within the enterprises)	3,6887	Load capacity utilization	2,4717	9
The degree of load capacity utilization	3,6262	The number of cyclists	2,4466	1
Vehicles used within the city	3,6065	The number of pedestrians	2,4466	21
Logistics costs	3,3908	Typical fuel consumption by vehicle category	2,4466	24
Home delivery services provided by stores	3,3908		2,4220	−15

(*continued*)

Table 3. (*continued*)

Data needs		Data having		Deviation (2–4)
Data category	Sum of resultant values	Data category	Sum of resultant values	
		The number of companies involved in urban freight transport		
The number of drivers of commercial vehicles and trucks	3,2890	Pollutant emissions by vehicle type	2,1500	17
The delivery transport cost in the overall cost of the supply chain	3,2518	The number of jobs in sectors related to the functioning of the urban freight transport	2,1005	7
The number of companies involved in urban freight transport	3,2393	Vehicle ownership	2,1005	3
Total number of car drivers	2,9840	The weekly schedule of accidents involving trucks	1,9597	15
Wages in urban freight transport	2,7819	Household size	1,8685	23
Vehicle ownership	2,6991	Energy consumption in urban freight transport	1,8536	3
Typical fuel consumption by vehicle category	2,6926	The number of kilometres travelled by commercial vehicle	1,8536	8
The number of vehicles by DMC and age	2,5205	Traffic/congestion level	1,5065	27
The number of accidents	2,4707	The average speed of vehicles	1,2973	24
The weekly schedule of accidents involving trucks	1,9315	Consumption of non-renewable fuels	1,2357	6
Share of freight vehicles in accidents	1,9256	Share of urban freight transport in exhaust emissions	1,2357	21
The number of jobs in sectors related to the functioning of the urban freight transport	1,6275	The share of trucks in total vehicular traffic	1,2101	10
The number of jobs in urban freight transport	1,6217	The noise level generated by the commercial vehicle	0,6178	33

(*continued*)

Table 3. (*continued*)

| Data needs | | Data having | | Deviation (2–4) |
Data category	Sum of resultant values	Data category	Sum of resultant values	
Household size	1,4686	The noise level when loading/unloading commercial vehicles	0,6178	9
The number of deaths	1,4628	Home delivery services provided by stores	0,6178	29
Population density and distribution of the population within urban areas	1,0001	The number of kilometres per capita	0,6178	24
Variance				353,8
Standard deviation				18,81

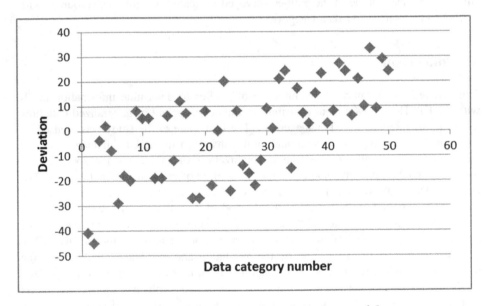

Fig. 1. Dispersion of surveyed population [own study]

The analysis showed a very high degree of mismatch between data needs and data resource having of the surveyed stakeholder groups. Stakeholders have reported the greatest need for data on:

- traffic/congestion level
- the noise level generated by the commercial vehicle
- days and hours of deliveries
- length of trips

- time spent within the city area
- pollutant emissions by vehicle type
- the volume of freight brought into the urban area
- total deliveries
- the places, where the trip starts
- travel times

However, adequate amount of available data resources could be indicated only with respect to data such as length of trips, the places, where the trip starts and travel times. Importantly, the greatest need for these data was reported by local governments, but for obvious reasons, these data can be shared mainly by users (carriers, logistics companies, customers/recipients). The availability of other data categories, in particular traffic/congestion level and the noise level generated by the commercial vehicle in the surveyed population is extremely minimal.

It should be emphasized that this study allowed to highlight significant deficits in terms of data having on the UFT functioning among local governments and even critical situation in respect of the entities that supervise infrastructure and road safety in cities (in this case, none of the entities surveyed indicated having data resources with respect to as many as 37 data categories).

4 Conclusion

This article is a summary of certain stage of studies to determine information needs within UFT. The study aims to contribute to the development of generalized data flow model in the integrated UFT management. The first stage of the study was focused on carrying out the questionnaires among entities involved in the UFT management and determining the significance of the hierarchization of the needs for specific data resources and their potential sources using the method of two-dimensional ranking. The authors made data systematization in terms of areas, indicators and categories. The indications obtained, with the use of the method of two-dimensional ranking, directly contributed to the presentation of the disproportion between data needs and data having among the studied population. This is a very important and significant effect. The determination of information gaps on the one hand, and the data redundancy on the other hand, provide accurate process image in the further development of the project.

It should be emphasized that the study has interdisciplinary nature. The target model of integrated data flows in the UFT management system will be supported by data flow diagrams and object-oriented methodologies. The authors are heading towards the development of analytical models (reality description level) and design models (interaction between the classes of the developed model), which will contribute to the modeling of the expected system.

Acknowledgements. This paper was financed under the project "Analysis of information needs of heterogeneous environment in sustainable urban freight" by the Polish National Science Centre, decision number DEC-2012/05/B/HS4/03818.

References

1. Mikulski, J., Kwaśny, A.: Role of telematics in reducing the negative environmental impact of transport. In: Mikulski, J. (ed.) TST 2010. CCIS, vol. 104, pp. 11–29. Springer, Heidelberg (2010). doi:10.1007/978-3-642-16472-9_2
2. Leonardi, J., et al.: Best practice factory for freight transport in Europe: demonstrating how 'good' urban freight cases are improving business profit and public sectors benefits. Procedia Soc. Behav. Sci. **125**, 84–98 (2014). Elsevier
3. Raicu, R., et al.: predictive models for routing in urban distribution. Procedia Soc. Behav. Sci. **125**, 459–471 (2014). Elsevier
4. Bekta, T., Crainic, T., van Woense, G.T.: From Managing Urban Freight to Smart City Logistics Networks. CIRRELT, Montreal (2015)
5. Browne, M., et al: Urban freight consolidation centres. Final Report, University of Westminster (2005)
6. Olsson, J., Woxenius, J.: Localisation of freight consolidation centres serving small road hauliers in a wider urban area: barriers for more efficient freight deliveries in Gothenburg. J. Transport Geogr. **34**, 25–33 (2014). Elsevier
7. Paddeu, D., et al.: Reduced urban traffic and emissions within urban consolidation centre schemes: the case of Bristol. Transp. Res. Procedia **3**, 508–517 (2014). Elsevier
8. Triantafyllou, M.K., Cherrett, T.J., Browne, M.: Urban Freight Consolidation Centers Case Study in the UK Retail Sector. Transp. Rese. Rec. TRB **2411**, 34–44 (2014)
9. Allen, J., et al: Freight quality partnerships in the UK – an analysis of their work and achievements. Report produced as part of the Green Logistics Project: Work Module 9 (Urban Freight Transport), University of Westminster (2010)
10. Lindholm, M.: Successes and failings of an urban freight quality partnership – the story of the gothenburg local freight network. Procedia Soc. Behav. Sci. **125**, 125–135 (2014). Elsevier
11. Taniguchi, E., Thompson, R.G. (eds.): City Logistics: Mapping the Future. CRC Press, Boca Raton (2015)
12. de Carvalho, J.M.C.: Systems theory, complexity and supply organizational models to erich city logistics: an approach. In: Taniguchi, E., Thomson, R.G. (eds.) Logistics Systems for Sustainable Cities, pp. 179–189. Elsevier, Amsterdam (2004)
13. Hillier, B.: The city as a socio-technical system: a spatial reformulation in the light of the levels problem and the parallel problem. In: Arisona, S.M., Aschwanden, G., Halatsch, J., Wonka, P. (eds.). CCIS, vol. 242, pp. 24–48. Springer, Heidelberg (2012). doi:10.1007/978-3-642-29758-8_3
14. Griffin, R.W.: Management (Polish Edition). PWN, Warszawa (2001)
15. Iwan, S.: Adaptative approach to implementing good practices to support environmentally friendly urban freight transport management. Procedia Soc. Behav. Sci. **151**, 70–86 (2014)
16. Public sector governance over urban freight transport. Report by Technical Committee B.4 Freight Transport and Intermodality, Paris (2012)
17. Iwan, S., Małecki, K.: Data flows in urban freight transport management system. In: Mikulski, J. (ed.) TST 2015. CCIS, vol. 531, pp. 1–10. Springer, Heidelberg (2015). doi:10.1007/978-3-319-24577-5_1
18. Taniguchi, E., et al.: City Logistics: Network Modelling and Intelligent Transport Systems. Pergamon, Oxford (2001)
19. Kaszubowski, D.: Application of benchmarking in the city logistics. In: Logistyka, vol. 5, pp. 1073–1082. ILiM, Poznań (2011)

20. Melo, S., Costa, A.: Definition of a set of indicators to evaluate the performance of urban goods distribution initiatives. In: Macharis, C., Melo, S. (eds.) City Distribution and Urban Freight Transport: Multiple Perspectives, pp. 120–147. Edward Elgar Publishing Ltd., Cheltenham (2011)
21. Iwan, S.: Implementation of Good Practices in the Area of Urban Delivery Transport. Scientific Publishing House of Maritime University of Szczecin, Szczecin (2013)

Design of M2M Communications Interfaces in Transport Systems

Malgorzata Gajewska[✉]

Faculty of Electronics, Telecommunications and Informatics,
Gdansk University of Technology, Gdańsk, Poland
m.gajewska@eti.pg.gda.pl

Abstract. In the paper the principle of M2M communications in transport systems is presented. The concept of M2M system architecture is considered. Characteristics of M2M communications for example data transmission, monitoring and other services for transport applications are presented. The problem of communications interfaces design for M2M applications is analysed. Next, the proposal of M2M communication interface for transport applications is considered. In the last part, the V2X communication is characterized as well as the 802.11p communication standard and the LTE-V system for implementation of this type of communication are presented.

Keywords: M2M · V2V · V2X · LTE-V · Road safety

1 Introduction

Automatic data exchange networks between devices, referred to as a communication M2M (Machine-to-Machine) or IoT (Internet of Things), is a technology which, according to specialists, in the coming years will be dynamically developed. The Ericsson company predicts that in 2020 year all over the world the 50 billion devices can use their own IP addresses, and they will communicate everyone to everyone as the M2M devices (in most cases). The other company predicts that in the year 2020, virtually any device will have access to a global network, and this applies to individual user equipment as well as all branches of the industry. This technique will certainly be used extensively in transport which may be one of the main beneficiaries of this technology. It is therefore important to analyse the interfaces M2M from the point of view of transport applications [8, 9, 11].

Generally, M2M is the technology of data transmission from one terminal to another terminal, such as machine to machine, sensor to sensor, mobile to machine but also man to machine. Especially, strongly we will see the development of M2M applications for use in transport system telematics. This application is often characterized by the absence of a human decision maker directly. But we have M2M applications in the transportation sector in which also require interaction with drivers. These M2M devices communicate with each other using great set of transmission technologies, e.g.: cellular networks, satellite networks or local networks (based on Wi-Fi, Bluetooth, ZigBee, and Radio

© Springer International Publishing AG 2016
J. Mikulski (Ed.): TST 2016, CCIS 640, pp. 149–162, 2016.
DOI: 10.1007/978-3-319-49646-7_13

Frequency Identification (RFID). The goal of M2M is to make machines capable of networking and communicating [8].

2   M2M-Based Model for Communication in Transport

Figure 1 shows the network communication without M2M between networks and the Internet (the dashed line) and new network structure with the M2M- based communication transport interface. A signal which is transmitted from the terminal side to the application side is sent by M2M platform [13].

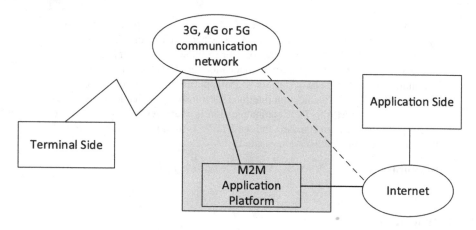

Fig. 1. M2M-based communication transport model [13]

If we use M2M communications we are able to adjust the operation of the network to the private needs of the user. Generally, the M2M-based communication mode enables efficient data transmission within the extensive monitoring system which can be implemented as a private network of high transmission rate. Traditional methods of connection, typically used in radio communication networks, are not acceptable for long-time connections with high transmission rate because of performance problems. And this is the main reason for the implementation of M2M platforms supporting future extensive monitoring applications.

M2M Application Platform enables independence from generally accessible networks and allows personalized services. Many users will be able to use the same IP address and simultaneously pursue a completely different service. As a result, it will be able a peer connection between more numbers of users than ever before. Additionally, the combination of their information is possible and its collection into one unit. This unit will again be used by someone else. In this regard, it will double the number of devices (e.g. GPS modules), and, as a result, this significantly decreases their price, which in turn causes more people to benefit from this.

Certainly, the M2M technology will repeatedly increase the amount of services that can be offered to individual customers, corporate but also expand the possibilities of action of the services responsible for our security. Unfortunately, the process also increases surveillance of people. However, this can't be avoided, which is confirmed by the assumptions for the implementation of the 5G network [11, 13].

3   M2M Transport Applications

The M2M may be used in transport systems telematics for different areas. Most significant are the applications as follows [8, 11]:

- monitoring of cars, trains, ships and aircrafts,
- vehicle fleet management by the company,
- system diagnostics - study of vehicle systems and detecting irregularities,
- system to the automatic reporting – e.g. system to the automatically detection of the fire in the car,
- control of equipment rented by leasing companies,
- vehicle-to-vehicle communications, especially for safety applications,
- systems to the security management in the train or other means of transport,
- automatic measurement and analysis of emission CO_2 – ecological problem,
- service management and holding of the transporting systems,
- remote measuring and controlling systems.

Moreover, M2M technology may be used to the control traffic lights or lighting cities, remote management of installations for the collection of fees for parking etc. For the another plays M2M may be used to the eCall, tracking stolen vehicle, remote diagnostics, fuel management, remote reading level generated fume but also to the transmitting information to the traveling about traffic jams, gasoline prices, the optimal route (after taking into account traffic information).

In Fig. 2, we can see the model of typical network with sensors and M2M platform. For M2M intelligent devices include sensors:

- devices to monitor the movement of vehicles,
- video cameras,
- meters energy consumption,
- smoke sensors,
- thermometers,
- flow meters,
- sensors presence,
- sensor to the monitoring of battery level,
- sensor to the measurement power consumption,
- sensor to the localization, monitoring vehicles speed and time of day (because in M2M we have the ability to run applications such as PAYD (Pay-as-you- drive) and PHYD (Pass-how-you-drive),
- sensor for industrial automation,
- sensor to the controlling of goods in containers on the ships and on the terminal,
- sensor to the automatically pay pass.

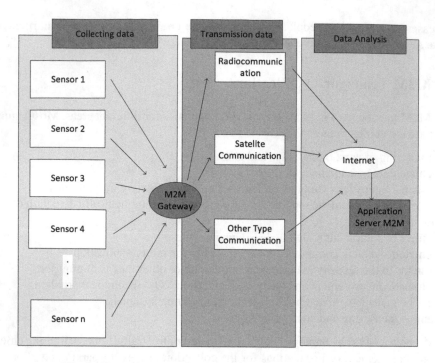

Fig. 2. Typical network with sensors and M2M [own study] [8]

As one can see, the M2M technology is a very valuable solution for transport systems. Many of the proposed solutions, M2M can help to improve road safety and help drivers. In addition, M2M solutions can indirectly contribute to an increase in crime detection because it will be possible to efficiently monitor multiple vehicles across the globe and the current analysis of acquired data [8, 11].

4   High Level Architecture for M2M

The high level architecture for M2M is presented in Fig. 3. This architecture includes two parts: the Network Domain and the Device and Gateway Domain. Network Domain is composed of the following elements [2, 3]:

– Access Network – is a network which allows the communication of Core Network with M2M Device and Gateway Domain. Communication interface between this layers can be implemented using different radio access networks as: UTRAN, eUTRAN (LTE), W-LAN, WiMAX, satellite network, GERAN and all future networks (5G).
– Core Network – is a network which provides, for example, IP connectivity, service and network control functions, interconnection functions (with other networks), and roaming.

- M2M Service Capabilities – this layer of the presented architecture could be shared by different applications and use Core Network functionalities as well as expose functions through a set of open interfaces.
- M2M Applications – application which run the service logic and use M2M Service Capabilities accessible via an open interface.
- Network Management Functions – this block consists of all functions required to manage the Access and Core Network.
- M2M Management Functions – this part of architecture consists of all the functions required to manage M2M Devices and Gateway uses a specific M2M Services Capability.

Device and Gateway Domain is composed of the following elements [2, 3]:

- M2M Device – is applied for using M2M Services Capabilities, and it is a device that runs M2M Application using M2M Service Capabilities. It must be noted that we have two methods for connect M2M Devices and Network Domain – for the first method the M2M devices are connected to the Network Domain via the Access Network (Direct Connectivity), for the second method this connection is realised by M2M Gateway (Gateway as a Network Proxy server).
- M2M Area Networks – provides connectivity between M2M Devices and M2M Gateways.
- M2M Gateway – a gateway which run M2M applications using M2M service capabilities.

As we can see in Fig. 3 high level architecture for M2M needs the use of M2M Service Capabilities in Core Network which is controlled by M2M Applications. These are the functions of M2M implemented in a network.

5   Communications Interfaces Design for M2M Applications

Communication between the M2M platform and application side is performed using special protocol dedicated to the M2M application – Wireless Machine Management Protocol. For the M2M communication the new kind of web application is defined which is needed to the realisation of transmission using the form of web service. Those services integrated in the web service can be discovered by any other web applications after it is deployed through standard internet protocols. Typical modern M2M network is composed of a group of devices that are equipped with the ability to transmit data. It is also necessary the linked realization of connections between the modules of the system and the host computer. Used program should provide an interface that allows the processing, storage and transmission of data. The data can be transmitted more frequently in one or two directions. Information can be transmitted from the individual network segments as control or broadcast information, and is transmitted through one central device. Additionally, this device can be used sometimes to update software. Transmission between the M2M network segments usually does not have a continuous - individual devices transmit data periodically, most of the time remaining on standby [13].

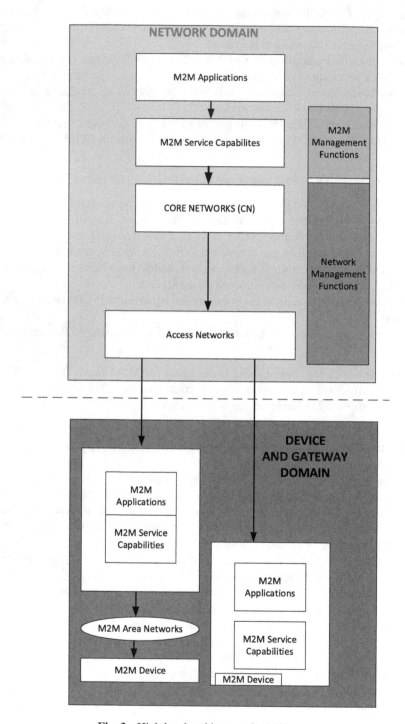

Fig. 3. High level architecture for M2M [2]

The proposal of communication interface design to the M2M application is presented in Fig. 4. For the first step, application starts with the connection. Users log to this application and get the session key number. On the next step, application side keeps this connection online and on the same time start listening data from the terminal side and call the web service to realized instructions sending to the target terminal [13].

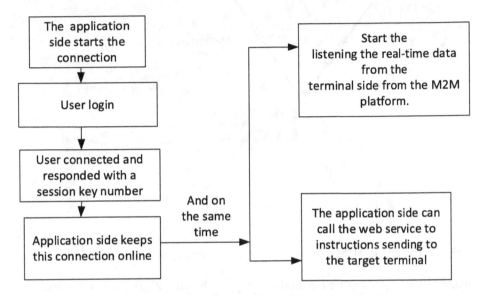

Fig. 4. Communication interface [own study] [13]

6   The Proposal of M2M Communication Interface for Transport Applications

Communication interface for transport systems should vary depending on the type of M2M communication. It seems necessary to develop different interfaces for communication machine to machine, sensor to sensor, mobile to machine but also man to machine. It seems to be reasonable to create separate interfaces for the case when we collect data from multiple users (e.g. the monitoring of fleet vehicles, gasoline prices, traffic jams), and otherwise when we will monitor the data to the one object (for example – one vehicle). In Fig. 5 the example of M2M communication interface for transport application is shown.

M2M Gateway control unit receives signals from sensors mounted in the truck through (e.g. GPRS). Then transmits this data to the server of M2M application, which analyzes them and, if necessary, announces an alarm and sends this information to the driver of a truck. To collect information from the sensors we can use e.g. ZigBee standard. The specification of this standard mainly includes simple networking solutions connecting small devices with low power consumption. The emphasis was also on the reliability of transmission, network configuration flexibility and to simplify the implementation of the protocol stack support. Transmission is used, among others, in 2.4 GHz

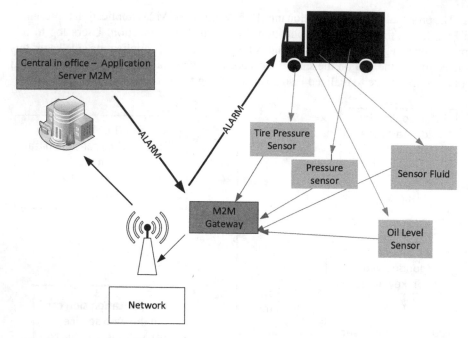

Fig. 5. Example of M2M communication interface in transport application [own study]

band, which does not require a central access point, because ZigBee network can create a mesh topology (the other type starry, P2P, etc.).

Note, that the information form sensors can be known in quasi-real time not only by a driver. This information is permanently monitored in a company what gives great benefit from the point of view of personal safety and the cargo safety.

7   V2X Communication as an Example of M2M

A special type of M2M application is communication using, as intermediary points in transmission, different modules of the road infrastructure e.g. cars. By doing so, you can develop systems for advanced traffic management vehicles, aimed at primarily improving of road safety and ride comfort. It becomes possible to implement control systems of vehicles (equipped with a number of different sensors and control systems) in such a way as to minimize the risk of collision. Also, the development of this technology allows wireless charging road tolls, traffic monitoring, and automatically modify the operation of traffic lights and most importantly - to warn motorists of the dangers of traffic.

The concept of such communication, using cars as intermediate nodes, was founded in the 90 s and is known under the name the AHS system (Automated Highway). Its rapid development begins form so-called Vehicle Infrastructure Integration Initiative in 2003, and from the act on the basis of which was to be the construction of automated intelligent system prototype for vehicles. This system should be equipped with the appropriate innovative solutions. Such a system would allow them to ride the highway in an efficient, safe and most importantly predictable [12].

At present, the types of this kind of radio communication, which are scheduled for execution until 2024, are known under the names: V2V radio communication (Vehicle-to-vehicle), V2I (Vehicle-to-Infrastructure), V2P (Vehicle-to-Pedestrian), V2H (Vehicle to Home), IN-V (In Vehicle Communication) and so-called Car-to-X communication. All of these ideas can also be found under the V2X heading (Automotive Vehicle-to-Everything), i.e. automobile communication vehicle with any element of infrastructure [1, 12].

The V2V communication means radio exchange information directly between vehicles. V2I is to use the possibility of exchanging information between the cars, and the surrounding infrastructure. The V2P communication is made between cars and pedestrians, while V2H is data transmission between vehicles and buildings. In the In-V mode is the ability to transmit information between various sensors and control devices, mounted on the vehicle. So that, it is possible e.g. parking the vehicle without a driver.

Very important from the point of view of road safety, is Car-to-X radio communication system, which helps inform the driver in advance of dangerous situations. For example, you can be warned about traffic jams on the road or a driver of another vehicle can be informed about vehicles approaching moving against the tide or upcoming privileged (including those that are close to our vehicle from the rear, and at the intersections of the side of the road). Network security is based on the principle that each vehicle equipped with this system can simultaneously and independently send and receive warning signals. These signals are transmitted, e.g. via WLAN-based IEEE802.11p/GSA protocol [6, 7].

V2V communication can prevent collisions at several common situations. For example in the situation two vehicles travel in the same direction. Then they can exchange information via V2V and if it occurs, too quickly approaching of the vehicle to a preceding individual, there will be sent warning signals. And in the absence of driver reaction starts emergency braking procedure. V2V communication is also important when there is not good visibility due to truck driving between small cars. Then both drivers of small cars do not see each other but V2V system send signals to him giving the knowledge of its existence. In this case the truck communicates with both two small cars. V2V communication is suited also to prevent collision in the situation when car wants to change a lane but second car travels at a neighboring lane, e.g. second car can be located in so-called "deadlock". Then, it could not be seen in a mirror. Figure 6 shows another situation, when you the V2V communication can be helpful. One can see that vehicles have not any information of their existence because they do not have direct visibility of each other. But using the V2V communications, this information can be sent between cars. Such knowledge is particularly important e.g. when one of vehicles is the emergency vehicle moving at a high speed. Each vehicle uses the information of the surrounding vehicles in order to determine the likelihood of a collision, and alert the driver if necessary.

Fig. 6. Example of V2V communication [own study]

V2V applications communicate using radio signals. The signals are transmitted via antennas of omni-directional characteristics of radiation (360°). Communication using these signals makes possible to obtain information about the presence of other vehicles (as well as e.g. their position or speed) on the road, even in the absence of direct visibility.

At the moment, it is assumed that the V2V communication should take place between vehicles within a range of up to 300 m. In this case you can identify potential threats that can lead to disaster if there is no any action of a driver or vehicle. In addition, the V2V system is intended to operate regardless of the weather, the way of lighting, but also regardless of the purity of vehicle sensors and the number of these vehicles. The latest solutions also provide information exchange between the vehicles on relative angle between them, regardless of turns, e.g., obstacles or buildings [10].

For communication between vehicles, as well as communication between vehicles and infrastructure, the communication standard IEEE 802.11p is defined (Wireless Access in Vehicular Environments - WAVE). It is an extension of 802.11 (Wi-Fi), dedicated to intelligent transport systems [6, 7].

The 802.11p standard defines communication channels of 10 MHz bandwidth, and the system works in the band of 5855 GHz to 5925 GHz. Additionally, it is assumed the 5 MHz guard band at the beginning of an allocated frequency band [4, 6, 7]. The plan of available radio communication channels is shown in Fig. 7.

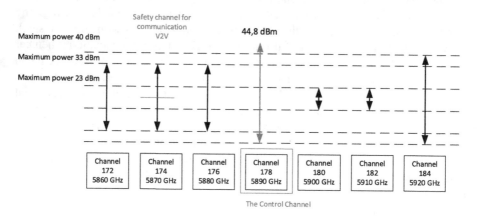

Fig. 7. Channel planning for the 802.11p standard [own study]

Available channels are different for very specific applications, e.g. 5.865–5.875 GHz channel is the channel of safety (Public Safety V-V Channel), in which we have the ability to transmit information about the accident or life-threatening road and 5.885–5.895 GHz is the channel to send control messages (CCH - Control Channel). It should be noted that the medium access control layer (MAC - Medium Access Control) requires that the control messages have priority defined - from 0 to 3 [4].

In addition, for each channel the maximum power level of signals transmitted is specified. Currently V2V communication uses two radio links. One of which is tuned to the 172 channel (see Fig. 7), which hosts a secure communication, and another is tuned to the 174 channel, where information is given to security. In addition, the 178 channel is used as a control channel. It allows the channels switching for channels used for transmission of some messages according to the provided services [4].

An important difference between 802.11p and other Wi-Fi standards is no need to establish so-called BSS connection (Basic Service Set), which allows transmission of signals in the channel right away, allowing for short-term communication with a relatively small delays. This is important because messages must be received by the drivers as soon as possible, so as to give them a chance to react in an emergency. It is also important that the device switches between channels in such a way that the channel informing about the risks can be monitored no less frequently than every 100 ms. The technique used to multiple channel access is OFDMA (Orthogonal Frequency Multiple Access), and used modulations are: BPSK, QPSK, 16-QAM and 64 QAM [5, 7].

Practical research and analysis, published in November 2015, carried out jointly by NHTSA (National Highway Traffic Safety Administration) and ITS JPO (Intelligent Transportation Systems Joint Program Office), indicate that the use of V2X communication system may help to prevent 80% of collisions involving two or more vehicles. At present, world automobile giants are interested in implementation of this technology in their vehicles, but also in searching of appropriate technologies and equipping its elements the cyclists, motorcyclists, and pedestrians. The main objective is the equipment we say in general road infrastructure, to eliminate minimum emergency situations (i.e. those in which do not work any element of infrastructure) and thereby reduce the number of accidents (especially as a result of which they are victims - injured or fatalities). Of course, the efficiency and effectiveness of V2X methods will depend inter alia on the speed of the moving vehicle.

8 LTE-V to V2X Communications

In December 2015, the 3GPP organization published technical report TR 25.885 (Release 14) [1] for use of LTE for the implementation of V2X communication services, namely to improve communication between vehicles, vehicles and parts of road infrastructure, as well as between vehicles and terminals carried by a pedestrian or cyclist. Table 1 shows the estimate of transmission parameters and performance of V2X communication for many different environments, which are present in moving vehicles.

There are distinguished: suburban environment, highways, motorways, urban environment (the lack of direct visibility), the intersection in the city and parking lots at shopping centers.

Table 1. Example for the V2X transmission parameters for different types of environments [1]

Environment	Effective distance	The absolute speed of device	The relative speed between two devices	The maximum acceptable delay	The probability that the recipient will get a message within 100 ms at an effective distance	Expected transmission reliability
Suburban environment	200 m	50 km/h	100 km/h	100 ms	90%	99%
Expressway	320 m	160 km/h	280 km/h	100 ms	80%	96%
Highway	320 m	280 km/h	280 km/h	100 ms	80%	96%
Urban environment (NLOS)	150 m	50 km/h	100 km/h	100 ms	90%	99%
Crossing urban	50 m	50 km/h	100 km/h	100 ms	95%	–
Retail park	50 m	30 km/h	30 km/h	100 ms	90%	99%
The inevitable disaster	20 m	80 km/h	160 km/h	20 ms	95%	–

The table identifies the probability of reliable transmission for the movement speed of vehicles, typical for each of these environments. The table also shows a set of transmission parameters at which it is possible to avoid collision of vehicles.

As shown in Table 1, the efficient operating distance is greater than the extent required to support TTC = 4 s (Time to Collision) at maximum relative speed. It results from the fact that it seeks to such a choice of parameters required for V2X communication what increase the reliability of transmission. In Table 1 are also set out the likelihood that the recipient will receive a message during 100 ms, at an effective distance.

LTE-V allows you to define well-outgoing transmission variants for a variety of services not-implemented to date, such as [1], e.g.:

– Adaptive cruise control - controlled by other road users, depending on the intensity of traffic and such weather conditions.
– Remotely generated warnings about traffic jams, which in many cases can be a potential threat.
– Automatic Parking System (APS) - includes up-to-date upgraded base station, in which real-time data is collected for vehicles in metropolitan area, e.g.: the availability of parking spaces on the street or in public garages. It is planned that the APS enables the driver seat reservation parking space, will lead him to this place for after-intermediary navigation application and able to make the parking fees.

Also, interesting is the use of LTE-V to perform V2X communication in areas beyond the reach of the network, as shown in Fig. 8. This type of communication is carried out in such a manner that both the A and B vehicles are equipped with V2V communication support. Both vehicles are in the area that is not supported by the access of E-UTRAN network (UMTS) supporting V2X communication.

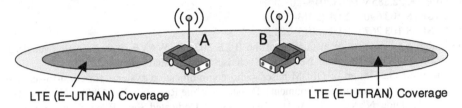

Fig. 8. Implementation of V2X transmission out of network coverage [1]

But V2X equipment in cars is preconfigured on the basis of parameters of existing E-UTRAN. First, V2X communication begins when both two cars are in E-UTRAN coverage area. At the moment, there are transmitted messages related to road safety through E-UTRA radio, using pre-configured parameters. If these two cars leave the area of E-UTRA coverage, then they lose their coverage and approach each other. They start direct V2X communication, and thus, they are able to detect the signals to each other, connected with road safety. So this is a completely new possibility of establishing communication between vehicles [1].

9   Conclusion

The M2M is the communication of the future, especially dedicated for transport systems telematics. It's obvious that the concept of M2M is very important from the point of view of industry development, in which the transport plays one of main roles. As we can see in this paper, the M2M communication is especially interesting for monitoring systems of great scale, in which a lot of information can be sent to different places. So, M2M gives possibility the simple realization of a number of additional services which cannot be used at this time. Important is that M2M communication interfaces in transport systems are particularly preferred because we have many different solutions contacted with this technology. The interest of people and companies these solutions will grow rapidly. So it's very interesting topic. Assumptions of the 5G radio communication system show that M2M communications will be a large group of proposed solutions. Moreover, the V2X communication is presented in which the M2M transmission is made between many vehicles and between vehicles and various other elements of road infrastructure. Using this method of communication will greatly improve traffic safety by eliminating a large part of vehicle collisions. The study also presents a new LTE-V standard for radio transmission to perform V2X communication, which is currently standardized and prepared for deployment. It is already known that it will be another revolution in communication because it will be possible shaping characteristics of the system to ensure the reliability of messaging warning and control, with a danger on the road and thus avoid a collision.

References

1. GPP TR 22.885 v 14.0 (2015–2012)
2. ETSI TS 102 690 v 2.1.1 (2013–2010)
3. ETSI TS 103 267 v 1.1.1 (2015–2012)
4. Eihler, S.: Performance evaluation of the IEEE 802.11p WAVE communication standard. Institute of Communication Networks, Technische Universität München
5. Gajewski, S.: Design of OFDM-based radio communication systems for coast-to-sea and coast-to-air propagation environments. Polish Maritime Research, No. 1(89), vol. 23 (2016)
6. Guo, J., Balon, N.: Vehicular Ad Hoc Networks and Dedicated Short-Range Communication (2015)
7. IEEE Standard for Information Technology – Telecommunications and Information exchange between systems – Local and metropolitan area networks – Specific requirements, Part 11 Wireles LAN Medium Access Control (MAC) and Physical Layer (PHY), Specification "Amendment 6: Wireless Access in Vehicular Environments" (2010)
8. Konreddy, R.: 4G+IPv6=Rewolucja M2M, Elektronika praktyczna (2013)
9. Liu, L., Gaedke, M., Koeppel, A.: M2M interface: a web services – based framework for federated enterprise management. In: Proceedings of the IEEE ICWS (2005)
10. Statement of Nathaniel Beuse: Associate Administrator for Vehicle Safety Research, National Highway Traffic Safety Administration, Before the House Committee on Oversight and Government Reform Hearing on "The Internet of Cars", November 2015
11. U.S. Department of Transportation: Connected Vehicle Insights, Trends in Machine-to–Machine Communications, Technology Scan and Assessment Final Report, October 2011
12. Vehicle-to-Vehicle Communications, Readiness of V2V Technology for Application. DOT HS 812 014, NHTSA August 2014
13. Yanlan, Y., Hua, Y., Shumin, F.: Design of communication interface for M2M-based positioning and monitoring system. IEEE (2011)

Practical Problems Within Safety Related Cryptography Communication Systems Assessment for Safety Critical Applications

Mária Franeková[✉], Karol Rástočný, and Peter Lüley

Faculty of Electrical Engineering, Department of Control and Information
Systems, University of Žilina, Žilina, Slovak Republic
{maria.franekova,karol.rastocny,
peter.luley}@fel.uniza.sk

Abstract. The authors of this article are focus on the analysis of safety-related (SR) communication system with open transmission system for safety-related applications with increasing SIL (Safety Integrity Level). The main part is orientated to summarization of practical problems within safety assessment of communications on the base of mathematical model of SR communication with using cryptography code of several lengths for keeping of message confidentiality and integrity, which is transmitted across open communication channel. The experimental part analyses SR communication in the end-to-end (between two nodes), with assumption of using cryptography and safety codes. Obtained results of probability of error of cryptography code are determined for model of BSC (Binary Symmetric Channel). In addition the cryptography degradation of block cryptography codes with using digital modulation techniques are discussed.

Keywords: Safety-related communication system · Open transmission system · Cryptography code · Safety code · BSC channel · Modulation techniques · Safety assessment · SIL

1 Introduction

During the development of the communication system which is part of the application participating in the control of safety-related processes it is necessary to take into account the requirements of generic standards for commercial - so-called COTS (Commercial Off-The-Shelf) systems (e.g. GSM, WLAN or Internet) but it is also necessary to consider standards for systems which failure can result in damage to human health, environmental damage or in significant material damage.

The design of safety-related communication systems must be based on the risk analysis (analysis of hazards and their consequences) related to the dangerous failure of the safety function [1]. Based on the risk analysis results there must be defined the safety requirements for the communication system (mainly the functional requirements and the safety integrity requirements) and based on those requirements there must be proposed measures to reduce the risks to an acceptable level. Records on such risk reduction must be performed. While the evaluation of the functional properties of the communication systems is based mainly on the qualitative methods (e.g. based on tests results evaluation

© Springer International Publishing AG 2016
J. Mikulski (Ed.): TST 2016, CCIS 640, pp. 163–174, 2016.
DOI: 10.1007/978-3-319-49646-7_14

or measuring's results evaluation), the safety integrity evaluation must be based mainly on the quantitative methods based on the probability theory [3, 4].

Existing standards (e.g. [2]) usually contain information about what needs to be done in order to demonstrate the safety properties of the communication systems but generally do not provide clear guidance on how it should be achieved. Mentioned procedures for the safety integrity evaluation are only informative and they usually recommend wide range of usable methods. What method respectively what combination of methods will be used is up to the evaluator. This decision depends on particular application and on evaluator's personal experience. Lack of uniformed methodology for the communication systems safety assessment creates opportunity for subjective interpretations of obtained results. At present, there is missing generally accepted theoretical apparatus for the communication systems safety integrity evaluation that will objectify the whole evaluation process. Because of this reason this issue receives attention in various international forums [5]. Mutual information exchange leads to gradual unification of views which has a positive impact on the mutual acceptance of evaluation results.

Solving the problem related to the absence of a uniformed methodology for the communication systems safety assessment is even more urgent for the use of very perspective open communication platforms in the industry or in different applications. If there is requested an open communication system with high safety integrity level (SIL) [1], then it is necessary to employ the cryptographic mechanisms to eliminate the risks caused by intentional attacks. During the methodology development for the safety integrity assessment of open communication systems with cryptographic mechanisms there must be considered the definition of open transmission system, types of attacks on messages transmitted via open communication systems as well as properties of existing safety mechanisms in the field of cryptography [6]. Good starting position can be provided by the experiences in the closed transmission systems safety integrity assessment, particularly in the field of railway applications, where has the use of systems with high SIL long tradition [7, 8] and it is supported by [2]. Many other standards intended for the industrial applications (e.g. IEC 61784-3 [9]) are based on the recommendations of [2]. According to [2] the open communication system may include an unknown number of users who work with safety irrelevant devices and also with safety-related devices. During the design phase of such communications system are the users unknown and the transmission system transfers unknown amount of information in unknown format. Transmission media and their characteristics are unknown to the users and transmission system is routing messages through one or more types of transmission media. The safety-related messages must be transferred with guaranteed authenticity, integrity and timeliness with the use of cryptographic techniques.

2 Approaches in the Communication Safety Analysis and Practical Problems

There exist standards for the qualitative safety assessment of the cryptography applied in the commercial (standard) areas such as banking sector or enterprise information systems. An example is FIPS 140-2 [10]. Methodologies in the field of cryptography applications for the safety assessment of systems with high safety integrity level such as railway transport or industry absent respectively there are none.

Application of cryptography in communication between safety-related devices in safety-related communication system can be designed in such a way that the cryptography technique is a part of the safety protection of particular safety-related device (it is implemented in the access protection layer) or the cryptography technique is a part of the safety protection of several safety-related devices that communicate within internal industrial network and it is implemented in a separate safety protection layer (more enforced approach, e.g. firewall included in the safety policy of the safety critical application).

As an example let's assume the safety-related communication in open communication system as shown in Fig. 1.

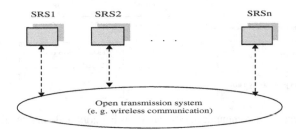

Fig. 1. Safety-related communication in open communication system [own study]

We assume the communication protocol of standard open transmission system supplemented with two safety-related layers. These layers are parts of every SRS (safety-related system):

- Transmission protection layer – performs safety functions related with the control of message integrity which can be corrupted during the transmission by the influence of EMI. This functions are in most cases performed by safety mechanism of the safety code;
- Access protection layer – performs safety functions related with the access control to the safety-related message. These functions are performed most often by the safety mechanisms of cryptographic techniques by which can be monitored the massage authentication or the message confidentiality.

This article is further focused on the issue of message confidentiality monitoring in the communication between two nodes as shown in Fig. 2. Attention is paid mainly on the calculation of the cryptographic transmission error probability. This is just a part of the issue related with safety integrity assessment of safety-related open communication systems.

If we want to perform communication system safety analysis for scenario arising from Fig. 2, first it is necessary to determine attributes in terms of safety analysis on which we want to focus.

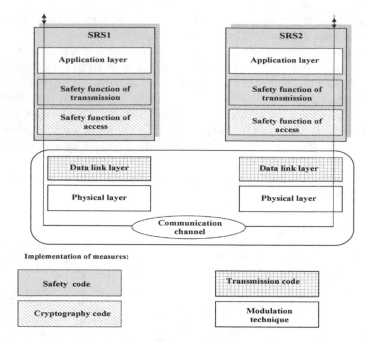

Fig. 2. Safety scenario of communication focused on integrity and confidentiality [own study]

In the development process (including the safety evaluation) of the safety-related communication system with open transmission system is important modeling [11]. Depending on the stage of the system development and on what development targets are followed it is necessary to use suitable modeling method and modeling tool or combination of several methods and tools. Usually it is build an abstract model which graphically or mathematically describes the important properties of the communication system. In most cases are during development of open communication systems modeled following topics:

Modeling of functional characteristics of communication protocol
It is recommended to create a functional behavior model of the system realized on the basis of semi-formal and formal methods (they are generally supported by software tools).

Modeling of interferences in the communication channel
EMI is the result of a number of different effects, that cannot be deterministically described, therefore the models of communication channels are based on probabilistic characteristics. Modeling should be based on the communication channel model. Preferred is the BSC (Binary Symmetric Channel) respectively AWGN (Additive Gaussian White Channel) model. It is also necessary to consider safety mechanisms of safety code (located according to Fig. 2 in the transmission protection layer) and transmission code (located according to Fig. 2 in lower parts of communication protocol).

Modeling of the random failures consequences on the safety of open transmission system

The aim of modeling of random failures consequences on the system safety is to create a model that allows calculation the probability respectively intensity of undetected corruptions of transferred messages. To analyze the consequences of random failures can be used BRD, FTA, FMEA, Markov processes, Petri nets respectively appropriate combination of these methods.

Authors were aimed to determine the mathematical apparatus to calculate the undetected error probability of used cryptographic code depending on the bit error rate of the communication channel (which is defined on physical layer by the noise characteristics of communication channel, by the type of modulation techniques and by its error rate). Calculation of cryptographic code error probability is in the article presented in combination with safety block code for two scenarios of transmission confidentiality: by the use of steam ciphers and by the use of block ciphers.

3 Calculation of Cryptographic Transmission Error Probability

Let's assume that the safety layers of transmission protection in SRS systems shown in Fig. 2 are realized with the use of safety code CRC-r, where r represents the number of redundant bits attached to the message. Let's assume the confidentiality issue of safety-related messages transferred between SRS 1 and SRS 2 is solved by the safety layer of access protection. This service is physically realized by massage enciphering that means by their transition from understandable form to their obscure form. According to standards [2] are for the safety critical applications recommended computationally secure block ciphers. Steam ciphers are not recommended for this purpose because of their lower safety. In the practical part of this article is therefore the cryptographic code error probability calculated for block ciphers but its determination is based on the derivation of cryptographic code error probability of steam cipher.

Let's mark the bit error probability of BSC model as P_b. Assuming that there is not used safety or cryptographic code in the transmission system it is possible to determine the word error probability as:

$$P_w = 1 - (1 - P_b)^k, \tag{1}$$

where k is the number of bits in transmitted word. In case the confidentiality of the transmitted word was secured by the use of steam cipher (where is enciphered steam of previous bits) all transferred bits in the word are affected by the same bit error probability P_b. Enciphered word error probability is the state when the output word contains one or more erroneous bits. Messages transmitted in SRS are (on the transmitting end) either enciphered continuously (in the case of steam cipher) or divided into code words (in the case of block cipher) but in both cases can the error occurrence lead to error chaining. Therefore in the event of error occurrence will be influenced also following n bits on the output of the cryptographic code decoder.

Cryptographic code decoder output world error probability when enciphered by the use of steam cipher P_{cw-s} (assuming BSC) can be according to [12] determined as:

$$P_{cw-s} = P(w,t) + P(w|\bar{t})P(\bar{t}), \tag{2}$$

where $P(w|\bar{t})$ is conditional error probability in transmitted word if there is at least one error in the transmitted word and $P(\bar{t})$ is the output word error probability when enciphered by steam cipher. Input word will be affected by the noise in case if at least one from $n-1$ input bits of previous input world has been erroneous due to noise in communication channel. If we assume the error independence then it is valid:

$$P(\bar{t}) = (1 - P_b)^{n-1} \tag{3}$$

and for the conditional error probability of transmitted word if there is an error affecting the transmitted word it is valid:

$$P(w|\bar{t}) = (1 - P_b)^k. \tag{4}$$

In case of burst errors with the length $tb = i$ can be $P(w|\bar{t})$ determined as:

$$P(w,t) = P(w|tb = k)P(tb = k) + \sum_{i=1}^{k-1} P(w|tb = i)P(tb = i). \tag{5}$$

If at least one from $n-k$ bits of previous input word was erroneous and the condition $n > k$ is fulfilled (marked as $tb = k$) then it is valid:

$$P(tb = k) = \begin{cases} 1 - (1 - P_b)^{n-k}, & n > k; \\ 0, n \leq k. \end{cases} \tag{6}$$

For $tb = 1$ and $1 \leq i < k$ is necessary that the error occur right in $n-1$ bits prior the monitored word but no further erroneous bits among next $n-i-1$ bits occur. Therefore for the condition $1 \leq i < k$ applies:

$$P(tb = i) = \begin{cases} P_b(1 - P_b)^{n-i-1}, & n > i; \\ 0, n \leq i. \end{cases} \tag{7}$$

By the substitution of (1) to (7) into (2) we obtain the formula for the cryptographic code error probability for steam ciphers:

$$P_{cw-s} = P(w|tb = k)\left[1 - (1 - P_b)^{n-k}\right]u(n-k)$$

$$+ \sum_{i=1}^{min(k-1, n-1)} P(w|tb = i)P_b(1 - P_b)^{n-i-1} \tag{8}$$

$$+ \left[1 - (1 - P_b)^k\right](1 - P_b)^{n-1},$$

where $u(n - k)$ expresses the step function. ($u(n - k)$ is 0 for $n < k$ and 1 for $n \geq k$). The upper bound of cryptographic code probability can be obtained by the assumption $P(w|tb = k) < 1$ and $(w|tb = i) < 1$.

$$P_{cw-s} \leq \left[1 - (1 - P_b)^{n-k}\right] u(n - k)$$
$$+ \sum_{i=1}^{min(k-1, n-1)} P_b(1 - P_b)^{n-i-1} \tag{9}$$
$$+ \left[1 - (1 - P_b)^k\right](1 - P_b)^{n-1}.$$

After simplification (according to [11]) we obtain equation for cryptographic code error probability for the steam cypher:

$$P_{cw-s} \leq 1 - (1 - P_b)^{n+k-1}. \tag{10}$$

In case the transmission confidentiality is provided by the use of block cipher the transmitted message with the length n bits is on the input of cryptographic code encoder usually divided into words (blocks) of the same length k bits and then each block enciphered independently – in general as a block with a total length of n bits. After the transmission and reception is the text restored in the cryptographic code decoder. Output words from cryptographic code decoder can be considered as correct when the received enciphered block is without any erroneous bits. Assuming the errors independence we can determine the cryptographic code error probability for block cipher P_{cw-b}:

$$P_{cw-b} = P(w|be)[1 - (1 - P_b)^n], \tag{11}$$

where $P(w|be)$ is conditional error probability of the output word if error occurred in the block on the output of the cryptographic code decoder. Assuming $P(w|be) = 1$ and by using Eq. (12) according to [11]:

$$(n + k - 1)P_b \geq 1 - (1 - P_b)^{n+k-1} \tag{12}$$

we obtain:

$$P_{cw-b} \leq 1 - (1 - P_b)^n \leq nP_b. \tag{13}$$

Because the deciphered block on the cryptographic code decoder output depends on the cryptographic code encoder output the error received in enciphered block certainly causes at least one erroneous bit in the output block (on the cryptographic code decoder output). That means, for all block ciphers with the size of n bits there exist $2^n - 1$ equally probable outputs block which contain erroneous enciphered block. Let's look closer at every fixed bit in these output blocks. In $2^{n-1} - 1$ possible output blocks is this bit correct that means it is in the same state as it would have been in if no error had occurred in the enciphered block. If the error in block occurred, we assume ensemble-average probability of bit correctness equal to $(2^{n-1} - 1)/(2^n - 1)$.

According to [11] by repeating the analysis for sufficient number of output bits can be concluded that for k – bits word in simple enciphered block is the average probability of its error on the output (in case there exist erroneous block on input):

$$\bar{P}(w|be) = 1 - \prod_{i=1}^{k} \frac{2^{n-1} - 1}{2^{n+1-i} - 1} = \frac{1 - 2^{-k}}{1 - 2^{-n}}. \tag{14}$$

By the combination of this equation with (13) we gain ensemble-average cryptographic world error probability for block cipher:

$$\bar{P}_{CW-b} = (1 - 2^{-n})^{-1}(1 - 2^{-k})[1 - (1 - P_b)]^n \tag{15}$$

If we use the suitable cryptographic code for the cryptographic communication system safety increase, then it is possible to change the Eq. (15) under the conditions stated in [11] to:

$$\bar{P}_{CW} = (1 - 2^{-n})^{-1}(1 - 2^{-k})\left[1 - (1 - P_W)^{n/k}\right], \tag{16}$$

where P_w is the un-enciphered code word error probability and ratio n/k is the number of words in the message. Taylor-series expansion gives:

$$\bar{P}_{CW} \approx (1 - 2^{-n})^{-1}(1 - 2^{-k})\frac{n}{k}P_W. \tag{17}$$

4   Obtained Results

The calculation of the average value of cryptographic code error probability for block cipher $\bar{P}_{cw-b}$ was realized for the example of safety-related communication system shown in Fig. 2. Calculation followed these conditions: in the transmission protection layer has been implemented safety code CRC-32 with maximal value of un-enciphered code word undetected error probability equal to $P_W = 2^{-r}$ (worst case according to [2]), it is assumed the BSC channel (worst case of bit error $P_b = 0, 5$).

Results of average value of cryptographic error probability $\bar{P}_{cw-b}$ were calculated for various lengths of messages on the cryptographic encoder input from ($n = 0,1$ kB, to 500 kB) and for practical used lengths of blocks k (64b 128b, 256b). We focused on block lengths for commercially used block ciphers such as DES modifications (2DES of 3DES) respectively AES. Results of average value of cryptographic error probability for blocks with lengths $k = 64, 128, 256$ bits depending on the length of transmitted messages is shown in Fig. 3. We are orientated to comparison of average value of cryptographic error probability $\bar{P}_{cw-b}$ for two formats of messages which are used in praxis: short format with length of 0,1 kB and long format with length of 500 kB. The result is illustrated in the Table 1.

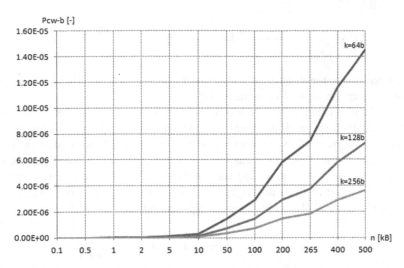

Fig. 3. Graphical representation of average probability of cryptographic code with the use of block cipher for various lengths of messages and various lengths of blocks [own study]

Table 1. Average value of cryptographic error probability for short and long formats of messages depending on the block length of cryptography code [own study]

Block of length k	$\bar{P}_{cw-b}$ Short format of message 0, 1 kB	$\bar{P}_{cw-b}$ Long format of message 500 kB
64 b	2,91E–9	1,46 E–5
128 b	1,46 E–9	7,28 E–6
256 b	7,28 E–10	3,64 E–6

Communication system is preferable to illustrate with graphical representation $P_b = f(\gamma)$, where P_b is bit error rate of communication channel and $\gamma = \frac{E_b}{N_0}$ is ratio of energy to spectral density of Gaussian noise per one bit. It means, it is also necessary to keep in consideration physical characteristic of communication channel within the type of applied modulation technique. The parameter γ is lower for defined bit error rate; the communication system is more effective.

Authors further assumed the untrusted open transmission system with the use of digital modulation technique M-QAM (*Quadrature Amplitude Modulation*) where the bit error probability of un-enciphered binary communication is a function of the average energy per one bit E_b.

$$P_b = f(E_b). \tag{18}$$

For M-QAM modulation can be the error rate of one bit calculated [12] according to equation:

$$P_{bM-QAM} = \frac{2}{log_2(M)} \left(1 - \frac{1}{\sqrt{M}}\right) erfc\left(\sqrt{\gamma \frac{3}{2} \cdot \frac{log_2(M)}{M-1}}\right), \tag{19}$$

where M represents the number of states, γ is the ratio E_b/N_0 [–] and *efrc* is the complementary error function of Gaussian noise.

If the calculated value of one bit error is substituted into the equation for the cryptographic code probability, then we gain the equations for the energy corresponding to a single bit E_b in un-enciphered and enciphered transmission calculation. By the comparison of these equations with (17) and (19) we can determine requested energy increase necessary for gaining the same cryptographic system error probability as provided by corresponding non-cryptographic system. This increase is expressed by qualitative parameter called cryptographic degradation (defined in decibels):

$$D = 10 \log_{10} E_{b1} - 10 \log_{10} E_b = 10 \log_{10} \left(\frac{E_{b1}}{E_b}\right), \tag{20}$$

where E_{b1} is energy necessary to produce a value of $\bar{P}_{cw-b}$ which is equal to the value of P_w if the energy value is equal to E_b. Bit error rate depending on E_b/N_0 for short (0,1 kB) and long (500 kB) formats of messages when used 64-QAM modulation is shown in Table 2.

Table 2. Results of probability of bit error rate when used 64-QAM (square constellation)

$\frac{E_b}{N_0}$[–]	5	15	25	40	50	100	150
P_{bM-QAM} [–]	8,41E–2	1,12E–2	2,27E–3	4,9E–4	1,41E–5	4,84E–8	2,7E–9

The worst case of average cryptographic world error probability for block cipher is calculated for $P_b = -2^{-1}$, when safety code error probability can be estimated by 2^{-r}, where r is number of redundant bits. If we use as safety code the CRC (Cyclic Redundant Check) code - type CRC-16, then $P_w = 2^{-16}$ and cryptographic world error probability for block cipher ($k = 128$ b) according to (17) is 1,75E–4 for short format of message ($n = 0,1$ kb) and for long format of message ($n = 500$ kb) it is 5,78E–2.

For the calculation of cryptographic degradation we assumed the most widely used type of wireless technology Wi-Fi. Standard IEEE 802.11n [13] uses more types of digital binary (B) or M-ary modulations for particular connections (BPSK, QPSK, 16-QAM, 64-QAM). The use of particular modulation depends on the parameter MCS (Modulation and Coding Scheme), which determines the modulation, coding and other signal and protocol parameters based on characteristic properties of every transmission such as interference, receiver sensitivity or others.

Results of cryptographic degradation depending on E_b/N_0 when used the 64-QAM modulation for the transmission of short message ($n = 2B$) and long message ($n = 250B$) is shown in Fig. 4.

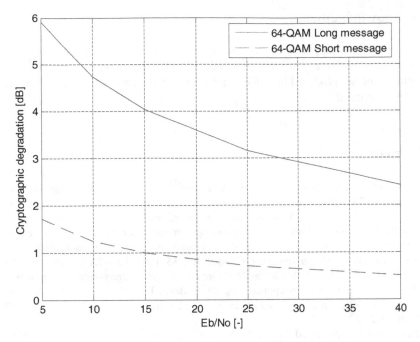

Fig. 4. Cryptographic degradation of 64-QAM modulation depending on E_b/N_o [own study]

5 Conclusion

The authors in this article dealt with methods for the quantitative evaluation of safety-related communication when is used cryptographic code in combination with safety code. The safety evaluation, limited to the services of confidentiality and integrity, was based on mathematical model of safety-related communication on the level of end devices. Based on the recommendations in the standard [2] was expected block cryptographic code with the length k (currently are preferred lengths $k = 64, 128$ bits) for which was expressed average value of cryptographic error probability depending on the length of transmitted message and on the communication channel error rate. From the results (Table 1) is clear that better values are achieved by the short format of message, what is more preferred model for the transmission of safety-related messages (short control commands with high priority). In addition, the authors calculated the impact of used modulation technique on cryptographic degradation. Presented results are for 64-QAM but it is possible to calculate results also for other types of modulation techniques.

6 Acknowledgements

This work has been supported by the Educational Grant Agency of the Slovak Republic (KEGA) Number: 008ŽU-4/2015: Innovation of HW and SW tools and methods of laboratory education focused on safety aspects of ICT within safety critical applications of processes control.

References

1. IEC 61508: Functional safety of electrical/electronic/programmable electronic safety-related systems, IEC (1998)
2. EN 50159: Railway applications: communication, signalling, and processing systems. Safety-related communication in closed and open transmission systems, CENELEC (2010)
3. Rástočný, K., et al.: Quantitative assessment of safety integrity level of message transmission between safety-related equipment. J. Comput. Inf. **33**, 1001–1026 (2014). ISSN 1335-9150
4. Franeková, M., Lüley, P.: Modelling of failure effects within safety- related communications with safety code for railway applications. Int. J. Mech. Transp. Commun. part VII, pp. 27–34, TRANSPORT 2015, Borovec, Bulgaria, 8–12 October 2015, Selected papers (2015). ISSN 1312-3823
5. www.unisig.org. Accessed 14 Dec 2015
6. Stallings, W.: Cryptography and Network Security. Principles and Practice. Prentice Hall (2011). ISBN 13:978 – 0-13-609-704-4
7. Karná, L., Klapka, Š.: Message doubling and error detection in the BSC model. In: 23rd International Symposium EURO - ŽEL 2015 - Recent Challenges for European Railways - Symposium Proceedings, Žilina: Tribun EU, pp. 52–57 (2015). ISBN 978-80-263-0936-9
8. Zelenka, J., et al.: Ratio counter – solution of relation between safety and availability of communication system. In: IEEE International Conference on Applied Electronics, Pilsen 2010, pp. 383–386, IEEE Catalogue Number CFP1069A-PRT (2010). ISSN 1803-7232, ISBN 978-80-7043-865-7
9. IEC 61784-3: IEC 61784-3: Digital data communications for measurement and control. Part 3: profiles for functional safety communications in industrial networks (2007)
10. FIPS 140-2: Security requirement for cryptographic modules. National Institute of Standard and Technology (2002)
11. Hristov, H., Hristova, M.: Modeling reliability of fault tolerant systems with homogeneous reservation. Mech. Transp. Commun. **11**(3) (2013). ISSN 1312-3823
12. Torrieri, J.: Principles of Secure Communication Systems. Artech House, Norwood (2009). ISBN 0-89006-555-1
13. IEEE 802.11n:2009: IEEE standard for information technology- Telecommunications and information exchange between systems- Local and metropolitan area networks- Specific requirements Part 11: Wireless LAN Medium Access Control (MAC) and Physical Layer (PHY) Specifications Amendment 5: Enhancements for higher throughput

Economical Feasibility of Eco-driving Induction in Road Vehicles

Mihaela Bukljaš Skočibušić[⊠], Hrvoje Vojvodić, and Pero Škorput

Faculty of Traffic and Transport Sciences, University of Zagreb,
Vukelićeva 4, 10000 Zagreb, Croatia
{mihaela.bukljas,pero.skorput}@fpz.hr,
vojvo0135@gmail.com

Abstract. With the use of telematics devices in cars, are shown in real-time data, such as the current vehicle speed, current fuel consumption, RPM etc. By analyzing the obtained data will be compared to eco-driving with normal driving. Also based on the analyzed data it is possible to identify the basic parameters of eco-driving and describe examples of the implementation of eco-driving with the economic aspect. In order to reduce vehicle operating costs, eco-driving permits monitoring of the above data, visualization and processing.

Keywords: Eco-driving · Telematics devices · Vehicle-operating costs

1 Introduction

Global efforts to reduce the emissions of motor vehicles powered by internal combustion engines have been implemented by the latest technical and technological solutions to maximize the ecological and energy efficiency use of vehicles.

Analysis of the dominant styles of driving by the driver, indicate on a low level of awareness among drivers regarding to the implementation of the rules of ecological and energy-efficient driving. One of the biggest reasons why car manufacturers invest considerable effort in the development and implementation of system supports for driver is achieving factory declared fuel consumption with the eco-driving style.

The telecommunications sector in the automotive industry develops very similar range of devices and applications designed for support of eco-driving styles. The development focus of these devices and applications is primarily to save fuel, which is also the strongest motivating factor for the widespread use of these styles of driving. Promotion and motivation of the eco-driving is generally conducted by motoring clubs and similar organizations.

Primarily, professional drivers, i.e. drivers passing high mileage per year are important for the motivation of using eco-driving, due to significant savings in the amount of fuel consumption, vehicle maintenance, and the like. Real time feedback for the driver in the system, such as the number of kilometers driven on 100 km can be helpful, but also supported with a detailed graphical view of monitored parameters of eco-driving on the geographical map.

© Springer International Publishing AG 2016
J. Mikulski (Ed.): TST 2016, CCIS 640, pp. 175–186, 2016.
DOI: 10.1007/978-3-319-49646-7_15

Such display is very useful in terms of raising awareness about the way you drive, and if carried out periodically gives a clear picture through graphic and stored data of the economic result for each driver separately.

Establishing communication with vehicle through OBD dongle is enabled via standardized OBD port, which is implemented in all vehicles from year 1998. The first vehicle that had OBD port also had ECU unit implemented to control the engine of the vehicle which is as well used to obtain concrete information on the current fuel consumption, the position of the accelerator pedal, etc.

2 In-vehicle Technologies

Modern vehicles contain a multiplicity of controllers that control vehicles primary and secondary functions. Primary functions are involved in maintaining safe control of the vehicle, that is the driving task, while the secondary functions [1] are available to the driver for comfort, control of communication, infotainment, navigation etc. Vehicles that are equipped with all of these systems allow deep interventions in the driving behavior of the vehicle thru its own communication network that controllers control. That kind of the communication network with several connected systems is called "Bus Communication" (Table 1) [2].

Table 1. Representative networks for bus communication [own study]

LIN ("Local Interconnect Network")	Serial communication for sensors and actuators
	Data rate up to 20 Kbit/s
CAN ("Controller Area Network")	Event-triggered controller network
	Operating even if some of nodes are defective
	Data rate up to 1 Mbit/s
FlexRay	Deterministic high-speed bus
	Error tolerant
	Data rate up to 10 Mbit/s
MOST ("Media Oriented System Transport")	Serial high-speed bus
	Data rate up to 24 Mbit/s (synchronous)
	Data rate up to 14 Mbit/s (asynchronous)
Bluetooth	Wireless radio data transmission
	Enables wireless ad-hoc networking of various devices
	Data rate up to 0.7 Mbit/s

Not like the common computer networks, car owners and authorized garage personal have full access to control units of primary and secondary functions with factory diagnostic tool shown in Fig. 1 [3].

Aftermarket tools like OBD dongles are not capable to have full access to the vehicles functions, but the basic information like the speed of vehicle, air temperature etc. is possible to read in a specific sampling time specified by the manufacturer.

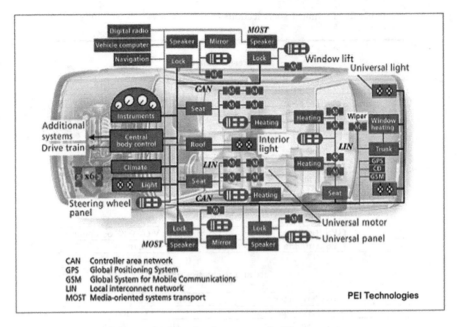

Fig. 1. Auto network [3]

3 Basic Functions of Test Equipment

For the purposes of this research testing were conducted by the telematics device for system support of eco-driving in real-time terms. The test configuration was mounted on the vehicle ŠKODA, ROOMSTER, 1.6 TDI (66 kw), 2014 year. For monitoring parameters of the vehicle and support of eco-driving, device C4 OBD2 dongle of manufacturer "Mobile Devices" was used. Manufacturer "Mobile Devices" also provides services of gathering pre-selected data or parameters during the trip. After the end of the trip, thru the vehicles CAN communication protocol OBD dongle stops with gathering data and send it to online database where it is processed on a way that the system is calibrated before implementing the OBD dongle. Technical specifications of dongle device are listed in Table 2.

Establishing communication with vehicle through OBD dongle is enabled via standardized OBD port (Fig. 2) [4], which is implemented in all vehicles from year 1998. The first vehicle that had OBD port also had ECU unit implemented to control the engine of the vehicle which is as well used to obtain concrete information on the current fuel consumption, the position of the accelerator pedal, etc.

Today's vehicles have built-in ECU, Comfort module, ABS and many other electrical units, without which the vehicle could not partially or fully work.

Accordingly, test assessment configuration was performed for driving in an urban area and high-speed road of Zagreb according to the parameters shown in Fig. 3.

According to Fig. 3 test system settings are set to monitor the parameters of eco-driving with monitoring additional parameters which include respecting traffic

Table 2. Technical specifications of dongle device [own study]

Performance	
Processor	ARM 11 – 500 MHz
RAM	64 MBytes (ext to 128 MB)
NAND Flash	256 Mbytes (ext to 2 GB)
Communication	
Modem	Quad-band GSM/GPRS
GSM Antenna	Internal (Quad band)
GPS positioning	
GPS Receiver	Sirf Atlas V (A-GPS on option)
GPS Antenna	Internal
Interface and telematics function	
3D Accelerometer	± 2g, ±4g, ±8g
3 axis Gyroscope	On option
Mini USB 2.0	Host/Device/UART
RTC	yes
LED	1 bicolor LED
Protocols	
CAN 2.0B interface	yes
RS232	on option with IBD Cable
ECM*	J1850 PWM, SAE J1850 VPW, ISO 9141 2ISO 14230 KWP2000, OBD II, ISO 15765 CAN
Power supply	
External	8-18V
External Voltage Measurement	8-18V
Li-ion battery & charger	Option (900mAh)
Type of physical connection, operating temperature and dimensions	
Connector	OBD2 connector
Operating Temperature	− 25 / + 60°C
Dimensions (mm)	27 x 48 x 49.5

regulations. For example, for the zone 5, where the optimum speed is between 30–40 km/h, it assigns 30 negative points if one of the defined values of eco-driving deviates from the optimal value. The system of assigning negative points in this way evaluates more monitored parameters.

PIN	DESCRIPTION	PIN	DESCRIPTION
1	Vendor Option	9	Vendor Option
2	J1850 Bus +	10	J1850 Bus −
3	Vendor Option	11	Vendor Option
4	Chassis Ground	12	Vendor Option
5	Signal Ground	13	Vendor Option
6	CAN (J-2234) High	14	CAN (J-2234) Low
7	ISO 9141-2 K-Line	15	ISO 9141-2 L-Line
8	Vendor Option	16	Battery Power

Fig. 2. OBD port [4]

Penalty scoring is separate for driving by night and in rush hours. Practice proves that in rush hours there is much bigger risk for traffic accidents, than in night. There for, penalties for driving in rush hours have been multiplied by four times, because of the much smaller number of traffic accidents happened by night. Number of traffic accidents was retrieved from Croatian statistics in traffic for a period of 10 years [5].

As parameters of eco-driving, following values are monitored: the amount of G force during accelerating, the amount of G force during braking, vehicle speed, engine speed, gear position. Monitored parameters were collect through mobile network SIM card in the device and the cellular network and stored on the external system, i.e. a central server system to control the fleet. Analytical procedures and techniques to visualize the data collected are shown in Fig. 4.

Warm colors show areas where the deviations of monitored values of eco-driving were the biggest. Similarly, it is possible to extract any parameter of eco-driving, whether for an individual vehicle or a fleet of vehicles.

Detailed view of monitored values lets you view the exact location with the description of deviation (Fig. 5), where exclamation marks are displaying exceeded G-force in cornering and accelerating, while the stop signs display exceeded G-force while deceleration. Such data are very useful for vehicle owners to see deviations from the optimal value while driving. One of the basic rules applying eco-driving are early shift into a higher gear (2000 and 2500 rpm's) to maintain a constant speed, anticipate the traffic situation, brake timely, regularly check the tire pressure and service the vehicle.

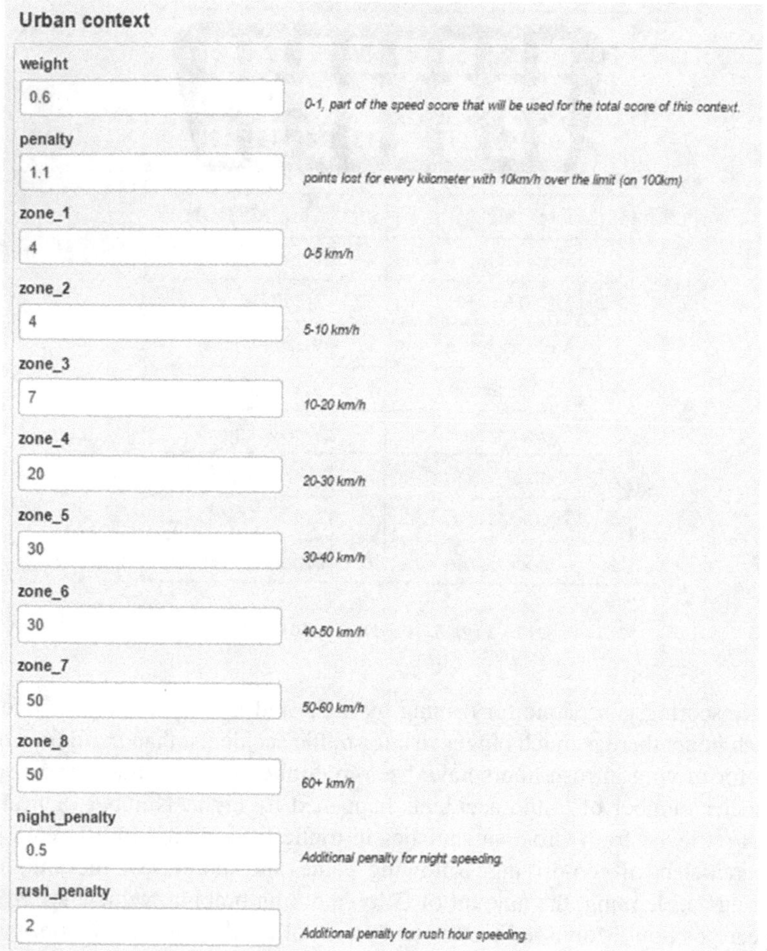

Urban context

weight

| 0.6 | *0-1, part of the speed score that will be used for the total score of this context.* |

penalty

| 1.1 | *points lost for every kilometer with 10km/h over the limit (on 100km)* |

zone_1

| 4 | *0-5 km/h* |

zone_2

| 4 | *5-10 km/h* |

zone_3

| 7 | *10-20 km/h* |

zone_4

| 20 | *20-30 km/h* |

zone_5

| 30 | *30-40 km/h* |

zone_6

| 30 | *40-50 km/h* |

zone_7

| 50 | *50-60 km/h* |

zone_8

| 50 | *60+ km/h* |

night_penalty

| 0.5 | *Additional penalty for night speeding.* |

rush_penalty

| 2 | *Additional penalty for rush hour speeding.* |

Fig. 3. Test configuration for grading eco-style driving [4]

4 Basic Functions of Test Equipment

For the purpose of this study two types of testing were conducted: testing without applying eco-driving, and testing with applying eco-driving on the same route. All methods of testing were repeated five times, and only trips with average number of alerts are considered, noting that testing without applying eco-driving was made without the knowledge of implementing device for tracking vehicle in purpose of eco-driving. Also, driver was transporting the same amount of cargo by the same vehicle to the same destinations in same order.

After five weeks, driver was presented with the data stored of his trips, as well as alert's in order to increase motivation for use of eco-driving. A short training on the basic rules of eco-driving [6], raised awareness of the benefits of this way of driving,

Fig. 4. Visualization parameters of eco-driving [4] (Color figure onilne)

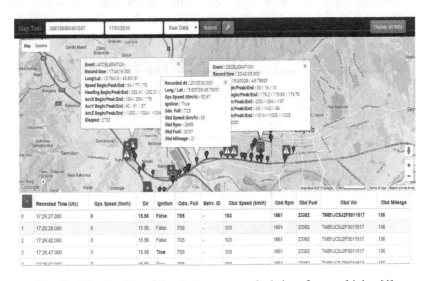

Fig. 5. Visualization of detailed parameter deviations for eco-driving [4]

for both the driver and the company. Basic rules of eco-driving are separated into two main groups shown in Table 3.

Table 3. Basic rules of eco-driving [own study]

Pre-trip	On the road
Tire pressure (check on a monthly base)	Change gears appropriately
Wheel alignment (check once a year, and when needed)	Anticipate traffic flow
Plan your trip (avoid peak traffic hours etc.)	Switch off the engine (waiting at railway crossing)
Service your vehicle by manufacturer service schedule	Close windows as much as possible
Minimize the payload and air resistance of your vehicle	Turn off the air conditioning as much as possible
Join a car pooling community scheme	Drive at the most efficient speed
Plan your trip	

4.1 Non-economical Driving

By following the usual way of driving, the foregoing vehicle stored 21 alerts of which 6 of them were for a sharp slowdown, with the remaining 15 for excessive acceleration shown in detailed Fig. 6. The driver that week on its frequent route had an average fuel consumption of 6.91 [L] to 100 [km] without applying the basic rules of Eco-driving.

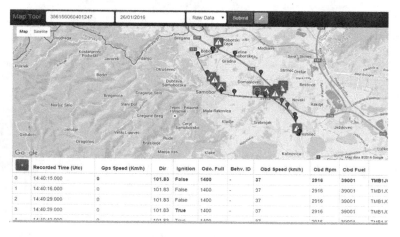

Fig. 6. Visualization of tracked parameters of non-economical driving [4]

A mistake that the driver was repeatedly doing was to excessive accelerating, noting that in the most cases it was completely unnecessary, with consequences of reducing traffic safety, increasing fuel consumption, tire wear, ware of the brake system, suspension components etc.

4.2   Eco-Driving

By viewing trips with the application of eco-driving with Fig. 7, it is possible to realize a significantly smaller number of alert's compared to driving without applying eco-driving. After familiarizing driver with the concept of eco-driving and its basic rules, the driver on the same route made only 12 alert's 2 of which are for an excessive slowdown, and the other 10 for excessive accelerating. Another significant mistake that the driver was doing was the lowered windows and turned on air-conditioning on not so hot days. That mistake was noticed after introducing the driver with concept of eco-driving and its basic rules. By applying eco-rides driver decreased fuel consumption for the significant 10.8%, or by 6.91 [L] to 6.16 [L] to 100 [km].

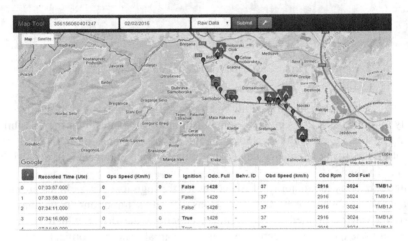

Fig. 7. Visualization of tracked parameters of eco driving [4]

5   Economical Analyze of Eco-driving

In order to reduce exhaust emissions of vehicles with internal combustion engines the effect of reducing costs is present. The Eco-driving has the potential sensible savings in fuel consumption and therefore CO_2 emissions. More and more countries through its transport policy and strategic recommendations to the EU encourage measures to reduce climate change.

For example, the European Climate Change Programme ("EPC") [7] as early as 2001 calculated that education of eco-driving within ten years could reduce emissions of CO_2 for a minimum of 50 Mt. In this connection, costs are reduced in fuel consumption, vehicle maintenance, insurance costs, costs of traffic accidents, as well as external costs. A 2006 study conducted by the Dutch research institute ("TNO") showed that the costs and benefits through program Eco-rides proved savings by using eco-rides 9 € per ton of CO_2 emission reduction. Figure 8. shows the cost (to costumer) per % CO_2 saved, and reveals eco-driving to be the most cost-effective approach to emission reduction on this basis [8].

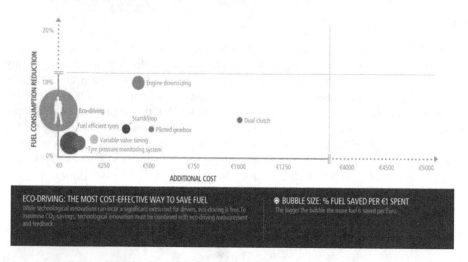

Fig. 8. Cost-effectiveness of eco driving [8]

It should be noted that in addition to the above effects of cost savings, more recently, although still in its infancy, some insurance companies offer the possibility of reducing insurance premiums vehicles based on assessments of individual driving styles of drivers using telematics device that monitors the parameters of eco-driving of the insured vehicle.

Although today the majority of EU countries through its fiscal policies promote the purchase of cars with lower CO_2 emissions [9], it is important to emphasize that eco-driving can be applied to any vehicle, coaching the driver which is far faster, simpler and cheaper way of achieving the same goal. The economical and environmentally friendly driving style opens up a huge scope for savings. Spotting the simplicity of eco driving, car industries started to implement eco-drive mode into their vehicles, one of the first was the Ford with their eco-drive mode [10].

6 Security Problems

All electronic units in vehicle could be divided into 2 groups, as external and internal communication threat points for unwanted entry from unauthorized personal shown in Fig. 9. As defined by Margaret Rouse [11], car hacking is the manipulation of the code in a car's electronic control unit (ECU) to exploit a vulnerability and gain control of other ECU units in the vehicle.

Every communication point [12] in vehicle is a security problem for drivers and car manufactures. Thru the internal communication, points it is possible physically establish communication with all of the electronic units for wrong and criminal purpose. With external communication points, it is evident that there is a bigger security problem. External communication points could be hacked from even a long distance like some other country [13]. Vehicle communication could be established from unwanted and unauthorized personal for unwanted actions like: remotely take over a

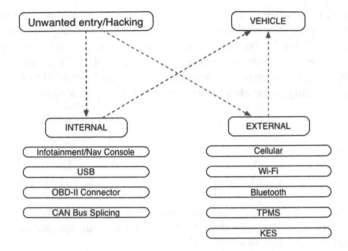

Fig. 9. Threat points [own study]

vehicle, shut down a vehicle, install malware on the vehicle, track a vehicle, thwart safety systems, steal a vehicle, unlock a vehicle and spy on vehicle occupants [14]. After establishing communication with the vehicle all of this actions could be executed at any moment, some actions even without our knowledge (install malware on the vehicle, spy on vehicle occupants etc.).

7 Conclusion

Every day more and more vehicles runs on road traffic routes and harmful exhaust emissions are becoming an increasing problem related to climate change and therefore also for the environment. Automotive industry develops various electromechanical assemblies to reduce CO_2 emissions and very successfully. Several years ago, the new vehicles reduced fuel consumption with the simplest components, such as tires, to more sophisticated solutions such as start-stop system and other similar systems. By using telematics devices such as OBD dongle it is possible with the reduced fuel consumption and emission to indicate the driver to drive more economically and what are the measures to reduce number of alerts while driving.

While researching different ways of driving, it was concluded by the shown ways that the way of eco-driving drastically reduces fuel consumption and emissions, and vehicle wear on tires, suspension, brake systems and engine.

While testing the system there were noticed potential security problems that arise with the possibilities of hacking the system via the OBD dongle, which can create undesirable safety effects.

Also, positive economic effects are shown that can be achieved by introducing eco-driving, of the strategic recommendations to the effects obtained on the basis of research on fuel efficiency, reduce CO_2 emissions, savings on consumables car parts, insurance costs and external costs.

Until today, unfortunately, there is no optimum security solution for OBD dongle devices that could eliminate all unwanted and ban from entering the said system in terms of its abuses, but on the positive side, using the same effects can be achieved flawlessly in the technical, environmental, safety, and applicative economic sense.

According to all of the above it, is expected that the application of new technologies in recent time is going to develop better security and simpler system of eco-driving.

References

1. Harvey, C., Stanton, A.N.: Usability Evaluation for In-vehicle Systems. CRC Press, Boca Raton (2013)
2. Pasricha, S., Dutt, N.: On-chip Communication Architectures. Elsevier, Amsterdam (2008)
3. The Development of Controller Area Networks for Cars. http://www.aa1car.com/library/can_systems.htm. Accessed 18 Nov 2015
4. Motor Vehicle Maintenance & Repair. http://mechanics.stackexchange.com/questions/23047/obd-ii-power-when-key-not-in-ignition. Accessed 18 Nov 2015
5. Bilten o sigurnosti cestovnog prometa: Ministarstvo unutarnjih poslova Republike Hrvatske (2013). http://www.mup.hr/UserDocsImages/statistika/2014/bilten_promet_2013.pdf. Accessed 18 Nov 2015
6. Yamabe, S., et al.: Physical fatigue comparison of eco-driving and normal driving. J. Syst. Des. Dyn. 5(5), 994–1004 (2011)
7. European Policy Center: Green revolution: making eco-efficiency a driver for growth. http://www.epc.eu/documents/uploads/pub_1401_green_revolution.pdf. Accessed 18 Nov 2015
8. FIAT corporation: FIATecoDrive, First study of eco-drive, Eco-driving uncovered.pdf, p. 16 (2010)
9. Gerlagh, R. et al.: Fiscal policy and CO2 emissions of new passenger cars in the EU, CPB Discussion Paper, 302 (2015)
10. Ford, News Release, Technology for efficiency-Ford ECOmode. http://media.ford.com. Accessed 18 Nov 2015
11. Tech Target Network. http://internetofthingsagenda.techtarget.com/definition/car-hacking. Accessed 18 Nov 2015
12. Wolf, M., Wiemerskirch, A., Paar, C.: Secure in-vehicle communication. In: Lemke, K., Paar, C., Wolf, M. (eds.) Embedded Security in Cars, Securing Current and Future Automotive IT Applications, pp. 95–109. Springer, Heidelberg (2006)
13. Smith, C.: Car Hackers Handbook, pp. 13–77. No Starch Press, San Francisco (2015)
14. Foster, I., et al.: Fast and vulnerable: a story of telematic failures. In: 9th USENIX Workshop on Offensive Technologies (WOOT 2015), Washington, D.C. (2015)

Preparing Reports on Risk Evaluation and Assessment in Rail Transport Based on the Polish Experience

Marek Jabłoński[✉]

OTTIMA Plus Sp. z o.o., Southern Railway Cluster,
40-594 Katowice, Poland
marek.jablonski@ottima-plus.com.pl

Abstract. The European Union Regulation on risk evaluation and assessment has resulted in the need for risk management in railway undertakings when they decide, based on analysis, that a change is significant and affects rail transport safety. Then it must be proven that the risk management process caused by this change is correct. The analysis of past experience in this field in Poland shows the need to improve this process, both in terms of cooperation between different entities involved in this process and in order to understand the objectives of the regulation and its key provisions. The fundamental problems and mistakes in reporting on risk evaluation and assessment include mistakes in defining the system undergoing change, mistakes in defining system boundaries, mistakes in identifying risks, improprieties in selecting risk acceptance principles, the improper use of codes of practice, ignoring safety requirements etc. These problems have triggered the analytical research on the assessment of the adequacy of risk management in rail transport as a result of a technical, operational and organizational change. The results of the analysis have been presented in this paper.

Keywords: Risk management · Risk evaluation and assessment report · Safety management system · Assessment body

1 Introduction

The issue of safety in rail transport is an issue that requires constant search for new, effective and efficient solutions, which aim to reduce risks in rail transport and improve the reliability of used devices. The role of a risk management process in rail transport is increasing [1]. The need for dealing with this issue results from the applicable European Union Directives [2–4]. According to A.M. Zarembski, J.W. Palese, in recent years, railways have turned to the discipline of risk management to improve safety and reduce the potential risk of accidents or derailments. Since accidents and derailments are very low probability of occurrence events, it is necessary to focus on the derailment causes themselves and develop risk management tools that quantify and analyse the "risk" for each key derailment or accident area [5]. It should be highlighted that railway undertakings are obliged to implement and maintain safety management systems when risk management is mandatory. It is not easy to apply the principles of risk management in rail transport due to the complexity of its transport and maintenance processes. It is a

© Springer International Publishing AG 2016
J. Mikulski (Ed.): TST 2016, CCIS 640, pp. 187–198, 2016.
DOI: 10.1007/978-3-319-49646-7_16

complex process that requires a holistic perspective based on very good knowledge of technical processes [6]. Railway undertakings are obliged to manage risk as part of their operating activities. This process should be part of organization operations and analysed in terms of all aspects of railway traffic safety. The risk should be evaluated when a new product is implemented, and when the railway system is designed, operated and maintained. These cases cause changes in the railway system and they should be analysed in terms of their impact on railway traffic safety. The awareness of railway undertakings managers of the need for efficient and effective risk management in the event of changes in the railway system is increasing. In Polish conditions, there are a number of shortcomings in applying the regulations related to this process. The Polish and European experience, due to the relatively new view of the risk management process, shows a significant need for further legislative work in this field, as well as changes in accepting responsibility for introduced changes. The process of risk assessment in the event of a significant change in the railway system entails a lot of necessary and required actions, which is often inconvenient for managers in terms of logistics. It results also from implementing new technologies and adapting them to new conditions [7]. The difficulties are mainly due to poor planning and lack of imagination of those who introduce changes as regards the impact of these changes on railway traffic safety. This leads to problems causing, for example, delays in performing investment processes and the need to find bad solutions resulting, for example, in unreliable risk analysis, and in particular, decisions that a significant change is non-significant in terms of safety. It is a widespread and reprehensible phenomenon. In addition to such situations, many other problems, mistakes and irregularities in applying legislation on risk management in rail transport can be observed. Key problems associated with implementing the requirements of Regulation 402/2013 [8], whose provisions govern the core assumptions of the risk management process, have been presented in the paper, based on the review of the relevant literature, including regulations and the author's own experience. The study is qualitative research based on the analysis of case studies.

2   Change Management in the Railway System

A change to the railway system is a factor creating new conditions for the functioning of this system. A change generates new circumstances, often affecting the level of safety. A change to the railway system, therefore, should be the subject of analysis in terms of safety. A railway undertaking that monitors changes in the designed, operated or maintained railway system increases the level of the culture of safety, at the same time eliminating any potential risks which, if there was no reaction, would increase the level of risks in rail traffic. A change, therefore, has been considered a very important element of the functioning of the railway system, the consequences of which are crucial for maintaining an acceptable level of safety. The identification of the types of changes that affect safety has been described in Regulation 402/2013. Such changes may be of a technical, operational or organisational nature. Such a division of changes makes it possible to monitor all the activities of railway undertakings in terms of an impact on safety. A technical change occurs when a railway system is introduced, modernised or renewed. An operational change is any changes affecting the rules of operating the railway system in terms of using its functionality. As regards organisational changes, only those changes which could impact on the operational or maintenance processes shall be subjected to consideration under the rules 402/2013. Change as a natural result of business activity in the railway sector has been subjected to monitoring and has become one of the key issues significantly affecting railway traffic safety. The examples of changes to the railway system in the areas of infrastructure, rolling stock and organization include:

Infrastructure:

1. The construction of a new railway line or extension of the existing one by another track.
2. The construction of the platform/station in a new location within the existing railway line.
3. The construction of a new or reconstruction of an existing railway subgrade resulting from changes in the geometry of the track system with a possible change in the embankment slopes/ditches tilt resulting in increasing or adjusting the maximum axle load to the designed category of line and/or increasing the design speed on the line.

Rolling stock:

1. The installation of ETCS on-board devices level 1.
2. The installation of ETCS on-board devices level 2 and higher.
3. The redevelopment of the gear system.
4. The redevelopment of the brake system.

Organization:

1. The introduction of new/modified rolling stock into operation.
2. An increase in the amount of rolling stock used.
3. The introduction of a new product on the rail market [9].
4. The presented changes show how wide the range of the discussed issue is. There is an infinite number of changes that may affect the rail industry and railway traffic safety, therefore this subject is now a key issue in the area of maintaining and increasing railway safety.

3   Risk Evaluation and Assessment in the Context of Significant Changes

The process of risk evaluation and assessment requires the selection of appropriate methods and techniques. When explicit risk is estimated, appropriate quantitative and qualitative methods are used. While the risk management process does not require that any specific tools should be applied, many of the more well-known techniques will be relevant, including:

- structured group discussions;
- checklists;
- task analysis;
- hazard and operability studies (HAZOPs);
- hazard identification studies (HAZIDs);
- failure mode and effects analysis (FMEA);
- fault trees; and
- event trees [10].

In addition to the above methods, it is reasonable to use comprehensive solutions for risk evaluation and assessment. Figure 1 shows the assumptions of such a comprehensive approach, that is the Risk Score Matrix model based on the assumptions of DIN V VDE V 0831-101 norm. The Risk Score Matrix (RSM) consists of the application of a risk matrix and score tables for assessment of the barriers, similar to RPN schemes. The final result consists of hazard rates (HR) related to the functional failures of the technical system and the assumptions on which the analysis rests, which may turn into safety-related application rules (SAR) [11].

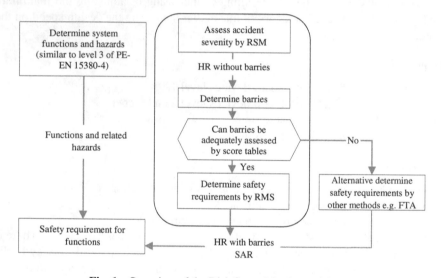

Fig. 1. Overview of the Risk Score Matrix model [11].

An important and widely used method for estimating risk is the ALARP principle.[1]

The ALARP principle ensures that the risks of any system with serious consequences in terms of human loss and injuries, is kept to a level which is As Low As is Reasonably Practicable. ALARP defines three risk levels:

Intolerable Risk, which cannot be justified or accepted, except in extraordinary circumstances Tolerable Risk, which can be accepted only if risk reduction is impractical or if the cost or risk reduction greatly exceeds the benefit gained.

---

[1] As low as reasonably practicable.

Negligible Risk, which is broadly acceptable and does not require risk mitigating measures. If risk is determined to be at the intolerable level, measures must be taken to reduce it immediately to tolerable level. If risk is found to be at tolerable level, risk mitigating measures should still be applied, provided that a cost benefit analysis is in favor of it [12]. The presented status of risk is important from the point of view of risk classification and identification of appropriate risk reduction measures to fulfil safety requirements. In Polish conditions, the FMEA method has been widely accepted and it is mostly used to determine explicit risk, that is a situation when the codes of practice or other reference system do not apply to an identified hazard. Despite many shortcomings, which include the high level of subjectivity of this method, it is still very popular. It seems reasonable to use solutions that combine several methods at the same time; then it is possible to reduce the number of mistakes that the FMEA method may entail.

When a report on risk evaluation and assessment is drawn up, it is necessary to extend the range of issues it covers. Based on the author's experience, the structure of the report on risk evaluation and assessment can be developed on the basis of the following subjects:

1. The impact of a change on safety (Yes, it matters - further action/No, it does not matter.)
2. The general description of the system before change (system objective e.g. intended purpose)
3. The description of a change - a type of change: technical, operational, organizational, the detailed description of the system:
    3.1. The functions and elements of the system after change (including e.g. human, technical and operational elements);
    3.2. The system boundary, including other interacting systems;
    3.3. Physical (i.e. interacting systems) and functional (i.e. functional inputs and outputs) interfaces;
    3.4. The system environment (e.g. energy and thermal flow, shocks, vibrations, electromagnetic interference, operational use);
    3.5. Existing security measures and the definition of safety requirements identified by the risk assessment process (at subsequent stages).
    3.6. Assumptions determining the limits for the risk assessment.
4. The criteria for assessing the significance of change:
    4.1. Failure consequences: a credible worst-case scenario in the event of failure of the system under assessment, taking into account the existence of safety barriers outside the system.
    4.2. Novelty used in implementing the change; this criterion includes innovation in both the entire railway sector, and what is new just for the organization implementing the change.
    4.3. Complexity of the change.
    4.4. Monitoring: the inability to monitor the implemented change throughout the system life-cycle and take appropriate interventions
    4.5. Reversibility: the inability to revert to the system before change; and

4.6. The assessment of the significance of change taking into account all recent modifications to the system under assessment and were not judged as significant.
5. The identification and classification of hazards (what might happen):
6. The choice of the risk acceptance principle (Using the codes of practice during risk evaluation, reference systems, assessment and evaluation of explicit risk).
7. The rules and scope of applying these risk acceptance principles in relation to defined hazards.
8. The identification of risk acceptance principles.
9. Confirmation that all defined hazards and risks fulfil the acceptance criteria.
10. Final conclusions and presenting the limitations of the document.

The presented issues do not exhaust the subject because in certain situations the scope should be expanded and clarified to match the context of the system undergoing change. Therefore, this description is illustrative, though quite detailed.

4   An Impact on Safety and the Evaluation of the Significance of a Change as a Milestone in the Risk Management Process

The greatest difficulty in the risk management process in the context of introduced changes is to assess their impact on safety and consequently, assess the significance of the change. In the first step it is necessary to assess the impact on safety in order to further assess the significance of the change according to specified criteria. If the proposed change has an impact on safety, the proposer shall decide, by expert judgement, on the significance of the change based on the following criteria:

a. failure consequence: credible worst-case scenario in the event of failure of the system under assessment, taking into account the existence of safety barriers outside the system under assessment;
b. novelty used in implementing the change: this concerns both what is innovative in the railway sector, and what is new for the organisation implementing the change;
c. the complexity of the change;
d. monitoring: the inability to monitor the implemented change throughout the system life-cycle and intervene appropriately;
e. reversibility: the inability to revert to the system before the change;
f. additionality: assessment of the significance of the change taking into account all recent safety-related changes to the system under assessment and which were not judged to be significant [13].

Figure 1 shows the logic of conducting the analysis leading to decisions about the impact of change on railway traffic safety. The presented diagram proves that first it is necessary to assess whether the change has any impact on safety. If it does, it must be determined whether it is significant. If it is significant for railway traffic safety based on the relevant assessment criteria, then a report on risk evaluation and assessment is drawn up, which is subsequently verified by an accredited assessment body. It should be noted that it is a railway undertaking or manufacturer that is responsible for deciding whether the change is significant or not. In any case, the decision should be documented and clearly identify specific individuals responsible for making a decision on the significance of the change (Fig. 2).

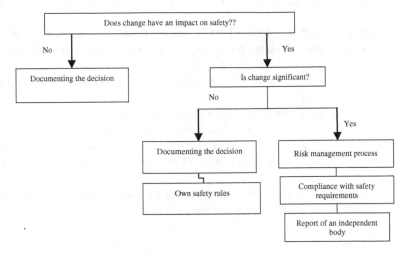

Fig. 2. The algorithm describing the decision-making process with regard to significant changes [own study]

The issue under discussion is particularly important as regards railway transport safety. Transferring responsibility for deciding that a change is significant and non-significant to railway undertakings, based on the subjectivity of the evaluation, has resulted in the situation that many changes are classified as non-significant, which reduces the role of accredited assessment bodies in participating in the process of change management and reduces the significance of changes to the railway system. Therefore, in order to decrease the number of situations when a group of railway undertakings has no responsibility, the control and supervisory role of the NSA (National Safety Authority) is important in supervising the implementation of changes in the railway system.

5   An Example of Application of the Method Risk Score Matrix in the Valuation of the Risks Associated with Rupture Axis of the Freight Wagon

Risk Score Matrix Method is a method which can be considered now as optimal from the point of view of technical risk assessment and the requirements of the Common Safety Method for Risk evaluation and assessment.

This method is based on the assumptions of the method of the event tree ETA (Event Tree Analysis) and allows the determination of the probability of occurrence with regard to the applied by the company railway protections.

It allows to confront the determined value of the risk to the safety requirements laid down in Regulation 402 and defined as: highly improbable "Means an occurrence of failure at a frequency less than or equal to 10 9 per operating hour and" improbable "Means an occurrence of failure at a frequency less than or equal to 10 7 per operating hour".

Existing methods such as. Method FMEA (failure mode and effects analysis) does not take into account the calculated probability of the occurrence of hazards from the point of view of a railway undertaking protections.

It is also no possible to designate a target of the risk in the context of the requirements posed by that regulation 402.

As part of the FMEA method calculated the number of priority risk is designated a value of not relating to the safety limits laid down in Regulation 402/2013.

Therefore, commonly used methods FMEA already adopted in the case of the requirements of Regulation 402 is not justified because of the limitations described above.

Figure 3 shows an example of the calculation of the risk of using the event tree for the risk of rupture of the wheelset axle in a freight wagon.

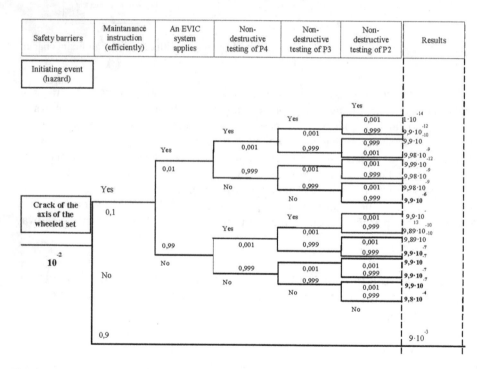

Fig. 3. An example of using the event tree ETA (Event Tree Analysis) for example cracks wheelset axle freight wagon [own study]

Were defined four safety barriers namely: Effectively was used documentation of maintenance, is used system EVIC (European Visual Inspection Catalogue), are carried out non-destructive testing axis of maintaining level P4 (major repair) and are implemented non-destructive testing axis of maintaining level P3 (review periodic chief) and are also implemented non-destructive testing axis of level P2 (review current). These four defined protections barriers led to the development of seventeen scenarios have been assigned a probability of risk in the context used a combination of protections.

The calculated probability values can be analysed with using the safety requirements laid down in Regulation 402.

Analysing the results of considerations it can be stated that the designated nodes in random probability estimates contribute to the final calculation of the probability. Important are used tools for safety protection.

The next step in this method is to apply the calculated value of the matrix Risk Score Matrix portfolio, taking into account the level of protections versus class of accidents.

Summing up the results of the calculation indicated in the case described, it is possible to use such a pattern of conduct with the risks for which the probability of danger will be included in the expected protections what is requirement of rules 402/2013.

6   The Preparation of Reports on Risk Evaluation and Assessment - Key Experiences and Problems

In scientific research six selected reports on risk evaluation and assessment were used, drawn up by infrastructure managers, railway construction investment companies and manufacturers, assessed by an accredited assessment body in terms of compliance with the ISO IEC 17020 norm [14]. Qualitative research based on the phenomenological paradigm was applied [15]. When analyzing selected cases of the examined reports, repeated irregularities were shown in the process of drawing up reports on risk evaluation and assessment and decisions about the significance of changes. Grouping the cases allowed for the assessment which subject areas are the most difficult for those who prepare documentation on documenting the impact of changes on rail traffic safety and they are often recommendations by the accredited assessment body. It should be noted that the experience in applying Regulation 402/2013 is not wide in Poland. This is due to the fact that the regulation has been in force for a short time. The paper is one of the first in Poland describing the study of selected aspects of conducting risk evaluation in the context of technical, operational and organizational changes in rail transport. As regards the scope of this study, six reports were assessed, out of which four related to operational changes introduced by the infrastructure manager, resulting from the introduction of new traffic control devices, including a control structural sub-system - trackside devices and two reports on the modernization of railway lines carried by railway construction investment companies, the scope of which included an energy structural sub-system. The reports were prepared in various forms. Sometimes the report on assessing the significance of change and the report on risk evaluation and assessment were a unified whole, and in other cases they were separate documents. Most reports (four) were prepared by a proposer, the other two by consulting companies assisting the proposer. All of the examined reports were ultimately positively assessed by an accredited assessment body, but all of them required prior alterations, which were collected and research findings were formulated based on them and the accompanying documentation. The following are key recurring problems that the authors of the reports on risk evaluation and assessment came across. They were identified by the accredited assessment body and described in the reports on safety assessment as part of the independent process of evaluating the adequacy of the risk management process in the context of the requirements of Regulation 402/2013.

1. The inaccurate description of the system being changed and the system before change. The authors do not accurately show what the subject of the change was, and do not describe how this change has influenced specific new hazards in rail traffic. They do not describe key interfaces between devices, either. To a large extent, these new interfaces resulting from the new functionalities of the system and changed configuration pose new risks, which should be covered by appropriate risk control measures. Some mistakes can also be observed in defining the boundaries of the railway system undergoing change. This is particularly important when a change affects a number of structural sub-systems according to TSI (Technical Specification of Interoperability), e.g. when a change relates to control sub-systems, energy and infrastructure.
2. It is imprecisely described which specific criteria influenced a decision about the significance of the change, as well as the reasons why a particular criterion occurred. There are also obvious factual mistakes in the interpretation of the significance of changes. Moreover, individual evaluation criteria are misunderstood.
3. The incorrect identification of the codes of practice within the scope of CSM RA (Common Safety Methods Risk Assessment) in the context of the defined hazards. Proposers often overlook this aspect and assess explicit risk, for example, by the FMEA quality method. Another problem is misunderstanding the definition of codes of practice and assigning a specific code of practice to specific hazards, as well as using codes of practice when a risk is only partially covered by the assigned code of practice.
4. The incorrect identification of the scope of using qualitative methods, e.g. FMEA compared to other risk acceptance principles. It should be also noted that the range of methods assessing explicit risk in Poland is reduced only to the uncritical application of the FMEA method, with no interest in other quantitative and qualitative methods.
5. The key problem is also failure to identify all possible risks, which in a given situation are important for the changed operation of the railway system. Industry experts from assessment bodies think this aspect demands improvements in particular.
6. Reports on risk evaluation and assessment are often drawn up by specialists who do not have sufficient experience in the railway sector, which is shown in the low quality of the description of technical issues as well as their detail. Another problem is insufficient knowledge of safety requirements with respect to the specific character of the railway system undergoing change.

The presented research findings indicate that the quality of reporting on the risk evaluation and assessment by accredited assessment bodies needs improving. The authors of these reports do not fully understand the seriousness of the impact of technical, operational and organizational changes on railway traffic safety.

7 Conclusion

The process of risk evaluation and assessment in terms of changes in the railway system is a complex process that requires the participation of many stakeholders. This issue is increasingly understood in Poland, however many mistakes have not been avoided, which the author has tried to highlight in the paper. Conducted qualitative research of selected reports on risk evaluation and assessment have helped to formulate the following conclusions:

1. The authors of the reports on risk evaluation and assessment make a series of mistakes of a methodological nature, which is related to the lack of knowledge of the correct interpretation of various issues such as: a change to the railway system, the codes of practice or safety requirements.
2. It is observed that decisions about the significance of a change are avoided. Railway undertakings are afraid of a long process of risk evaluation and assessment and the costs of participation of accredited assessment bodies in the risk management process.
3. The reports are often too general and do not present technical aspects in detail, as well as functionalities resulting from the changes, especially in the context of inter-faces with other rail devices or systems.

The presented findings should provide a foundation for improving risk management processes by railway undertakings. The paper highlights the need to initiate a broad discussion on the quality of reports on risk evaluation and assessment and the problem of classifying changes as non-significant in order to avoid the complicated procedure for documenting changes to the railway system and the reliable assessment of their impact on its safety.

References

1. Jabłoński, A., Jabłoński, M.: Implementation and managing of innovation in the conditions of legal and economic constraints on the base of rail transport. In: Mikulski, J. (ed.) Telematics in the Transport Environment. CCIS, vol. 329, pp. 423–432. Springer, Heidelberg (2012)
2. Directive 2004/49/EC of the European Parliament and of the Council of 29 April 2004 on safety on the Community's railways and amending Council Directive 95/18/EC on the licensing of railway undertakings and Directive 2001/14/EC on the allocation of railway infrastructure capacity and the levying of charges for the use of railway infrastructure and safety certification (Directive on rail safety)
3. Directive of the European Parliament and Council Directive 2008/110/EC of 16 December 2008 amending Directive 2004/49/EC on safety on the Community's railways (Directive on rail safety)
4. Directive of the European Parliament and Council Directive 2008/57/EC of 17 June 2008 on the interoperability of the rail system within the Community (Recast)

5. Zarembski,, A.M., Palese, J.W.: Managing risk on the railway infrastructure. In: WCRR (2006)
6. Jabłoński, A., Jabłoński, M.: Key challenges and problems in conducting independent evaluations of the adequacy of the risk management process in rail transport. In: Mikulski, J. (ed.) Tools of Transport Telematics. CCIS, vol. 531, pp. 311–321. Springer, Heidelberg (2015)
7. Jabłoński, A., Jabłoński, M.: Transfer of technology in the field of rail transport through cluster initiatives management, communications in computer and information science. In: Mikulski, J. (ed.) Activities of Transport Telematics. CCIS, vol. 395, pp. 58–66. Springer, Heidelberg (2013)
8. Commission Implementing Regulation (EU) No 402/2013 of 30 April 2013 on the common safety method for risk evaluation and assessment and repealing regulation (EC) No 352/2009
9. An expert opinion on the practical application of requirements of a common safety method for risk assessment (CSM RA) - a guide, UTK, Warszawa (2015)
10. Common Safety Method for risk evaluation and assessment, Guidance on the application of Commission Regulation (EU) 402/2013, March 2015
11. Braband, J.: Rapid risk assessment of technical systems in railway automation. In: Proceedings of the Australian System Safety Conferrence, ASSC (2012)
12. As Low As Reasonably Practicable
13. Berrado, A., et al.: A framework for risk management in railway sector: application to road-rail level crossings. Open Transp. J. **5**, 34–44 (2011)
14. Commission Implementing Regulation (Eu) No 402/2013 of 30 April 2013 on the common safety method for risk evaluation and assessment and repealing Regulation (EC) No 352/2009, Art 4
15. PN-EN ISO/IEC 17020:2012, Conformity assessment. Requirements for the operation of various types of bodies performing inspection
16. Collins, J., Hussey, R.: Business Research, A Practical Guide for Undergraduate and Postgraduate Students, 2nd edn. Palgrave Macmillan, New York (2003)

Multi-option Model of Railway Traffic Organization Including the Energy Recuperation

Marianna Jacyna[1]($\boxtimes$), Piotr Gołębiowski[1], and Michał Urbaniak[2]

[1] Faculty of Transport, Warsaw University of Technology, Warsaw, Poland
{maja,pgolebiowski}@wt.pw.edu.pl
[2] Faculty of Civil and Environmental Engineering,
Gdansk University of Technology, Gdansk, Poland
michal.urbaniak@wilis.pg.gda.pl

Abstract. The article presents the issue of organization of railway traffic within the station, taking into account energy recuperation. We identified a number of factors which affect the energy efficiency of recuperation, including the issue of transfer of energy technology between a few vehicles. We presented some aspects of the decision problem of graphic train timetable construction with regard to recuperation. As an indicator of quality solution assessment we proposed vector objective function consists of three sub-objective function.

Keywords: Energy recuperation · Organization of train traffic problem · Railway energy consumption optimization · Graphic train timetable

1 Introduction

Transport means overcoming space as obstacles resulting of a distance. To realize the movement of people and/or cargo it should take into account many of the factors that determine this process. These can include for example: infrastructure, superstructure or economic factors as the demand for transportation. Transport service should be implemented an adequate level of, which can be expressed using four metrics:

- punctuality – services should be implemented at the scheduled time,
- rapidity – services should be implemented appropriate mode of transport that can move with appropriate speed, and the infrastructure which is in good working order and has the appropriate technical parameters to realize the service in minimal time,
- certainty – transport service should be organized in such a way that the customer was sure its implementation,
- safety – the service should be carried out in a secure manner - respecting rules and regulations in force in any area.

Because of the need to organize the transport service in a comprehensive way it should be solve many of problems which are interrelated, if not outright it indirectly. Among these issues can distinguish the following issues [10]:

© Springer International Publishing AG 2016
J. Mikulski (Ed.): TST 2016, CCIS 640, pp. 199–210, 2016.
DOI: 10.1007/978-3-319-49646-7_17

- distribution of the stream of traffic on the network (models of movement organization) [12, 13],
- adapt the infrastructure to the reported demand for transport in models of transport system development, including problems of allocation of resources necessary to perform the tasks [10],
- development of the transport network, including the location of objects in the transport network [21],
- choice of technological equipment in transport and logistics buildings [3],
- minimizing the lead time of transport service [12],
- the problem of choosing option modernization of infrastructure [2].

These issues concern all road, air and sea transport. Of course, it would be the other scale of these issues and the complexity of these problems and ways to solve them. The authors have focused on rail transport. One of the major problems in rail transport is to organize traffic on the network.

Railway traffic organization is a complex decision-making problem which requires analysis of many issues relating to, for example [18]:

- to satisfy freight and passenger modal,
- guaranteeing the continuity and safe service in the following areas:
  - regulation of trains service at railway stations
  - regulation of transport capacity and bandwidth line and station,
  - driving the train in accordance with the documentation,
  - cooperation with different types of operators,
- minimize transport time,
- providing to the passengers comfort of travel.
- ensure minimizing the cost of carriage,
- achieving the best possible value indicators.

This involves optimization:

- number and size of trains,
- circulation of vehicles and traction and conductors teams,
- construction of the train timetable and the plan for their collating,
- organization of traffic in station areas.

This means that it should take into account the interests of many users and contributors of all the transportation process (customers, operators, passengers, rail carriers, maintenance providers, energy providers, et al.). Their aims are generally conflicting.

One aspect of organization railway traffic is to prepare a rational train timetable. The construction of train timetable for passenger traffic takes place in two steps [11]: generating the transport offer and constructing the traffic chart of the trains. Forming carriage offer it is a definition of rational course of communication train lines and the allocation for each line set train type and frequency of services. Whereas the construction of a train traffic graph it is routing trains on the graph in a coordinate system: time - road. In Poland, for preparing a graph of train traffic responsible is administrator of infrastructure.

With the process of rational design of train schedules is related to the organization of movement within the station in terms of energy savings [22]. Energy Recovery saves about 14% of the demand for energy. Using an electric vehicles with electronic traction power supply system gives the possibility to recover some of the energy and use it to acceleration. The recovered energy must be stored or transmitted to the global energy system. As a "no-cost" alternative, it can be the energy recovery directly to the overhead line, and then use to acceleration another train. The effectiveness of this solution is closely linked to the appropriate organization of railway traffic. It is necessary that when a vehicle loses energy by the recuperative braking, another one was able to use it to acceleration.

Proper organization of railway traffic on the railway station area needs many elements which contains in the development of train traffic diagram [5, 7, 9, 11], for example:

- directing traffic after the occurrence of the obstacles on the railway network by using a diagram of train traffic [16, 23],
- using the simulation models to drive the operation of trains on the railway [23],
- mathematical modelling of the railway traffic organization on the railway network [3, 8, 11, 12],
- using cyclic [20] and no-cyclic [6] train timetables,
- using energy-efficient driving strategies [1, 3, 4, 14].

The article presents an approach to the organization of railway traffic having regard to the many subcriteria. The paper presents a mathematical formalization of the decision making problem for the construction of train traffic diagram on a selected area of the railway network having regard to energy recuperation. Also partial indicators of the solutions quality comprising the vector objective function and the boundary conditions that define the set of feasible solutions was presented.

2   The Essence of the Recuperation in Railway Traffic Organization Problem

2.1   Costs Related to the Volume of Electricity Consumption

The costs of railway transport, related to transport activities, can be divided into two basic groups: internal costs, understood as the operating costs of rail transport operators and external costs [2] defined as the costs incurred by the environment or society (e.g. environmental costs, the cost of air pollution and accident costs). Among the internal costs necessary to distinguish own cost of the railway enterprise and the cost of access to infrastructure. Both groups of costs are dependent upon an important factor - the availability of electricity for traction purposes.

In rail transport, to power the traction vehicles are currently being used two basic types of traction: petrol and electricity. The subject of this article is electric traction, so topics of petrol traction will be skipped. Supplying energy for rail transport (both for the purposes of traction and non-traction) provided by specialized bodies, whose tasks include trading and distribution of electricity and the provision of electricity service such as maintenance, modernization and restoration of traction devices.

In Poland, there are two billing systems for electricity: the settlement flat rate and the settlement based on energy counter. In the system of settlement flat rate, carriers pay a fee for the used energy in the form of contractual fees calculated on the basis of the algorithms depend on the volume of transport. In the settlement based on energy counter through the introduction of technology to measure the actual energy consumption of the vehicles (using electricity meters). It can be used for the settlement actually used electrical power by the particular vehicle. Comparison of settlement systems described on the example data published by one of the railway company - Mazowieckie Railway- shown in Table 1.

Table 1. Comparison of traction energy costs of the settlement flat rate and the settlement based on energy counter [15]

Kind of the settlement	Year	Performance of freight [train km]	Settled energy [kWh]	Cost of energy [PLN]	Cost of energy per unit [PLN/kWh]
The settlement flat rate	2010	13905586	186552562	91708479,9	0,493
The settlement based on energy counter	2011	14573986	162846537	77580154,99	0,478
	2012	16571919	181818159	87363012,5	0,480

The Mazowieckie Railway Company at the turn of 2010 and 2011 has changed the way to the energy settlement from settlement flat rate on the settlement based on energy counter. The size the performance of freight in 2011 has increased compared to the previous year by 4.8%, while the settled amount of energy decreased by as much as 12.7%. By reducing the unit price of energy and changing the way the settlement, cost of energy decreased by 15.4%. In the context of optimizing the costs traction energy in the railway company only the settlement based on energy counter has financial no justification for allowing full use of the latest rolling stock, newest technology and organizational solutions.

The second group of energy-related costs represent charges for access to infrastructure. The size of the rates for access to infrastructure consists of [19]:

-   basic fees:
-   for minimal access to railway infrastructure, expressed in PLN/trainkm,
-   for access to equipment related to the operation of trains, like: unit for refuelling, train stations and stops, freight terminals, hump, unit train formation, stabling tracks and ramps and loading yards,
-   additional fees, which include: assistance in running abnormal trains, providing supplementary information about train movements, allocation and preparation of a train path outside the annual timetable, sharing of regulations and list of railway stations and retail outlets as a print, sharing lifts from the regulations and other documents which require incurring the costs.

It is important that the cost of access to the lines, where infrastructure electric traction is available is higher by approx. 16% (depending on the category of line) than the lines that do not have such access. Due to the fact that the cost of electricity is a significant cost of transport activity should endeavour to rationalize the energy consumption.

2.2 The Strategy of Optimize Energy Consumption for Traction Purposes

The development of traction systems and transportation caused the increasing importance of energy recuperation that means recovery energy during regenerative braking. An important problem is the energy exchange directly between different vehicles on the route. Energy, which is in the process of regenerative braking captures the catenary by train approaching the station could be used by another train in acceleration process at the same station. This solution does not require the use of on-board or stationary energy storage. There are also no needed specialized equipment and infrastructure. It is essential to proper coordination of traffic on given line in order to possibility cooperation between the processes of consumption and return of energy. Suitable coordinating vehicles on the network in such a way, that the energy given back in the regenerative braking process was entirely used for acceleration purposes other vehicles is not possible [17]. It should be noted that the change scheduled for individual trains of about 10 s can change the energy balance of close to 100%.

The simplest situation for the introduction of time coordination between braking and the time of the acceleration, so that was possible transfer of energy between trains is on lines dedicated to the urban, suburban and underground traffic (Fig. 1). This is mostly influenced conditions such as:

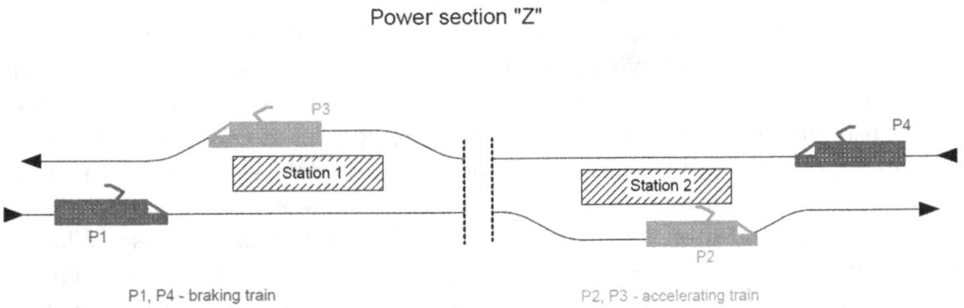

Power section "Z"

P1, P4 - braking train   P2, P3 - accelerating train

Fig. 1. Diagram of coordination the times of an acceleration and braking of trains on the station on suburban line [own study]

- the number of acceleration and braking cycles associated with huge traffic on the line,
- use the same type of rolling stock,
- the same speed of trains
- cyclic timetable,

- relatively short distances between vehicles on the network and low loss of transfer energy between vehicles,
- relatively uncomplicated track system,
- no needed transport connection between trains moving in opposite directions.

It is possible to introduction to the mathematical model supporting the timetables construction additional restrictions. This restrictions should cause that, the time of start braking by the vehicle P1 (Fig. 1) was as close as possible to the time of acceleration of the vehicle P3 (similarly for vehicles P2 and P4). It should be noted that the situation should be considered from the point of view of the whole power section so that an optimum coordination processes of acceleration and braking includes not only station area – Fig. 2. In this case, the optimal solution will be coordinated at the same time three trains The time of acceleration P5 vehicle should be as close as possible to the time of start braking the both P6 and P7 vehicle, seeking to maximize use of the transfer of energy from recovery.

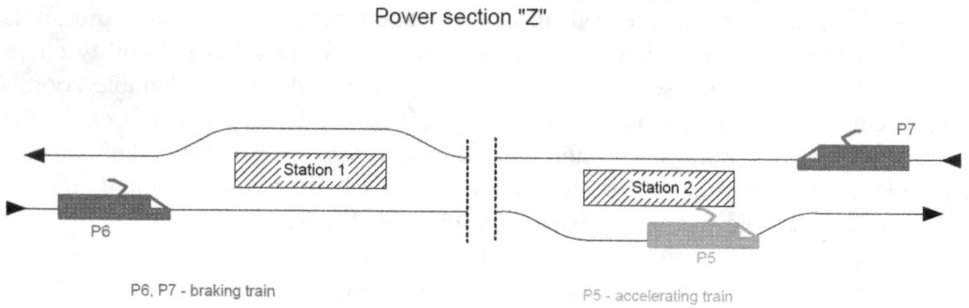

Fig. 2. Diagram of coordination the times of an acceleration and braking of trains on the whole power section of suburban train network [own study]

Much more complex situation occurs in the areas of nodal stations. This follows the extensive track system, variety operated trains, increased rotation of vehicles and the need to consider changes from one train to another. In this case, the optimization of energy costs will be subordinate aspect of the integration of the train connections.

With the restrictions of this strategy should be discussed the technical aspects of the design and construction of overhead power lines. Spansion isolated and isolators sectional are most often on the sections where trains are in braking phase. When in adjacent power sections will be various voltage and a vehicle, which passing through an isolated power section with consumption of energy or give back power to the grid, can be generate an electric arc between the pantograph and the power cable. This leads to a faster degradation of catenary and the pantograph.

3   Methodology of Multi-criteria Support Decision

One of the basic problems that appears when choosing the best solution (variant), is to establish a criterion that will allow to assess the quality of the solution. When the problem is not complicated, then to choose a solution is enough to choose one criterion.

It has place e.g. in some tenders where the solution quality is assessed using only the cost of completion of the service. The issue becomes more complicated when problem is complex and difficult to choose unequivocally the best solution. Then it should be consider simultaneously many criteria for assessing the quality of solutions.

Multi-criteria decision-making problems can be divided into four groups [10]:

- problems of multiple-choice, where the decision problem involves the determination of one embodiment among a set of options,
- problems of multi-criteria organizing, where the decision problem involves to organization of options,
- problems of multi-classification which based on defining classes and allocation the different options to these classes,
- problems of multiple-description.

Multi-criteria optimization problems are characterized by two elements: description of the set of feasible solutions and a set of functions (criteria) depict the set of solutions in a set of quality assessment. Multi-criteria problems searching for the best solution x^* minimizing the global criterion function F, composed of many sub-criteria $f_k(x)$, can be represented in the following form:

$$F = \langle f_1(x), f_2(x), \ldots, f_k(x), \ldots, f_K(x) \rangle \rightarrow \min \tag{1}$$

with restrictions:

$$a_i(x) \geq b_i \quad i = 1, \ldots, m \tag{2}$$

The set of feasible solutions D^{dop} is defined as a set of elements x (where a vector of decision variables is $x = \langle x_1, \ldots, x_n \rangle$) form:

$$D^{dop} = \{x : a_i(x) \geq b_i \quad i = 1, \ldots, m\} \tag{3}$$

and the corresponding space criteria D_f is defined as:

$$D_f = \{F : F = \langle f_1(x), f_2(x), \ldots, f_k(x), \ldots, f_K(x) \rangle, \quad x \in D^{dop}\} \tag{4}$$

Depending on the analytical form of particular the test function $f_k(x)$ and restrictions $a_i(x) \geq b_i$. it can be talk about different types of multi-criteria mathematical programming. In an optimal solution of organization of railway traffic, taking into account of energy recuperation it should be defined sub-criterion of function.

4 The Model of Train Timetable Graphic Construction Taking into Account Energy Recuperation

4.1 General Assumptions

The problem is to search organization of the railway traffic, which is represented by routes of trains on the graphic timetable, taking into account various interests, which are contrary to oneself:

- minimization of differences between the real time of a train ride and the model time,
- minimization of trains stops time in forwarding offices,
- minimization of differences between the amount of energy taken from the catenary and delivered onto her.

Knowing the numbers of operating control points wk, $wk \in WK^E(twr)$ and linear elements lk, $lk \in LK^E(twr)$ railway network can be presented by graph $GK^E(twr) = <WK^E(twr), LK^E(twr)>$. Additionally route twr of train on the graphic timetable so that $twr \in TWR^E$ and trains poc to run in the form of set $poc \in POC^E(t_{kat})$ were defined. Trains have defined routes of communication lines which include electrified segments of the network $t_{kat} \in T^E_{kat}$. Communication lines are prepared individually for each sections of the demand for transport $kat \in KAT^E$. It is also necessary to determine expected frequency of running of individual trains on each communication lines $f(t_{kat}, poc)$.

Having defined railway network and parameters of the transport offer which will be should be carried out on its, it is assumed that the canvas of the graphic timetable is described by graph $GR^E(twr) = <WR^E(wk, twr, poc), LR^E(wk, wk', twr, poc)>$, in which hours of arrivals and departures of trains are being represented by points with numbers $wr(wk, twr, poc) \in WR^E(wk, twr, poc)$, however paths of trains written as lr $(wk, wk', twr, poc) \in LR^E(wk, wk', twr, poc)$. Additionally parameters written as follows are assumed:

- $lrns(wk)$ – length of station time spacing,
- $lrnp(wk)$ – length of passenger stop time spacing,
- $lrnsz(lk, poc)$ – length of open line time spacing,
- $lrsk(wk)$ – length of waiting time for defined station.

In the article the problem of the railway traffic organization in the area of station taking into account energy recuperation is being analysed. In this scope, apart from earlier mentioned parameters, it is necessary to establish the number and the location of traction substations $pt(twr)$, $pt(twr) \in PT(twr)$ located on the given route of the graphic timetable and determination of activity areas of individual traction substations $obs(pt(twr))$, $obs(pt(twr)) \in OBS(PT(twr))$.

4.2 Decision Variables and Boundary Conditions Imposed on Them

The problem is to prepare organization of the railway traffic in order to ensure, from the one hand safety and fluidity of the movement, and from the other expectations of buyers of transport services.

Thus, assuming markings presented in the point 4.1 we are seeking the value $y^E(lr$ $(wk, wk', twr, poc))$ determining, whether the train path $lr(wk, wk', twr, poc)$ put on the graphic timetable is appropriate. We are establishing in addition, that $y^E(lr(wk, wk', twr, poc))$ is a binary variable so $y^E(lr(wk, wk', twr, poc)) \in \{0, 1\}$. The variable $y^E(lr(wk, wk', twr, poc))$ assumes value 1 when the outlined path is appropriate and 0 otherwise. The variables $y^E(lr(wk, wk', twr, poc))$ wrote in the form of vector $\mathbf{Y^E}(twr)$.

Outlining the appropriate path requires taking into account the number of boundary conditions:

- moment of the appearance of the train on the graphic timetable (initial hour, hour of the departure from the first forwarding office) $wr(wk, twr, poc)$ can have only one hour of the arrival to the next forwarding office $wr(wk', twr, poc)$,
- moment of the start or finish the state of the train for the indirect forwarding office $wr(wk, twr, poc)$ can have only one following moment of the start or finish the state of the train $wr(wk, twr, poc)$,
- moment of the disappearance of the train on the graphic timetable (final hour, hour of the arrival to the last forwarding office) $wr(wk', twr, poc)$ can have only one hour of the departure from the preceding forwarding office $wr(wk, twr, poc)$,
- for each operating control point difference between the moment of the arrival of the next train poc' and the moment of the departure of the train poc must be greater than or equal to the length of station time spacing $lrns(wk)$,
- for each passenger stop difference between the moment of the arrival of the next train poc' and the moment of the departure of the train poc must be greater than or equal to the length of passenger stop time spacing $lrnp(wk)$,
- for each operating control point moment of the arrival of the next train poc' must be greater than or equal to the moment of the departure of the train poc enlarged about length of open line time spacing $lrnsz(lk, poc)$,
- for each passenger stop moment of the arrival of the next train poc' must be greater than or equal to the moment of the departure of the train poc enlarged about length of open line time spacing $lrnsz(lk, poc)$,
- for each operating control point difference between the moment of the arrival of the next train poc' and the moment of the departure of the train poc must be greater than or equal to the required frequency of running of trains $f(t_{kat}, poc)$,
- for each passenger stop difference between the moment of the arrival of the next train poc' and the moment of the departure of the train poc must be greater than or equal to the required frequency of running of trains $f(t_{kat}, poc)$,
- for each operating control point, where trains can terminate and start the run, difference between the moment of the departure of the next train poc' and the moment of the arrival of the train poc must be greater than or equal to the length of waiting time for defined station $lrsk(wk)$.

4.3   Vector Objective Function

The problem of searching appropriate organisation of the railway traffic taking into account energy recuperation is connected with making a decision, for which from the one side the safety and the fluidity of the movement will be ensured, and from the other expectations of buyers of transport services. So the decision cannot be taken based on one assessments criterion of the solution quality. It should be applied a few fragmentary criteria which have a measurable degree of the accomplishment of the purpose. Because criteria will be expressed in different units, it will be difficult to make an explicit decision, which variant of the organization is the best one. Therefore a global objective function which will let for the comprehensive evaluation of the problem will be exploited. Thus, the set of acceptable solutions D_{dop} is a joint set of every considered in the model measurable purposes (see point 2).

For the purposes of multi-criteria optimization of the traffic organization on the railway network taking into account energy recuperation was introduced three fragmentary criterion functions $f_1(\mathbf{Y^E}(\mathbf{twr}))$, $f_2(\mathbf{Y^E}(\mathbf{twr}))$ and $f_3(\mathbf{Y^E}(\mathbf{twr}))$, constituting components of the vector objective function:

$$F\left(\mathbf{Y^E}(\mathbf{twr})\right) = \left[f_1(\mathbf{Y^E}(\mathbf{twr})), f_2(\mathbf{Y^E}(\mathbf{twr})), f_3(\mathbf{Y^E}(\mathbf{twr}))\right] \rightarrow \min \qquad (5)$$

where:

– function $f_1(\mathbf{Y^E}(\mathbf{twr}))$ with the interpretation of the minimization of the amount of differences $p(lr(wk, wk', twr, poc))$ between the real time of the realization of the trains' path $t_{rz}(lr(wk, wk', twr, poc))$ and the model time $t_w(lr(wk, wk', twr, poc))$ has a form:

$$f_1\left(\mathbf{Y^E}(\mathbf{twr})\right) =$$
$$\sum_{lr(wk,wk',twr,poc)\in LR^E(wk,wk',twr,poc)} p(lr(wk, wk', twr, poc)) \cdot y^E(lr(wk, wk', twr, poc)) \rightarrow \min \quad (6)$$

function $f_2(\mathbf{Y^E}(\mathbf{twr}))$ with the interpretation of the minimization of the amount of stop times in forwarding offices $t_{post}(lr(wk, wk', twr, poc))$ has a form:

$$f_2\left(\mathbf{Y^E}(\mathbf{twr})\right) =$$
$$\sum_{lr(wk,wk',twr,poc)\in LR^E(wk,wk',twr,poc):wk=wk'} t_{post}(lr(wk, wk', twr, poc)) \cdot y^E(lr(wk, wk', twr, poc)) \rightarrow \min$$

$$(7)$$

function $f_3(\mathbf{Y^E}(\mathbf{twr}))$ with the interpretation of the minimization of differences $e(lr(wk, wk', twr, poc))$ between the amount of energy taken from the catenary by train $ep(lr(wk, wk', twr, poc))$ and delivered onto her $eo(lr(wk, wk', twr, poc))$ has a form:

$$f_3\big(\mathbf{Y^E}(\mathbf{twr})\big) =$$
$$\sum_{lr(wk,wk',twr,poc)\in \mathbf{LR^E}(wk,wk',twr,poc):wk=wk'} e(lr(wk,wk',twr,poc))\cdot y^E(lr(wk,wk',twr,poc)) \to \min$$

$$(8)$$

Searching of the minimum of the vector objective function $F(\mathbf{Y^E}(\mathbf{twr}))$ is held under fulfilling the number of boundary conditions described in the point 4.2.

5 Conclusion

Organization of the railway traffic is a complex decision-making problem. The complexity results from the number of problems and the number of factors which should be considered. Many of them have a relation to the preparation of the work schedule of railways, according to which the traffic on the railway network is held. Apart from that, it is necessary to consider interests of many users and participants of the whole transport process which have conflicting goals in general.

A correct trains timetable is taking into account both of regulations and safety rules determined by requirements of control command and signalling, as well as principles with leading the trains traffic on the railway network. The organization of the railway traffic should from considerations of energy savings take into account the rationalization of the energy consumption, i.e. exploiting the energy from recuperation. Usage of electric traction vehicles with power electronic drive systems, gives the possibility to recover some of the energy as electricity, which during braking is returned to the catenary, and during the start can be reused. The effectiveness of this solution is closely associated with a respective management of the railway traffic.

Prepared mathematical model of the graphic trains timetable construction taking into account energy recuperation allows for its preparation for trains using the electric traction equipped into devices for recovering the energy during braking. We considered a fact of running of trains in the interval of the time spacing - where it is possible it should be station time spacing, where it is not possible it should be open line time spacing.

References

1. Albrecht, T., Otteich, S.: A new integrated approach to dynamic schedule synchronization and energy saving train control. In: Allan, J., Hill, R.J., Brebbia, C.A., Sciutto, G., Sone, S. (eds.) Computers in Railways VIII, vol. 61, pp. 847–856. WIT Press, Southampton (2002)
2. Ambroziak, T., et al.: Wariantowe rozłożenie potoku ruchu w zadanej sieci przy uwzględnieniu kosztów zewnętrznych. Logistyka **4**, 1605–1616 (2014)
3. Ambroziak, T., et al.: Analysis of the traffic stream distribution in terms of identification of areas with the highest exhaust pollution. Arch. Transp. **32**, 7–16 (2014)
4. Bocharnikov, Y.V., et al.: Optimal driving strategy for traction energy saving on DC suburban railways. IET Electr. Power Appl. **1**, 675–682 (2007)

5. Caprara, A., et al.: Passenger railway optimization. In: Handbooks in Operations Research and Management Science, vol. 14, pp. 129–188 (2007)
6. Carey, M., Lockwood, D.: A model, algorithms and strategy for train pathing. J. Oper. Res. Soc. **46**, 988–1005 (1995)
7. Faryna, P. (eds.): Elektroenergetyka kolejowa. Railway Business Forum, Warszawa (2011)
8. Gołębiowski, P.: Mathematical model of shaping the railway transportation offer. In: CLC 2013: Carpathian Logistics Congress – Congress Proceedings (reviewed version) [CD-ROM], 1st edn., pp. 397–402. Tanger Ltd., Ostrava (2014)
9. Hansen, I.A., Pachl, J.: Railway Timetabling & Operations. Analysis - Modelling - Optimisation - Simulation - Performance Evaluation. Eurailpress, Hamburg (2014)
10. Jacyna, M.: Modelowanie i ocena systemów transportowych. Oficyna Wydawnicza Politechniki Warszawskiej, Warszawa (2009)
11. Jacyna, M., Gołębiowski, P.: An approach to optimizing the train timetable on a railway network. In: Brebbia, C.A., Miralles i Garcia, J.L. (eds.) Urban Transport XXI, vol. 146, pp. 699–710. WIT Press, Southampton (2015)
12. Jacyna, M., Gołębiowski, P.: Traffic organization on the railway network and problem of construction of graphic train timetable. J. KONES Powertrain Transp. **22**, 79–87 (2015)
13. Jacyna-Gołda, I., Żak, J., Gołębiowski, P.: Models of traffic flow distribution for various scenarios of the development of proecological transport system. Arch. Transp. **32**, 17–28 (2014)
14. Kasprzak, J., Mysłek, J., Podoski, J.: Zasady trakcji elektrycznej. Wydawnictwa Komunikacji i Łączności, Warszawa (1980)
15. Koleje Mazowieckie: Management Report, Warszawa (2010–2012)
16. Meng, X., Jia, L., Qin, Y.: Train timetable optimizing and rescheduling based on improved particle swarm algorithm. Transp. Res. Rec. J. Transp. Res. Board **2197**, 71–79 (2010)
17. Pazdro, P.: Koncepcja ruchowej optymalizacji efektywności hamowania odzyskowego. TTS Technika Transportu Szynowego **1–2**, 62–64 (2003)
18. Pielas, C.: Organizacja ruchu kolejowego. Wydawnictwa Komunikacji i Łączności, Warszawa (1968)
19. PKP Polskie Linie Kolejowe S.A.: Cennik stawek jednostkowych opłat za korzystanie z infrastruktury kolejowej zarządzanej przez PKP Polskie Linie Kolejowe S.A. obowiązujący od 14 grudnia 2014 r.. PKP Polskie Linie Kolejowe S.A., Warszawa (2013)
20. Serafini, P., Ukovich, W.: A mathematical model for periodic event scheduling problems. SIAM. Discrete Math. **2**, 550–581 (1989)
21. Szczepański, E., Jacyna-Gołda, I., Murawski, J.: Genetic algorithms based approach for transhipment hub location in urban areas. Arch. Transp. **31**, 73–83 (2014)
22. Urbaniak, M., Jacyna, M.: Zarządzanie kolejowym ruchem podmiejskim z uwzględnieniem możliwości rekuperacji energii. In: Materiały konferencyjne VIII Międzynarodowej Konferencji Naukowo-Technicznej Systemy Logistyczne Teoria i Praktyka, pp. 279–280. Oficyna Wydawnicza Politechniki Warszawskiej, Warszawa (2015)
23. Woch, J.: Podstawy inżynierii ruchu kolejowego. Wydawnictwa Komunikacji i Łączności, Warszawa (1983)

Risk Analysis of Railway Workers Due to Interference into GSM-R System by MFCN

Marek Sumiła[(⊠)]

Warsaw University of Technology, 00-662 Warsaw, Poland
sumila@wt.pw.edu.pl

Abstract. The article presents a method of evaluating the range of area covered by interference for railway employees equipped with the GSM-R handheld mobile stations (MS). Previous studies covered only the on-board MS e.g. Cab radio and EDOR. EIRENE defines three other types of terminals used by railway workers. They are: GPH, OPH and OPS. Research in this area is yet unknown. The article presents a comparative analysis of these classes of MS and the method allows to evaluate the area affected by interference. A valuable addition is a comparative analysis of the extent of the impact of interference on the real parts of railway lines in Poland for which an implementation of GSM-R system is planned.

Keywords: GSM-R · Interference · Risk analysis · Handheld mobile station

1 Introduction

GSM-R was developed as a unified railway radio communication system based on public GSM system 2.5G in the core but expand to fulfil special railway requirements in the fields of voice and data transmission to provide a high level of availability and safety [11, 12]. This condition is connected with uninterrupted service and high availability [25].

The standard is a result of common collaboration between many various European railway companies to achieve interoperability by develop single communication platform that cover previous radio communication system in Europe and enable higher train speeds and traffic density. It fits in with the general trend of creation of telematics systems, where any mobile equipment need to work with a base station and control equipment of each network manufacturer. Some other examples for such unified systems are presented in [1, 21, 22]. To fully introduce and use this interoperability, GSM-R became a part of the European Rail Traffic Management System (ERTMS) standard and carries the signalling information directly to the train driver.

To provide essential requirements and functionality for this system UIC established a special task force EIRENE to standardize GSM-R in the field of specification (SRS) [16] and the field of functional (FRS) [17] and MORANE (Mobile Radio for Railways Networks in Europe) specifications which guarantee performance at speeds up to 500 km/h (310 mph), without any communication loss.

Based on those specification and ETSI EN 300 919 [10] specified five different types of mobile stations. Each of them allows to establish mobile communication for

© Springer International Publishing AG 2016
J. Mikulski (Ed.): TST 2016, CCIS 640, pp. 211–222, 2016.
DOI: 10.1007/978-3-319-49646-7_18

operation and maintenance among groups of railway employees i.e. station controllers, dispatchers, train engineers, shunting team members, drivers and train crews to fulfil their duties and has special functionality to allow functional numbering and location dependent addressing, shunting mode and direct mode for voice communication, data transmission for ETCS and call pre-emption in case of an emergency (REC).

Among the types of terminals we can distinguish:

- General Purpose Radio GPH (General Purpose Handheld) is a type of mobile radio that can be used by wide kind of railway men and is similar to the public mobile phones,
- Cab radio is a special type of terminal to be used by the driver to voice communication and is installed on-board the trains and has modular construction; the Cab radio is mandatory for interoperability because the same unit on the train has to be used to communicate in different networks operators,
- Operational radio OPH (Operational Purpose Handheld) is a mobile radio dedicated for the use by railway personnel involved in train operations such as trackside maintenance,
- Shunting radio OPS (Operational Purpose Handheld for Shunting) is a mobile radio for use by railway personnel involved in train operations such as shunting and is a variant of the OPS but with specific shunting features,
- EDOR (ETCS Data Only Radio) is a special kind of on-board radio in the train used for data communication to work with the ETCS train control application; it is also as the Cab radio mandatory for interoperability to work with ETCS 2 of higher.

The experience of western Europe countries, where the GSM-R has already been implemented, indicates that the transmitters (TRX) of public MFCN (Mobile/Fixed Communications Networks) can cause interference and directly affect the operation of GSM-R receivers. Results of different measurement campaigns performed since 2007 [8, 18, 23, 28, 29] have been used to understand the nature of majority causes of interference i.e. blocking and intermodulation effects.

The operational performance of GSM-R service and the behaviour of GSM-R Cab radios and EDOR's in the presence of GSM, UMTS and LTE signals in the adjacent spectrum above 925 MHz were published in [18]. Detailed analysis of the UIC database published in 2014 contains more than 660 cases of interference [29]. Among those, 194 cases were classified as serious influence on railway safety and put under detailed analysis. European Commission, DG Communications Networks Content & Technology issued in 2013 in a working document [23] reported the issues of registered GSM-R networks' interferences and actions taken into account to correct the problem. These activities resulted in an elaborate CEPT Report ECC no. 229 [7] as complement earlier work described in reports [2–6]. Activities undertaken by ETSI led to elaborate two technical specifications TS 102 933-1 [13] and TS 102 933-2 [14] to improve the performance of GSM-R receivers and to be more resistant to interference from MFCN. Research in the area of influence of MFCN refers to Cab radio and EDOR (ETCS data only cab radio). These are specific types of terminals are mandatory for interoperability but do not cover all types of defined terminals. Only mentioned ETSI technical specifications in the last issues consider the requirements and tests for

2 Watt's mobile stations. According to mentioned documents is necessary to perform further research and risk analysis of interference for handheld mobile station.

The following parts of the article present a comparative analysis of two major classes of mobile stations (MS) and an evaluation of the range of interference from MFCN transmitter and at the end will be invoke the results of interference simulation.

2 Problem of Interference in GSM-R Receivers

Early studies [3] conducted by CEPT (European Conference of Postal and Telecommunications Administrations) suggest that the victim of interference between GSM-R and public networks is receiver's module of GSM-R terminal. In regard to this, it should be considered the signals transmitted in the downlink GSM-R band (921–925 MHz) and the public network MFCN (Mobile/Fixed Communications Networks) band i.e. 925–960 MHz. The direct neighbouring bands of those systems can produce out-of-band signals and lead to the phenomenon of blocking and appearance of intermodulation products in GSM-R receivers. ETSI TS 137 104 [15] defines out-of-band emissions as emissions that are unwanted emissions appear immediately outside the channel bandwidth and as a resulting of the modulation process and non-linearity part of signal processing in the transmitter but excluding spurious emissions.

The blocking of a receiver (sometimes also called de-sensitization of a receiver) is defined in most harmonized European standards (including those for GSM/GSM-R) as a measure of the ability of the receiver to receive a modulated wanted input signal in the presence of an unwanted input signal, on frequencies other than those of the spurious responses or the adjacent channels, without exceeding a given degradation [7, 14]. The phenomenon of blocking can be caused by narrow-band and broadband signals and appear when the received wanted signal and from an off-channel unwanted signal not exceed the blocking level. In ETSI TS 102 933-1 [13] where was described requirements for radio reception the performance shall be met when the continuous sine wave signal id set to a level of −40 dBm. The same Technical Specification shows that for the out-of-band emissions of the interfering signal, the maximum noise signal level shall not exceed −113 dBm/200 kHz on wanted ARFCN and its (bi)adjacent channels.

The presence of narrowband signals occur for GSM 2.5G network. Broadband technologies are used in GSM 3G, LTE and WiMAX networks. The reports CEPT [6, 7] describe the consequences of the disruption caused by these systems and have greater impact on GSM-R band because of introduction the wideband signals in adjacent bands.

A preliminary analysis of the impact on MFCN to GSM-R receivers is shown in [27, 28] and a wider impact of this phenomenon on the safety and reliability of the rail system can be found in [1, 24, 26].

3 Comparison of GSM-R Terminals

Essential requirements for GSM mobile stations covers ETSI norm EN 300 919 [10]. Requirements for GSM operation on railways are detailed in ETSI EN 301 515 [12] and ETSI TS 102 281 [11]. Detailed analysis indicate two class of mobile station:

- eight Watt on-board terminals, e.g. Cab radio and EDOR,
- two Watt handheld terminals, e.g. GPH, OPH, OPS.

Cab radio and EDOR are installed on the trains and they are quite large and heavy. They can have modular construction and have not difficulty in expanding of new elements that can improve the parameters of the receiver. In addition to that, the fact that the Cab-radio and EDOR as train radios they have an extend antenna mounted on the train's roof. It gives about 4 m received signals above ground [16].

The GPH, OPH and OPS are different types of terminals. They are much more mobile than previous ones. They are light and have solid construction, most often with solid antenna. In those types of terminals much more difficult to add extra filters or sophisticated radio modem without changing whole radio's construction. As a handheld mobile stations they work, during normal operation, on about 1–2 m the high depends on the circumstances: standby (wearing a mobile on a belt) or usage (making a call).

Further research indicate that EIRENE in its specifications [16] invokes an ETSI TS 145 005 [9] which specifies further differences. For Cab-radio and EDOR the sensitivity is −104 dBm, and for small mobile stations is not less than −102 dBm[1]. The mobile station maximum output power and lowest power control level shall be, according to its class. The Cab radio and EDOR are in the 2nd class and transmit up to 8 W (39 dBm). The handheld mobile stations are in the 4th class can transmit with power 2 W (33 dBm)[2]. Both kind of mobile stations work with tolerance form extreme conditions ±2.5%. It should be specified that the radiated power is also dependent on antenna gain and the loss of afferent signal cables. A typical isotropic antenna has a gain of the antenna at the level of 2.15 dBi. Transportable radio antenna can present a greater the antenna gain.

The comparison shows that there is no direct correlation between transportable and handheld terminals when we evaluate influence of interference on terminal working in GSM-R system. Summary comparison presents the Table 1.

In March 2016 the UIC together with INFRABEL organizes the international conference "UIC ERTMS WORLD CONFERENCE" to get knowledge about implementation status of ERTMS system and both its components ETCS for train control and GSM-R for radio-communications. It has become a worldwide reference for all decision-makers and railway professionals involved in the development of rail systems across the world. Table 2 presents data about numbers of subscribers each type of terminals used to GSM-R communication until 2014 in the UE countries.

---

[1] ETSI 145 005 table 6.2-1a Reference sensitivity level for MS.

[2] ETSI 145 005 table 4.1-1 MS maximum power at GMSK modulation.

Table 1. GSM-R types of terminals parameters comparison [own study]

Parameter	Cab radio and EDOR	GPH, OPH, OPS
Movement	Up to 280 km/h	~6 km/h
Power	8 W (39 dBm)	2 W (33 dBm)
Sensitivity	−104 dBm	−102 dBm
Antenna high mounted	4 m	1–2 m
Antenna	Extend	Built-in
Construction	Modular	Solid
Expandability	Possible	Impossible

Table 2. Conference ERTMS Atlas 2016 – data updated until 2014 [30]

Country	Mobile subscribers (activated until 2014)					
	Cab radios	EDOR	GPH	OPH	OPS	Dispatchers
Austria	1840	-	162	1827	645	87
Belgium	2296	tests	96	291	0	315
Bulgaria	0	0	150	0	50	22
Croatia	0	0	0	0	0	0
Czech Republic	1200	3	600	350	tbd	340
Finland	few	-	>4400		10	103
France	7200	450	110	450	0	-
Germany	16361	407	17744	6929	2942	3938
Hungary	1500	?	?	?	0	0
Italy	4100	700	55000	500	-	200
Netherlands	2500	550	100		?	0
Norway	1093	2 (tests)	1735	4152	0	207
Poland	0	0	10 (tests)	0	0	2
Portugal	4	0	20	2	0	2
Romania	0	0	14	15	1	3
Slovenia	10 (tests)	0	0	0	0	0
Spain	531	1062	1180	22	6	36
Sweden	2800	220	1550	290	0	?
Switzerland	3447	1288	6444	184	50	566
United Kingdom	4998	54	600	8000	0	569

Based on the collected data it can be said that 2014 is activated almost 178000 subscribers. Number users with handheld terminals was two times greater than the transportable terminals. This is a result that confirms the validity of the study.

4   Analysis of Interference Areas

Risk analysis for handheld mobile stations lead to a new look at the problem area covered by interference. In general the GSM-R receiver can be blocked if the public network signal level exceeds −40 dBm [6, 13] or −35 dBm as it was proposed in the UIC report [28].

$$P_{po} = P_{GSMR} - L(d_{GSMR}) - (P_{MFCN} - L(d_{MFCN})) \; [dBm] \qquad (1)$$

where:

P_{po} – signal level at the point of railway, P_{GSMR} – GSM-R transmitter power [dBm], $L(d_{GSMR})$ – attenuation at distance from GSM-R transmitter, P_{MFCN} – MFCN transmitter power [dBm], $L(d_{MFCN})$ – attenuation at distance from MFCN transmitter.

An assessment should be considered as a comparison the signal from MFCN transmitter and the signal from the GSM-R base station (Fig. 1).

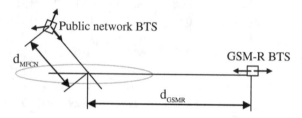

Fig. 1. Determination of the C/I at the point of the railway line [own study]

One should use information about carrier type (2G, 3G, LTE) and bandwidth, EIRP, antenna's characteristics, antenna's high and tilt and also azimuth and distance to the railway area can be used to calculate a signal strength in the main signal beam in vertical and horizontal axis. Assessment of size of the area covered by the interference depends on the radiation characteristics of antennas. In this article for simplicity, we consider the characteristic of MFCN BTS antenna's and assume that the level of wanted (GSM-R) signal is constant.

For precise calculation it should be considered the horizontal and vertical propagation of MFCN antennas. In many installations have been used sector antennas because of their radiation patterns. A typical GSM antenna radiation patterns has horizontally 120° and vertically about 7°.

The impact area of interference in the horizontal plane is approximately the same for all classes radios. It estimation illustrates by the example described later in this article. More interesting is the issue of vertical propagation of signals from MFCN. For calculation of the impact area we can use the formula

$$h = H - L \cdot tg\alpha \qquad (2)$$

where:

h – the height of the main beam above the track, H – height of antenna mounting, L – distance between BTS and railway track, α – tilt of the antenna.

In cases where tilt is equal to zero degree the range of signal propagation depends and the half power beam patterns ribbons. For the most vulnerable in analysing the case of radio receivers Cab radio and EDOR it should be taken as a constant the height of 4 m but form the handheld mobile stations the height is average 1,5 m. The height of the antenna determines the area of influence of the main beam signal calculating the minimal distance L_2, and the maximum distance L_3 of influence. Therefore we can assume that

$$L_1 = \frac{H - 4}{tg\,\alpha}, \ L_2 = \frac{H - 4}{tg(\alpha + \varphi)}, \ L_3 = \frac{H - 4}{tg(\alpha - \varphi)} \tag{3}$$

where:

L_1 – main beam of the signal, L_2, L_3 – min. and max. distance covered by signal, H – height of antenna mounting, α – tilt of the antenna, φ – vertical deviation of the main beam signal

To illustrate presented formulas was introduced Fig. 2.

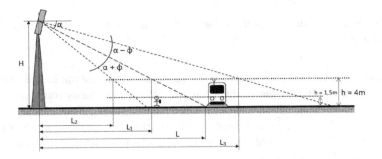

Fig. 2. Vertical deviation of the main beam signal and its propagation [own study]

Simple comparison of range of area affected by interfering signals changes according to the antenna suspension. The speed of workers movement is an additional factor that unwanted signals influent longer on their terminals than a driver's terminal i.e. a Cab radio or EDOR.

4.1 Case Study

The proposed method of risk assessment was used to evaluate a real influence on railway areas. For this purpose a place was chosen on the railway line E-65 in Poland. E-65 is under implementation ERTMS system and GSM-R network as its subsystem [19]. It was selected the 97th km of railway line at railway station Ciechanów. In the area near the railway station public BTS was identified. In the area near the railway station public GSM transmitter was identified. An on-line BTS search tool to specified distance, azimuth, and the owner of the BTS was used. The result of research shows Fig. 3.

Fig. 3. The location of the MFCN transmitter near Ciechanów railway station Śmiecińska 10 Str. [own study]

Base station is located at the distance of 170 m to the railway track. The azimuth of the antenna is unknown for the Authors but the railway area partly rounding the BTS because from the station leaves a railway siding (about from 120 to 370 m). There is high probability that at the area will present staff performing shunting or other railway workers equipped with a GSM-R handheld. The expected area of influence of the transmitter MFCN shown in Fig. 4.

If we assume that the transmitter works in 5 class (63.19 dBm) in the band 948.1–953.1 MHz and the attenuation is calculated according to ITU-R P.525-2 [20] we can see that the signal strength is change from the level −8.75 dBm (for the distance 100 m) to −28.75 dBm (for the distance 1000 m) (see Fig. 5). The value is about 11 dB higher than was set in [13].

When we invoke that the minimum values of the coverage of GSM-R signal for probability of 95% are based on a coverage level of −98 dBm for voice and non-safety critical data and not less than −95 dBm for speed up to 220 km/h for the ETCS 2 (the level is planned for this railway line) it can be noted that the difference between levels of wanted and unwanted signals reaches the value 66 dB at the distance of 1 km. In evaluated area of Ciechanów station and its sliding the values are about 10 dB higher.

The impact area of interference in the vertical plane can be calculated if we assume that TRX antenna is mounted at 30 m and the tilt is equal to 3°. We can use formula (3) and is equal about:

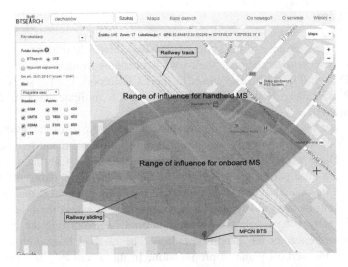

Fig. 4. Area covered by the MFCN transmitter for different class of MS [own study]

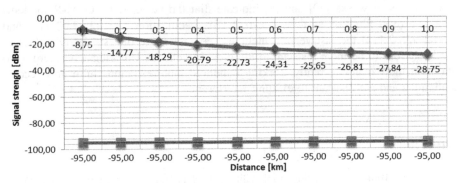

Fig. 5. Area covered by the MFCN transmitter for different class of MS [own study]

- from 147 to 845 m for on-board mobile stations,
- from 161 to 925 m for handheld radio.

As it was expected, changing the height of signal's reception influences the covered area by the interference but it has not significant influence on attenuation of MFCN signals. In the discussed location the difference in the range is approx. 70 m. For closer areas there are additional ribbons covering this areas. It can be assumed that the abovementioned case is not secluded. Confirmation of this statement requires further research and analysis.

5 Conclusion

The problem of interference of the system GSM-R is relatively little known in Poland. In Europe some countries elaborate their own procedures to improve the coexistence between MFCN and GSM-R. The annex to the ECC Report 229 [7] shows activities undertaken by Germany, Netherlands, Norway and UK to mitigate impact of MFCN on GSM-R networks.

The previous research discussed in the first and second part of the article shows a small number of works allow to assess the phenomenon of handheld mobile stations' interference. These works focused mainly on-board radios e.g. Cab-radio and EDOR. As shown in the third part of the article, the number of handheld radios are two times higher than the on-board radios and they are fundamentally different in their parameters and way of using by the railway staff. It was necessary to assess them individually.

The main part of the article presents the method for evaluating the extent of the impact of interference on the handheld radios. The method takes into account the horizontal and the vertical plane of propagation. In the discussed case, the handheld radios the vertical plane may be particularly important because of the range of area has change due to on-board radios.

The study of works [29] and [8] indicate that the newer radio GSM-R modems using improved filtration systems due to ETSI specification TS 102 933-1 [13] and TS 102 933-2 [14] the last version includes separate range of tests for this class of mobile stations to be more resistant to the blocking effects. Unfortunately, there is no data about how many currently working handheld mobile stations fulfil new recommendations and are resistant to existing interference.

References

1. Bester, L., Toruń, A.: Modeling of reliability and safety at level crossing including in Polish railway conditions. In: Mikulski, J. (ed.) TST 2014. CCIS, vol. 471, pp. 38–47. Springer, Heidelberg (2014). doi:10.1007/978-3-662-45317-9_5
2. CEPT Report 41: Compatibility between LTE and WiMAX operating within the bands 880-915 MHz/ 925-960 MHz and 1710-1785 MHz/ 1805-1880 MHz (900/1800 MHz bands) and systems operating in adjacent bands
3. ECC Report 096: Compatibility between UMTS 900/1800 and systems operating in adjacent bands, Krakow (2007)
4. CEPT ECC Report 127: The impact of receiver standards on spectrum management, Cordoba (2008)
5. CEPT ECC Report 146: Compatibility between GSM MCBTS and other services operating in the 900 and 1800 MHz frequency bands (2010)
6. CEPT ECC Report 162: Practical mechanism to improve the compatibility between GSM-R and public mobile networks and guidance on practical coordination (2011)
7. CEPT ECC Report 229: Guidance for improving coexistence between GSM-R and MFCN in the 900 MHz band (2015)
8. CG-GSM-R(13)035 Bundesnetzagentur; Coordination between UMTS and GSM-R

9. ETSI TS 145 005 Digital cellular telecommunications system (Phase 2+); Radio transmission and reception (3GPP TS 45.005 version 12.5.0 Release 12). Sophia Antipolis Cedex (2015)

10. ETSI EN 300 919: Digital cellular telecommunications system (Phase 2+) (GSM); Types of Mobile Stations (MS) (GSM 02.06)

11. ETSI TS 102 281 (V2.0.3): Railways Telecommunications (RT); Global System for Mobile communications (GSM); Detailed requirements for GSM operation on Railways

12. ETSI EN 301 515: Global System for Mobile communication (GSM); Requirements for GSM operation on railways. V2.3.0 (2005)

13. ETSI TS 102 933-1: (V2.1.1) Railway Telecommunications. GSM-R improved receiver parameters. Part 1: Requirements for radio reception (2015)

14. ETSI TS 102 933-2: (V2.1.1) Railway Telecommunications (RT). GSM-R improved receiver parameters. Part 2: Radio conformance testing (2015)

15. ETSI TS 137 104: E-UTRA, UTRA and GSM/EDGE; Multi-Standard Radio (MSR) Base Station (BS) radio transmission and reception

16. EIRENE System Requirements Specification. European Integrated Railway Radio Enhanced Network. GSM-R Operators Group. UIC CODE 951, version 15.4.0, Paris (2014)

17. EIRENE Functional Requirements Specification. European Integrated Railway Radio Enhanced Network. GSM-R Functional Group. UIC CODE 950, version 7.4.0, Paris (2014)

18. FM(13)134r2: Compatibility measurements GSM/UMTS/LTE vs. GSM-R (with 8 annexes); version 3 (2015)

19. http://www.kurierkolejowy.eu/aktualnosci/11637/Podpisano-umowe-na-ERTMS-na-E65. html. Accessed 15 Dec 2015

20. ITU-R P.525-2. ITU-R Recommendation. Calculation of free-space attenuation (1994)

21. Perzyński, T., Lewiński, A., Łukasik, Z.: Safety analysis of accidents call system especially related to in-land water transport based on new telematic solutions. In: Mikulski, J. (ed.) TST 2015. CCIS, vol. 531, pp. 90–98. Springer, Heidelberg (2015). doi:10.1007/978-3-319-24577-5_9

22. Perzyński, T., Lewiński, A., Łukasik, Z.: The concept of emergency notification system for inland navigation. In: Weintrit, A., Neuman, T. (eds.) Information, Communication and Environment. CRC Press Taylor & Francis Group, London (2015)

23. RADIO SPECTRUM COMMITTEE: GSM-R Interferences – Contributions from delegations and ERA on issues. Statistics and best practices as a follow-up to the discussion in RSC#42. Working Document. European Commission. DG Communications Networks Content & Technology, Brussels (2013)

24. Rosiński, A.: Modelling the Maintenance Process of Transport Telematics Systems. Publishing House Warsaw University of Technology, Warsaw (2015)

25. Rosiński, A.: Reliability-exploitation analysis of power supply in transport telematics system. In: Nowakowski, T., Młyńczak, M., Jodejko-Pietruczuk, A., Werbińska-Wojciechowska, S. (eds.) Safety and Reliability: Methodology and Applications - Proceedings of the European Safety and Reliability Conference, ESREL 2014, pp. 343–347. CRC Press/Balkema (2015)

26. Stawowy, M., Dziula, P.: Comparison of uncertainty multilayer models of impact of teleinformation devices reliability on Information Quality. In: Podofillini, L., Sudret, B., Stojadinovic, B., Zio, E., Kröger, W. (eds.) Proceedings of the European Safety and Reliability Conference, ESREL 2015, pp. 2685–2691. CRC Press/Balkema (2015)

27. Sumiła, M., Miszkiewicz, A.: Analysis of the problem of interference of the public network operators to GSM-R. In: Mikulski, J. (ed.) TST 2015. CCIS, vol. 531, pp. 253–263. Springer, Heidelberg (2015). doi:10.1007/978-3-319-24577-5_25

28. UIC O-8736-2.0. UIC Assessment report on GSM-R current and future radio environment (2014)
29. UIC O-8740. Report on the UIC interference field test activities in UK (2013)
30. http://www.ertms-conference2016.com. Accessed 14 Dec 2015

Railway Radio Communication Systems in the Context of an Emergency Call

Mirosław Siergiejczyk[(✉)]

Faculty of Transport, Warsaw University of Technology, Warsaw, Poland
msi@wt.pw.edu.pl

Abstract. This article presents selected issues of cooperation between the VHF 150 MHz radio communication system and the GSM-R system. Both of them will be used in Polish railways. Particular attention was paid to issues associated with linking RADIOSTOP emergency signals of the VHF 150 MHz radio communication system and the REC (Railway Emergency Call) of the GSM-R system, as well as eREC (enhanced Railway Emergency Call) for the GSM-R standard. The RADIOSTOP railway emergency signals of the VHF 150 MHz radio communication networks were analysed. The REC emergency connection in the GSM-R system was characterised. Followed by a discussion of technical problems associated with generating emergency signals in the period of system migration. In addition to the problems associated with the technical aspect of connecting the two alarm signals, the issues arising from their radio coverage were also analysed.

Keywords: Railway · Radio communication · Emergency call · GSM-R system

1 Introduction

Interoperability of a railway system means the ability to allow the safe and uninterrupted movement of trains which meet the required efficiency level of these lines. Adapting a standardised traffic control system was necessary to ensure interoperability. Supported by the European Union, a project of a unified rail traffic control system is the ERTSM (European Rail Traffic Management System). The ERTMS guarantees the chance for the trains to freely move over rail networks of individual countries without the need to stop at the borders [2]. The ERTMS consists of a traffic control system (ETCS) and a radio connection system (GSM-R). Thanks to the ERTMS, the train driver receives detail data on permissible movement parameters for a given section of the railroad and all messages are sent by the system to the driver's desktop. GSM-R is the transmission medium for the ETCS system (level 2 and 3), and transmits, i.e., permits to move issued by RBC (Radio Block Centre) [16]. GSM-R replaces a number of national rail radio communication systems and its implementation results in not only an increase of the radio communication quality, but also allows to eliminate barriers stemming from different standards of voice communication and data transmission used in Europe [6, 7].

© Springer International Publishing AG 2016
J. Mikulski (Ed.): TST 2016, CCIS 640, pp. 223–234, 2016.
DOI: 10.1007/978-3-319-49646-7_19

Every day, thousands of trains enter rail routes in Europe. One of the most important prerequisites for a train to be permitted for traffic is equipping a traction unit with radiophone in working order. Currently, the railway network in Poland works in the 150 MHz band and that system is the basis for the implementation of the GSM-R and ETCS systems. After correct implementation of these systems, the 150 MHz radio will be disabled in the assumed time periods. However, replacing a 150 MHz radio with the GSM-R standard required a simultaneous operation of these networks during the migration period. It needs to be noted that there are certain problems associated with linking the RADIOSTOP alarm signal of the VHF 150 MHz radio communication and the REC of the GSM-R system.

The problem of coexistence of GSM-R and VHF 150 MHz, and in particular the cooperation signals RADIOSTOP and REC, first described in a document adopted by the Council of Ministers in 2007 "National Programme for the Implementation of the European Railway Traffic Management System in Poland" [6]. This document indicates that the function RADIOSTOP buildings on lines GSM-R, will be used until the complete dismantling of the infrastructure of analog system on these lines. Therefore, it is necessary to develop a technical solution and organization that enable the migration period to tie together the two systems. This will allow the smooth operation of the train without any ambiguity that might impede the maintenance of the current level of safety.

One of the key issues in the implementation of the GSM-R system is the problem of organizing the usage of the radio network, taking into account both, the requirements for the continuity of telecommunication services, their safe transmission and the cooperation with the previously operated analogue VHF 150 MHz radio communication network.

2   Analysis of the RADIOSTOP Signal

The radio communication system used in Poland was developed around 1965 in order to ensure communication between train drivers and traffic orderlies on the route. The Centre for Research and Development of Railway Technology (currently the Railway Institute) prepared a document "Study of the PKP radio communication concept". It was decided that the system was to have an option of selective group call in order to more rapidly call a train in case of emergency. Finally, after analysing the scope of interference and propagation of radio waves, the VHF 150 MHz band was chosen for the needs of radio communication. This radio communication system is not compliant with the recommendations of the UIC 751-3 sheet and does not provide interoperability required by the EU.

The VHF 150 MHz is primarily connected with the driver on duty traffic, train conductor and connectivity drivers on the rail lines. According to the instructions of PKP Polish Railway Lines SA concerning the use of radio communication devices traction, radio communication is designed to ensure communications between:

- train dispatcher and train driver,
- train dispatcher movement of adjacent traffic posts - in the event of a total interruption in the wired communications or security threats rail traffic,

- train drivers located on the same route,
- lineman rounds of track and train driver,
- drive traveling on the trail or between lineman rounds and train dispatcher check-points restrict movement of the trail - only in cases of danger or safety incidents railway (railway accidents, railway accidents)
- support the position of the terminal devices detect faults rolling stock and train driver,
- conductors team and train driver,
- conductors team and train dispatcher - in case of security threats traffic and railway incidents and the need to communicate passenger trains.

Existing system is not intended for the staff stations communicate with one another setting. Nevertheless, in case of interruption of telephone communications can be used for the radiotelephone announcing trains. If so used tables of different identities for each route, which are known only to the railway control rooms. Such arrays consist of two rows of figures. In order to prevent turning on the communication to third parties control room impressive given any two adjacent numbers on the top row, and the control room answering given numbers underneath them.

The currently operated analogue radio communication system was defined as a class B communication system. According to the Decision of the EC 2012/88/EU [1], class B systems cover a limited set of existing, national communication and train control systems that were in operation before April 20, 2001. A list of these systems can be found in the technical document published by the European Railway Agency, titled "List of class B Control Systems" (ERA/TD/2011-11, version 2.0). These systems greatly hinder the interoperability of locomotives and traction units but play an important role in maintaining a high safety level in the area of the trans-European railway network. In the ERA/TD/2011-11 document, one of the indicated class B communication systems was a system defined as "SHP PKP radio system with a RADIOSTOP function". There-fore, an analogue VHF 150 MHz system with the RADIOSTOP function provides class B railway communication in Poland. In order not to allow the creation of additional obstacles for interoperability, the Member States should ensure further functioning of existing class B systems or their interfaces, according to the current specifications, unless modifications are necessary, aimed at removing safety-related flaws of these systems.

An analogue train radio communication network is equipped with an area braking (HO) function, otherwise known as the RADIOSTOP function and used for providing safety of railway traffic. According to the guidelines of the Company PKP Polskie Koleje Państwowe S.A. (PKP Polish National Railway), all traction units intended for train operation need to be equipped with devices of the RADIOSTOP system [10].

The RADIOSTOP system allows immediate halt of moving traction units, in the place where there is a hazard for traffic safety. Therefore, the area braking function is used to stop all trains being within the range of the broadcast distress signal. The trains stop after receiving the RADIOSTOP signal on the channel, on which the receiving station of the train is set at. Analogous in meaning, i.e. the order to stop all moving traction units, is the alarm signal communicated verbally.

A radio-phone unit of the train, apart from the typical systems, contains an automatic transmitter and receiver of the HO alarm signal. The RADIOSTOP signal is activated by the train driver after pressing the "Alarm" button on the radio-phone keypad, which results in the special sound signal being sent by radio. The "Alarm" signal is a combination consisting of three successive short tones that are diverse in terms of frequency and periodically repeated. As a result, the carrier wave's signal is modulated with a defined sequence (3 × 100 ms) of three acoustic signals f1 = 1,160 Hz, f2 = 1,400 Hz and f3 = 1,670 Hz, followed by a 500 ms interval. The signal is broadcast continuously, until being turned off. An employee that broadcast the "Alarm" signal should immediately use the alarm channel of the radio-phone to inform the nearest traffic orderly about the cause of broadcasting the signal. It does not apply to employees who need to evacuate the control cabin in an emergency. The traffic orderly should also immediately notify the dispatcher about that fact. Activation of the RADIOSTOP function causes immediate, automatic braking of trains being within the range of a given radio-phone. Automatic braking is activated when the radio-phone on the traction unit is connected with ATP (Automatic Train Protection) equipment. The devices used by PKP are not dependent on semaphore indications. The RADIOSTOP function itself is installed in a traction unit, on the second pneumatic duct.

The button activating the alarm signal should be of red color and ergonomic shape. Its design and location should guarantee the user easy access, as well as protect against accidental use. The ALARM signal itself is a combination consisting of three successive short tones that are diverse in terms of frequency and periodically repeated. The RADIOSTOP function on line covered by the GSM-R system will be used until the 150 MHz radio system is disassembled on these line and was taken into account in the STM module for the ATP system.

A procedure for using the alarm system, developed in the PKP Polskie Koleje Państwowe S.A. (PKP Polish National Railways), has been in force since 2004 on PKP railway lines. This procedure was thoroughly described in the "User manual for the Ir-5 train radio communication devices", in the part of the Radio-telephone alarm signal. The knowledge of principles described in the manual is essential for both the drivers and the traffic orderlies [9].

3   The REC/e-REC Railway Emergency Call Feature in the GSM-R Network

Railway emergency connections are used to communicate information to railway personnel about dangers in a given area that require specific actions or halting a moving train. The REC (Railway Emergency Call) function is one of the most important functions of the GSM-R system. It is, in part, an equivalent of the RADIOSTOP function implemented in the VHF 150 MHz system. The REC function was thoroughly described in the following documents [4, 5]:

– UIC Project EIRENE – Functional Requirements Specification – Version 7.3.0,
– UIC Project EIRENE – System Requirements Specification – Version 15.3.0.

The REC feature is a special type of a group call feature with the highest priority (0). Due to a short time of setting up the connection, the REC signal is not coded.

According to the system specification contained in EIRENE documentation, there are 2 types of the REC connection:

- Train Emergency Call,
- Shunting Emergency Call.

In their specifications, EIRENE points out that a train emergency call should be sent to all drivers and traffic orderlies within a defined area and, additionally, the information about a REC connection should be passed to a proper RBC. In contrast, a shunting emergency call (SEC) should be sent to all users engaged in actions associated with shunting in a given area and should automatically assume connection ensuring priority.

The call type initiation is determined automatically, depending on the radio operation mode. If a moving object is in the "shunting" operation mode, then the emergency call button should initiate a SEC and in another case, the call should be a standard REC emergency call. The functions associated with initialising an emergency call are mandatory.

The connection type is determined automatically, on the basis of the mode of operation of the terminal that initiated the connection. It is necessary that a railway emergency connection is forwarded to all drivers and traffic orderlies within the area of operation, where the danger occurred.

A REC alarm call, according to the eMLPP (enhance Multi-Level Priority and Preemption) function, has the highest priority. Therefore, initiating an alarm will break all current connections and set-up a distress channel. Unlike the RADIOSTOP system, that feature does not halt trains but only allows the transfer of information about a danger. The dispatcher makes the decision to stop a train/trains, after getting information regarding the cause that triggered the alarm from the driver who generated the notification.

According to the REC feature specification included in the EIRENE documentation, the REC feature should also enable configuring the areas of a group emergency call in such a way, so that it contains combinations of cells controlled by one or more MSC units in one or more networks. In such a case, when the REC areas are controlled by more than one SMC, it is necessary to define a parent MSC for each area of a group call. In case of international connections, in order to minimise the emergency call set-up tie, the parent MSC directly controls the cell, in which a REC call was initiated.

According to EIRENE requirements, an alarm connection is received automatically. It is a duplex connection. The user initiating a REC call is marked, with the aim to distinguish him from other participants of a railway emergency call in the dispatcher terminal. The procedure for train crews is stopping the train after receiving a REC signal – the journey may be continued after a permit from an employee authorised to control traffic in the area, through a point-to-point connection with the train. Initiating a REC call should be confirmed by the base station with a feedback code number 1612 – with the aim to prevent making an emergency call with the use of an unauthorised radio-phone.

The REC feature is one of the most important GSM-R features. The alarm call has a high priority for calls informing train drivers, dispatchers and other persons about traffic stoppage due to a hazard in a given area. It is activated by pressing a special, distinctive alarm button on radios and in radio cabins. The emergency call is set-up within 2 s from pressing the button and the REC reception is automatically confirmed (confirming feedback information contains the locomotive number or train number).

This connection is defined for three phases [7]:

– phase 1: warning,
– phase 2: information,
– phase 3: alarm end.

The eREC service is an advanced emergency connection service. Similarly to the eLDA function, in order to determine a train's location, GPS data and data from near-track detection devices are used. It prevents situations where a train, being on the border of two sections, connects to the wrong dispatcher and, consequently, sends an alarm message to incorrect trains.

The eREC (enhanced Railway Emergency Call) service is an advanced emergency connection service. Similarly to the eLDA (enhanced Location Dependent Addressing) function, in order to determine a rain's location, GPS data and data from near-track detection devices are used. It prevents situations where a train, being on the border of two sections, connects to the wrong dispatcher and, consequently, sends an alarm message to incorrect trains.

The eREC system, thanks to the introduction of the concept of sectors, allows to distinguish a REC signal in a given telecommunications cell. A sector is an area defined for a given railway route or line. The concept of a sector is not identical as radio coverage, therefore, a single GSM-R cell may contain several eREC sectors. Figure 1 presents a sample cell diagram of the GSM-R network. In the case of the eREC function, each of these cells is divided into sectors. A single cell of the GSM-R system may contain from 1 to 9 sectors.

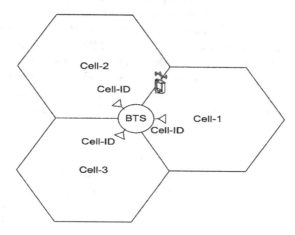

Fig. 1. Sample scheme of GSM-R network cells (own study on the basis of [16])

Each of the sectors of the eREC feature receives an ID number (eREC Sector Identity). This attribute is associated with the distress signal and allows to distinguish users within a given area. A user who wants to use the eREC function must register the terminal in the GSM-R network, according to a procedure specified by EIRENE. Only after the registration is complete, the radio will receive updates of the sector ID number. Updates of the sector ID number are possible through:

– eREC radio interface through USSD messages (ID number is obtained from the train location information database),
– a balise.

Once the sector ID number is granted, it is necessary to confirm it. The carrier enabling confirmation of the ID number is the SMS-CB (SMS Cell Broadcast).

Management of a connection with a eREC feature is based on the following requirements:

– call initiation – the network grants a relevant sector ID number during the call set-up,
– setting up a network connection – the network forwards call data, together with an eREC sector ID number to the group call register (GCR),
– receiving the call – the user is added to the conversation, if the sector ID number registered in the GCR is identical to the active sector number in a given area.

Moreover, the interface between the eREC and a dispatcher must remain unchanged, i.e., procedures for setting-up and ending the call are identical, as in the case of the REC feature and are based on the group call register (GCR).

4   Evaluation of Cooperation of Radio Communication Systems Taking into Account RADIOSTOP and REC Signals

According to a provision in the NPW ERTMS, the RADIOSTOP function in Poland will be used on lines covered by the GSM-R system until the VHF 150 MHz system is dismantled on these lines. That function has been included in the STM module for the ATP system and is to guarantee proper interaction of the RADIOSTOP function with traction units equipped with ECTS and ATP STM going through lines equipped with SHP and a 150 MHz radio. Therefore, in the migration period, the vehicles moving along the lines equipped with the 150 MHz and GSM-R standards are required to answer and properly interpret the RADIOSTOP signal. As a result, inconveniences associated with the use of two systems will arise. They will be felt by train drivers, traffic orderlies, as well as maintenance crews. This involves, i.e. the need to acquire proficiency in operating two types of equipment from two different radio communication systems and equipping workshops with a higher number of specialized testers and measuring instruments. This will lead to an increase in the scope of spare parts and equipment. In addition, the driver should switch from the analogue network to GSM-R in places marked with a W33 sign and from the digital network to the VHF 150 MHz network in places marked with a W34 sign [6, 11].

Moreover, for each railway line equipped with GSM-R, as long as it is equipped with the 150 MHz radio, the trackside devices have to guarantee that:

– generating of a RADIOSTOP signal by the trackside devices will be accompanied by automatic generating of the GSM-R emergency signal (phase 1),
– the RADIOSTOP signal received by trackside devices (e.g. generated by the train driver) will automatically cause a GSM-R emergency signal to be generated (phase 2),
– generating of a GSM-R emergency signal by the trackside devices will be accompanied by automatic generating of a RADIOSTOP signal (phase 3),
– a GSM-R emergency signal received by trackside devices (e.g. generated by a train driver or an employee moving along the tracks) will automatically cause a RADIO-STOP signal to be generated (phase 4).

The drawing Generating alarm signals during the migration period of the VHF 150 MHz and GSM-R railway radio communication systems, is presented in Fig. 2 systems in relation to the LCS.

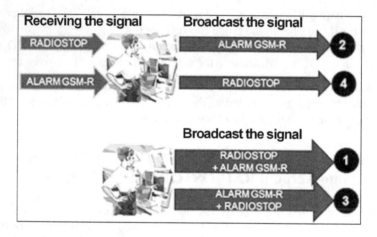

Fig. 2. Generating alarm signals during the system migration period [17]

In regard to the above information, it is assumed that the traffic orderly will be in possession of an integrated terminal that will simultaneously send RADIOSTOP and REC signals. Such an integration of a terminal is possible through adding appropriate interfaces between the dispatcher side that operates a 150 MHz radio and the dispatcher side of the GSM-R standard.

Apart from the problems associated with the technical aspect of the call of both alarm signals, we need to remember about the differences arising from their range. It is not possible to program the GSM-R network in such a way that the REC signal covers exactly the same range as the RADIOSTOP signal. Figure 3 presents the range of both.

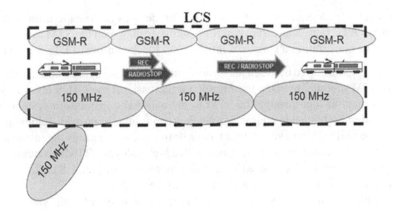

Fig. 3. The range of GSM-R and VHF 150 MHz systems in relation to the LCS [17]

Two cases may be considered:

- activation of the RADIOSTOP signal in the LCS area,
- activation of the REC signal in the LCS area.

In the first case, when a RADIOSTOP signal will be broadcast in the LCS area, an immediate braking of trains using the VHF 150 MHz standard will take place. In contrast, interoperable trains, using the GSM-R radio communication system will not be halted; only the automatic connection setup of the REC channel for all trains covered by the RADIOSTOP signal will occur.

In the second case, broadcasting the REC signal will cause a minimally delayed braking of class B trains (using the analogue standard) in the LCS area; in class A trains the automatic connection setup of the REC channel will occur. Both of the above cases do not favour efficient management of traffic and introduce ambiguities that make it harder to ensure safe train traffic. Perhaps a more advantageous option would be to add an interface to the radio communication terminals that would allow processing REC signals to RADIOSTOP and vice versa, thus ensuring halting all trains on setting up a communication channel on all trains in the LCS area.

When analysing the efficient ranges of both systems, it needs to be noted that the RADIOSTOP signal covers the whole infrastructure over a radius of about 10 km, while REC, most of all, the infrastructure covered by a given base station. Therefore, in order to counter uncontrolled emission of the RADIOSTOP signal, it would be necessary to develop a new grid of train radio communication channels in the 150 MHz band, taking into account the implementation of Local Control Centres on railway lines in Poland.

Because the RADIOSTOP function will be used until the VHF 150 MHz system is fully dismantled, a Specific Transmission Module (STM) was develop, which allows proper interaction of the analogue emergency call with traction units equipped with ECTS. As a result, for the purposes of description completeness of the procedures of the ETCS system, 2 additional track equipment levels were defined:

- zero level – lack of equipment constituting a source of information for the on-board ETCS system,

– STM level – track equipped with national ATP devices (automatic train protection system) or ATC (automatic train control system), from which the STM module could gather data constituting a source of information for the on-board ETCS system.

There is a possibility to use such hardware configuration, so that, by updating through a specific STM module, manage train traffic on the basis of the ETCS level 1 system. Update information sent through the STM to the national transmission system "track-vehicle" is used by trackside equipment. Such a configuration ensures achieving technical interoperability. However, it does have a drawback – it requires the installation of an on-board STM device on units equipped with a higher-level ERTMS system devices. What is more, there is also a need to install a domestic on-board STMD device on vehicles of other carriers, entering a given railway network [14].

The simplest solution to the problem arising from the range of both systems, would be to implement the GSM-R system along the entire line. Such a concept would ensure the avoidance of perturbations associated with the operation of two systems during the migration.

Using two emergency signals on one line requires developing special procedures and principles of managing train traffic in emergency situations, due to the fact that picking up a RADIOSTOP signal by an analogue on-board radio cause an emergency halt of all trains equipped with the analogue communication system that would pick up the signal, while in the case of the REC signal being received by a GSM-R on-board digital radio should trigger the driver's reaction that is described in a relevant procedure [3].

In the case of lines managed by the PKM Pomorska Kolej Metropolitalna (PKM – Pomeranian Metropolitan Railways), one of the requirements of access to PKM lines is equipping trains with radio communication devices working in a 900 MHz band. The relation between the VHF 150 MHz analogue network and the GSM-R system occurs only on the contact point of lines managed by PKP Polskie Linie Kolejowe S.A. (PKP Polish Railway Lines) and the PKM lines. Of course that relation only concerns the possibility of the "Alarm" signal of the analogue network, broadcast automatically by the RADIOSTOP system installed on a train, to be received by radio-phones of the analogue network or two-system radios, despite their voice function being turned off. According to the manual PKM-04, which regards the use of radio communication devices, in the case of a train being stopped by the RADIOSTOP system, one should follow the "User manual for train radio communication devices" Ir-5 (R-12) PKP Polskie Linie Kolejowe S.A. (PKP Polish Railway Lines) [12, 13].

5   Conclusion

A very important factor influencing the railway transport safety is the efficient and uninterrupted communication, allowing rapid and efficient flow of information necessary to manage and control traffic and handle passengers. In Poland, the currently radio communication VHF 150 MHz system, stopped being developmental a long time ago. However, the most important drawback, from the point of view of the European Union, is the inability to provide interoperability required in the EU. It makes safe and uninterrupted train traffic on the territory of EU Member States impossible. Thus, it is necessary to quickly upgrade the current railway network up to the European standards and replace the VHF 150 MHz radio with the GSM-R standard.

The technological leap involved with the implementation of a modern radio communication system brings along also many threats. From a technical point of view, the small insufficient number of frequencies is a very large threat. This problem might arise especially on big hubs and railways stations, where shunting communication is additionally required. In order to solve that issue, the International Union of Railways (UIC) submitted to the European Telecommunication Standards Institute (ETSI) a request to extend the GSM-R in the trunk band (so called E-GSM-R).

The problem of cooperation systems, VHF and GSM-R is limited to cooperation signals RADIOSTOP and REC. A prerequisite to maintain security at the appropriate level is to develop operating procedures and conversion of both signals, so that it is uniform for both analog radio and digital. Unfortunately, there are on any railway line in Poland there is no cooperation between the two systems. The infrastructure manager is at the stage of testing various technical solutions that allow the retransmission of both alarm signals. Only after analyzing the results of tests he submits a way of cooperating systems in the area of emergency call. Please note that the modernization of the railway radio system is a process that requires the use of appropriate organizational procedures during the migration period to ensure safe monitoring of drive trains equipped with radio cabin of the old and the new type.

Another challenge is the operation of the GSM-R system during the migration period, i.e. cooperation with the VHF network – especially in the scope of emergency calls. Differences stemming from the technology of both systems constitute a considerable challenge for the engineers that will have the task to configure the devices of both networks, so that they are able to "get along". A much bigger system is the parallel operation of two emergency systems, i.e., RADIOSTOP and REC. Moreover, we need to remember that the migration period is a period of substantially increased maintenance and operation costs of both systems and significant organisational difficulties for basic participants of the transport process; train drivers and traffic orderlies that need to be familiar with and operate two considerably different radio systems. Therefore, the implementation of the new system will result in a lack of trained personnel. Simultaneously, there might be a problem with creating appropriate qualifications of the workforce that has been operating analogue devices for a few dozen years.

Summing up, the GSM-R standard replaces a number of national railway radio communication systems and its implementation results in increasing the quality of railway radio communication and allows to eliminate barriers stemming from different communication standards operated in Europe. Therefore, it is necessary in Poland to replace the current, analogue VHF 150 MHz system with the GSM-R standard and to smoothly pass through the system migration period.

References

1. Committee Decision no. 2012/88/EU of 25 January 2012 on technical specification for interoperability in the scope of "Control" subsystems of the transEuropean railway system (Dz. Urz UE L 51 z dnia 23.2.2012 r str. 1, z późn. zm.) (in Polish)
2. Directive of the European Parliament and the Council 2008/57/EC of 17 June 2008 on the interoperability of the railway system in the European Communite
3. Gago, S., Siergiejczyk, M.: Operational issues of railway radio communication systems in the migration period. Logistyka, April 2014 (in Polish)
4. EIRENE, Functional Requirements Specification Version 7.4.0, April 2014
5. EIRENE, System Requirements Specification Version 15.4.0, March 2014
6. National Programme for the Implementation of the European Railway Traffic Management System in Poland, Warsaw (2007) (in Polish)
7. Pawlik, M.: Polish National Programme for Implementation of the European Railway Traffic Management System ERTMS. Technika Transportu Szynowego, January 2007 (in Polish)
8. PKP Polskie Linie Kolejowe S.A.: Technical standard vol. VII: Telecommunication, Warszawa (2009) (in Polish)
9. PKP Polskie Linie Kolejowe S.A.: Attachment to the ordinance no. 17/2004, of 27 December 2004, User manual for the Ir-5 train radio communication equipment (in Polish)
10. PKP Polskie Linie Kolejowe S.A.: Attachment to the ordinance no. 22/2004, of 27 December 2004, User manual for the Ie-14 train radio communication equipment (in Polish)
11. PKP Polskie Linie Kolejowe S.A.: Manual Ie-1. Signalling manual, Warszawa (2015) (in Polish)
12. PKP Polskie Linie Kolejowe S.A.: Attachment to the ordinance no. 4/2012, of 10 January 2012. Technical-operational requirements for the Ie-105 train radio-phone (in Polish)
13. Pomorska Kolej Metropolitalna S.A.: PKM-04 User manual for radio communication equipment, Gdańsk 2015 (in Polish)
14. Siergiejczyk, M. (ed.): High Speed Railways in Poland. Wydawnictwo Instytutu Kolejnictwa. Railway Institute Publishing House, Warsaw (2015) (in Polish)
15. Siergiejczyk, M., Mierzejewska, A.: Issues of linking railway radio communication systems in the context of an emergency call. Logistyka, April 2015 (in Polish)
16. Winter, P.: International Union of Railways, compendium on ERTMS, Eurail Press, Hamburg (2009)
17. www.gsmrail.pl – Ryszard Markowski's website on the GSM-R system (in Polish). Accessed 12 Dec 2015

The Impact of the Tolling System for Road Traffic Safety

Maria Michałowska and Mariusz Ogłoziński[✉]

Department of Transport, Faculty of Economics,
University of Economics in Katowice, Katowice, Poland
mariamic@ue.katowice.pl, oglozinski@o2.pl

Abstract. The article concerns the issue of application of ITS solutions in toll collection and their impact on road safety. The aim of the article is to present an electronic tolling system as the most effective formula of payment for the use of road infrastructure in the context of ensuring road safety. In the article there are presented results of a study about traffic incidents in the area of toll station Żernica on the A4 motorway near Gliwice Sośnica junction, where the implementation of manual tolling system resulted in many dangerous situations and deterioration of safety indicators in the test area. The article refers to the IV European road safety programme, and the positive effects of implementation of telematics in transport are shown. The results of a research in the effectiveness of ITS solutions in improving road safety were summoned, pointing to their broad capabilities in this area.

Keywords: Tolling system · Road safety · Effectiveness

1 Introduction

The use of high-tech information and communication solutions in road transport can contribute significantly to increase its efficiency and safety. The role of intelligent transport systems (ITS) in enhancing transport safety has been identified in the IV European road safety programme [2], and past experience suggests telematics solutions as one of the most efficient tools to improve road safety. Progressive development of telematics technologies and the possibility of their integration, make that this is the direction of the future, though now the potential of intelligent transport systems in Poland is not widely used. Extensive use of any smart solutions in road transport requires a change in the philosophy of road transport management, including its safety. ITS systems, allowing better traffic management, improve safety and traffic flow, as well as contribute to reduce exhaust and noise emissions [6]. The positive effects connected with application of telematics are also the targets achieved for safer road infrastructure, safer vehicles, driver's support systems, improving services in emergency situations and the protection of vulnerable road users [5].

© Springer International Publishing AG 2016
J. Mikulski (Ed.): TST 2016, CCIS 640, pp. 235–242, 2016.
DOI: 10.1007/978-3-319-49646-7_20

2　ITS in Road Safety

The results of research carried out in Europe, the United States, Canada, Japan and Australia on ITS systems in road transport and their effectiveness in improving road safety indicate that the use of intelligent transport systems can result in:

- reduction in expenditure on transport infrastructure by 30 % while maintaining the same functionality of the system,
- higher efficiency of the transport system, and better use of the infrastructure objects (for example, higher bandwidth of the road network, shorter waiting time for passage at intersections) by 20 %,
- increasing the level of safety in road traffic (reducing the number of road accidents and their victims) by 40 %,
- reducing the negative impact of transport on the environment (less exhaust emissions and noise) by 10 %,
- easier integration of different modes of transport and connections with other systems through greater intermodality of transport and minimizing the negative effect of bottlenecks [7].

The capabilities of such systems seem to be endless, and examples of ITS solutions in road transport are shown in Table 1.

There is a great variety of possible use of ITS systems in road safety:

Table 1. Examples of applications of ITS solutions in road traffic (own study based on [7])

Areas of application	Implemented services
Traffic management and travel	Information prior to travel
	Information for motorists while driving – sudden events, traffic jams, road works, weather
	Traffic control
	Exhaust emission control
	Control of intersections of roads and railways
	Assistance to traffic law enforcement
	Infrastructure maintenance management
	Keeping the line and navigation
	Secure parking lots
	Toll collection
Accident management	eCall accident notification service
	Management of emergency services vehicles
	Management of emergency actions
Vehicle safety systems	Preventing collisions in traffic (sections and junctions)
	Video anti-crash systems
	Safety watch
	Anti-crash safety
Safety systems	Safety of public travels (including pedestrians)
	Safety of disabled road users
	Smart intersections

- inside vehicles systems,
- road infrastructure systems,
- mixed systems (cooperation between vehicles and infrastructure).

Systems within the vehicles collect data relating to the movement of the vehicle and its surroundings and transfer it to the driver in order to optimize the decision-making processes. In the extreme case the systems shall take the actions necessary to avoid collision or minimize its effects. ITS road infrastructure systems collect data on the road infrastructure and provide it to the users. Mixed systems combine the possibilities of both of them [4].

Use of ITS systems to improve road safety has been presented in White Paper of the European Commission in 2001 as a desirable action to change the transport from traditional to more intelligent [1]. The overall aim of these activities was to create a clean and safe road transport, as well as the integration of ITS systems in all modes of transport, which should contribute to better management of transport infrastructure. In the White Paper published in 2011, ITS systems are presented as the most effective tools for improvement of road safety, which use could change the current approach to these issues. The document specified, that development of transport sector should be based e.g. on increasing its efficiency through better traffic and information management systems, what in relation to road transport, refers unambiguously to ITS systems. The increase of efficiency of use of road transport and infrastructure is not possible without increasing the level of safety in road traffic, and this can be achieved through wide use of telematics solutions. Increasing the efficiency of use of transport was set out in the White Paper of 2011 as one of ten main objectives for the creation of a competitive and resource efficient transport system. The European strategy of research, innovation and implementation in the field of transport, assumes that the development of ITS systems will contribute to better use of the road network, its greater safety and a more rational exploitation. Presented in this document, a list of initiatives in the field of transport safety aims at important reduction in number of road accident victims through the harmonisation and introduction of new technological solutions in the field of road safety, such as supportive systems for drivers, intelligent speed limitation devices, the eCall service, systems of cooperation and vehicle-infrastructure interfaces (V2I) [3].

The possibilities related to the use of ITS technology in improving road safety, according to estimates of the authors of the OECD report, may result in reduction in the number of road accident victims in the OECD countries for about 47.000 a year and savings associated with that of 73 billion dollars per year, as well as reduction in the total number of victims and wounded in road accidents by 40 % a year and savings associated with that of 194 billion dollars per year [8]. Included in the report estimates of reduction in the number of accidents, victims and people injured are based on the assumption that ITS technologies are fully implemented in all the member countries of OECD. However, due to varying degree of progress in implementation of ITS systems in individual countries, it is estimated that the condition might not be reached even by another 20 years. The implementation of the ITS technologies in road transport therefore is not a one-time action, but a long-term process. The authors of the report also highlight the problems and challenges relating to the implementation of ITS systems in the service of road safety (Table 2).

Table 2. The challenges of a wide implementation of ITS systems in the service of road safety: (own study based on [8])

Challenge	Description	Action
Financing	1. The high cost of modern road traffic safety systems 2. Many countries are struggling with budget problems, which largely affects investments in road safety 3. Unwillingness to invest in costly and unproven technologies	It is necessary to promote ITS technology, in particular highlighting the advantages of its use and creating a good climate around it
Assessment of the solution	The possibility of malfunction of one or several ITS subsystems on a domino effect (one defect causes the next)	Current assessment of the implemented ITS solutions is necessary in order to monitor their performance and quality
Users	1. Telematics solutions may distract the driver, give a false sense of safety and encourage risky behaviour on the road 2. Users/drivers should be aware of the possibilities and limitations of ITS systems	1. The need to regulate the law of using ITS solutions in vehicles, since some of them can distract the driver and reduce the level of road safety 2. The need for training the users/drivers in the use of ITS solutions
Responsibility and legislation	1.It is needed to decide on the role of a human in ITS system: weather one fully obey an autonomous ITS system, or only the system provides the user data on the basis of which one shall take decisions 2. Who is liable in the event of a collision	1. Implemented ITS solutions should be subjected to detailed quality tests to ensure maximum confidence and responsibility for their operation 2. The need to certify new ITS solutions
Infrastructure	Should forgive errors, be safe, high-quality, ready to work, interoperable, act on the basis of the developed standards	Steps to develop common standards for the use of ITS systems infrastructure should be taken, which will affect the efficiency of the transport – it is the main argument for the use of ITS solutions
The staff	It is necessary to ensure a well prepared staff dedicated to road safety and ITS	In order to make the full use of the ITS systems potential, preparing adequate personnel should be made in the public administration and the private sector
Research and development	In order to further development of ITS technology in road safety R&D sphere should concern its weak and strong sides, human issues, as well as the legal aspects of its use and the possibility of extending its application	It is required to collect and store detailed data about the road safety performance and the causes of road accidents, including the participation of ITS systems – it is necessary for the purpose of continuous improvement of ITS technology

The potential benefits in the area of road safety, relating to the implementation of ITS systems are also indicated in the directive 2010/40/EU of the European Parliament and of the Council on the framework for the deployment of intelligent transport systems in the field of road transport and for interfaces with other modes of transport. The development of ITS applications related to safety and protection of road traffic has been found to be one of the four main priority areas. It was assumed that innovation will play a major role in solving the problems of transport in the European Union. Implementation of solutions based on ITS will provide innovative services related to various modes of transport and traffic management, and will help to better inform users and will provide a safer, coordinated and intelligent use of transport networks.

3 Tolling System in Poland

The constant development of ITS solutions in road transport gives the ability to technological advancement. The use of this opportunity in Poland is conditioned with approach perspective to these issues by policy makers at all levels of administration (national, regional and local). The existing toll collection system on motorways for vehicles up to 3.5 tonnes of gross maximum weight is an example of organizational torpor and lack of openness to new technological solutions. Until recently, a small proportion of the motorways in the network of roads in Poland and the limited availability of telematics technology, caused that ITS charging systems solutions for the use of road infrastructure have not been widely used. There was a real chance to launch, together with the development of the motorway network, one nationwide electronic toll system with ITS solutions, which would be consistent with the functioning since 2011 toll collection system for trucks and buses (ViaToll). In the opinion of the authors of the literature of the subject, the construction of traditional – manual charging points on the motorways A1, A2 and A4 resulted in completion of Poland to the group of old technology countries [5]. Installation of the manual charging system can be seen in different aspects (Table 3).

Many of the negative aspects above have a significant impact on road safety. One of the most important is that the drivers have to stop their vehicles on the toll station. Often it makes long queues, aggression and irritation of drivers, what is particularly dangerous in road traffic.

4 Road Safety Indicators on A4 Motorway

The research was carried out in the central part of the A4 motorway in southern Poland near Gliwice Sośnica node. The motorway was not tolled for passenger cars since it was built until 2012. On the 1st of June 2012 there was introduced a manual toll collection system for drivers of vehicles up to 3.5 t of gross maximum weight. Before that happened, the toll station was under construction for one year. The research was based on data delivered by the Regional Police Headquarters in Katowice. At first the data for the "toll station in operation" period was prepared. This included: the data for the toll station, and the area of 1 km before and after the toll station on the motorway in

Table 3. Negative and positive aspects of introduction of the manual charging system on motorways in Poland [own study]

Manual charging system				
Negative aspects				Positive aspects (disputable)
Organizational	Economic	Environmental	Road safety	
Different payment systems depending on the operator of the motorway – the inability to use the viaAuto device on motorways belonging to concessionaires	1. Costs of construction and maintenance of the toll stations 2. Costs of employment the staff of the toll stations, working in conditions of exposure of health and life 3. Costs of printing, distribution and disposal of tickets 4. The need to stop the vehicle and often multiple starting of the vehicles, increasing fuel consumption 5. The artificial generation of congestion 6. The time lost while waiting in line for toll stations	1. Printing, distribution and disposal of tickets with magnetic strip 2. Lack of recycling – the tickets are not used again 3. Exhaust emissions connected with waiting in line for toll stations	1. The threat of road safety connected with the need to stop the vehicle and often repeated starting off, the increased possibility of the crash 2. Stress, irritation, aggression of the drivers resulting with the necessity of waiting in line for toll stations 3. The desire to make up for lost time after leaving toll stations - increased probability of risky behavior on the road	1. Closed system – no posting services to control and the enforcement of the obligation to pay the toll 2. Increased fiscal revenues with increased fuel consumption and operation of the vehicles

both directions (Katowice and Wrocław). Then the data for the same distance was prepared for the remaining two periods: "toll station under construction", and "no toll station".

The results of the research indicate that introducing a manual toll collection system on the toll station Żernica on the A4 motorway, resulted in deterioration of road safety indicators compared with the same period before the construction of the toll station Żernica has started. The study referred to three periods:

- 01.04.2010–31.03.2011, no toll station, the traffic was running smoothly without the need for stopping vehicles in order to pay the toll,
- 01.04.2011–31.05.2012, toll station under construction, what was associated with the impediment of the movement of vehicles and greater probability of emergence of dangerous situations,
- 01.06.2012–31.05.2013, toll station in operation, what was associated with the necessity of stopping vehicles to retrieve the ticket and pay the toll, and more likely the emergence of dangerous situations.

Table 4 summarizes the list of traffic incidents in the test area during 2010–2013.

Table 4. Traffic incidents around toll station Żernica on the A4 motorway [own study]

The studied period	Collisions	Accidents	Deaths	Injuries
01.04.2010-31.03.2011 no toll station	27	1	0	1
01.04.2011-31.05.2012 toll station under construction	46	2	0	3
01.06.2012-31.05.2013 toll station in operation	59	3	2	2

A graphical presentation of the content of the table is shown in Fig. 1.

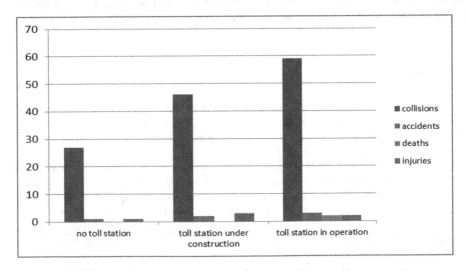

Fig. 1. Traffic incidents around toll station Żernica on the A4 motorway [own study]

As it is clear from the content of the presentation, the noticeable deterioration of road safety indicators was during the building of the toll station. The test results indicate that during this period there has been an increase in the number of collisions by

70.3 %, the number of accidents has doubled, and the number of injured has tripled. So a significant increase in the number of collisions, accidents and persons injured due to the creation of obstacles for the movement of vehicles was related to the beginning of construction the toll station. The information about the greater number of collisions, accidents, injuries, and even deaths, after construction – during operation of the toll station Żernica – testify to the greater probability of emergence of dangerous situations arising from the necessity to stop the vehicles in toll station, in order to get a ticket and pay the toll.

5 Conclusion

The obtained results of the research show that there is a great influence of the kind of tolling system on road safety performance. The introduction of automatic tolling system based on ITS would not cause an increase in the number of traffic incidents related to the necessity to pay the toll. At the same time, this solution would allow the road administration (in this case General Directorate for National Roads and Motorways) task, which is charging for the use of road infrastructure, while maintaining the current level of road safety in the test area. Findings provide a basis for the recognition of the automatic tolling system as highly efficient in the field of improvement of road safety. Also one cannot skip the positive image effect associated with openness to new technological solutions, as well as the organizational, economic and environmental aspects. In a long term, the automatic tolling system, in comparison to a manual one, is cheaper, increases road traffic bandwidth, does not make queues and dangerous situations, and above all does not affect the road safety.

References

1. CEC, White Paper, European transport policy for 2010: time to decide, 12 September 2001 COM (2001) 370 final, Commission of the European Communities, Brussels (2001)
2. EC, Towards a European road safety area: policy orientations on road safety 2011-202, {SEK (2010) 903}, Brussels, 20 July 2010
3. EC, White Paper, Roadmap to a Single European Transport Area – Towards a competitive and resource efficient transport system, Brussels, 28 March 2011, KOM (2011). 144 final
4. Krystek, R. (ed.): Zintegrowany system bezpieczeństwa transportu, t. 1, Diagnoza bezpieczeństwa transportu w Polsce, WKŁ Warszawa, PG Gdańsk (2009)
5. Markusik, S. (ed.): Infrastruktura logistyczna w transporcie. t. III, cz. 1, Infrastruktura liniowa - wodna, transportu lotniczego oraz telematyka transportu, Wydawnictwo Politechniki Śląskiej, Gliwice (2013)
6. Michałowska, M. (ed.): Efektywność transportu w warunkach gospodarki globalnej. Wydawnictwo Uniwersytetu Ekonomicznego, Katowice (2012)
7. Nowacki, G. (ed.): Telematyka transportu drogowego. ITS, Warszawa (2008)
8. OECD, Road Safety – Impact of new technologies (2003)

Implementing High Density Traffic Information Dissemination

Petr Bures[1(✉)] and Jan Vlcinsky[2]

[1] Faculty of Transportation Sciences, Czech Technical University in Prague,
Konviktska 20, 110 00 Praha 1, Czech Republic
bures@fd.cvut.cz
[2] TamTam Research, Slunecnicova 338, Karvina, Czech Republic
jan.vlcinsky@tamtamresearch.com

Abstract. The paper focuses on the utilization of RDS-TMC channel in a more efficient way to carry more traffic information. This approach is called TurboTMC is based on ALERT-C and ALERT-Plus protocols. The paper outlines the idea of the service and further discusses the implementation issues with location table and message prioritization and practical benefits. The TurboTMC is media independent, so it can be used with other media like RDS 2.0 or DAB.

Keywords: Traffic and travel information · Distribution · RDS-TMC · Turbotmc · High density · Multi media

1 Introduction

Information about current traffic situation, provided by traffic information services (TIS), have important role in safety of road users and also allows increasing fluency of road traffic. Therefore the TIS is an integral part of the national ITS architecture [1, 2]. To allow European users benefit from those services when going from one country to another the European Union decided to harmonize them and make them to some extent mandatory. This was set forth by the ITS Directive (2010/40/EU) and its delegated acts (thoroughly analyzed in [3]), namely:

- *885/2013* "provision of information services for safe and secure parking places for trucks and commercial vehicles" and
- *886/2013* "data and procedures for the provision, where possible, of road safety - related minimum universal traffic information free of charge to users".

Current TIS disseminates traffic information via several channels, ranging from broadband to narrowband. This paper focuses on enhancing the efficiency of traffic information transmitting in a narrowband RDS (-TMC) channel and keeping the backwards compatibility allowing use of ALERT-C service on the same radio frequency.

Even though RDS is rather old technology, it has great advantage of widely available infrastructure and end-user devices. It makes possible to deploy the service in multiple countries in Europe in harmonized manner and making it accessible also for users from abroad. Since RDS is already there and for free, users do not have to tackle

© Springer International Publishing AG 2016
J. Mikulski (Ed.): TST 2016, CCIS 640, pp. 243–256, 2016.
DOI: 10.1007/978-3-319-49646-7_21

difficulties with getting internet connectivity or differences in used DAB technologies in different countries.

1.1 RDS-TMC Basics

RDS or Radio Data System allows transmission of information services over an analogue (UHF FM) radio channel (CEN ISO 14819). The services are carried over by so called RDS groups. RDS standard (EN50067) defines 32 possible group types (0A-15A, 0B-15B), reserved for different types of services. TMC or Traffic Message Channel service uses 2 group types, 3A (TMC service information) and 8A (actual traffic messages).

RDS has very limited capacity to transmit information, just 1187.5 bits per second, including almost 40 % overhead. All RDS services including TMC must be very space efficient. Each group has 104 bits which gives the rate of 11.4 groups per second. Usable payload of each group is much smaller, only 37 bits, see Fig. 1. Since other RDS services need certain amount of space in RDS channel, the TMC service is limited to use maximally 25 % of the channel bandwidth (under normal circumstances) which results in 2.85 TMC groups/s.

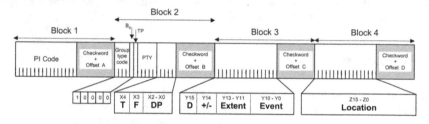

Fig. 1. Structure of an 8A RDS (TMC) group [8]

To minimize misinterpretation of messages due to channel errors, a TMC group is considered valid only when one or two of its exact copies were received with no other TMC group in between, this mechanism is called immediate repetition. This effectively halves possible channel capacity for TMC.

TMC allows continuous, "silent" delivery of live traffic information feed suitable for processing by a navigation device, which then might offer to a driver proposal for alternative routes to avoid traffic incidents or other unusual traffic situations. TMC introduces set of rules and lookup tables that are used together to encode and to decode traffic information at TIS center and navigation device. The lookup tables, namely event codes table and locations table must be present in same version at the center and in navigation device.

RDS-TMC is standardized by European norm and in European context called ALERT-C service, there are other services in RDS-TMC, the proposed TurboTMC and the obsolete ALERT-Plus. Even though RDS-TMC and ALERT-C are often used as synonyms the latter shall be perceived as a sub group of the earlier.

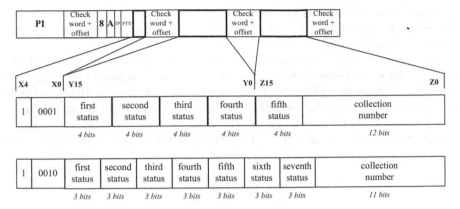

Fig. 2. ALERT-Plus RDS-TMC user message structure [7]

1.2 Channel Capacity Limitation

Channel capacity is a number of messages transmitted in reasonable time scope (window) to a navigation device. There are two possible view points to this problem. One is based on implicit duration types of information, second on user expectation about the timeliness of the service.

Traffic information could be categorized into *dynamic* and *static* and *event oriented* and *status oriented*. They have different implicit expected life, so called duration.

- *static* information is usually event driven and could be represented by planned roadworks,
- *dynamic* information could be both event driven or statuses and could be represented by accidents or travel time.

Dynamic messages have implicit duration of 15 min, since the situation is likely to change in that time scope. Static messages have implicit duration of 60 min. The implicit duration gives a time frame in which each message shall be repeated or it would be deleted from a receiver memory. Another perspective for message repetition is a driver's willingness to wait for the information (latency), which counteracts implicit durations. The acceptable latency for the dynamic information is in most cases around 5 min.

Traffic situations description can range from simple to complex, the complexity projects into number of RDS groups used for such message. Simple message contained in one RDS group has most important information:

- WHERE situation happened described by location (16 bits, extent 3 bits and direction 1 bit),
- WHAT actually happened described by event code (11 bits) and
- WHEN or how long it will be relevant described by duration (3 bits).

Adding more information, like more related events, diversion instructions, quantities, results in more RDS groups, of up to 5 RDS groups for message containing rich information about the traffic situation.

Less complex messages allows to transmit more messages over the RDS-TMC, therefore enhances capacity of the channel. Providers, however tend to give as precise description as possible, which leads to messages with more than one RDS group. In measurements done by authors in 2013 in Prague more than 80 % of the TMC messages consisted of more than one group [4].

The statements above lead to Eq. 1 that gives potential number of messages N that can be effectively presented via a RDS channel.

$$N = 11.4/rate/repetition/msg\_size * time\_window \qquad (1)$$

Where *rate* is channel capacity used up by TMC, *repetition* is the immediate repetition of TMC groups, *msg_size* is average number of RDS groups per message and *time_window* is accepted time of renewal of the messages. Next Table (1) summarizes maximum number of messages based on different parameters.

Table 1. Number of messages that can be effectively presented via an RDS channel [own study]

TMC capacity [%]	25 %	25 %	25 %	25 %	25 %	25 %
Immediate repetition [–]	2	2	2	2	2	3
Iverage message size [groups]	1	1.5	2	2.5	3	3
Time window [min]	5	5	5	5	5	5
Number of messages	427.5	285	213.8	171	142.5	95

For ideal conditions, where Provider uses approximately only 1.5 groups per message and 5 min repetition window, we get capacity of 285 messages per TMC channel. That corresponds with 300 messages limitation set for terminals by ALERT-C standard (CEN ISO 14819).

The number of 285 messages could be enough for a service covering whole country, but only in case that the service is based mainly on event driven situations such as accidents and roadworks really affecting traffic, e.g. not informing about every small side walk works that do not really render any traffic congestion. However, if the service shall inform about the traffic statuses, like travel times, parking occupancy or level of service, for each important road segment in the area, then this number becomes a limitation.

1.3   Overcoming Channel Capacity Limitation

Modern techniques of traffic data collection (e.g. using fleets) are able to automatically cover large areas and to generate many status messages (likely going to thousands). Therefore we needed to find a way how to convey more information to the receiver through RDS-TMC, using shared infrastructure with already present ALERT-C service. Some RDS channel capacity (at least 40 %) also has to remain for basic RDS information. This could be done by enhancing ALERT-Plus standard and adding new features to preserve compatibility and smooth transition to this service for navigation device vendors [5].

In this paper we present TurboTMC approach to increase quantity of traffic status information that could be transmitted over RDS-TMC channel. This approach is based on, but not limited to ALERT-Plus which has been proposed as European standard (ENV 12313) in 1998 but never reached any practical implementation.

In ALERT-Plus, the quantity increase is done by concept that allows one RDS group to carry 5 or 7 status information (see Fig. 3). This can be 5 successive travel time statuses on predefined segments or 7 parking occupancy statuses related to 7 different car parks. Structure not so different from TPEG [6]. Since ALERT-Plus allows only single group messages, we get 7:1 or 5:1 quantity enhancement ratio. Transmitted messages contain statuses which can only be interpreted with reference to locations. To facilitate broadcasting, locations are grouped together in blocks called collections. Both transmitter and receiver must have table with collections and all their related information or it is not able to interpret received messages. Table must be delivered to receiver prior to interpretation attempt via other means than RDS-TMC channel.

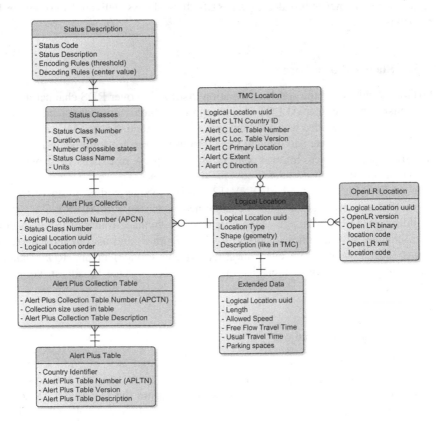

Fig. 3. Structure of TurboTMC look up location table, with logical location concept [own study]

In TurboTMC approach ALERT-C and ALERT-Plus services coexists, allowing operation of a universal free to all event oriented service as well as an added value status messages oriented service on the same transmitter.

2 Methodology and Standards

In classical ALERT-C (CEN ISO 19819) each traffic incident encoded as a TMC message consists of an event code and a location code in addition to expected incident duration, extent and other details and is carried over by one or more RDS groups. Event code and location code are stored in look up tables together with description and special features that allow their interpretation on the receivers map.

ALERT-C event look up table contains a list of up to 2048 event phrases. Some phrases describe individual situations such as a crash, while others cover combinations of events such as a crash causing long delays. ALERT-C location look up table represents simplification of road network, assigns numerical codes to locations (typically major junctions) on the road. Location tables are then integrated in the maps provided by in-vehicle navigation system.

ALERT-C event and location look up tables, which allow real interpretation of TMC message, are never broadcast over RDS. It is always delivered into device by other means.

2.1 Pre Standard as a Base

TurboTMC stands for traffic status information distribution over RDS channel in much higher density than regular ALERT-C. The approach builds on a pre-standard called ALERT-Plus, since the idea was good but just never matured. TurboTMC retains good parts and reworks obsolete and difficult parts into usable service for modern navigation devices. In situations where Internet connectivity is a sparse or expensive resource, the RDS channel comes with no related cost for the end user.

The ALERT-Plus pre-standard (ENV 12313-4) specifies coding protocol for distribution of status information using the same media as standard ALERT-C.

Building on ALERT-Plus gives the possibility to enhance traffic distribution done by RDS channel several times (theoretically 5 or 7 times) and also to separate status messages (added value) from event messages (basic service) in one RDS channel. Which opens a possibility to monetize on added value service broadcast along free basic service.

2.2 Building a New Standard

The ALERT-Plus standard was created a too early, initial lack of wide scale status information and lack of supporters led to the fact, there was no real long term deployment thus there was not enough time to fine tune the technology. Since the introduction of ALERT-Plus, the situation has changed dramatically, the original concepts with which was standard drafted did not withstand the technology evolution. The situation changed at least in following aspects:

- Status information are now available in large quantities.
- Long term use of ALERT-C allowed to simplify original design, this has impact on ALERT-Plus as well.
- End user devices has changed:
    - text based interface turned into graphical user interfaces with maps,
    - built-in navigation devices changed into, mostly, portable ones, even to mere applications,
    - RDS decoders are not any more expensive separate units but are built into most of today's smart phones,
    - navigation device processing power is much higher and
    - data set (maps, look up tables) updates changed from CD/DVD to update over the Internet.

2.3   TurboTMC, Removing the Obsolete

The ALERT-Plus standard, has (had) a lot of unnecessary features, and some unused options, list of most prominent ones with proposed solution of handling them in TurboTMC follows:

- transmission options in 1A 5A RDS groups (to be removed),
- definition of location table (to be redefined),
- format of management messages (to be redefined),
- transmission of management messages (to be redefined),
- message geographical scope features (to be removed),
- broadcast strategy (to be removed),
- message identity by country, service, location table and collection id (to be redefined).

Originally ALERT-Plus lacked some features, like:

- extensibility for new possible status information (fuel prices, etc.),
- compatibility with ALERT-C message management structure and
- possibility of geocoding location using different location referencing methods.

The standard ALERT-Plus implied using its own location tables (different from ALERT-C). That would be difficult to achieve, so the ALERT-C location codes should be primarily used in TurboTMC location look up table to derive physical locations. This will not add any additional work to map data and navigation device (software) vendors. We also proposed to extend TurboTMC location look up table by other location referencing methods like AGORA-C, TPEGloc or OpenLR which would in future remove dependency on static ALERT-C location tables.

3 TurboTMC Requirement Analysis and Proposal

This chapter describes TurboTMC implementation issues. We start with requirements a follow with proposal and actual implementation issues.

3.1 Requirements

Based on analysis of current and previous standards and state of the art, following statements points out main requirements for TurboTMC service.

- Should be used for status information only, and designed to be a service complementing regular ALERT-C (event oriented) service.
- Should be backwards compatible, receivers not capable of decoding the content should simply ignore it.
- Should use in ALERT-C unused tuning variants of 8A groups to carry status and management information.
- One RDS group should carry FIVE 4-bit statuses or SEVEN 3-bit statuses, or in future TEN 2-bit statuses (where 4 states are enough).
- Different variants (FIVE SEVEN) of status message should be carried by different variants of 8A tuning message (now unused variants).
- Management messages that carry information about the look up location table should be broadcast within the same 8A tuning variant as status information messages.
- Should use regionalization to maximize relevant information throughput. Each transmitter should transmit statuses related to its region of coverage, the status information transmitted by one transmitter are referenced to ONE collection table that could hold several thousands of collections.
- Should have minimal requirements on updating navigation device/software functionality. Message should be, in the device, separated into several regular ALERT-C messages handled in a standard way.
- Should use static look up location table, referencing predefined physical segments where status measurements are collected. For status information this is possible since they always comes from known locations (links equipped with detectors).
- Look up location table should be primarily based on segments built from existing ALERT-C locations, just reordered and collated into sets (locations > collections > collection tables > location table).
- Advanced look up location table should use extended handling of location description where collections could be built from segments referenced by dynamic method (TPEGloc, OpenLR). Since network is known in advance - the look up table would contain these dynamically created references in advance - it removes dependency on ALERT-C location tables.
- Should allow to reference almost any location (know before). Could be done by employing previous requirement.
- Location table updates should be handled via Internet.

3.2   Service Proposal

Status information are contained in so called tuning variants of the 8A group, specifically the variants 1–3 (bits X3-X0 in Fig. 3). These variants are not used in regular ALERT-C, thus are not decoded by a not updated receiver, this is for backwards compatibility. Tuning variant 1 is used to carry FIVE 4-bit statuses and variant 2 for SEVEN 3-bit statuses.

Previous Table (2) shows an example of status codes and their processing thresholds for 3 bit status used for information about parking occupancy. The thresholds are needed to encode specific occupancy into status code and center value to decode status code into specific occupancy. Thresholds (center values) must be defined by service operator and shared as a part of location table with end user devices. For example occupancy 78 % will yield to status 3 (very high car park occupancy) and will be decoded in end user device as a center value of status 3 interval 80 %.

Table 2. Example of parks occupancy status codes implementation [own study]

Status code	Description	Threshold	Center value
0	No information available	–	–
1	Car park full	100 %	125 %
2	Car park full soon	90 %	95 %
3	Very high car park occupancy	70 %	80 %
4	High car park occupancy	50 %	60 %
5	Parking spaces available	20 %	35 %
6	Car park almost empty	0 %	10 %
7	Use discouraged or car park closed	–	–

Five or seven statuses used in the message are linked with locations grouped in a referenced collection, 1st status to the location at first position in collection, 2nd status to second position, etc. Figure (3) represent proposed structure of TurboTMC look up location table, with its main feature, the logical location concept.

The look up location table consists of collection tables that holds individual collections of locations. Since only the collection ID together with collection table number is transmitted over RDS, the Location ID of one location inside a collection could be large unique number. This allows us to detach locations from any prescribed form of representation and assign them globally unique ID (GUID). The location itself then could hold different forms of predefined static (ALERT-C) or dynamic (OpenLR, TPEGloc…) location references.

Because of regionalization concept, some logical locations can be part of one or many collections in one or many collection tables, also some collections may be repeated as whole in different collection tables. In previous Fig. (3), the logical location could link to separate files containing:

– ALERT-C location description (country + location table number + primary location + extent + direction),
– other location reference like OpenLR description in XML or binary format.

TurboTMC is meant to be used as complementary to standard ALERT-C service and could increase information output several times. The increase is shown in next Table (3), where

Table 3. Number of information carried by mixed TurboTMC service [own study]

Service	8A/s	% TMC	grp/msg	I/msg	Repetition	Carousel	TOTAL
A-C	0,6	5 %	1.5	1	2	5	57
tT	2,3	20 %	1	7 (5)	2	5	2394 (1710)
A-C	1,1	10 %	1.5	1	2	5	114
tT	2,9	25 %	1	7 (5)	2	5	2993 (2137)
A-C	1,1	10 %	1.5	1	2	5	114
tT	3,4	30 %	1	7 (5)	2	5	3591 (2565)

- one service is two rows combination of A-C (ALERT-C) and tT (TurboTMC),
- 8A/s is number of groups per second devoted to given service,
- % TMC is number of 8A/s divided by 11.4 (total number of RDS groups per second),
- grp/msg is average size of ALERT-C and TurboTMC message in RDS groups,
- I/msg is number of useful information per one message,
- repetition is number of immediate repetitions for TMC groups accepted by receiver,
- carousel is broadcast repetition interval in minutes and
- TOTAL is received information from service during carousel time.

In mixed service 114 ALERT-C messages per 5 min interval is sufficient - we inform only about events, that has effect on the traffic. For 40 % ratio TMC groups to all RDS groups we would be able to transmit 114 (A-C) + 3 591 (tT) so 3 705 information.

4   TurboTMC Implementation

4.1   Setting up the Service

In the ALERT-Plus, there were several options how to handle message management, we have decided to use 8A tuning RDS groups. The management information are carried in the same group as statuses (tuning variant 1 or 2), the distinction between management and normal status message is done by using a special collection number. The management information consist of TurboTMC Location table number and collection table number.

Other relevant information like Service ID and ALERT-C Location table number, that is in TurboTMC used to identify country relevant to a service, are carried by 3A group since standard ALERT-C service is present in the channel. The ODA number, used to identify what service is being carried by 8A groups is set to CD46, the same as for ALERT-C, since the special ODA number defined for mixed service was withdrawn.

4.2   Location Table Creation

Main concept of location table is a logical location with unique GUID referring to physical location (see Fig. 3), table structure than allows to "plug into" this location different location referencing method and to link it to any number of collections and collection tables. Logical locations can be street segment, parking spot, parking place, point of interest (petrol station).

During creation of initial location table we have faced the problem of how to fill the collections. Collection shall in ideal world be put together from adjacent locations, this however was not possible since many roads did not have exact number of segments necessary to fill in whole collection.

Status information analysis of segments (locations) with high status variability revealed that in reality the segments with "high" statuses are not geographically grouped, they were fragmented. The original assumption that by filling the collection with adjacent segments we maximize information throughput was not true. Therefore the collections could be put together even randomly, good option also is putting locations that change the most together into collections. This is however analytically difficult, and we have decided for initial set up to fill collections with locations just with regard to their general area.

Collection ID is 11 or 12 bit number that allows to address 2 048 or 4 096 collections in a collection table, collecting 2 048 * 7 = 14 336 or 4 096 * 4 = 16 384 locations (see Fig. 3). Using Czech ALERT-C location table version 5.0 for logical locations creation led to 25 512 simple segments.

Therefore TurboTMC location table had to be regionalized into **3 overlapping regions**, collection tables (see Fig. 4). Regions were defined by transmitter coverage area. Collection tables of all regions held same set of high importance locations, like highways, and even used same collection ids for addressing them in different collection tables. Two neighboring regions also held some amount of same locations where the service area overlapped, this was however without effort to have them in collection with ids.

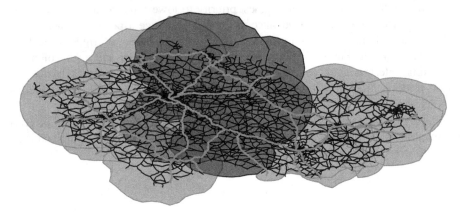

Fig. 4. Example of regionalization of TurboTMC location table [own study]

4.3   Transmitting Data

Fixed amount of monitored segments or points leads to fixed amount of statuses in a full status space updated each interval. As capacity of RDS channel is limited (3 500 in 30 % utilization), it is obvious that not all messages could be updated in 5 min interval. To decide which status information send and which not we used prioritization. Priorities are assigned by statuses collated in collection since all statuses in one collection are transmitted together. Individual statuses do contribute to whole message priority. Transmitting of TurboTMC messages is than based on priority that is recalculated at every window start (5 min interval). Following are aspects, participating in priority calculation (Each priority parameter can have different weight):

- **Age:** How long time ago was the message broadcast last time, limited to some maximum time T i.e. 15 min. Relative age is calculated as time since last transmission, expressed in seconds rounded down to value of T if grater and divided by T. Relative age will be always in range [0–100 %].
- **Severity:** Overall severity of statuses in the message. The more severe or urgent statuses are reported by a message, the more important it is to transmit it. Aggregated status is calculated as sum of squared status codes for all locations. The value is then normalized to range [0–100 %]. The assumption here is that higher status number means higher severity of the information conveyed by a status.
- **Change:** How much has the message changed since last transmission. The more rapid changes are introduced by a new message compared to lastly transmit one, the more relevant it is to notify an end user about these changes. Collection message Change is sum of squared changes (current status last non zero broadcast status) in particular statuses, normalized to range [0–100 %].
- **Status Class:** Different types of statuses might have different urgency. Currently we do not use this, but may become handy as soon as we introduce status classes with different durations and importance.

The above aspects are combined together by weighted average, where individual characteristics could be assigned different weight based on its importance for prioritization process. Each time the messages (collections) are prioritized they are ordered into a FIFO buffer from those with highest priority to lowest and broadcast. Broadcast message changes its priority value since its Age and Change aspects decreases.

If the end of the queue is reached a priority recalculation is triggered and the process starts again. Figure 5 shows in each line a collections and each column is passed time interval. Dark cell background means low priority, light background is highest priority, drop from light to dark means that the message was transmitted in previous window.

With this prioritization we have achieved to transmit all relevant images in given time interval, those with low priority got transmitted at least once per 15 min and those with high every transmitting window (initially set to 5 min).

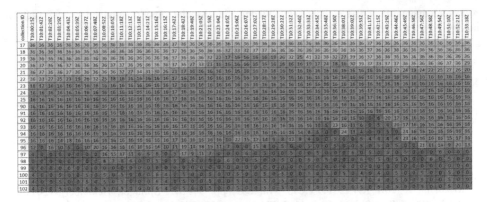

Fig. 5. Example of growing priorities of messages in subsequent windows (columns) [own study]

5 Discussion

5.1 TurboTMC in DAB and RDS 2.0

The proposal how to carry RDS-TMC in DAB was published as standard (ETSI TS 101789) in 2002. TMC information was just repackaged into data chunks of size of 37 bit (see Figs. 1 or 2), carried by blocks in fast information data channel FIDC. TurboTMC data could be fitted into DAB channel using the same technique.

Using TurboTMC in RDS 2.0 (IEC 62106) is also very simple, the RDS 2.0 is fully compatible with existing RDS and just extends the RDS from one channel to 4 channels with same capacity, encoding and framework as the original channel spaced next to each other. 1st channel remains the same for compatibility reasons, other 3 channels might be dedicated for specific use, i.e. full channel (1187.5 bits) for TurboTMC. That would mean in 100 % utilization 11 970 status information carried in 5 min interval.

6 Conclusion

In this paper we presented implementation of a traffic information service that allows to carry in normal RDS channel status information along with a standard RDS-TMC service. To enhance multiple times channel capacity and also to make possible to monetize status information as an added value service.

The basis for this service was pre standard ALERT-Plus, which was used for structure of RDS group. Location table concept with logical locations and message prioritization queues were original ideas. This service in ideal conditions allows to transmit more than 3 500 information in 5 min time frame using only 35 % of the channel capacity.

Acknowledgements. The authors acknowledge the financial support provided by the Technology Agency of the Czech Republic through project TurboTMC (TA03031386).

References

1. Bures, P., Belinova, Z., Jesty, P.: Intelligent transport system architecture different approaches and future trends. In: Duh, J., et al. (eds.) Data and Mobility: Transforming Information into Intelligent Traffic and Transportation Services, Proceedings of the Lakeside Conference 2010, vol. 81, pp. 115–125. Springer-Verlag, Berlin, Berlin (2010)
2. Belinova, Z., Bures, P., Barta, D.: Evolving ITS architecture – the Czech experience. In: Mikulski, J. (ed.) TST 2011. CCIS, vol. 239, pp. 94–101. Springer, Heidelberg (2011). doi:10.1007/978-3-642-24660-9_11. ISSN 1865-0929, 978-3-642-24659-3
3. Algoe, Rapp Trans: Study regarding guaranteed access to traffic and travel data and free provision of universal traffic information: D8 – Final report. 1st edn., Brussels (2011). http://ec.europa.eu
4. Bures, P., Vlcinsky, J.: Monitoring of live traffic information in the Czech Republic. In: Mikulski, J. (ed.) TST 2011. CCIS, vol. 239, pp. 9–16. Springer, Heidelberg (2011). doi:10.1007/978-3-642-24660-9_2
5. Bureš, P.: Quality and quantity improvement for current RDS-TMC. In: Mikulski, J. (ed.) TST 2013. CCIS, vol. 395, pp. 410–417. Springer, Heidelberg (2013). doi:10.1007/978-3-642-41647-7_50
6. Bures, P.: The architecture of traffic and travel information system based on protocol TPEG. In: Proceedings of the 2009 Euro American Conference on Telematics and Information Systems: New Opportunities to Increase Digital Citizenship, Prague, Czech Republic (2009)
7. prENV 12313-4 Traffic and Travel Information (TTI) - TTI Messages via traffic message coding - Part 4: Coding Protocol for Radio Data System - Traffic Message Channel (RDS-TMC) - RDS-TMC using ALERT Plus with ALERT-C
8. EN ISO 14819-1 Traffic and Travel Information (TTI) - TTI Messages via traffic message coding - Part 1: Coding protocol for Radio Data System - Traffic Message Channel (RDS-TMC) - RDS-TMC using ALERT-C

V2X Technology – Cooperative Systems as the Development Direction and Safety Improvement for the Transport

Katarzyna Bołtowicz[✉] and Przemysław Staśkowiak

Kapsch Telematic Services Sp. z o.o, Poleczki 35/A1, 02-822 Warsaw, Poland
`{katarzyna.boltowicz,`
`przemyslaw.staskowiak}@kapsch.net`

Abstract. In this article the authors describe the principles of the potential possibilities and directions of development of innovative systems – V2X (Vehicle to All). They present some examples and the effects of the use of such systems in European countries and in the USA. The authors show also the statistics of road events and/or incidents on which usage of the systems of this type has/can have a positive or even decisive impact. The second part of the article contains the ideas and perspectives of further development of V2X systems in terms of possible use case spectrum of such systems as well as the possible extent and location of use. In this part the authors indicated i.e. few concrete examples of the systems' (possible) usage for Poland.

Keywords: V2X technology · Intelligent transport systems · Traffic optimization

1 Introduction

Basing on different analysis the collisions on the crossing roads determine 40–60% of all the accidents.

As written by Gene Carter only one year in US was enough for people to drive over 3 trillion miles, to cause more than 6 million accidents, to lose 35,000 people and to suffer over 3 million damages.

It's enough to have a look on the drawing below to understand on a very simple example how it is possible. The number of collision points on different basic types of crossings presented below is already enough to imagine possible situations, especially during heavy traffic flow (Fig. 1).

The vehicle's manufacturers try to protect constantly their innovative ideas as well as already released prototypes and solutions of connected vehicles. However there are several different technologies designed as confidential and properly protected from the beginning. Those are commonly known as: vehicle to vehicle (V2 V), vehicle to infrastructure (V2I) and jointly specified as vehicle to everything (V2X). The solutions are built on the short range wireless communication technology and were designed most of all to stave off accidents and thereby to protect humans' life [2].

© Springer International Publishing AG 2016
J. Mikulski (Ed.): TST 2016, CCIS 640, pp. 257–267, 2016.
DOI: 10.1007/978-3-319-49646-7_22

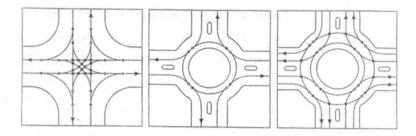

The legend:

- ▪ - crossing,
- ● - exit
- ○ - entry

Fig. 1. Car accidents critical points on the crossing roads: (a) 4-lines with no traffic lights, (b) 1-line small roundabout, (c) 2-lines small roundabout [4]

V2X systems belong to the newest generation of the solutions classified as the ITS (Intelligent Transport System). They allow to exchange information directly between vehicles, between vehicles and the road infrastructure, meaning i.e. the vehicle and the road signs or the traffic lights. The data exchange is an immediate one and uses the radio waves communication. This is why it is much quicker than using GPS satellites or GSM mobile networks.

How do the systems work in practice? The V2 V devices installed in our vehicles and connected to the on-board computer send information to other vehicles informing i.e. of activation of Acceleration Slip Regulation system (ASR). The communicates are being displayed in the vehicles which approach the spot of the inconvenience. That way the next/coming drivers are also warned of the traffic jams, traffic congestion or the accidents. V2I systems work on similar bases. Valid information could also be generated manually by the system users or operators. Thanks to the communication with the road infrastructure (using a proper onboard equipment) the vehicles' drivers are being informed of the speed limits, hard weather conditions or many other difficulties on the road. Such solutions are being tested in Europe in Vienna and Prague. In the future they are going to cover all roads being a part of so called European Corridor [1]. These kind of messages are already defined as DENM (Decentralized Environmental Notification Message) and some basic use cases are already standardized within ETSI (*European Telecommunications Standards Institute*), i.e. ETSI EN 302 673-3 V1.2.2: (Table 1).

C-ITS (Cooperative Intelligent Transport System) is the "systems network" which allows all of the communication partners (vehicles, transport infrastructure and service suppliers) to exchange information as the base for reaching a new level of the traffic security and improvement of its effectiveness[1] (Fig. 2).

---

[1] In accordance to MoU (Memorandum of Understanding) C2C CC.

Table 1. ETSI's basic set of applications – active road safety – driving assistance – RHW [3]

Applications class	Application	Number	Use Cases
Active road safety	Driving assistance – road hazard warning	UC005	UC005 Emergency electronic brake lights
		UC006	UC006 Wrong way driving warning
		UC007	UC007 Stationary vehicle - accident
		US008	US008 Stationary vehicle – vehicle problem
		UC009	UC009 Traffic condition warning
		UC010	UC010 Signal violation warning
		UC011	UC011 Roadwork warning
		UC012	UC012 Collision risk warning
		UC013	UC013 Decentralized floating car data – Hazardous location
		UC014	UC014 Decentralized floating car data – Precipitations
		UC015	UC015 Decentralized floating car data – Road adhesion
		UC016	UC016 Decentralized floating car data – Visibility
		UC017	UC017 Decentralized floating car data – Wind

Kapsch contributes in many different standardization and research projects and initiatives for V2X. One of them is E-Co AT ((just mentioned) European Corridor – Austrian Testbed for Cooperative Systems) where it cooperates on the system's specification (in general). The example has been visualized below (Fig. 3):

The major examples of the current development and deployment driving forces are:

- Europe: CAR 2 CAR Communication Consortium (C2C-CC) – a non-profit organization; EU C-ITS Platform – since November 2014
- USA: V2V implementation supported mainly by the government agencies, US Department of Transportation (USDOT) and National Highway Transportation Safety Administration (NHTSA)

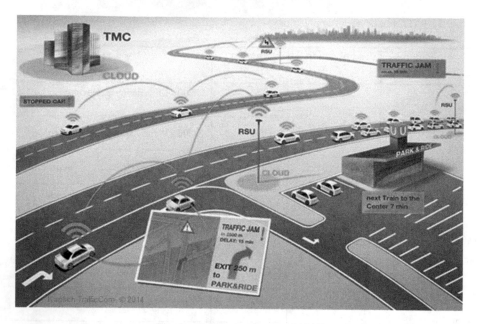

Fig. 2. C-ITS visual example [11]

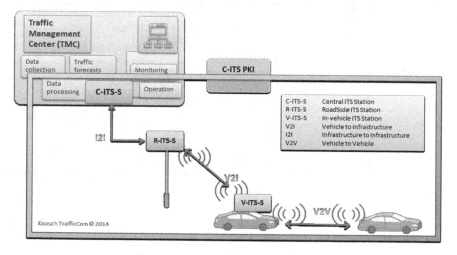

Fig. 3. E-Co AT C-ITS based on V2X architecture [11]

– MoU (Memorandum of Understanding) between OEMs (Original Equipment Manufacturers), implementation expected in 2017. V2I infrastructure on the road should be available by this time for at least a part of the road network (Fig. 4).

As an example of V2X system we can consider the one developed by Kapsch TrafficCom. The system is called Wave in U.S. and G-5 Standard in Europe. It uses the onboard unit (OBU) as presented on the Fig. 5 below. The information is being taken

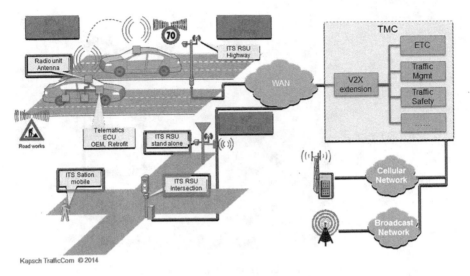

Fig. 4. The V2X landscape [11]

directly from the roadside and communicates with a device in the vehicle on the Bluetooth. For that purpose a tablet PC, a navigation system or any other hand held device can be used. That way the driver receives information while travelling along about road conditions, accidents or even about flight delays. The equipment is weather resistant. The OBU works also as a mobile transceiver. Thanks to that the information from several traffic sources can be displayed in a real time. It contains also many other useful functions such as multi modal transport informing how many free parking spaces are there in the parking lots, etc.

Kapsch participates in many standardization, research and test initiatives related and acts as a member or partner in different International ITS Community associations. It was for instance one of the fourteen project's partners as well as for several different initiatives for the subject solution. Except for the already mentioned it took part in the Testfeld Telematik (2011–2013) – the testing platform in the vicinity of Vienna for the cooperation services (V2X Demo Tour from Kapsch AG) based on the components as described above and pictured below (Fig. 5).

A real pioneer in V2X systems usage are however the Japanese and especially Toyota concern. The Japanese scientists have integrated V2 V and V2I systems in frames of the ITS Connect solution. Its' major goal was to avoid collisions on the cross-roads. One of the major functions of the system is to support a difficult turn left maneuver; seems trivial but for quite some drivers can really pose considerable problems and stress.

What is interesting ITS Connect is not a test system any more. Beginning October 2015 it's a part of the Toyota Crown equipment which is very popular sedan in Japan. By the end of 2016 it is going to be a part of IV generation of Toyota Prius equipment as well.

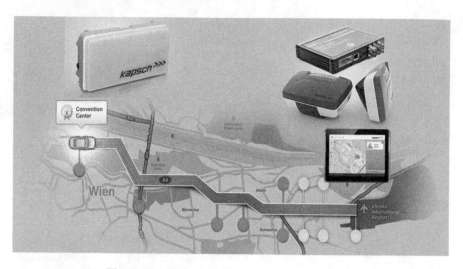

Fig. 5. Testfeld telematik system components [11]

V2X systems are being also tested in USA. Between March and April 2015 automatically steered vehicle Delphi drove all U.S. from San Francisco on the west side to New York on the east. The distance was 5.5 thousand km long and took 9 days. 99% of time the vehicle was steered automatically. It drove with no issue passing (as other regular road traffic participants) the round-abouts, tunnels, bridges and road works. The scientists collected more than 3 terabytes of data which will be the base for further project's development. The Delphi car had different V2 V and V2I systems installed that allowed a wireless communication with other vehicles and the infrastructure and extended the possibilities of the radar, visual cameras system and other systems supporting the driver [1].

One of the first practical tests of the V2X technology was executed in Europe on the 1300 km distance at the end of 2014 and it took 7 days. Specially prepared vehicles visited then Frankfurt, Vienna and Rotterdam in an attempt to connect to vehicle technology, not only with each other but also with the road infrastructure. The test was carried out thanks to the cooperation of several partners: German TUV Sud, NXP, Cohd Wireless, Siemens and Honda.

During the journey the vehicles communicated to each other to warn i.e. on a danger situations that a driver was not able to perceive well in advance (before they appear). The drivers checked also the receiving of the signals of the intelligent road signs, traffic lights and other elements of the road infrastructure. What is more, within the trials the communication in different weather conditions was tested as well as during differentiated traffic intensity and extra obstacles such as road works, collisions and similar events [5].

2 Advantages of Using the Solutions Based on the Subject Technology

United States Department of Transportation evaluated that the usage of V2X solutions could help to avoid even 80% of accidents. Such an equipment can simply warn the driver of unexpected and invisible dangerous situations which are not detectable by regular onboard systems. Dividing data, in approx. 0,5 km range, by a vehicle that is equipped in the V2X system the driver is able to receive information on the current road situation (i.e. causes of accidents that happen most commonly). That helps him to take a proper decision and action needed in time. Different types of possible connected cars' use cases have been presented in the table below [2] (Table 2).

Table 2. Connected-car exemplary use cases [2]

V2I Safety	V2 V Safety	Convenience
Red light violation warning	Emergency brake light warning	Eco driving
Curve speed warning	Forward collision warning	Smart cities
Spot weather	Red light violation	Parking information
Work zone safety	Slow traffic ahead	Truck platooning
Bridge height	Aggressive driver warning	Speed harmonization
Pedestrian in crossing signal	Emergency vehicle notification	Queue warning
Stop sign gap assist	Road hazard detection	Insurance pricing

As in the case of i.e. ITS Connect system described above it monitors the situation on the cross-road and informs the driver with the sound and light signal of any disturbances on the turn's way when only the brake pedal is being released. The system emits also light and sound signals when only the vehicle approaches the red light and the braking process hasn't been initiated yet. The system improves also the traffic flow's fluency informing the driver how much time is left to the traffic light change into green so the starting point is as ably as possible.

The next element of the same system is an active cruise controller which reacts on the speed change of the vehicles driving nearby and adjusting the speed in accordance to the road situation. ITS Connect warns also of coming emergency vehicles informing additionally i.e. on their direction and the distance from our vehicle.

As already shown the systems can inform of the different road events in advance like changing traffic lights, traffic jams, possible obstacles for a different reason and many other major functionalities which is only the beginning as they are supplemented by a full range of supporting solutions not necessarily connected directly to the road events as such. Basically, all together are making our life easier and more secure and the roads additionally simply more predictable.

3 Conclusion

Even though we have few different solutions already or soon available, as presented above on just a few examples, of course we have to wait for the practical benefits of the V2X systems use. First they must be placed as standard in most cars then they must receive adequate investment in the infrastructure. In practice this means that we'll be able to talk about universality of such solutions in decades. Undoubtedly the development of this type of technology is an important step on the way to a significant improvement in road safety and traffic improvement [1].

To operate such an ingenious system a software that is needed is constructed on fifteen times more lines of code Boeing 787 Dreamliner. Its' location on many processors makes it a very inviting object for hackers. Probably this is a huge defiance for the entire industry – to balance adequately comfort and the functionally of the systems, what is really important from the end user perspective, and securing its' blind-spots on the other hand [2].

There are some examples of such challenges listed below that should be addressed before real launching and making such solutions publicly available [2]:

- The technology is a concrete additional cost to the total cost of a car, although as with the new technologies it will change in some time for sure
- Before we'll be able to see any benefits of the solutions a critical mass of cars equipped with the systems is needed on roads
- Probably the legal regulations in different countries will be providing finally kind of demand for the manufacturers to equip new vehicles in V2X systems however current share of unequipped vehicles and related life cycle should be considered in parallel
- What has been already mentioned before - the safe security systems, ensuring end users' privacy are the critical issue that have to be still developed.

Telefonica company prepared a report presenting broadly defined solutions enabling collaboration between cars' component systems and the mobile devices.

The report was created as and outcome of the survey's analysis led on 5 thousands of adult consumers having the driving license and living in 5 following countries: Spain, Germany, Great Britain, Brazil and U.S. The key conclusions were as follows [6]:

- There is sufficient global demand for connected cars solutions type – 71% of respondents said that they are interested in one or are already using them.
- 80% of consumers expect that the technology of connected cars will bring the same ease, as in the case of the use of modern technologies at home or at work and beyond via mobile phone
- In all markets surveyed three basic issues connected with cars have been pointed out: enhanced security, the use of early warning systems and creation of more intelligent navigation. Nearly three-quarters (73%) of respondents indicated security features and diagnosis as the most important, giving to understand that they expect in the future the emphasis to these areas

- 35% of respondents declare that they will not have their own car before the year 2034. Respondents intend to use the services while the short-term and ultra-short term car rental and joint trips
- The central panel in the car is the preferred place of access to the combined mobile services and manufacturing – says more than 60% of respondents
- There is no single ideal solution for paying for services related to connected cars. For example, most Spanish drivers would prefer to pay the appropriate amount at one time, while the Americans, Germans and Britons are in favor of the basic charge with the possibility of counting the cost when choosing services.

Not waiting for the target solutions in general we have a possibility to use and take advantage of several solutions available already. Two general directions are to use existing solutions both for city transport and traffic optimization and improving the outside traffic control which means also safety.

In general, the V2X systems are being developed to deliver the drivers an extra functionalities for their comfort and protection a that can be conjugated with data included in other onboard devices. Obviously the decision making process is still on the driver nevertheless V2X is a technology that makes possible in the future that autonomic vehicles will be much dependable as it supports them to practically predict and adequately address an unexpected and imperceptible road situations [2].

To be publicly adopted V2X needs to be trusted. It it's not successfully established no spectacular profits can be expected. This is why the major challenges for todays' systems are security and privacy. The systems that are nowadays approachable are designed to secure the communication between vehicles and ensure that it's certified as legitimate instead of a driver or a specific vehicle identification. The systems usually ignore also signals coming from poorly calibrated or compromised systems [2].

The systems should be treated as top priority for the transport; both for the cities and outside i.e. with the following dedication to public transport, road works warning, speed limits, emergency vehicles and school bus or simply for information purposes for the travelers.

A possibility to be more safe on the long distance journey and to have a possibility to connect and monitor the traffic (vehicles) around is also a huge advantage. There is a lot more of course – zero accidents vision, autonomic vehicles, advanced driving assistance multimodal solutions and new mobile quality is definitely the direction of the industry growth.

So called "connected car" was one of the hottest trends in 2014 according to Connected Life by Nielsen report from January 2014; 63% of respondents were looking for the solution and considering it as a significant decisive factor when choosing and buying a new car. Nowadays this is the trend that can't be stopped and which direction is clear: the autonomic vehicles. The direction doesn't surprise if we consider all the advantages mentioned already above like reducing the number of accidents on the roads, reducing the number of fatalities but also will significantly reduce the number of spent fuel. It is estimated that the introduction of self-propelled cars can increase the number of parking spaces in cities up to 10%.

Not only that is the reason. Any company today regardless of what it does can be considered as a technology company. It's been known for quite some time already.

Hence our quest for innovation and the huge interest in new technologies. Today we have reached the stage where we naturally deal with self-learning computers, artificial intelligence, virtual assistants, etc. There are a lot of R&D projects in progress on a different type of systems being able to recognize emotions, pictures, gestures and many others that leads us to higher level technology [7].

The Global Mobile Economy Report (GSMA, CC Forecast, May 2013) estimates that the global connected vehicles market's value in 2018 will reach €39 billion. That means it will be tripled within 6 years. Intel forecasts 152 million of new cars with originally built-in technology by 2020 – sevenfold increase. IHS Automotive predicts that the autonomic cars will appear in the mass market around 2025 and by 2035 will constitute 9% of the sales of all cars (Fig. 6).

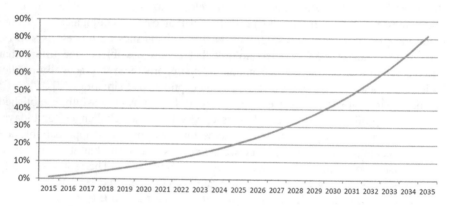

Fig. 6. Vehicles equipped with V2X in Europe – predictions [11]

Many countries all over the world have already led tests and trials of autonomic vehicles. Many of such projects are currently in progress: driverless cars will be allowed on Britain's motorways next year, Göteborg has just released a pool of autonomic taxies, A9 between Berlin and München will be a test road section for such a solution, and many others. The majority of such vehicles and solutions seems to be still not precise enough to be fully trusted however this is the top priority for the manufacturers and they declare full responsibility for their products.

There are still few additional factors that cannot be forgotten which in general can be classified in two major categories – social trust and economics/politics, however as the time flows and brings constantly new improvements same will be reflected naturally in these both areas.

The trend in general is expected as most fundamental change to transport since the invention of the internal combustion engine. What we could see so far as a vision in science – fiction films becomes simply a reality. We are facing currently a disruptive innovation defined in 1995 by Clayton M. Christensen – "an innovation that creates a new market and value network and eventually disrupts an existing market and value network, displacing established market leaders and alliances" [10] (Fig. 7).

Fig. 7. Autonomic vehicles visualizations [8, 9]

References

1. autoDNA. http://www.autodna.pl/blog/nowatorskie-systemy-bezpieczenstwa-v2x-juz-na-jap onskich-drogach/. Accessed 28 Dec 2015
2. Carter   Gene.   http://embedded-computing.com/guest-blogs/v2x-technology-makes-cars-safer/. 10 Apr 2015
3. ETSI EN 302 673-3 V1.2.2, Part 3: Specification of Decentralized Environmental Notification Basic Service, September 2014
4. Michalski,   L.:   http://www.obserwatorium.word.olsztyn.pl/index.php?option=com_k2& view=item&id=53:charakterystyka-infrastruktura&Itemid=81&tmpl=component&print=1& lang=pl&jjj=1455717075588. Accessed 15 Feb 2016
5. Okurowski,   T.:   http://www.lokalizacja.info/pl/rynek-masowy/nawigacja-samochodowa—connected-car/praktyczny-test-technologii-v2x-w-europie-test-na-trasie-1300-km.html#.Vsr 6g032ZaS. Accessed 10 Nov 2015
6. Telefonica S.A.: Connected Car Industry Report (2014). https://m2m.telefonica.com/ multimedia-resources/connected-car-industry-report-2014-english. Accessed 15 Jun 2015
7. Natalia, H.: TrendBook (2016). Accessed 10 Mar 2016
8. http://www.rinspeed.eu/aktuelles.php. Accessed 29 Jan 2016
9. http://www.robertdee.pl/. Accessed 29 Jan 2016
10. https://en.wikipedia.org/wiki/Disruptive_innovation. Accessed 01 Sept 2015
11. Kapsch TrafficCom. Accessed 29 Jan 2016

Traffic Analysis Based on Weigh-In-Motion System Data

Wiktoria Loga and Jerzy Mikulski[✉]

University of Economics in Katowice, 40-287 Katowice, Poland
wiktoria.loga@edu.uekat.pl, jerzy.mikulski@ue.katowice.pl

Abstract. The subject of the paper is the study of the data derived from Weigh-in-Motion (WIM) system while paying special attention to the traffic analysis nearby the measurement point. The attempt is to determine the changes and tendencies in drivers' behaviour and the traffic density based on the data delivered by the devices of dynamic weighing system. Frequent traffic flow analysis is necessary to identify the trends in road traffic, dangerous phenomena and to effective management of the transportation politics, traffic organization and road network planning. The dynamic weighing systems can be the source of complex information about the traffic conditions on the communication route due to continuity and the range of measurements.

Keywords: Weigh-in-Motion systems · WIM system · Traffic management · Traffic engineering

1 Introduction

Growth of street traffic volume is one of the reasons the level of road safety and travelling comfort is decreasing. The field responsible for planning, designing and using the transportation system is traffic engineering. To achieve its goal there is the need for complex data about conditions and trends in the traffic at the selected area. The source of the data are traffic counts. Usually the expense in terms of time, money and human recourses involved in those counts are high. Significant help for reducing those numbers and measurement errors is the usage of modern technologies that automatically count traffic parameters. The collected data in combination with wireless communication, enables to shorten data processing markedly. Additionally automatic and continuous measurements allows to conduct the research more frequently. In the context of road safety and environmental protection becoming more and more significant nowadays there is a need for constant observation of dynamic changes and new or seasonal trends in analyzed data. Precise verification of those variables requires carrying out the traffic counts at least every 5 years. It seems to be particularly relevant with reference to national roads, as fundamental elements of road network. Moreover all already implemented automatic systems such as WIM system may be solid backup for the measurements [1].

© Springer International Publishing AG 2016
J. Mikulski (Ed.): TST 2016, CCIS 640, pp. 268–279, 2016.
DOI: 10.1007/978-3-319-49646-7_23

2   Weigh-In-Motion Systems as a Useful Tool for Traffic Analysis

Weigh-in-Motion system components may be divided between overrun part (usually permanently connected with road surface) and measuring, recording instruments. Besides measuring apparatus indispensable element of the system is the application layer. Complete data are presented through Graphical User Interface (GUI), which is possible due to dedicated software. The design of the interface, type and way of presenting data may be adjusted to final user's requirements. The application displays information about vehicle parameters in real time and forwards the data about possible overload to suitable public authorities [2] (Fig. 1).

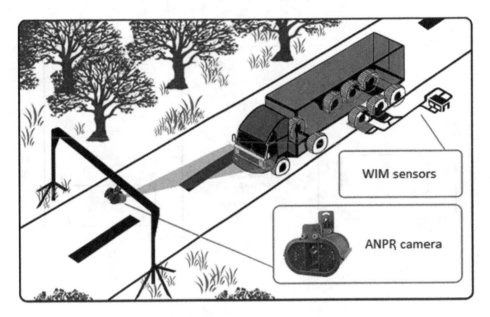

Fig. 1. Weigh-in-Motion system implementation [11]

Conducting traffic counts, research and analysis are the necessary part of planning and managing transport infrastructure and vehicles flow. The basic purposes of the traffic studies are as follows [4]:

- identification and formulation of the laws underpinning traffic flow along with its models,
- data acquisition for specifying movement needs and trends at analyzed area
- acquiring data essential for designing and evaluation of the effectiveness of applied solutions for transportation systems
- analysis of changes and trends in traffic flow from a sociological point of view.

Because the effects in traffic flow are under constant changes, the control and monitoring of those transformations becomes necessary. To this end, traffic counts should be periodically repeated. The example of recurring measurements is the analysis of Annual Average Daily Traffic (AADT), in Poland conducting every 5 years by General Directorate for National Roads and Motorways, as a part of General Traffic Measurement. Usage of data registered by WIM system enables running so-called continuous measurements. It allows to identify all traffic fluctuations. The variations may be examined paying special attention to long or short time periods, e.g. traffic fluctuations weekly, hourly or seasonally. Information about variability of traffic conditions in time enables to formulate additional prognosis concerning for instance traffic volume of heavy goods vehicles, road capacity, noise level or possible environmental impact [4] (Fig. 2).

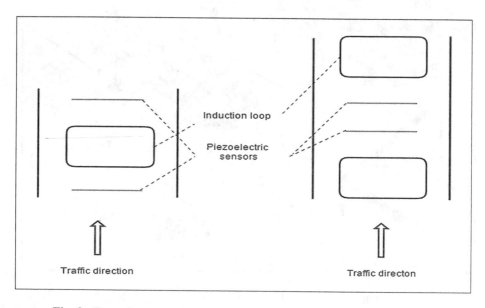

Fig. 2. Exemplary mounting of piezoelectric sensors (own study based on [3])

Currently most traffic analysis in General Traffic Measurement is made in cooperation with traffic observers/pollsters. For those purposes it is necessary to employ from dozen to even 20 thousands individuals each time. Additionally the full cycle of implementation, up to publication of research results takes about 2 years. In manual traffic measurements information must be affixed to the special forms by each observer. Due to this process collected traffic parameters are analyzed mostly using paper documentation. To complete the research it is needful to convert data to electronic format, that requires great amount of time and human recourses. As the consequence, financial investments are increasing while the measurement accuracy is declining, caused by possible human failure (incorrect filling of the from, miscalculations, incorrect vehicle classification [1].

3 Measuring Points Characteristics

In traffic analysis based on WIM data special attention was attached to Silesian Voivodeship, because the measuring point is located in this area. The location choice was based on high AADT volume (highest among all WIM stations included in General Directorate for National Roads and Motorways infrastructure). The AADT value was recognized as advantageous in the view of conducted research, identification of changes and trends in the traffic. The data used in analysis were obtained from WIM station situated on National Road no.1 and National Road no. 44. Mentioned stations are 2 of 94 dynamic weighing points in Poland. One of the test-benches is functioning in Siewierz and it supports two lanes in the same direction (City of Katowice). To compare the traffic conditions in the region, the analysis also involved second WIM station in Silesia, located in Mikołów-Śmiłowice and supporting all traffic lanes [6] (Fig. 3).

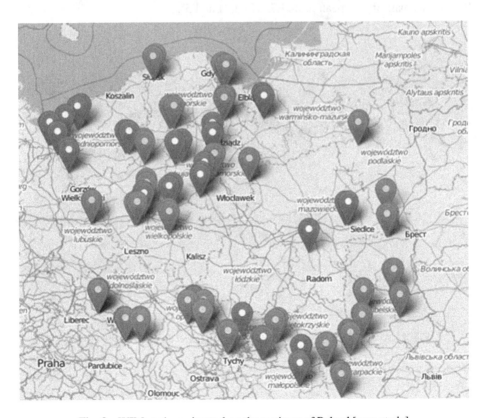

Fig. 3. WIM stations situated on the territory of Poland [own study]

As part of Weigh-in-Motion stations in city of Siewierz and Mikołów Śmiło-wice, two sets of measurement apparatus were mounted. That solution enabled to acquire higher accuracy of conducted procedures. Two separate sets of sensors provide independent data, thereby increasing the number of obtained test samples. Each measurement apparatus is composed of two piezoelectric sensors, inductive loop and temperature sensor. Piezoelectric sensors exhibits responsiveness on tempera-ture fluctuations, the control of this parameter facilitates the correction of the received results. Additionally the stations were equipped in laser sensor that monitors vertical clearance violation. The sensor was mounted on 4 m jib aside from traffic lanes. The video registration devices (ANPR camera and CCTV camera) were placed on 6-m gantry above the road [6].

Both measuring points were implemented and set in motion in the similar time. As part of weighing in motion system, piezoelectric sensors with the B+7 class of accuracy (according to COST 323 Specification) were mounted in the road surface. Apart from the measuring apparatus the stations were equipped in induction loops and video regis-tration sphere. The sphere includes ANPR cameras that are able to read vehicle regis-tration plates automatically and day & night camcorder that registers every passing vehicle. It must be also said that National Road no. 1 as national north-south corridor is also a polish part of European route E75 (Figs. 4 and 5).

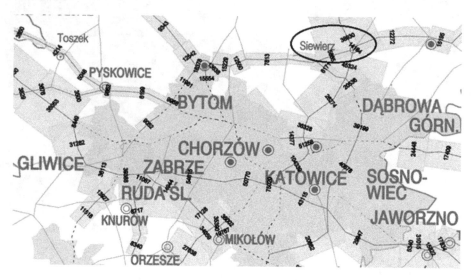

Fig. 4. Annual Average Daily Traffic - WIM station located in Siewierz [7]

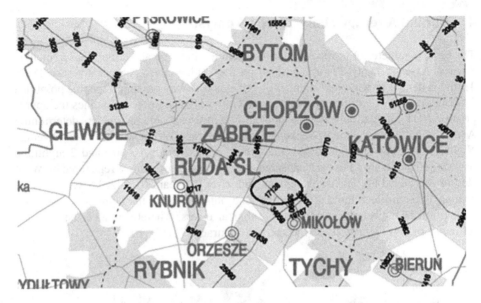

Fig. 5. Annual Average Daily Traffic - WIM station located in Mikołów [7]

The Weigh-in-Motion is using dedicated software that allows to send from the weighing station data concerning [5]:

– gross weight of wheel,
– gross weight of axle,
– total number of axles
– axle spacing,
– gross vehicle weight (GVW),
– vehicle length,
– vehicle height,
– potential vehicle over weight,
– maximum gross vehicle weight,
– vehicle speed,
– traffic line,
– vehicle classification,
– number of the measurement,
– exact date and time of measurement.

Additional data transmitted [5]:

– photos of potentially oversized or overweight vehicles,
– pictures of license plates,
– data read from registration plate saved as text file.

4 Annual Average Daily Traffic Analysis

The basic parameter of traffic calculations included in General Traffic Measurement is Annual Average Daily Traffic (AADT) on the studies area. Annual Average Daily Traffic is defines as a average number of vehicles two-way passing a specific point in a 24-hours period. According to the General Traffic Measurement guidelines the appropriate AADT calculation way was adopted. The indispensable parameters determining AADT volume are: Average Weekday Traffic, Average Weekend Traffic and Average Night Traffic. The study included 9 daytime periods (6 am–10 pm) and 2 nighttime periods (10 pm–6 am). For each analyzed day the volume of vehicle registered by WIM station was added up. The measurement dates over the years 2012–2014 were selected primarily to eliminate the risk of invalid read-outs, lapses in the communication with the station or measurement apparatus technical errors. Thereby reduce the negative influence of those situations on the analysis accuracy [7] (Fig. 6).

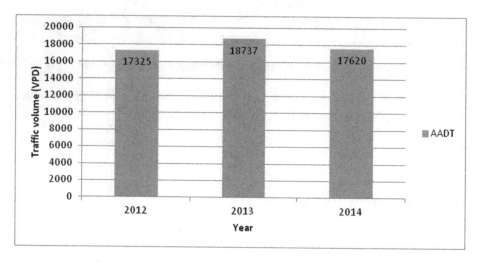

Fig. 6. Annual Average Daily Traffic registered by Siewierz WIM station [own study]

In case of Siewierz station obtained data presents the results only for one traffic direction. Assuming the traffic on the discussed location distributes in 50/50 proportion, the received AADT data can be doubled. Analyzing the AADT changes based on WIM data, the ratio value decreased by 11% over the years 2010–2012. Then 8% growth of ratio was observed in 2013, comparing year 2013 with 2014 another 6% decline. Conclusion to the calculation may be finding that dynamics of AADT changes on National Road no. 1 segment has halted and ratio fluctuation is noticeably lower. In case of Mikołów Śmiłowice station the AADT ratio variations has been already slight between year 2005 and 2010, the growth reached only 0.7%. Putting together the 2010 AADT value with 2012 value (based on WIM station data) the 1% decrease was observed. Comparing year 2012 with 2013 percentage decline was identified again, this time about 9%, while during the next year (2013/2014) small increase (about 4%) was noticed. Also in this location the AADT changes dynamics was relatively small [6–8] (Fig. 7).

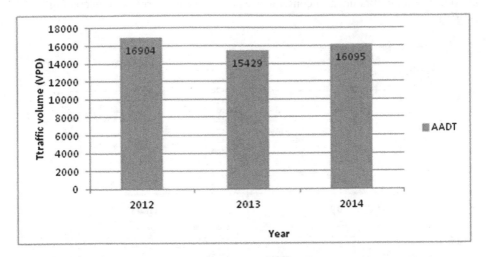

Fig. 7. Annual Average Daily Traffic registered by Mikołów Śmiłowice WIM station [own study]

5   Measuring Point Traffic Characteristics

Traffic volume analysis is a part of research conducting in order to identify current traffic conditions on selected road network area. To calculate traffic volume characteristics and trends correctly, the precise variation over time intervals analysis is necessary. Significant help in capturing traffic fluctuations are continuous measurements. Possibility of conducting such studies enables the use of WIM system data. WIM stations capture and record every vehicle passing the measuring point, constantly 24 hours a day regardless of weather and traffic situation on the road. Because of those WIM advantages, the research time may be loosely selected (any season, day or hours) and adjusted to traffic engineers needs.

Data used in the study was obtained from Weigh-in-Motion station located on National Road no. 1 in Siewierz (two lanes, direction Katowice). The time range of used data embraced 7 consecutive days, from 9th February 2015 until 15th February 2015. The road sensors were calibrated to gain the highest possible reliability level of the measurements. The first step in identifying traffic characteristics was examining traffic volume changes in selected location, based on hourly distribution for each day. The results of the analysis were presented graphically on figure.

In the figures, a significant difference can be observed in traffic volume between weekdays and weekends. During the weekdays, so called rush hours outstand in the figures; this distribution is typical for urban and suburban areas. That means that in those days the traffic volume remains high, also number of vehicles increases in the commuting hours. The morning peak is characteristically higher than the afternoon peak, which on the other hand is more prolonged. In analysed time also higher traffic volume exists in afternoon hours on Friday, that phenomenon is standard for this day of week and it is caused by weekend journeys. On Sunday the vehicle accumulation is clearly higher in afternoon and night hours as a consequences of weekend journeys homecoming [6, 9] (Fig. 8).

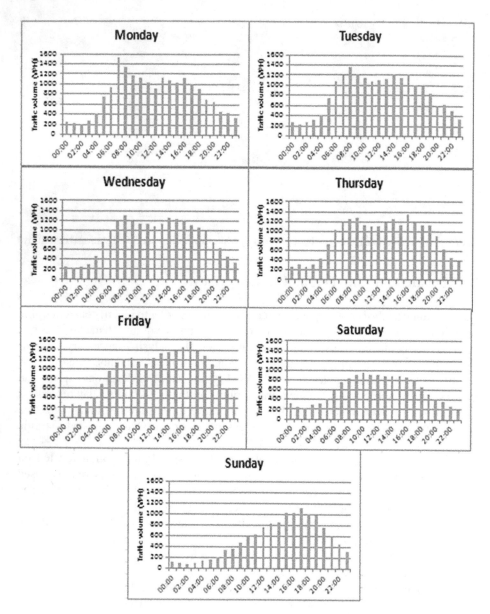

Fig. 8. Hourly distribution of traffic volume changes day by day [own study]

Important element that allows to define the vehicle in the traffic is the classification. This term means assigning the vehicle to one of the predefined groups. The classification systems are rarely implemented for autonomous functioning, they usually are components for bigger systems. Also as part of Weigh-in-Motion station, the classification subsystem was implemented. Ordinance of the Infrastructure Minister demands using classification based on schematic vehicle parameters. Vertical force sensors enable to obtain very precise measurements on the grounds of vehicle axis, the solution provides high selectivity, clarity and effectiveness. Currently several vehicle classification schemes are used. Class division according to European COST 323 Specification is presented in the Table 1 [6, 9].

Table 1. European COST 323 Specification [10]

Category	Silhouette	Description
Category 1	Cars, vans (< 35 kN)	Cars, cars+light trailers or caravans
Category 2		Two axle rigid lorry
Category 3		More than 2-axle rigid lorry
Category 4		Tractor with semi-trailer supported by single or tandem axles
Category 5		Tractor with semi-trailer supported by tridem axles
Category 6		Lorry with trailer
Category 7		Busses
Category 8		Other vehicles

Percentage changes of different vehicle categories in the traffic volume in daily distribution was presented on the Figure. Definitely the biggest share in vehicle movement was observed for passenger cars. Depending on the day of week the share fluctuated in ranging from 62% (Tuesday) to 87% (Saturday). Following class in terms of vehicle volume were tractors with semi-trailer supported by tridem axles (Category 5). Next class regarding percentage values were two axle rigid lorries and tractors with semi-trailer supported by single or tandem axles (Category 2 and Category 4). In weekly distribution the highest volume of goods vehicles were reported at the begging of the week (Monday, Tuesday, Wednesday), following days the shares were significantly decreasing [6] (Fig. 9).

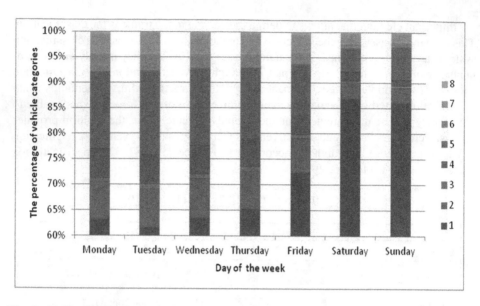

Fig. 9. Daily distribution of different vehicle categories in the traffic volume (10.02.2014–16.02.14) [own study]

6 Conclusion

Weigh-in-Motion systems may be significant support in discussed measurements. The preselecting stations provides essential data to calculate ratios and values used in General Traffic Measurement. Additionally continuity of actions allows to conduct research in different time intervals. Frequent and complex traffic parameters analysis registered by WIM system enables early identification of trends in road traffic and dangerous phenomena. Knowledge about those processes is essential in relevant managing the transportation politics at defined area, implementing adequate traffic organization and also is support in road network planning. Traffic analysis is a tool used in traffic regularity observation. The study should take into consideration complexity and features describing vehicle movement. A vital part of analysis are observations of traffic volume fluctuations. Information about law-breaking committed by drivers or transport companies is also important. All those characteristics may be gained using Weigh-in-Motion systems data.

References

1. Celiński, I., Sierpiński, G., Staniek, M.: eGPR- Generalny Pomiar Ruchu z wykorzystanie urządzeń mobilnych. Inżynieria Ruchu Drogowego, 3/2012
2. Leśko, M.: Wybrane zagadnienia diagnostyki nawierzchni drogowych. Wydawnictwo Politechniki Śląskiej, Gliwice (1997)
3. Mitas A.W., et al.: Elektroniczne narzędzia pomiarowe w transporcie: wagi preselekcyjne. Elektronika: konstrukcje, technologie, zastosowania, nr 12/2011

4. Gaca, S., Suchorzewski, W., Tracz, M.: Inżynieria ruchu drogowego. WKŁ, Warszawa (2011)
5. Dostawa dwóch preselekcyjnych systemów ważenia pojazdów w ruchu oraz ich instalacja w ciągu drogi krajowej nr 1 w miejscowości Siewierz oraz w ciągu drogi krajowej nr 44 w miejscowości Mikołów- Śmiłowice, Specyfikacja Istotnych Warunków Zamówienia, Generalna Dyrekcja Dróg Krajowych i Autostrad, Katowice (2011)
6. Loga, W.: Using weigh-in-motion system data in evaluation of the traffic conditions, Engineering Thesis, University of Economics in Katowice
7. Metoda przeprowadzenia Generalnego Pomiaru Ruchu w roku 2015, Generalna Dyrekcja Dróg Krajowych i Autostrad, Warszawa (2014)
8. Synteza wyników GPR 2010, Generalna Dyrekcja Dróg Krajowych i Autostrad, dostępne online: http://tnij.org/iu4b2vg. Accessed 5 Mar 2015
9. Gajda, J., et al.: Pomiary parametrów ruchu drogowego. Wydawnictwa AGH, Kraków (2012)
10. COST 323 Weight-in-Motion of Road Vehicles (1999)
11. Mikłasz, M., et al.: Ważenie pojazdów w ruchu. Rozwiązania i praktyka w Polsce. Magazyn Autostrady nr 5, Elamed, Katowice (2011)

Supporting the Process of Information Visualisation in Transport Systems

Piotr Lubkowski[✉] and Dariusz Laskowski[✉]

Military University of Technology, Gen. S. Kaliskiego 2, 00-908 Warsaw, Poland
{piotr.lubkowski,dariusz.laskowski}@wat.edu.pl

Abstract. The process of ensuring the security of monitored objects and material goods requires the access to the information from sensors located in different points of monitoring and data acquisition systems. Operational use of monitoring system supports the preventative action, observation and emergency operations. The possibilities of the monitoring system associated with a continuous registration of events in real time enable their storage and further processing, which can in turn be used to analyses the actions taken and the training of transport system crews. An important element of supporting the work of transport system crew, except technical aspects, is the opportunity to obtain a direct and easy access to information from the sensors for a person in charge and to have a software application that allows visualization of data from the monitoring system on the screen of the mobile data terminal. The paper presents an application that offers the possibility of supporting the process of events visualization that uses data captured from sensors of video surveillance system.

Keywords: Visualization of information · Surveillance services · Transportation systems

1 Introduction

Increasing the level of safety of transport operations requires more and more frequently use of advanced technical means implementing the transfer of information from stationary and mobile video surveillance systems. Video surveillance systems are used for many years in both public and national utility facilities as well as in commercial establishments, factories and businesses, warehouses or in vehicles. The information from the monitoring systems is traced back increasingly by armed forces in order to secure activities in military operations, traffic and fire rescue teams or crisis management services.

A properly functioning video surveillance system that uses elements of stationary and mobile infrastructure can thus be an essential element supporting the process of events visualization also in transportation systems [1, 2]. Operational use of monitoring system supports the preventative action, observation and emergency operations. The possibilities of the monitoring system associated with a continuous registration of events in real time enable their storage and further processing, which can in turn be used to analyse the actions taken and the training of transport system crews. It should be also

© Springer International Publishing AG 2016
J. Mikulski (Ed.): TST 2016, CCIS 640, pp. 280–292, 2016.
DOI: 10.1007/978-3-319-49646-7_24

pointed out that the role of the monitoring system is not limited only to the operational application. The monitoring system can also be used in the prevention activities and as part of the evidence procedure. In each of these applications an extremely important issue is the reliability and quality of the realized monitoring process.

To ensure the effective functioning of the monitoring system in supporting the work of transport system crews in the first instance such technical aspects should be taken into account:

- the size of monitoring area,
- arrangement of visualization sensors,
- characteristics of sensors as well as registration and visualization equipment's,
- the level of technical knowledge of operating personnel.

An important element of supporting the work of transport system crew, except those mentioned technical aspects, is the opportunity to obtain a direct and easy access to information from the sensors for a person in charge and to have a software application that allows visualization of data from the monitoring system on the screen of the mobile data terminal [3]. In a number of cases the services do not have this type of information, as well as not have simple-to-use mobile visualization applications. The paper presents an application that offers the possibility of supporting the process of events visualization that uses data captured from sensors of video surveillance system and that is designed for the Android mobile platform. A characteristic feature of the application is the possibility of combining sensors with a map of the area in which are conducted the activities of transport system crews. The proposed solution is part of the monitoring and data acquisition system designed for the INSIGMA (*Intelligent Information System for Global Monitoring, Detection and Identification of Threats*) project [4].

2 Information Visualization Services

Monitoring systems are a combination of video recording (sensors), transmitting, storing and reproducing devices in one integral unit. They enable observation of people or objects in real time as well as event recording for later analysis. Monitoring systems represent, therefore, an executive element of information visualization services, whose primary goal is to provide information in digital form, represented by video or audio data, for further processing [4, 5]. In modern visualization systems data recording is carried out using sensors in the form of IP cameras equipped with image and sound sensors. System is complemented by such components as network hubs, video recorders and telecommunications infrastructure. Depending on the approach taken, these components are combined into a single monitoring network using a wired (based on Ethernet 10/100 Base-T standard) or wireless (using IEEE 802.11 standard) medium.

The service of information visualization (IVS) allows identification and perception of elements (objects, events) of the surrounding reality in a function of variable time and space. It also allows to support the decision making process under dynamically changing conditions of the mission or operations. This service through the acquisition and analysis of information (in the form of e.g. digital data gathered from sensors) supports the process of assessing the effects of the events on their objectives and goals. Lack of or insufficient information is identified as one of the main factors influencing the action taken during ongoing operations and is generally equated with increased risk caused by the human factor. Therefore, visualization service is particularly important in areas where volume of available information is large and wrong decisions can lead to serious consequences.

In the case operations carried out by road or railway rescue services, but not only their success depends on efficient operation of whole team [6]. Therefore, the service of visualization information should be characterized by the possibility of providing specific information to individual team members related to the nature of the tasks performed by them, as well as common situational picture that allows obtaining information common to the whole team. The significance of the visualization information service for various applications is widely analysed in the literature. Examples of such applications are shown in [7–9]. This publication presents a solution that combines the functionality of information visualization services from the data made available in a video monitoring system.

3   Risk Assessment of Information Visualization Services

Monitoring system's ability to correctly detect and identify threats is one of the essential factors determining the efficiency, effectiveness and reliability of its operation. Hence, in the process of information visualization additional risks and caused by them consequences that have a direct impact on this process should be taken into consideration. This so-called risk analysis is part of the applications validation process which is presented later in this paper.

Nowadays video surveillance networks are characterized by such features as frequent changes in topology, the mobility of users and providers, the use of wireless links and limited processing power and capacity of network nodes. Proper operation of the sensor is influenced by factors such as lighting, distance between object and the sensor or the presence of background objects that make identification process difficult. Not without significance are also the already mentioned technical parameters, in particular the sensitivity of the sensor, focal length, resolution and compression used. Hence, the following threats related to reliable data identification should be taken into account: lack of proper object lighting (insufficient sensitivity of the transducer) or no infrared operation mode, too long distance between the object and the camera (no proper selection of the focal length), no possibility of object specification (insufficient sensor resolution) or no extraction possibility (low resolution, sensitivity). These threats can lead to the following problems related to visualization process (Fig. 1):

- false rejection - an object that has its model in the database is unrecognized and rejected due to the fact that it does not have its counterpart,
- misclassification - an object that has its model in the database is not properly assigned to other model in the database,
- false acceptance - an object that does not have its model in the database is assigned to a model that already exists in the database.

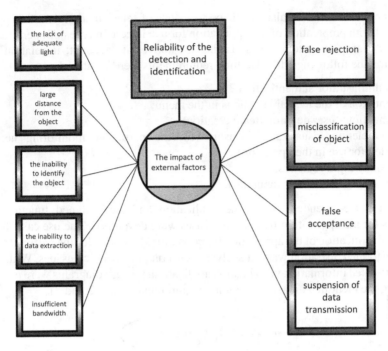

Fig. 1. The Bow Tie threat analysis and problems of visualization in terms of the influence of external factors [own study]

4   Requirements for Information Visualization Application

The information visualization system is dedicated for IT support of transport operations by presentation of data from video monitoring system. Functional requirements for the application result from its intended purpose and include the possibility of parallel display of situational data from several sensors, located in the place of operation, on a mobile terminal (terminals) with which the transport system crew is equipped. The application supports video visualization service realization shared in the IT system developed for the time of operation and/or functioning in the place of its performance. Therefore, it should comply with a set of requirements generally referred to as client requirements that include [10]:

– user authentication;
– detection of the availability of IT network resources and login to identified resources in the wireless communication mode;
– presentation of the available visualization sensors (identification of cameras in the supported video monitoring system) in connection with the site map;
– full screen video data preview from the selected sensor in the on demand mode;
– image scaling;
– modifications of user authentication parameters.

The application should also enable performance of selected administration projects connected with preparation of the application for use in certain types of operations and by certain transport system crews. Taking the above into account, in relation to the application the following administration requirements were identified:

- user/administrator authentication;
- presentation of the available options in the menu;
- user creation - user deletion (team creation);
- adding/deleting visualization sensors (for the particular monitoring networks intended for use in the operation);
- analysis of login registers;
- automatic verification of input data.

Considering the intended use of the application and the assumed functionalities, diagrams for the application function use cases were developed. The use case diagram enables identification of the application functionalities available for the particular actors of the performed operation without analysing the complex technical issues. With regard to the presented information visualization application, two use diagrams were proposed, reflecting the identified areas of functional requirements (Figs. 2 and 3).

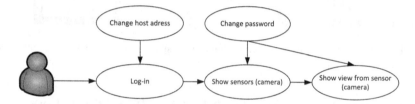

Fig. 2. Diagram of use cases in the client functionality [own study]

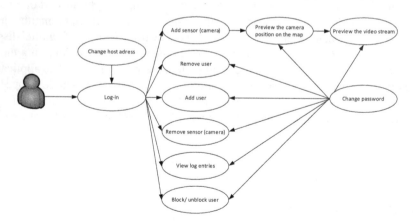

Fig. 3. Diagram of use cases in the administrator functionality [own study]

The service of information visualization in the video monitoring system assumes the existence of three main actors whose actions have impact on its proper use in transport operations:

– user – authorized to use the client application;
– administrator – authorized to use the administrator application;
– system,

The difference between the first two actors consists in different level of access to the system. In other words, they have a different level of access to take specific actions connected with the use of the visualization service application. The above actors have certain common features, as the scope of user rights is included in the set of the administrator rights. Each of them is provided with the possibility of login, logout and viewing the cameras. Thus, the scenarios of use cases for these functionalities are similar. An example of use scenario for the user in the mode of video downloading from the camera is presented in Table 1.

Tab. 1. Results of conducted experiments [own study]

Camera preview	
Description	The user views the list of the available cameras
Actors	User/system
Initial condition	The user must be logged in
Main scenario	1. The system detects whether the user device has access to the video monitoring network and Internet 2. Google map is activated in the subscriber device 3. The system collects information on the cameras located in the database defined by the administrator 4. The system applies markers on the map, which symbolize the cameras identified in the area of operation
Supplementary scenario	A. In the device without the Internet access, the system collects information on the cameras from the database prepared by the administrator before the operation B. Based on the collected information it creates a list of buttons which are a reference to the data stream describing the possibility of data download from the camera

5   Implementation of the Visualization Application

The information visualization application, which is the basis for the visualization service, was developed on the basis of Android platform using the project environment composed of Android SDK, Java Development Kit, MySQL database, support package for PHP language and the formalized JSON language. Android SDK includes a set of libraries responsible for communication between the system components and the application. In turn, the MySQL database is an element storing information on the parameters and location of sensors. The database is integrated with the application using the PHP package which, in this case, is an intermediary in communication between the database and the application. The application was designed to operate with Android system 4.0.3 (API 14) or later. During initialization, the application performs the procedure to verify the access to the Internet resources and checks the connection based on the defined set of source and configuration files.

For communication with the user, a graphical interface was developed, built on two activities related to the supported functionalities: user and administrator. The user has only the right to preview the visualization shared by the system. The administrator has not only the user rights but also other, additional functions resulting from the need to manage the visualization system. Owing to the Navigation Drawer mechanism, it is possible to access the administration and visualization menu which is realized in the form of a strip of options emerging on the left of the application home screen. In order to improve video preview, the mobile application was provided with the function of changing the image view position using gestures. They are used in two cases - when checking the marker position in the administrator application and when scaling the video image of the mobile application. The application view in the login mode is presented in Fig. 4.

Fig. 4. Login screen of the application for visualization of information from the monitoring system [own study]

The possibility of choosing the camera preview through tags included in Google maps [10] is one of the main assumptions of the visualization system. In order to implement this functionality in the visualization application, the following projects should be implemented:

- generation of the key enabling the use of Google maps;
- adding the key and OpenGL based on *AndroidManifest.xml* file modification;
- detection of the Internet connection;
- implementation of map support in the program.

Implementation of the above projects results in displaying the map with included positions of sensors in the monitoring network. Selection of the marker symbolizing position of the camera allows for calling the camera description and subsequently displaying the camera image. It is possible for the administrator to define the area displayed by the application when it is running (at the stage of operation preparation) or to limit the area during mission by the user. Implementation of these operations is possible owing to *GoogleMap* type object. The administrator defines also the container in which information on the cameras available in the database is stored. This method of storage facilitates using the application in the data preview mode and connecting the cameras with the map image. The application screen view in the map preview mode is presented in Fig. 5.

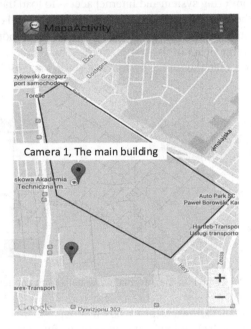

Fig. 5. Application screen in the mode of presentation of sensors on the site map [own study]

The application may be launched using physical hardware terminals (e.g. tablet, smartphone) or using a virtual device, the so called emulator shared, for example, in Android Studio. In each case it is necessary to generate an installation file. The application installation file (.apk) is generated after calling the *GenerateSigned APK* option from the Build menu of the Android Studio program. The generated file, after being copied to a mobile device or to an emulator, is installed and the application shortcut is provided on the device home screen.

6   Implementation of the Visualization Application

In order to perform the development and functional tests, the measurement station was developed composed of the subscriber device and computer including the MySQL database with data on the sensors and XAMPP website server located in the Division of Communication Systems in the Faculty of Electronics of the Military University of Technology [11]. The subscriber device was configured to operate in the user mode. The purpose of the performed tests is to present the functional possibilities of the developed application.

The tests were performed in a mobile environment, at a rate reflecting a moving pedestrian subscriber in an urban area representing the place of the operation. Owing to the use of a subscriber device in the form of a smartphone equipped with the GPS module, it is also possible to track the moving subscriber. During the tests, one measurement terminal was used, switched to work in IEEE 802.11 network guaranteeing access to the video monitoring system and Internet access to load the site map. The tests were performed on the premises of the Military University of Technology (Fig. 6). Products of renowned suppliers of hardware and software for both the systems and applications are the components of the test platform. Therefore, it appears reasonable to conclude that the specified measuring system is a correct and highly reliable testbed.

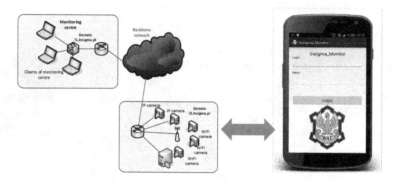

Fig. 6. Diagram of the testbed for the information visualization application [own study]

Functional tests are also known as black box testing, and are aimed at detection of the error resulting from the lack of implementation of the functionality described by the requirements. Functional tests cover both the functions available to the users through the application interface and back-end operations, such as system security. Test cases concerning the presented visualization application were implemented to check whether it behaves as it should. In the first part, it was focused on functional tests of the client application. This group of tests covered:

- user authentication;
- detecting the Internet connection and displaying the window with a list of cameras;
- presenting the camera image stream;
- video scaling including the reset option;
- changing the user password;
- changing the settings through the change of the website address.

In the second part, functional tests of the visualization system administrator application were performed. It was assumed that this component should be tested according to the following diagram:

- administrator authentication, adding and deleting the user/administrator;
- blocking/unblocking, adding and deleting the camera, login history viewing;
- automatic verification of data input in the text fields.

Based on the performed tests, it was found that the application functions correctly in relation to majority of the performed tests. It should be stated that the application is resistant to attempts of intended or unintended interference in its correct operation. In the case of login, all attempts of unauthorized access were detected and blocked. It was noted that the camera image presentation and image scaling functioned properly. Based on the available markers, the user was able to access all monitoring points. However, it was also noted that it was needed to colour the markers in order to clearly connect them with the sensors (Fig. 7).

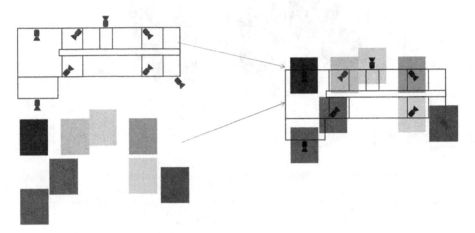

Fig. 7. Connection of markers with dislocation of sensors [own study]

The above problem resulted from restrictions of the applications related to the necessity of providing an exact address of the network sensor and sensor coordinates for the site map.

The results of qualitative tests are presented below. The goal of qualitative tests was to validate the IVS application and the quality and reliability of IVS in surveillance network. The results showing the impact of the selected threats related to reliable data identification. Such factors as light intensity distance between the object and the camera as well as sensor resolution were analysed. In the process of quality and reliability evaluation that we call user satisfaction, single stimulus method compliant with ITU-R BT 500-11 recommendations was used [12]. During the tests three types of objects were identified which differs in the size and shape. Tests were performed using the series of IP Axis cameras.

The next figures (Figs. 8 and 9) present the impact of light intensity and distance from object on the user satisfaction. As can be seen, in the absence of artificial lighting, there is an increase of user satisfaction. This is due to the fact that the tested camera has a function of night mode resulting in better visibility of objects. The results indicate also a relationship between the quality and reliability of the identification process and the distance of the object from the camera.

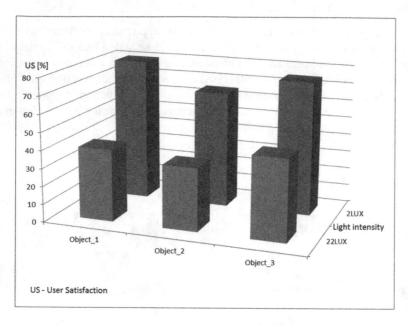

Fig. 8. Impact of light intensity on user satisfaction [own study]

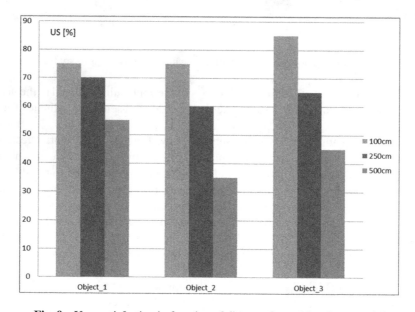

Fig. 9. User satisfaction in function of distance from object [own study]

The impact of sensor resolution on the quality and reliability of IVS is presented in Fig. 10. During the tests the camera worked in low resolution mode (QCIF: 176 × 144), and then was switched to the mode of high resolution (4CIF: 704 × 576). The results show that even at low camera resolution the IVS work correctly. However, the use of this variant of work does not provide the 100% guarantee of IVS quality and reliability.

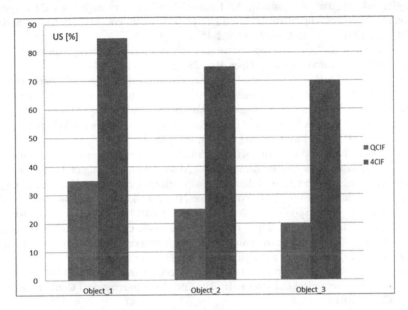

Fig. 10. Impact of sensor resolution on IVS quality and reliability [own study]

7 Conclusion

The paper presents results of works connected with implementation and launching of the application for visualization of information from the monitoring network of the Telecommunications Institute in the Faculty of Electronics of the Military University of Technology. The results confirm the possibility of using the application for video visualization in the process of supporting selected functions of transport processes.

It should be also mentioned that the presented application, at the current stage of development, does not aspire to be a professional application, but its purpose is to estimate the possibility of using it in specific environmental and technical conditions, in order to obtain conclusions for further development work. The obtained results are determinant for further studies in the area of monitoring systems.

Acknowledgments. The work has been supported by the European Regional Development Fund within INSIGMA project no. POIG.01.01.02,00,062/09

References

1. Siergiejczyk, M., Paś, J., Rosiński, A.: Application of closed circuit television for highway telematics. In: Mikulski, J. (ed.) TST 2012. CCIS, vol. 329, pp. 159–165. Springer, Heidelberg (2012). doi:10.1007/978-3-642-34050-5_19
2. Siergiejczyk, M., Paś, J., Rosiński, A.: Evaluation of safety of highway CCTV system's maintenance process. In: Mikulski, J. (ed.) TST 2014. CCIS, vol. 471, pp. 69–79. Springer, Heidelberg (2014). doi:10.1007/978-3-662-45317-9_8
3. Kowalski, M., et al.: Exact and approximation methods for dependability assessment of tram systems with time window. Eur. J. Oper. Res. **235**(3), 671–686 (2014)
4. Łubkowski, P., Laskowski, D., Maślanka, K.: On supporting a reliable performance of monitoring services with a guaranteed quality level in a heterogeneous environment. In: Zamojski, W., Mazurkiewicz, J., Sugier, J., Walkowiak, T., Kacprzyk, J. (eds.) Theory and Engineering of Complex Systems and Dependability. AISC, vol. 365, pp. 275–284. Springer, Heidelberg (2015). doi:10.1007/978-3-319-19216-1_26
5. Łubkowski, P., Laskowski, D.: Test of the multimedia services implementation in information and communication networks. In: Zamojski, W., Mazurkiewicz, J., Sugier, J., Walkowiak, T., Kacprzyk, J. (eds.) Proceedings of the Ninth International Conference on Dependability and Complex Systems DepCoS-RELCOMEX. June 30 – July 4, 2014, Brunów, Poland. AISC, vol. 286, pp. 325–332. Springer, Heidelberg (2014). doi:10.1007/978-3-319-07013-1_31
6. Werbiska-Wojciechowska, S., Zając, P.: Use of delay-time concept in modelling process of technical and logistics systems maintenance performance. Case study, Eksploatacja i Niezawodność - Maintenance and Reliability **17**(2), 174–185 (2015)
7. Hamzah, M., Sobey, A., Koronios, A.: Supporting decision making process with information visualization: a theoretical framework. In: The 2nd IEEE International Conference, ICIME 2010, China (2010)
8. Mathias, W., Burneleit, E., Albayrak, S.: Tactical information visualization for operation managers in mass casualty incidents. In: Proceedings of the Workshop on AmI for Crisis Management, Italy, vol. 953. CEUR-WS.org (2012)
9. Shrinivasan, Y.B., van Wijk, J.J.: Supporting the analytical reasoning process in information visualization. In: Proceeding of the Twenty-Sixth Annual SIGCHI Conference on Human Factors in Computing Systems, CHI 2008, Italy (2008)
10. Kwiatkowski, P.: Development, implementation and testing of mobile applications for the visualization of information from video surveillance, Master thesis, WAT, Warsaw (2015)
11. Lubkowski, P., Laskowski, D., Pawlak, E.: Provision of the reliable video surveillance services in heterogeneous networks, In: Safety and Reliability: Methodology and Applications - Proceedings of the European Safety and Reliability Conference, ESREL 2014, pp. 883 – 888. CRC Press, Balkema (2014)
12. ITU-R Rec.BT 500-11 Methodology for the subjective assessment of the quality of television pictures, Geneva (2006)

Deep Learning Approach to Detection of Preceding Vehicle in Advanced Driver Assistance

Paweł Forczmański[✉] and Adam Nowosielski

Faculty of Computer Science and Information Technology,
West Pomeranian University of Technology, Szczecin, 52 Zolnierska St.,
71-210 Szczecin, Poland
{pforczmanski,anowosielski}@wi.zut.edu.pl

Abstract. In paper we propose a detection method for objects in video stream taken in front of a car by means of deep learning. The successful detection of preceding cars is a part of the analysis of current road situation including emergency and sudden braking, unintentional lane change, traffic jam, accident, etc. We include the results of preliminary experiments employing video stream captured by camera installed behind frontal wind screen. The detection and classification are performed using Convolutional Neural Network preceded by road lane detection. We performed several experiments on real-world data in order to check the accuracy of the proposed algorithm.

Keywords: Deep learning · Convolutional neural network · Vehicle detection · Vehicle classification · Road lane detection

1 Introduction

1.1 Motivation

Nowadays, fast evolution of intelligent transport systems is observable. One of the most interesting areas of such progress is the development of autonomous and self-driving cars. Autonomous car looks like the vehicles used today, having a steering wheel and several seats that face the driving direction. The driver is necessary to supervise such a car, which takes control only in certain situations. Currently, some elements of autonomy are already offered to the customers, i.e. adaptive cruise control, self-parking, and automatic braking [1]. The next, or some say, parallel stage of development is a self-driving vehicle, which potentially takes control over the driving in case of heavy traffic or on highways. Such evolution may lead to the total disappearance of the steering wheel making vehicle drive using the same set of sensors as autonomous vehicles. In such case, the car has to know its exact position, identify objects nearby, and constantly calculate safe and optimal route. This real-time-based situational and contextual responsiveness requires a powerful visual computing algorithms that combine data from all available sensors, while also planning the safest path [2].

Both concepts, however, are founded on the same idea of recognizing environmental conditions and taking decisions about driving. In case of autonomous car the decision

© Springer International Publishing AG 2016
J. Mikulski (Ed.): TST 2016, CCIS 640, pp. 293–304, 2016.
DOI: 10.1007/978-3-319-49646-7_25

is only a suggestion for a driver, who is eventually responsible for the car, while self-driving car does not assume any human actions. Hence, both car types should profit from artificial intelligence and machine vision methods.

The most common problem related to both autonomous and self-driving cars addressed in the literature is the recognition of traffic signs while driving [3]. Fortunately, such an issue is well described and solved in many cases. There are successful implementations in many middle-class cars also. The other problem, which we focus on, is the detection of cars on the road. While this is crucial in terms of driving safety, contemporary car crash-avoidance systems and experimental self-driving have also such systems implemented. However, they often rely on radar and other sensors to detect cars and various obstacles on the road. Hence, the expected improvement is a vehicle/obstacle detection system that can perform in close to real-time based on visual cues only. Such vision-based detection could make systems for recognizing vehicles both cheaper and more effective.

In our algorithm we employ recently proposed deep learning approach. It is associated with a specific branch of machine learning that solves the task of mapping input data such as image features, e.g. brightness of pixels, to certain outputs, e.g. abstract class of object. The deep learning model, in a form of a hierarchical representation of the perceptual problem, compose functions into sets of input transformations, through transitional representations, to output. Such composition constitutes a deep, multi-layered model encoding low-level, elementary features into high-level abstract concepts. The strength of such an approach is that it is highly learnable. The idea of learning comes from a classical neural networks, i.e. we feed an input of a deep network iteratively, and let it compute layer-by-layer in order to generate output, which is later compared with the correct answer. The output error moves backward through the net by back-propagation in order to correct the weights of each node and reduce the error. This operation improves the model during computations. The scheme of processing is presented in Fig. 1.

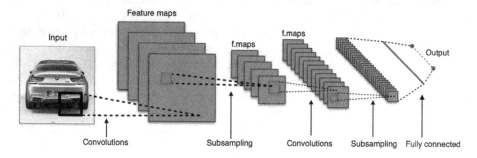

Fig. 1. Data flow in convolutional neural network [own study]

In this paper we propose to use convolutional neural network to solve a problem of multi-class vehicle recognition based on video analysis. While the idea is not purely original, we provide a complete processing flow, including road lane detection, region of interest extraction and final vehicle classification. We propose several improvements in order to make it possible to execute our algorithm close to real-time.

The paper is organized as follows. First we present related works, especially in the field of vehicle detection and classification, from science and industry. Then, we introduce the most important elements of out algorithm including lane detection, region-of-interest definition and vehicle extraction/classification. Finally, we present some exemplary results of numerical experiments and we finish it with some conclusions and future plans.

1.2 Existing Solutions

The general task of detecting certain objects in static scenes or video streams implicates the determination of the image portion containing the investigated object. The selection of appropriate characteristic features used to build an object model is crucial in this case. Additional aspects to be considered are the mechanism of feature matching and the method of scanning input scene. In most cases, the process of detecting specific objects does not rely on an information about the detected object, i.e. object pose, size and position. Further difficulties may result from the scene illumination changes, environmental conditions and specific look of objects [4].

The solution of the problem of detecting vehicles in images can be based on various assumptions related to scene and observer characteristics. If the scene is static and the camera observes moving vehicles, the solution can benefit from background modelling methods [5, 6]. In such an approach, moving objects are segmented from background and then classified [7]. That improves the performance, since the detection can be the most time-consuming task. On the other hand, when the camera is mounted on a moving vehicle, the background modelling cannot be applied, since the scene changes constantly and the extraction of moving objects is complex. Hence, the algorithm have to scan whole scene area with a sliding window and compare extracted regions with a template [5]. In case when the initial information about interesting object is absent, it is required to perform searching in all possible locations and using all probable window (or image) scales. It may significantly increase the computational complexity. Therefore, it is profitable to limit the scanning area for a classifier and the size of sliding window basing on some heuristics. It can employ the knowledge about road area, horizon level and the distance of observation.

The scientific literature contains many proposals of visual features used as a determinant of vehicle presence. Most of them use Local Binary Patterns, Histogram of Oriented Gradients, Spectral Features from Discrete Cosine Transform etc. In [8] an exemplary method for vehicle detection is proposed. It models an unknown distribution of vehicles by means of higher order statistics (HOS). Finally, the classification is performed using HOS-based decision measure. Another interesting algorithm is presented in [9]. It project all input pixels' colors into a new feature space. Bayesian classifier is applied to classify pixels belonging to the vehicles. Finally, Harris corner detector is applied to find extreme corners, and mark a vehicle. The authors of [10] presented a detection method that extracts rear-views vehicles. It is important, that it does depend on the road boundary or lane position. The segmentation of the region of interest employs shadow area under the vehicles and simple low-level features such as edges and symmetries. Vehicle detection combines statistics-based and knowledge-based methods. The number of false detections is reduced using the a'priori knowledge about detected vehicles. Finally, Support Vector Machine is used for two-class classification problem. A similar application of an ensemble of several one-class SVMs is presented in [11].

Many recent solutions for the vehicle detection, e.g. [12, 13], employ extractors based on Histogram of Oriented Gradients (HOG), Local Binary Patterns (LBP), and Haar-like features, passed to the certain cascading classification algorithm, e.g. AdaBoost. The HOG represents direction of brightness changes in some area, while LBP labels pixels by thresholding, resulting in a binary sequence. In case of simple rectangular Haar-like features, the difference of the sum of pixels in areas inside the rectangle is calculated. All of them are rather easy to calculate yielding fast computations and real-time operation.

The alternative direction in research is associated with a use of deep learning methods to create robust classifiers without a need of carefully crafted low-level features. An interesting study is presented in [14]. The authors collected a large volume of road data and applied computer vision and deep learning algorithms to solve problems related to car and lane detection. They showed that convolutional neural networks (CNNs) may be used to perform such tasks in real-time. As it was shown, deep learning can be applied in autonomous driving. Another exemplary approach based on deep learning is presented in [15]. The authors propose to extract the frontal view of a car to recognize the maker and the model. After that, they transform the frontal view of a car to its feature mapping using principal component analysis (PCA). Finally, they use deep learning with three layers of restricted Boltzmann machines (RBMs). Another novel approach is presented in [16], where a deep learning based vehicle detection algorithm with two-dimensional deep belief network (2D-DBN) is proposed. The authors developed a 2D-DBN architecture that employs second-order planes as input and bilinear projection yielding enhanced accuracy in vehicle detection. According to the authors, the experimental results showed that their method works better than provided state-of-the-art algorithms.

Besides scientific-only approaches, there are also several commercial products available on the market. They are based on various principles, however most of them works in a similar way. The earliest solutions to preceding vehicle detection were algorithms implemented in Lexus LS (2006), Volvo s80 (2007), Lincoln MKS/MKT (2009) and Audi A8 (2010). Nowadays, more other carmakers introduce similar systems, both relying on radar and CCD cameras. An independent company - Mobileye [17] - proposed to use an information about detected lane and an additional image information to give a trustworthy threat assessment and collision anticipation. It provides an early detection based on optic flow analysis. Unfortunately, the details are not publicly available and the performance of the developed system is not easy to evaluate. Nevertheless, one of the most advanced systems is the development platform for autonomous cars offered by Nvidia Corp. – Drive PX [18]. It provides driver assistance technologies powered by deep learning, fusion of sensor data, and surround vision. According to the developers, Drive PX can work with video streams captured by 12 cameras, together with lidar/radar, and ultrasonic data. Such information is processed by the appropriate algorithms (based on Deep Neural Networks) in order to precisely analyze the whole environment around the vehicle, including both dynamic and static objects.

As it can be seen, most of presented systems are rather complex and very specialized solutions, dependent on many additional sensors, closely integrated with the car subsystem, hence not to be applied in general purpose, low-end devices, like smart-phones or tablets. Therefore in the paper we present an algorithm, that uses only a simple monocular camera and a dedicated software, making proposed solution widely applicable. Thanks to software-only architecture presented system can be easily extended or modified, mainly in the field of vehicle classification.

2   System Overview

2.1   Algorithm Outline

Developed system works on a video stream, assuming a camera is placed behind the windscreen, looking forward. The view area covers a road in front of the host vehicle, with a horizon line roughly in the middle of a picture. The scheme of processing is presented in Fig. 2. It consists of two main modules: road lane detection and object classification. The first module is intended to extract a region of interest (road lane occupied by the host vehicle), while the second one detects vehicles within this region. The algorithm works in a loop, where each iteration ends with a detection/classification.

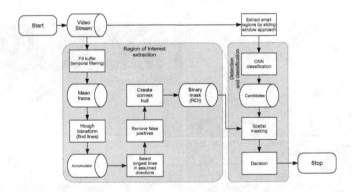

Fig. 2. Scheme of processing [own study]

2.2   Road Lane Detection

The first stage of processing involves the detection of road lines and, thus, the area (Region of Interest), in which we identify vehicles [19]. We assume, that the road in front of the host vehicle is constantly observed. For our purposes, a mobile device equipped with a simple camera (e.g. smartphone or tablet) is appropriate, hence it is very popular and general-purpose. In order to make the calculations easier, we assume that the horizon line is detected with help of a typical accelerometer and g-sensor, however there are many algorithms that are based on visual features only [20]. Our lines detection is based on a well-known Hough transform, however there are other possible algorithms to employ [21].

In order to detect road lines constituting a road lane, we propose to use the following set of operations:

1. Calculate an average of 10 successive frames in order to remove image artefacts and noise;
2. Convert the result into grayscale;
3. Detect edges using Canny detector and create binary image;
4. Remove values above calculated horizon level;
5. Detect lines using Hough Transform (HT) and leave only detected lines that extend at an angle range of -65 to 65 degrees;
6. Select long lines by locating the highest peaks in the HT matrix;
7. Fill gaps in detected lines;
8. Create convex hull based on the endpoints of the line segments;
9. Create binary mask – road lane in front of a vehicle.

Further, obtained region of interest is grown, in order to cover all possible locations of vehicles in front of the host vehicle. Several exemplary detections are presented below (Fig. 3). As it can be seen, the method is resistant to the illumination changes, camera position and horizon level. However, if the road changes its direction in a rapid manner (e.g. sharp curve occurs) the area cannot fully approximate the area in front of the car, since it assumes rather long line segments.

Fig. 3. Examples of road lanes detection in different road conditions [own study]

2.3   Deep Learning Based Vehicle Classification

At the stage of detection/classification we use Convolutional Neural Network, originally known as the NeoCognitron [22] and then advanced by LeCun et al. as LeNet [23]. An algorithm applied in this case is based on a successful AlexNet [24].

The key computational problem in a CNN is the convolution of a feature detector with an input signal. In such an approach, the features flow from single pixels to elementary primitives like vertical and horizontal lines, circles, and color areas. In contrary to the traditional filters that work on single-channel images, CNN filters process all the input channels. Since convolutional filters are translation-invariant, they produce a strong response wherever a specific feature is discovered. In our approach we employed pre-trained network (based on ImageNet) and the MatConvNet toolbox [25] for implementing convolutional neural networks. The CNN used in our algorithm consists of 37 layers. The input layer (which actually contains input image) is 224 × 224 elements. The structure is presented in the Tab. 1, where 'type' represents the type of a layer (*cnv* - convolution; computes the output of neurons that are connected to local regions, *relu* - Rectified-Linear and Leaky-ReLU; an elementwise activation function; *mpool* – Max Pooling, performs downsampling operation along the spatial dimensions; *sftm* – performs SoftMax regression) (Table 1).

Table 1.   The structure of CNN used for detecting vehicles [own study]

layer	1	2	3	4	5	6	7	8	9	10	11	12	13
type	cnv	relu	cnv	relu	mpool	cnv	relu	cnv	relu	mpool	cnv	relu	cnv
support	3x3	1x1	3x3	1x1	2x2	3x3	1x1	3x3	1x1	2x2	3x3	1x1	3x3
stride	1	1	1	1	2	1	1	1	1	2	1	1	1
padding	1	0	1	0	0	1	0	1	0	0	1	0	1
out dim	64	64	64	64	64	128	128	128	128	128	256	256	256
filt dim	3	n/a	64	n/a	n/a	64	n/a	128	n/a	n/a	128	n/a	256
rec. field	3	3	5	5	6	10	10	14	14	16	24	24	32

layer	14	15	16	17	18	19	20	21	22	23	24	25	26
type	relu	cnv	relu	mpool	cnv	relu	cnv	relu	cnv	relu	mpool	cnv	relu
support	1x1	3x3	1x1	2x2	3x3	1x1	3x3	1x1	3x3	1x1	2x2	3x3	1x1
stride	1	1	1	2	1	1	1	1	1	1	2	1	1
padding	0	1	0	0	1	0	1	0	1	0	0	1	0
out dim	256	256	256	256	512	512	512	512	512	512	512	512	512
filt dim	n/a	256	n/a	n/a	256	n/a	512	n/a	512	n/a	n/a	512	n/a
rec. field	32	40	40	44	60	60	76	76	92	92	100	132	132

layer	27	28	29	30	31	32	33	34	35	36	37
type	cnv	relu	cnv	relu	mpool	cnv	relu	cnv	relu	cnv	sftm
support	3x3	1x1	3x3	1x1	2x2	7x7	1x1	1x1	1x1	1x1	1x1
stride	1	1	1	1	2	1	1	1	1	1	1
padding	1	0	1	0	0	0	0	0	0	0	0
out dim	512	512	512	512	512	4096	4096	4096	4096	1000	1000
filt dim	512	n/a	512	n/a	n/a	512	n/a	4096	n/a	4096	n/a
rec. field	164	164	196	196	212	404	404	404	404	404	404

Employed, pre-trained network can distinguish between 1000 general classes of various images (objects). In our case we focused on a subset of 30 objects that can be found on roads, namely (the number in the brackets is the original class number): ambulance (408); amphibious vehicle (409); beach wagon (437); tandem bicycle (445); taxi (469); convertible (512); fire truck (556); freight car (566); garbage truck (570); go-kart (574); jeep (610); limousine (628); minibus (655); minivan (657); motor scooter (671); mountain bike (672); moving van (676); passenger car, coach, carriage (706); pickup truck (718); police van, police wagon (735); racing car (752); recreational vehicle (758); school bus (780); sports car (818); streetcar, tram (830); tank, armoured combat vehicle (848); tow truck (865); tractor (867); trailer truck (868); trolleybus (875).

The scheme of processing the final results from CNN classifier is presented in Fig. 4. In our algorithm we scan every video frame with a sliding window and present each extracted sub-image to the CNN classifier. We take into consideration 30 particular outputs from the network – each one representing probability of a relevant class (listed above). Each single answer falls within a range (0,1>, while the sum of all answers in each location cannot exceed one. Hence, the next stage is voting, which creates accumulated results map that is further superimposed over a region of interest obtained at the previous stage of processing. The voting works as follows. If the sum of all answers is higher than 0.5, then we assume a vehicle in the analysed location. Next, we look for the highest activation and check, if it is larger than 0.1, then we can decide about predicted vehicle class. The binary mask from lane detection stage serves as a filter, which tells vehicles that are on the same road lane from vehicles that are on the neighbouring lanes. Moreover, object that are detected over a horizon line are rejected as a false positives.

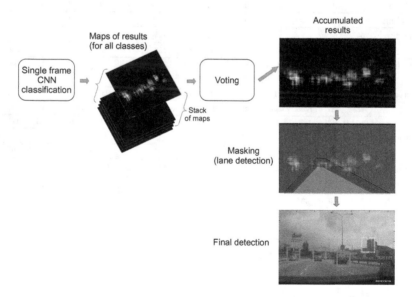

Fig. 4. Final detection stages: classification, voting and masking [own study]

3   Experiments

We performed experiments on data collected from driving recorders working in various lighting conditions and surroundings. The frame-rate was equal to 30 frames per second, with a spatial resolution of a single frame equal to 640 × 360 pixels, 24-bit RGB. A simulation was performed using a MATLAB prototype.

Below in Figs. 5 and 6 one can see some exemplary results of our algorithm. The road lane is marked yellow and the detected/classified vehicles are marked red or blue, depending on the lane position. The vehicle being on the same lane as the host car is marked red, while the other cars are marked blue. In each case, the class number is provided above each marking. The probability level above each extracted vehicle is also shown.

Fig. 5. Exemplary detections in good lighting conditions [own study]

In most cases vehicles are detected and the class is predicted correctly, however, when detected car is far from the camera the classification can be ineffective. It is caused by not ideal representation of samples in the learning set, which, as it should be stressed out, was created for general purpose object classification only.

As it can be seen, poor lighting conditions do not influence the detection of road lane and the detection of vehicles, however, the classification accuracy in this case is degraded (e.g. a bus was misclassified as a trailer truck, a van was recognized as a sports

Fig. 6. Exemplary detections in poor lighting conditions [own study]

car and a taxi cab was recognized as a sport car, too). Nevertheless, the detection accuracy stands at a very high level, which makes proposed solution ready for a practical implementation.

4   Conclusion

In the paper, the problem of vehicle detection in real driving conditions was analysed. The purpose of the performed experiments was to verify the possibility of employing deep learning approach for the processing on visual-only data that come from monocular camera. Proposed algorithm does not rely on any dedicated hardware and can be successfully implemented in general-purpose simple computer system, e.g. mobile device. The experiments were performed on video sequences captured during driving in different outdoor conditions. As it was shown, the vehicle detection based on the proposed assumptions is possible and quite successful. The classification fails in case of large distance from the camera and not-perfect representation of a learning set. The future works would include increasing the quality of learning set as well as more precise lane detection stage (especially in case of road bends).

As it was mentioned, many prototypes, like Google's experimental self-driving cars currently rely on a wide array of radar/lidar, and other sensors to detect vehicles and other objects on the road [2]. Elimination of some of that sophisticated equipment could make such cars cheaper and easier to design. On the other hand, presented system may be implemented in low-end mobile devices e.g. smartphones and tablets. Moreover, not just driverless cars that would benefit; modern crash-avoidance systems found in existing cars could also potentially make use of such an algorithm.

References

1. Berger, C.: Continuous integration for cyber-physical systems on the example of self-driving miniature cars. In: Bosch, J. (ed.) Continuous Software Engineering, pp. 117–126. Springer, Heidelberg (2014)
2. Google Inc: Google self-driving car project. https://www.google.com/selfdrivingcar. Accessed 10 Feb 2016
3. Forczmanski, P., Malecki, K.: Selected aspects of traffic signs recognition: visual versus RFID approach. In: Mikulski, J. (ed.) Activities of Transport Systems Telematics. CCIS, vol. 395, pp. 268–274. Springer, Heidelberg (2013)
4. Mazurek, P., Okarma, K.: Application of Bayesian a priori distributions for vehicles' video tracking systems. In: Mikulski, J. (ed.) Transport Systems Telematics. CCIS, vol. 104, pp. 347–355. Springer, Heidelberg (2010)
5. Frejlichowski, D., Gościewska, K., Nowosielski, A., Forczmański, P., Hofman, R.: Detecting parked vehicles in static images using simple spectral features in the 'SM4Public' system. In: Kamel, M., Campilho, A. (eds.) ICIAR 2015. LNCS, vol. 9164, pp. 489–498. Springer, Heidelberg (2015). doi:10.1007/978-3-319-20801-5_54
6. Forczmański, P., Seweryn, M.: Surveillance video stream analysis using adaptive background model and object recognition. In: Bolc, L., Tadeusiewicz, R., Chmielewski, L.J., Wojciechowski, K. (eds.) ICCVG 2010. LNCS, vol. 6374, pp. 114–121. Springer, Heidelberg (2010). doi:10.1007/978-3-642-15910-7_13
7. Okarma, K., Mazurek, P.: Application of shape analysis techniques for the classification of vehicles. In: Mikulski, J. (ed.) Transport Systems Telematics. CCIS, vol. 104, pp. 218–255. Springer, Heidelberg (2010)
8. Aarthi, R., Padmavathi, S., Amudha, J.: Vehicle detection in static images using color and corner map. In: International Conference on Recent Trends in Information, Telecommunication and Computing ITC 2010, pp. 244–246 (2010)
9. Rajagopalan, A.N., Burlina, P., Chellappa, R.: Higher order statistical learning for vehicle detection in images. In: Proceedings of the 7th IEEE International Conference on Computer Vision, vol. 2, pp. 1204–1209 (1999)
10. Wen, X., et al.: A rear-vehicle detection system for static images based on monocular vision. In: 9th International Conference on Control, Automation, Robotics and Vision, pp. 1–4 (2006)
11. Cyganek, B., Woźniak, M.: Pixel-based object detection and tracking with ensemble of support vector machines and extended structural tensor. In: Nguyen, N.T., Hoang, K., Jędrzejowicz, P. (eds.) ICCCI 2012. LNCS (LNAI), vol. 7653, pp. 104–113. Springer, Heidelberg (2012). doi:10.1007/978-3-642-34630-9_11
12. Sun, D., Watada, J.: Detecting pedestrians and vehicles in traffic scene based on boosted HOG features and SVM. In: Proceedings of the IEEE 9th International Symposium on Intelligent Signal Processing (WISP), pp. 1–4 (2015)
13. Tang, Y., et al.: Vehicle detection and recognition for intelligent traffic surveillance system. Multimedia Tools Application (online) (2015)
14. Huval, B., et al.: An Empirical Evaluation of Deep Learning on Highway Driving, arXiv: 1504.01716v3 (2015)
15. Gao, Y., Lee, H.J.: Deep learning of principal component for car model recognition. In: Proceedings of the International Conference on Image Processing, Computer Vision, and Pattern Recognition (IPCV), pp. 48–51 (2015)
16. Wang, H., Cai, Y. Chen, L.: A vehicle detection algorithm based on deep belief network. The Scientific World Journal, vol. 2014, Article ID 647380 (2014). http://dx.doi.org/10.1155/2014/647380. Accessed 1 Dec 2015

17. Mobileye, N.V.: Headway Monitoring and Warning. http://www.mobileye.com/technology/applications/vehicle-detection/headway-monitoring-and-warning/. Accessed 10 Feb 2016
18. Nvidia Inc.: Drive PX, Accelerating the Race to Self-driving Cars. http://www.nvidia.com/object/drive-px.html. Accessed 10 Feb 2016
19. Chang, H.Y., Chih-Ming, F., Huang, C.L.: Real-time vision-based preceding vehicle tracking and recognition. In: 2005 Proceedings of IEEE Intelligent Vehicles Symposium, pp. 514–519 (2005)
20. Kumar, A.M., Simon, P.: Review of lane detection and tracking algorithms in advanced driver assistance system. Int. J. Comput. Sci. Inf. Technol. (IJCSIT) 7(4), 65–78 (2015)
21. Lech, P., Okarma, K., Fastowicz, J.: Fast machine vision line detection for mobile robot navigation in dark environments. In: 8th International Conference on Image Processing and Communications (IP and C), Advances in Intelligent Systems and Computing, vol. 389, pp. 151–158 (2016)
22. Fukushima, K.: Neocognitron: a self-organizing neural network model for a mechanism of pattern recognition unaffected by shift in position. Biol. Cybern. 36(4), 193–202 (1980)
23. Lecun, Y., et al.: Gradient-based learning applied to document recognition. Proc. IEEE 86(11), 2278–2324 (1998)
24. Krizhevsky, A., Sutskever, I., Hinton, G.: Imagenet classification with deep convolutional neural networks. Adv. Neural Inf. Process. Syst. 25, 1106–1114 (2012)
25. Vedaldi, A., Lenc, K.: MatConvNet: CNNs for MATLAB. http://www.vlfeat.org/matconvnet/ (2014)

The Scope and Capabilities of ITS – the Case of Lodz

Remigiusz Kozlowski[✉], Anna Palczewska, and Jakub Jablonski

Faculty of Management, University of Lodz, Lodz, Poland
remigiusz.kozlowski@gmail.com,
anna.m.palczewska@10g.plgmail.com, jablon.jak@gmail.com

Abstract. ITS play a very important role in improving traffic in the cities. Several implementations of such systems in many countries led to positive effects both, for the traffic improvement as well as for reducing of exhaust emissions which, consequently, increase the quality of life. Some of Polish cities have just started to implement ITS and those systems cover only a part of their surface. This is mostly due to the high costs of introducing such systems. The capabilities of ITS are constantly developing therefore each subsequent implementation offers a greater range of possibilities such as: better control of traffic flow, general safety and environment protection. The authorities of Lodz, on the occasion of significant reconstruction of the city, decided to install ITS which includes 236 intersections. This fact makes it the biggest ITS in Poland. This article examines the scope and capabilities of ITS, which has just been launched (December 2015) in Lodz.

Keywords: ITS · Implementation process · Advanced technologies · Transport telematics

1 Introduction

Mass production of cars, which began in the 20's of the nineteenth century [3] led to the emergence of a large number of cars on the streets of American cities. In addition to the numerous benefits of the development of the automotive industry has brought a series of negative effects. Which could include, among others: "accidents caused by operation of transportation systems (loss of life, medical care for the disabled and so on.), congestions and the others (air pollution, toxic emissions for health, environment and buildings, climate change, emissions of greenhouse gas ($CO2$), which have a lasting impact on the earth's climate, noise)." [1] Possibilities to negate those negative effects were sought. Among them is the current use of ITS systems. The results of studies on this topic have been published, among others, by M. Litwin, J. Oskarbski and K. Jamroz. They showed: [4].

- Increase of the road network capacity by 20–25 %;
- Improved road safety (reduce of the number of accidents by 40–80 %);
- Reduced travel time and the energy consumption (by 45–70 %);
- Improved comfort of travel and traffic conditions for drivers, public transport users and pedestrians;
- Reduced road fleet management cost;

© Springer International Publishing AG 2016
J. Mikulski (Ed.): TST 2016, CCIS 640, pp. 305–316, 2016.
DOI: 10.1007/978-3-319-49646-7_26

- Reduced costs associated with the maintenance and renovation of roads;
- Improved quality of the environment (emissions reduced by 30–50 %);
- An increase in economic benefits in the region.

There is no doubt that installation of such systems is worthwhile. In Polish conditions, the implementation of ITS is particularly difficult because of the backwardness of transport infrastructure in comparison to other European countries. The reason is the need for a set of large urban investments not directly related to transport, such as the revitalization of large parts of the city (this is the case in Lodz). These investments very often require reconstruction of transport infrastructure in the city.

In such situations, the problem is the incorporation of appropriate deployment of ITS in conjunction to other works carried out in the city. It is also important that the city residents, tired by the hardships of revitalization, modernization of transport infrastructure should be prepared to accept yet more the inconveniences associated with the installation and startup of ITS system. Certainly it is worthwhile to make the effort, because of the contribution that type of investment brings to the development of the city and the region. [9] Lodz is an example of a city that, struggling with a number of problems, took up the challenge of both city revitalization, modernization of transport infrastructure and construction of the system ITS.

2   Characteristics of Lodz Transport System

Transport infrastructure in Lodz and the region in the last 10 years have undergone significant change. Lodz is a city located in the centre of Poland, with significant transport movement in the directions of both north-south and east-west. Both in the city centre and on its outskirts, in the past few years a number of new investments were implemented. Lines constituting to the city ring road were built: motorways (A2 and part of A1) and expressways (S-8, part of S14) (Fig. 1).

Further investments are planned - the completion of the S-14 (tender for the construction of the so-called Western ring road was put in 2015), which will led the stream of vehicles at the entrance to Lodz Reymont airport without having to go through the city centre, the completion of the A1 (eastern ring road), which is set to finish in mid-2016 (Fig. 2).

However, apart from these routes additional access roads from the ring road to the city centre are needed. Many routes were rebuilt, some of the extensions were planned thus allowing the creation of access roads. It was also necessary to reconstruct the routes which are the main thoroughfares in the city – thus the reconstruction of WZ route. The tunnel, which was built on the most crucial stretch accelerates communication between the western part of the city – 'Polesie' and eastern 'Widzew' (two large city districts, Polesie with 138 thousand inhabitants and Widzew with 135 thousand) [7] settlements situated on the two ends of the city. What was most important during this investment was that all solutions implemented in W-Z road tunnel could be connected and managed as a part of city ITS system (Fig. 3).

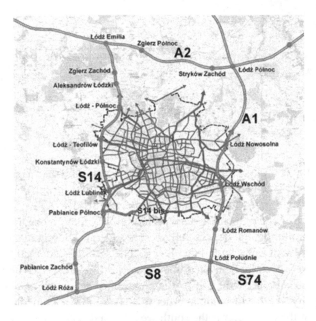

Fig. 1. Lodz ring road system [2]

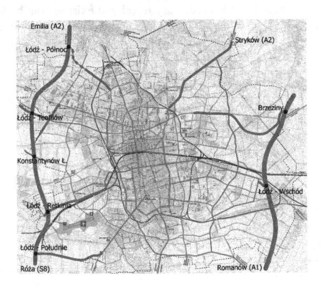

Fig. 2. Lodz ring road access routes: expressway S14 and motorways A1 and A2 [2]

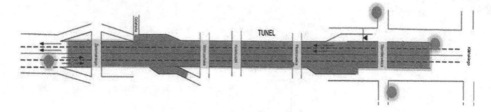

Fig. 3. W-Z tunnel map with cameras location [2]

In this way, the communication within the city was improved. 'Trasa Górna' road was also established channelling movement in the direction of north-south (seaports, Silesia) from the centre of the city on to the main artery road.

In 2012 reconstruction of the Łódź Fabryczna Railway Station began. In place of the old station, new multimodal node was built, where several different means of transport are integrated - railway, bus, car and taxi levels. The station is adapted to move High-Speed Rail and has specially dedicated tracks just for this purpose. The next stage of construction of the station is going to be the creation of a railway tunnel between the two parts of the city, north and south, in order to lead out the movement of High-Speed Rail further out of the city towards the south west Poland. The station is located in the city centre. [5] Through this investment, it is possible to reconstruct the 70 hectares in the centre and rebuild it from the beginning by the newly existing standards. In connection to this investment in it is also necessary to rebuild existing and build some new streets. On the old, overhead railway line new road will be constructed to feed traffic to

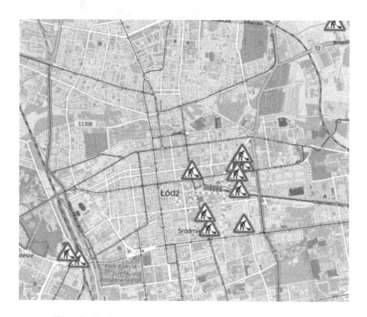

Fig. 4. Lodz road investments map in year 2016 [13]

the station. All other streets around the station are being rebuilt. Station is estimated to serve up to 200 thousand [6] passengers per day (Fig. 4).

Further investments and upgrades are planned. As a key in the near future, connections with the A1 (Ofiar Terroryzmu street connected with Rokicińska street, extension of Puszkina street to Pomorska street, extension of Łodz-Górna road) and S-14 (extension of Włókniarzy street, access from Szczecińska street to Aleksandrów Łódzki node, access road to Konstantynów Łódzki node). Access roads to S14 are in the process of planning, and some roads on the route require construction or reconstruction of access points. The current difficulty with which one can meet in the Lodz is the fact that all the work is conducted in the city, while the A1 has not yet taken over the stream of cars in the direction of north-south. This is particularly troublesome when the certain passages are modernized in the busiest hours.

It should be also noted that, as a result of construction of new infrastructure, the attractiveness of the city location is significantly improved and the traffic starts to increase. It is estimated that such situation will take place at approach to the new railway station. Despite numerous modernizations and building a new road in the wake of old railway tracks, the amount vehicles, due to the increase in the number of passengers using the new train station, will be much greater than the capability of the road infrastructure being prepared for these purposes. In the recent years construction plans (for 2012 and 2013 [12]), a new street 'Nowotargowa' was planned, with two lanes in each direction. Later it was reduced to only one lane, currently these plans were postponed and only part of the road, from Piłsudzkiego Street (W-Z road) to the station is constructed (Fig. 5).

Fig. 5. Lodz central public transportation interchange centre [own study]

3 Lodz ITS System Characteristics

Lodz is the third largest city in Poland with 710 thousand citizens living inside the city borders and over 1,1 million inside Lodz agglomeration. [8] The city is important transportation hub, with its location in the central Poland, proximity to main North-South A1 and West-East A2 highway intersection and the existence of Logistics Centre in the macro region. There are almost 360 thousand registered vehicles and 1030 kilometres of public roads. [11] There are 76 Public Transportation lines (17 by tram and 59 by bus). [6] All those factors influence Lodz transportation system. Road congestion and, as a result, Public Transportation ineffectiveness is serious problem. Starting after 2010 major investments were undertaken, and among them complex, Poland largest ITS system, to be introduced.

System was finished and launched in December 2015, spanning over 236 out of almost 280 city intersections. [11] Main goal of the investment was oriented around public transport communication and improvement of the road traffic. By giving the priority to public transport modes, their effectiveness should improve, overall travel time will be shortened and the comfort and safety will be enhanced. The cost investment was about 80 mil PLN. Realization took only 16 months, since the first phase of planning, through project making, ending with the ITS system launch in city. During this time required technical documentation was prepared, intersections were equipped with special communication devices, which connected them to the system, and optical network data transmission and Traffic Management Centre were constructed.

Lodz ITS system is accompanied by several other investments, some already launched, some in planning stage. Such as public transport Interchange Centre, situated on the main West-East road, finished in October 2015, and from the beginning integrated into ITS system – first of several planned. Another example is Lodz Regional Tram system, also integrated with ITS.

ITS in Lodz was equipped with several subsystems:

- Traffic control system - SCATS,
- Public transport management system - MUNICOM,
- Camera system - CCTV, ANPR,
- Tunel management system - SCADA.

The traffic control system covers an area optimization of traffic lights, priorities and tunnel management system, which use SCADA. Thanks to this, it gives a possibility to, for example, in case of slower traffic in the tunnel can automatically change the settings on other intersections, in turn "repositioning" the flow of vehicles.

The public transport management system offers dynamic information for passengers, who use public transport. Parts of this system are: tables with information for passengers, http://rozklady.lodz.pl/ web portal for passengers, where they can check information from chosen stops using the virtual timetables. System gives possibility of setting and changing priorities from Traffic Management Centre. For example, a tram, that will be arriving early, is going to be delayed to match the timetable [2] (Fig. 6).

Fig. 6. Virtual table of public transport information [14]

Monitoring system is built on two different types of devices: ANPR cameras and CCTV cameras.

For these systems to work properly a lot of modernization works and installation of new equipment throughout the city was done. Modernized were 236 traffic lights, built 40 km of cable ducts, laid 555 km of new wiring, installed 130 passenger information boards, 9 boards VMS for drivers, 176 CCTV cameras and ANPR, 700 vehicles of the City Public Transport Company were modernized and the system in the tunnel was constructed [2].

In addition, the system was provided with: a module, providing mobile information, information for drivers, passenger information and management of road infrastructures thanks to eDIOM [2].

The architecture of the system allows future integration with other systems. For example: GDDKiA system operating at interchanges.

4   Lodz ITS System Capabilities

The main component of Lodz ITS is System of Area Traffic Management (pl. System Obszarowego Zarządzania Ruchem) or SOZR - using shortened polish name. It is based on two GIS applications. First is GIS eDIOM browser application - fully functional database based on ORACLE database system, integrated with some of SOZR subsystems. Second application is Central (integrating) Application - a desktop system that integrates all subsystems, among them:

SOZR system has modular, multilayer structure, where all subsystems are integrated into one. System will enable a real time, on-line visualization, gathering and archivization of data about traffic. It is also possible to monitor the status of traffic control equipment and components, like traffic lights. In case of malfunction or any undesired incident – alarm is sent to the central (Table 1).

Table 1. Components of Lodz ITS [10]

ANPR – traffic intensity	Vehicle localization subsystem
SCATS – system SOSR (Area Traffic Control System)	Public transportation subsystem
CCTV	Tunnel subsystem
VMS – driver information	Parking subsystems
Passenger information board subsystem	Street lighting subsystem
Chosen eDIOM layers	Winter maintenance subsystem

First of the subsystems is Area Traffic Control System (SOSR - pl. System Obszarowego Sterowania Ruchem). It is based on SCATS (Sydney Coordinated Adaptive Traffic System) used in 265 cities, 27 countries in the world [11] and provides constant analysis of traffic condition. Main functions of this system are:

- Traffic light control,
- Light supervision,
- Data gathering,
- Strategic management.

Version installed in Lodz has capability to maintain up to 300 intersections. It will enable priority passage for city public transportation and maximization of traffic bandwidth on controlled junctions. All public transport vehicles are equipped with system detectors supported with on-site detectors (bus lines, tram tracks etc.). System detects approaching tram or bus, and based on the data, give the priority by changing the traffic lights cycle time.

Among other components, there is the Driver Information System, which is based on road signs and interactive VMS tables. They inform the drivers about traffic congestion, and provide alternative routes. In basic work mode they also inform about real travel times between selected points of the road network of the city. System functioning is guaranteed by a network of detectors located at the road intersections. ANPR cameras are used to calculate average travel time. VMS tables are constructed to withstand temperatures between -40 and $+800°C$, they are also equipped with automatic adjustment of illumination level. At this point, there are nine tables of this kind, located at key point of the city. It is possible for drivers to look at those tables remotely, on-line by using web site of the system: www.its.lodz.pl (Fig. 7).

Passenger Information System – SDIP – is another part of Lodz ITS system, similar to Driver Information System, where bus and tram stops are equipped with interactive tables. Those tables inform about departure time of nearest vehicles. They are also equipped with voice system, allowing voice messages about incidents to be provided, if needed. System uses data gathered by SOZR system, to provide accurate times and delays. Mobile application "myBus online" is going to be integrated into Lodz ITS system, giving mobile users similar, real time capability. Existing Lodz Regional Tram network was integrated into new system.

CCTV Camera System is one of vital elements of SOZR system. By gathering video data it supports main traffic operator decision making, road threats and incidents detection and overall transport network safety level. Overall, almost 80 new or already

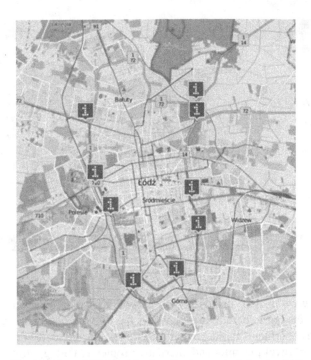

Fig. 7. Map of VMS tables localisation around the city [13]

existing cameras are connected into system. City Police, Fire Brigade, Municipal Police and Lodz Public Transport Company own additional workstations connected to main system.

Another module, available to traffic operator, is the Public Transport Management System. It gives the possibility to adjust the planned timetables, control the punctuality and to support dispatcher activities, including fleet management, monitoring and vehicle computer management.

Automatic Number Plate Recognition System – ANPR – is another camera based subsystem. It provides the data about vehicles on city roads. One of the main functions is to calculate travel times, to send the information to other subsystems. Data is gathered at the beginning and at the end of measured sector, both day and night. System works at effectiveness level of 95 %. 121 [10] ANPR cameras are situated in city.

New websites were created, operating on the basis of data provided by the ITS system. These tools provide many options. Functionalities of 'passenger information system' – www.rozklad.lodz.pl include virtual boards with timetables for each of the stops and travel planner that takes into account the stops and various forms of public transport (trams, buses). In addition, in real-time, website presents vehicles that are 'on the move'. It also allows their identification through virtual plates. All of that makes it easier to plan the way to the stop and eliminate the time wasted on waiting for a tram or bus (Fig. 8).

Fig. 8. Lodz ITS system webpage [13]

Through the www.its.lodz.pl website you can watch current road events or accidents. The page visualizes traffic in real time. It allows the users to obtain information on the road where there are difficulties, caused both by congestion and road works. Additionally there is a tool for travel planning and alternative routes planning, allowing bypassing traffic congestion for individual transport users, which in the city are currently caused by the large number of investments carried out in recent years.

Another feature provided by system through the ITS website is access to the images from CCTV cameras - 54 pieces. Every 5 min the system provides images from cameras located at intersections (FIg. 9).

Fig. 9. Internet website of passenger information implemented within Lodz ITS system [14]

In the future the system will be enlarged by connecting more intersections and constructing more cameras. System will be also expanded by adding 'city bike' programme that will be launched in Lodz at the beginning of May 2016.

5   Conclusion

City of Lodz has taken a very courageous decision to implement ITS simultaneously with many other large investments such as: revitalization processes of the city, construction of new 'Fabryczny' railway station and a number of other very large investments, such as W-Z road. This solution enabled the matching of the ITS system and other investments in order to achieve synergy effect for all investments. The scope and opportunities presented in this article are therefore the result of mutual adjustment of many urban projects to each other. In practice, it was a huge organizational, financial and even political effort of the Lodz authorities, city institutions, responsible for carrying out these investments and companies engaged in these projects.

In the presented case, there is a good chance that respectively received scope and capabilities of ITS will provide, in the future, a number of benefits and rewards for both residents and visitors moving around the city, after many years of difficulties, due to realization of all those municipal investments. By August 2016 first results of system functioning were acquired by analysing the travel time from before and after the implementation of the system. The results show that time for public transport was reduced by 9,6 % and for individual transport reduced by 29,46 %. [2].

Acknowledgements. We would like to thank the Director of Roads and Transport in Lodz, Mr. Grzegorz Nita and Mr. Michał Sarnacki, the Head of the Department of Traffic Engineering and Traffic Control Board of Roads and Transport in Lodz for help in preparing empirical materials used in this article.

References

1. Kalašová, A., Mikulski, J.: Calming traffic - a modern way of improving safety in cities. Arch. Transp. Syst. Telematics **7**(4), 16 (2014)
2. Department of Traffic Engineering and Traffic Control Board of Roads and Transport in Lodz. http://zdit.uml.lodz.pl. Accessed 12 Dec 2015
3. Hounshell, D.: From the American System to Mass Production, 1800–1932: The Development of Manufacturing Technology in the United States, p. 1. The John Hopkins University Press, Baltimore (1985)
4. Litwin, M., Oskarbski, J., Jamroz, K.: Inteligentne Systemy Transportu –Zaawansowane Systemy Zarządzania Ruchem, w: I Polski Kongres Drogowy Lepsze drogi -lepsze życie: referaty. 1st Polish Road Congress Better roads - better life: proceedings, Polski Kongres Drogowy, Warszawa, pp. 167–174 (2006)
5. Łódź Fabryczna Railway Station Portal. www.nlf-b2.pl. Accessed 12 Dec 2015
6. Lodz Public Transport Company portal. www.mpk.lodz.pl. Accessed 12 Dec 2015

7. Main Statistical Office Lodz – Population – Data about Lodz 2014, Table 30, Population Based on Balances by Former Office Agencies of the City of Lodz Office. www.lodz.stat.gov.pl. Accessed 12 Dec 2015
8. Main Statistical Office of Poland portal. www.stat.gov.pl. Accessed 12 Dec 2015
9. Marczak, M., Kozłowski, R.: Budowa inteligentnych systemów transportowych jako szansa dla zrównoważonego rozwoju regionów, Ekonomia i Zarządzanie, Numer 2, Tom 6, Białystok, pp. 40–41 (2014)
10. SOZR system Technical Documentation, Sprint S.A, October 2014
11. SPRINT company portal. www.sprint.pl. Accessed 12 Dec 2015
12. www.lodz.naszemiasto.pl. Accessed 12 Dec 2015
13. http://its.lodz.pl. Accessed 12 Dec 2015
14. http://rozklady.lodz.pl. Accessed 12 Dec 2015

Modeling Information Spread Processes in Dynamic Traffic Networks

Rafał Kucharski[1]([✉]) and Guido Gentile[2]

[1] Department of Transportation Systems,
Cracow University of Technology, Kraków, Poland
rkucharski@pk.edu.pl
[2] SISTeMA ITS, PTV Group, DICEA,
Università di Roma "La Sapienza", Rome, Italy

Abstract. We propose the probabilistic information spread model to represent the spatiotemporal process of becoming aware while traversing the traffic network. In the contemporary traffic networks drivers are exposed to multiple traffic information sources simultaneously. Traffic managers look for a realistic estimate on when, where and how many drivers become informed about the actual traffic state (e.g. about the event). To this end we propose the probabilistic Information Spread Model (ISM) representing the process of spreading information to the drivers via multiple information sources (radio, VMS, on-line information, mobile applications, etc.). We express the probability of receiving information from a given information source using specifically defined spreading profile (formalized through the probability density function) and market penetration of respective source, with a novel information spreading model for on-line sources (websites, mobile apps, social networks etc.). Moreover, by assuming the information sources are mutually independent, the simplified formula for the joint probability can be used so that the model becomes practically applicable in real-time applications. Model is designed to work within the macroscopic dynamic traffic assignment (DTA) as a part of the network flow propagation model. Thanks to that, the informed drivers can be traced as they propagate through the network towards their destinations. We illustrate the model with the simulations on Dusseldorf network showing how information is spread in several ATIS scenarios (VMS, radio news, online sources, and simultaneous sources).

Keywords: Dynamic traffic assignment · Information spread models · Advanced traveler information systems · ITS

1 Introduction

In contemporary traffic networks number of sources spread the information to recipients. The informed drivers, aware about the event, act differently than unaware drivers, they can adapt to the actual traffic conditions and change their routes. Advanced Traveler Information System (ATIS) aims to provide the actual information to the travelers and for the traffic management action it is crucial to understand and quantify the information spreading processes in the traffic networks. Most importantly how many drivers are

© Springer International Publishing AG 2016
J. Mikulski (Ed.): TST 2016, CCIS 640, pp. 317–328, 2016.
DOI: 10.1007/978-3-319-49646-7_27

informed with both the spatial and temporal dimension (when and where they become informed and how they propagate through the network). The efficiency of ATIS can be, in turn, measured with the information: range (number of informed travelers), timing (when the drivers receive the information) and targeting (did the drivers affected by the event got informed). We identify following information sources as the most important in the contemporary urban traffic networks: radio news, variable message signs (VMS), on-line information and mobile applications. Since the driver can be exposed to all of them simultaneously, all of them need to be considered jointly to fully define process of becoming informed. Driver can pass-by VMS sign while listening to the radio news and using mobile apps. To comprehensively approximate the number of informed travelers and the time when they are informed, we propose in this paper the Information Spread Model (ISM). It allows to determine number of travelers receiving the information while travelling throughout the traffic network and exposed to several sources of information. Moreover, thanks to applying the model in the DTA framework we can model the propagation of the informed travellers through the network.

ATIS are one of core elements of ITS with goal to inform and/or guide travellers so that they can intelligently adapt to the actual traffic situation and improve the overall transport system efficiency (see [3] for the throughout review). Since the first ATIS concepts were formulated [12, 17], number of information sources became widely accessible. Hundreds of VMS have been installed, traffic news are broadcasted with actual information, actual traffic data is available on number of online platforms and millions of mobile users can update their information at no cost. This leads to a complex background in contemporary traffic networks with multiple sources of information and various content they provide. ATIS can provide to the driver awareness on the current (instantaneous) or forecasted state of the network [13]. The information can be either the location of the event on which driver autonomously builds his route [5], full forecast arising from the event [1], or a routing suggestion provided by the system [8].

The stated- and revealed-preference studies of rerouting provide a good insight on the information spreading processes in the traffic networks. Emmerink [6] conducted a survey to see the impact of VMS and radio broadcasts on route choice. Sixty percent respondents claimed that their route-choice would be influenced by radio broadcasts, and forty percent by VMS signs. Schlaich [20] got much less optimistic results with the revealed-preference data from floating mobile data. His research on how the VMS information affects route-choices showed ca. 30% compliance.

The most challenging part of the proposed ISM model was to properly represent and understand the way the information is spread on-line. The proposed concept originated from the evacuation models [22] where information about event spreads in time successively reaching recipients. Fortunately, recent data from Twitter [16] led to better founded understanding of information spread processes. Numerous studies were conducted on the Twitter data. [5, 10, 19] provide valuable insight on the two key aspects: dynamics and range of the information spread process. The dynamics are observable through the 'tweets' posted after emergency situations: earthquake, hurricane, riots, etc. [18] made an outstanding research on how fast the information dissipates though the communication network of Twitter. They investigated the twitter traffic related to the false news (i.e. "Rioters released wild animals from London ZOO") and showed how they spread to the recipients. The analysis was made time dependent so that speed of spreading is observable. Several examples analysed by Procter revealed similar relation between number of twitters and time – fitting Rayleigh-like distribution, which was in-line with previous assumptions [4, 11, 22]. Apart from the temporal dimension, the information spreading strongly relies on other aspect: virality. Ghosh [7] showed that information in the communication network is either completely negligible and forgotten very fast, or contrary: it spreads like a virus through communication network, reposted forward with exponential probability [19]. The observations on virality proved that the virality strongly affects both the speed and range of the information. Although the process of becoming viral is random and there are no efficient methods to determine if the information will become viral [14], there are monitoring techniques to identify that the information is spreading virally [19, 21]. In general, we suggest to use the external tools to determine the virality and market penetration (range) of the online information if they are available.

The paper is organized as follows: after above literature review we formally introduce the information spread model in the next section. First we introduce the generic theoretical definition of the information source and formulate the probability of receiving the information via one source and multiple sources. Then we introduce the respective formulas for the most common sources of information. Finally we and show how the model can be applied in the macroscopic DTA environment. Subsequently, we present the simulation results for the Dusseldorf network showing information spreading process in several ATIS scenarios. The paper is concluded with the summary and further directions.

2 Information Spread Model (ISM)

Let's propose the Information Spread Model (ISM) as the macroscopic, probabilistic model. ISM estimates the number of drivers which are getting informed, further applying it within the DTA model allows to propagate the informed drivers throughout the network. Output of the model is (1) number of drivers becoming informed while traversing a given network arc at a given time $\iota_a(\tau)$ and (2) number of informed drivers at a given arc at a given time $\bar{q}_a(\tau)$.

Being informed is treated as a binary variable: the individual driver can be either informed or not informed about the event at a given time. The driver is not informed as long as he did not receive information from any of the sources and he becomes informed after receiving the information from any source. Since the ISM is designed to work with the macroscopic DTA, instead of modelling the process of becoming informed at the level of the individual drivers, we propose to handle it probabilistically and determine the probability of becoming informed. Thus, the ISM computes $\iota_a(\tau)$ – the probability of becoming informed while traversing arc a and exiting it at time τ. Such output can be conveniently used to determine the expected number of informed drivers, who are further propagated throughout the network.

To compute the probability of becoming informed let's define $\iota_a^S(\tau)$ – the probability of being informed by means of a generic information source S while traversing arc a at time τ. We compute it from the market penetration P^S and the so-called spreading profile ι^S of the source S. Market penetration is the share of the drivers which have access and utilize the information source S. Spreading profile determines how the information reaches drivers and is formalized with a probability density function (PDF). The spreading profile yields the probability of becoming informed as a function of time, space and any other significant variables. Let's, by definition, assume that the spreading profile is integrable, so that the probability of becoming informed while traversing arc a can be obtained as an integral of the spreading profile ι^S from the arc's tail a^- and entry time $-t_a$ to the arc's head (a^+) and exit time τ. Practically, since the spreading profile is a PDF, the integration can be substituted with the cumulative density function (CDF) denoted $I^S(\tau)$. This way the probability of becoming informed while traversing arc a can be equivalently computed with a difference between CDF at head and at tail (using travel time of arc ta as a time during which driver is exposed to the information source). Finally, the probability of becoming informed $\iota_a^S(\tau)$ is given as (1), with the second term being much more efficient as the CDF usually have a closed form and can be directly evaluated.

Since the driver is exposed to several information sources, we need a joint formula for $\iota_a(\tau)$ - probability of becoming informed from at least one source. It can be conveniently expressed with the complementary event, namely that the driver did not become informed through any source. Yet to compute the joint probability of not becoming informed through any source we either need to determine the correlations between the various information sources, or assume they are independent. To be able to introduce practically applicable form of the model, we assumed that all sources are mutually independent. Thanks to this we can propose the simple formula for the joint probability and determine the total probability $\iota_a(\tau)$ computed as the product of a complementary probabilities $(1 - \iota_a^S(\tau))$ over all the available sources $S \in \mathbf{S}$ (2).

$$\iota_a^S(\tau) = \int_{\tau-t_a}^{\tau} P^S \cdot \iota^s(\tau)d\tau = P^S \cdot \left(I_{a^+}^S(\tau) - I_{a^-}^S(\tau - t_a)\right) \tag{1}$$

$$\iota_a(\tau) = 1 - \prod_{S \in \mathbf{S}} \left(1 - \iota_a^S(\tau)\right) \tag{2}$$

The above general concept is designed to work with the most popular sources of information which are present in the contemporary traffic network. In particular, we formulate the equations for the following spreading paradigms:

1. fixed time information, i.e. news broadcasted through the radio;
2. fixed place information, i.e. message shown at a road-side VMS;
3. on-line information, i.e. published at a webpage, accessible via the mobile application, or posted on a social network.

2.1   Fixed-Time Information

The traffic information (*NEWS*) broadcasted at a fixed time (i.e. the radio news) reaches the recipients wherever they are. It can be assumed that everyone who is listening to the news (equal to the radio market penetration P^{NEWS}) at the broadcast time will become informed, regardless the location. Therefore, the probability of becoming aware through the radio news $\iota_a^{NEWS}(\tau)$ depends only on time when the arc a was traversed. $\iota_a^{NEWS}(\tau)$ is expressed with (3) and equals the market penetration P^{NEWS} if the news were broadcasted during traversing arc a at time period $(\tau - t_a, \tau)$ and is null otherwise. For consistency, the broadcast time needs to be limited to a single time instant τ_{NEWS}.

$$\iota_a^{NEWS}(\tau) = \begin{cases} P^{NEWS} & \tau_{NEWS} \in (\tau - t_a, \tau) \\ 0 & \text{otherwise} \end{cases} \tag{3}$$

2.2 Fixed Place Information

The information broadcasted at a fixed location, i.e. through a Variable Message Sign (VMS) can be reached only by those who traverse the arc equipped with VMS while it broadcasted the information. Therefore, the spreading profile is expressed with the similar form as above, yet being function of not only the time, but most importantly of the space. Namely, the probability of becoming informed by means of VMS is positive only for the drivers crossing a VMS-equipped arc during its broadcast time and is null otherwise (for other arcs and when the information is not broadcasted). Since the VMS is visible for everyone, P^{VMS} is the probability of noticing the VMS while passing by, rather than the market penetration. Consequently, the probability of becoming informed while traversing arc a is given with (4). In this case the broadcasting period do not need to be limited to a single time instant.

$$\iota_a^{VMS}(\tau) = \begin{cases} P^{VMS} & \text{for arc with VMS broadcasting info at } (\tau - t_a, \tau) \\ 0 & \text{otherwise} \end{cases} \tag{4}$$

2.3 On-line Sources

Representing the process of spreading information available on-line (-marked with superscript @), which can be checked by the driver at any time is way more challenging than the above sources. Online sources are commonly used nowadays and include: mobile applications, social networks, online services, etc. Let's consider a generic information published on-line at a given time $\tau^@$ after which drivers can receive it. Depending on the information source drivers can either be notified (via alert), come across it while browsing (via the general news feed of a social network) or see it while checking the web-site (e.g. when they look at the online traffic map). In all cases the cumulative probability of being informed can be expressed as a function of time past the broadcast time gradually increasing from zero to the total market penetration when the information spreading process is over. As advocated below, we express the probability of receiving information (PDF) with a Rayleigh distribution, yet any similar PDF can be applied if revealed in the field observations (exponential, log-normal, Erlang, Weibull, etc.). The Rayleigh shape fits the empirical findings from the *Twitter* revealed in numerous studies. Procter [18] studied how the tweets are re-tweeted and reach the new recipients, Fig. 1 shows how the twitter information reached the recipients in time (information about London riots in 2012) compared with the theoretical Rayleigh distribution approximating the spreading process. The revealed spreading process supported the assumptions of the early researchers (i.e. [22]) that the Rayleigh distribution well estimates the probability of receiving the information in time. The PDF of a Rayleigh distribution is given with (5) and its CDF with (6) where $\Delta\tau$ is the time past the information is published $\Delta\tau = \tau - \tau^@$ parameterized with the spreading pace σ. The Rayleigh distribution is by definition null prior the information is published online.

$$\imath^{@}(\tau) = P^{@}\frac{\Delta\tau}{\sigma^2}e^{-\Delta\tau^2/2\sigma^2}, \quad \Delta\tau > 0 \tag{5}$$

$$I^{@}(\tau) = P^{@}\left(1 - e^{-\Delta\tau^2/2\sigma^2}\right), \quad \Delta\tau > 0 \tag{6}$$

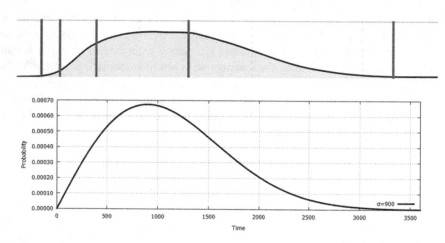

Fig. 1. Empirically observed Twitter posts frequency - top [18] and Rayleigh PDF with $\sigma = 900$ s [own study]

For the sake of realism, let's further extend the above formulas to cover the, so-called, *virality* of the information. As shown in [10] some news spread in the communication networks like viruses, while other remain unnoticed and do not reach bigger audience. In general, the viral information reaches more recipients and reaches them faster. In the traffic networks the information about the significant information are likely to become viral. To include the virality in the proposed model we will express both the market penetration $P^@$ and spreading pace σ of the Rayleigh distribution as a functions of the information virality, so that the viral information will reach bigger audience and will reach it faster. The market penetration can be then reformulated as a total share of drivers having access and utilizing the on-line source multiplied with the range within the source. E.g. the total share of social network users multiplied by the share of users noticing the information in their news feed. Both parameters can be read from the external web traffic monitors working in the real time (see i.e. [14]). In this paper we simply reformulate the basic Rayleigh PDF (5) so that both market penetration and spreading pace are time-dependent (7). Finally, using the cumulative of the spreading profile (8), the probability of receiving information on-line while traversing arc a at time τ is given with (9). Such formulated probability can be used directly as one of information sources in (2).

$$\iota^{@}(\tau) = P^{@}(\tau) \cdot \frac{\Delta\tau}{\sigma(\tau)^2} e^{-\Delta\tau^2/2\cdot\sigma(\tau)^2}, \quad \Delta\tau > 0 \tag{7}$$

$$I^{@}(\tau) = P^{@}(\tau) \cdot \left(1 - e^{-\Delta\tau^2/2\sigma(\tau)^2}\right), \quad \Delta\tau > 0 \tag{8}$$

$$\iota_a^{@}(\tau) = I^{@}(\tau) - I^{@}(\tau - t_a) \tag{9}$$

3   Illustrative Example

To illustrate how the proposed model works with multiple sources of information we show (in Table 1) the joint probability computed with (2) for multiple information sources in various scenarios. The joint probability is sensitive to introducing additional sources and their probabilities (1) and it changes when the market penetration, or spreading profile of the component source changes.

Table 1. Joint probability of being informed via multiple sources in various ATIS configurations [own study]

Scenario	Information probability of a source (1)					Joint probability (2)
	On-line #1	Radio	On-line #2	VMS	Radio #2	
1	.2	–	–	–	–	**.20**
2	.2	.2	–	–	–	**.36**
3	.2	.2	.2	–	–	**.49**
4	.2	.2	.2	.2	–	**.59**
5	.2	.2	.8	.2	–	**.90**
6	–	.2	–	–	.2	**.36**

To summarize the model Fig. 2 is presented. It depicts the cumulated probability of becoming informed ι for a generic driver traversing the network in the following cases:

- in the default setting (thick black line), the information is available from on-line source and it reaches 40% of drivers ($P^{MAX} = 40\%$).
- if the information is broadcasted also through the radio at 10:35 (thin black line), the joint probability of being informed rises up to 60%, as the two independent sources amplify due to (2).
- if driver passes the VMS sign (at 11:12) with the information broadcasted, the joint probability reaches 90%.
- if the information is viral (dotted line) it spreads fast and reaches more drivers (to 80%).

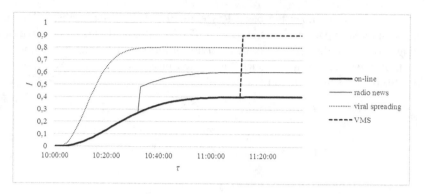

Fig. 2. Probability of receiving the information from multiple sources as a function of time in several ATIS scenarios [own study]

3.1   Applying ISM Inside DTA

The proposed model is designed to work within the Dynamic Traffic Assignment (DTA). DTA determines the traffic flows on the network satisfying the demand [2]. It is done through the assignment methods, typically following the 'user-equilibrium' concept of balancing travel costs of all drivers [23]. The results of DTA are the network performances (i.e. temporal profile of travel costs c and times t) and the demand pattern. Demand pattern of DTA is either set of OD paths defined with specific temporal profile of flows q, or, alternatively, set of local routing decisions (arc conditional probabilities p) defined for each node. Arc conditional probability tells what share of drivers will use arc a subject to being at its tail at time τ which, coupled with origin demand, become equivalent to explicit paths [15].

ISM can be applied inside the DTA thanks to the partition of the propagated flow into two states: uninformed $\hat{q}$ and informed $\bar{q}$. The process of becoming informed can be seen as a transition from the uninformed to informed state in Markov sense with the transition probability computed in ISM with (2). Flows in both states are propagated simultaneously in the trajectory-based Network Flow Propagation (NFP) procedure, adapted for the ISM (Algorithm 1). In the typical NFP, the demand flow from the origin to the destination d is propagated along the trajectories using the arc conditional probabilities p, calculated in a DTA route-choice model. The NFP is computed subject to the destination d in reverse topological order of nodes (along the shortest tree trajectory). It cumulates the node flows from the backward star of the node and the demand flows at the origins with (11) to propagate the flow to the forward star with the respective arc conditional probabilities (10).

To apply this framework for the ISM, the above formulas are extended to propagate the two states of flow (uninformed and informed) and utilize the transition probability (ι obtained from the ISM model via (2)). At each node the uninformed flow $\hat{q}$ is reduced with the part ι which became informed at the arcs of the backward star (12) and is propagated forward with (13) – which includes the demand flows D. Respectively, the informed flow is increased with the uninformed flow which became informed at the backward arcs (14). Such cumulated informed flow is propagated towards destination d with arc conditional probabilities p, yet the demand flows are not included anymore (15).

$$q_a^d(\tau) = \left(D_{a-}^d(\tau) + q_{a-}^d(\tau)\right)p_a^d(\tau) \tag{10}$$

$$q_i^d(\tau) = \sum_{a\in i^-} q_a^d(\tau) \tag{11}$$

$$\hat{q}_i^d(\tau) = \sum_{a\in i^-} \hat{q}_a^d(\tau) \cdot \left(1 - \iota_a^d(\tau)\right) \tag{12}$$

$$\hat{q}_a^d(\tau) = \left(D_{a-}^d(\tau) + \hat{q}_{a-}^d(\tau)\right)p_a^d(\tau) \tag{13}$$

$$\bar{q}_i^d(\tau) = \left(\sum_{a\in i^-} \bar{q}_a^d(\tau) + \sum_{a\in i^-} \hat{q}_a^d(\tau)\cdot \iota_a^d(\tau)\right) \tag{14}$$

$$\bar{q}_a^d(\tau) = \bar{q}_{a-}^d(\tau)p_a^d(\tau) \tag{15}$$

Above formulas allow to consistently apply the ICM model inside the NFP computed along the trajectories towards a single destination. Such formulation allows to propose the ISM algorithm as follows:

Algorithm 1. DTA Network Flow Propagation with Information Spreading Model [own study]

```
function NFP with ISM
    input:
        c,t,p,d              * travel costs, times, arc conditional probabilities, de-
mand
        for τ = 1 to T                  * for each time instant chronologically
            for d = 1 to D              * for each destination
                for each i in Reverse TO    * process nodes in reverse topological
order
                    q̂ᵢ += dᵢ           * load the demand
                    q̂ᵢ = ∑ₐ∈ᵢ₋ q̂ₐ·(1−ιₐ)   * compute the node flows:- unaware flow
                    q̄ᵢ = ∑ₐ∈ᵢ₋ (q̄ₐ+q̂ₐ·ιₐ)   *                         -aware flow

                    for each a∈i⁺               * for each arc of the forward star
                        for each S∈S            * compute the information probability
                            compute ιₐˢ with (1) * for each source
                        next S
                        compute ιₐ with (2)      * and joint for all sources
                        q̂ₐ += q̂ₐ₋·pₐ      * propagate forward the uninformed flow
                        q̄ₐ += q̄ₐ₋·pₐ      * propagate forward the informed flow
                    next a
                next i
            next d
        next τ
        return q̂,q̄
end function
```

4 Numerical Examples

To illustrate the ISM we applied it on the DTA model of Dusseldorf. We simulated
event taking place at 9 pm and showed how the drivers are informed about it in various
ATIS scenarios. At Fig. 3 below, we present number of informed drivers (green bar)
every 10 min in the following scenarios. In the first scenario the information is spread

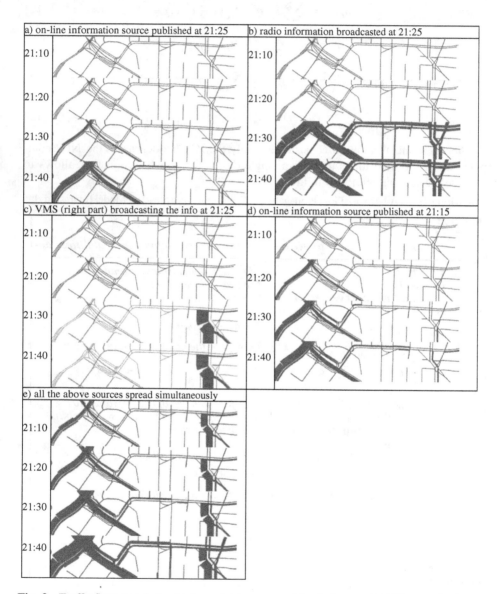

Fig. 3. Traffic flow snapshots of informed drivers every 10 min in several ATIS scenarios [own
study] (Color figure online)

online from 21:25; second shows the case of radio-news broadcasted at 21:30; in the third scenario VMS broadcast the information at single arc from 21:30; in fourth scenario the on-line information is published earlier, at 21:15; the last scenario shows how the information spreads when all the above sources are available simultaneously.

5 Conclusion

We proposed the model which can work with multiple sources of information present in contemporary traffic networks. We identified three generic information spreading paradigms: fixed-time, fixed-place and on-line data and proposed the respective formulas. We believe that all the major sources of information in contemporary traffic networks (online data, VMS, radio news) can be modelled with one of those paradigms with respective probability formulas. The joint probability for multiple sources computed with (2), although seems simplified is highly efficient at representing the various ATIS configurations and is sensitive to changes in component information sources as well as introducing another information sources. As shown in the numerical examples, the model can realistically represent various ATIS scenarios and provide a dynamic forecast of informed drivers in traffic networks. Thanks to computing the probability at the single arc level, proposed model can be handled within the DTA and applied practically representing the actual ATIS configuration. Model provides valuable output: traffic flows of informed drivers propagating through the network in space and time. Such outputs are of needed in contemporary traffic management centres of ITS systems and due to the efficient implementation model can be used in real-time for real-size networks. Such model can be further used e.g. to evaluate the efficiency of ATIS or as a starting point to model rerouting phenomena when informed drivers change their paths to avoid the event [9].

References

1. Bifulco, G., Di Pace, R., Simonelli, F.: A simulation platform for the analysis of travel choices in ATIS context through stated preferences experiments. In: EWGT Conference, Padua, 2009
2. Cascetta, E.: Transportation Systems Analysis. Springer, Heidelberg (2009)
3. Chorus, C.G., Moilin, E., Van Wee, B.: Use and effects of advanced traveller information services (ATIS): a review of the literature. Trans. Rev. **26**(2), 127–149 (2006)
4. Cova, T.J., Johnson, J.P.: A network flow model for lane-based evacuation routing. Transp. Res. Part A: Policy Pract. **37**(7), 579–604 (2003)
5. Earle, P., Bowden, D., Guy, M.: Twitter earthquake detection: earthquake monitoring in a social world. Ann. Geophys. **54**(6), 708–715 (2012)
6. Emmerink, R.H.M., Axhausen K. W., Rietveld P.: Effects of information in road transport networks with recurrent congestion. Research Memorandum (1993)
7. Ghosh, R. et al.: Time-aware ranking in dynamic citation networks. In: IEEE 11th International Conference on Data Mining Workshops (ICDMW) (2011)

8. Güner, A.R., Murat, A., Chinnam, R.: Dynamic routing under recurrent and non-recurrent congestion using real-time ITS information. Comput. Oper. Res. **39**(2), 358–373 (2012)

9. Kucharski, R.: Rerouting phenomena modelling of unexpected events in dynamic traffic assignment. Ph. D. thesis, Cracow University of Technology (2015

10. Leskovec, J., Adamic, L., Huberman, B.: The dynamics of viral marketing, ACM Trans. Web **1**(1) (2007)

11. Lindell, M.K.: EMBLEM2: an empirically based large scale evacuation time estimate model. Trans. Res. Part A: Policy Pract. **42**(1), 140–154 (2008)

12. Mahmassani, H. S. et al.: Dynamic traffic assignment with multiple user classes for real-time ATIS/ATMS applications. In: Proceedings of the Advanced Traffic Management Conference Large Urban Systems (1993)

13. Mahmassani, H.S., Yu-Hsin, L.: Dynamics of commuting decision behaviour under advanced traveller information systems. Transp. Res. Part C: Emerg. Technol. **7**(2), 91–107 (1999)

14. Mathioudakis, M., Koudas, N.: Twitter monitor: trend detection over the Twitter stream. In: Proceedings of the 2010 ACM SIGMOD International Conference on Management of data. ACM (2010)

15. Meschini, L. et al.: An implicit path enumeration model and algorithm for dynamic traffic assignment with congestion spillback (1999)

16. Milstein, S. et al.: Twitter and the micro-messaging revolution: communication, connections, and immediacy.140 characters at a time (2008)

17. Polydoropoulou, A., Ben-Akiva, M.E.: The effect of Advanced Traveler Information Systems (ATIS) on travelers behavior. Transportation Research Board (1998)

18. Procter, R., Vis, F., Voss, A.: Reading the riots on Twitter: methodological innovation for the analysis of big data. Int. J. Soc. Res. Methodol. **16**(3), 197–214 (2013)

19. Sakaki, T., Okazaki, M., Matsuo, Y.: Earthquake shakes Twitter users: real-time event detection by social sensors. In: Proceedings of the 19th International Conference on World Wide Web, WWW 2010 (2010)

20. Schlaich, J.: Analyzing route choice behavior with mobile phone trajectories. Transp. Res. Rec. J. Transp. Res. Board **2157**, 78–85 (2010)

21. Terpstra, T. et al.: Towards a realtime Twitter analysis during crises for operational crisis management. In: Proceedings of the 9th International ISCRAM Conference (2012)

22. Tweedie, S.W., et al.: A methodology for estimating emergency evacuation times. Soc. Sci. J. **23**(2), 189–204 (1986)

23. Wardrop, J.G.: Some theoretical aspects of road traffic research. Proc. Inst. Civ. Eng. **1**(3), 325–362 (1952)

Telematics in Sustainability of Urban Mobility in Silesian Agglomerations

Ryszard Janecki[✉]

University of Economics in Katowice, 1 Maja 50, 40-287 Katowice, Poland
ryszard.janecki@ue.katowice.pl

Abstract. Positive importance of a modernized and sustainable urban transport for the development of cities and agglomerations of the region is a key prerequisite to intensify activities to improve the sustainability of users of transport systems. Shaping contemporary balanced as much as possible urban transport systems in the region, should take advantage of the opportunities created by the solutions of the Intelligent Transport Systems (ITS). The considerations contained in the article confirms the importance of telematics as one means of sustainability of urban mobility.

Keywords: Objectives of sustainable mobility · Urban agglomeration · Intelligent transportation systems ITS · Transport telematics · Challenges of urban mobility · Concept of ITS

1 Introduction

Among many actions postulated in the recent years in the transport policy of cities and agglomerations, more and more often there are projects which aim to improve the sustainability of mobility in their area. Formulation and implementation of tasks this area is apparent from two basic premises. The first is the acceptance of the concept of balancing urban mobility as one of the priorities of the common transport policy of the Member States of the European Union in the current decade. Thus, working towards sustainable mobility in cities and urban areas has become an important area of Community cohesion policy and its funds[1].

The second of the premises are the benefits, that can be generated by the modernized urban and agglomeration transport in line with the principles of sustainable mobility. These include among others:

- improving the living conditions due to a healthy environment,
- ensuring the safety of the urban community in the wider sense
- improving the accessibility of cities and agglomerations,
- higher and more effective economic efficiency of these centres.

---

[1] The issue was heavily accentuated in [21]. This strategy involves the implementation of a competitive transport system that will increase mobility, remove barriers in key areas will increase employment. It will be significantly reduced Europe's dependence on imported oil and reduced CO_2 emissions.

© Springer International Publishing AG 2016
J. Mikulski (Ed.): TST 2016, CCIS 640, pp. 329–341, 2016.
DOI: 10.1007/978-3-319-49646-7_28

A particularly important benefit is to ensure the realization of the needs of society in terms of mobility. This means better, according to the common expectations, availability of inhabitants and visitors of cities and agglomerations to services, facilities and economic and social activities.

In the solutions leading to the improvement of sustainable urban mobility, urban and agglomeration transport becomes in this effort one of the main tools aimed at the qualities of an effective and efficient instrument for achieving rational and sustainability of urban mobility [9]. At the same time the effects of transport operations are clearly visible, in terms of cooperation with stakeholders in sustainable urban mobility they achieve a high degree of social acceptance.

Considering the above context, the article presents solutions for intelligent transport systems (ITS) used as a tool in the process of balancing mobility in the Bielsko-Biala agglomeration and metropolis of Upper Silesia. Cities and agglomerations of the region are at the beginning of the road leading to sustainable mobility today corresponding to their inhabitants.

2 Objectives of Sustainable Urban Mobility and Activities Leading to Their Achievement

The concept of sustainable urban mobility has its origin in the doctrine of sustainable development [1, 5, 16], as illustrated in Fig. 1.

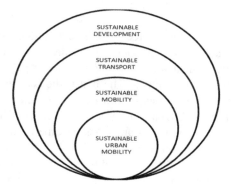

Fig. 1. Sustainable urban mobility as a component of sustainable development [own study]

Its basic assumptions relate to the conditions to be met by the development control process. These include taking into account at every stage of operations, environmental restrictions, the need for self-limitation and system management needs and support development in the name of intergenerational justice [2, 14].

The sectoral allocation of the doctrine, thus creating sustainable transport in line with sustainable development, includes actions aimed at ensure a balance between: technical, functional, spatial, economic, social and environmental factors, which characterize the transport system and subjected to balancing [6].

Instrument in the form of a modernized urban and agglomeration transport allows for sustainable urban mobility. From a practical point of view, sustainable urban mobility is the ability to meet the needs of residents of agglomeration or city and other transport users in terms of freedom of movement, access to key destinations and promotion good economy in the absence of danger to human health and safety, preserving the environment for future generations and the effective functioning of the urban transport system [12, 20]. Included in the definition the subject of sustainable urban mobility and its attributes is shown in Fig. 2.

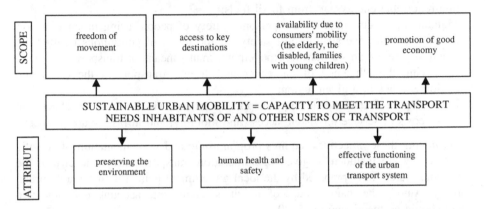

Fig. 2. The concept of sustainable mobility [own study]

Here can be seen that the idea of sustainable urban mobility includes two basic functional areas of modern towns and cities. The first is defined by the development policy of the agglomeration, city or region, and relates to the health of residents, congestion mobility, transport security and its users, participation of residents in the changes, strategic planning and climate problems. The second covers the wider sphere of the economy and its sustainable development, and support the diverse needs by the urban transport system [13, 20].

With the given areas of functioning of cities and agglomerations are related the objectives pursued under the concept of sustainable urban mobility. They should be seen as a response to the challenges coming from the spheres of social and economical of urban and agglomerations centres. With the extensive catalogue of the most important goals may be included [20]:

- The creation for agglomeration or city residents in the given region and to all other transport users, access to key destinations (home, work, education, etc.) and services, as well as meet the quantitative and qualitative needs of travel;
- Promotion the good economy, it is such an economic activity whose aim is continuous improvement of the quality of life now and in the future, taking into account existing financial constraints and environmental conditions, and under the condition of active involvement of all participants in the economic processes in the agglomeration or the city of a given region;

- Preserving the environment in the state characterized by a constant reduction of pollution of air, soil and water, and noise level, reduction of greenhouse gas emissions and energy consumption, as well as the ability to ensure residents meeting the needs related to recreation, physical culture and active recreation, should also be taken into account the needs of the economic environment and the community as a whole, which will allow for treatment the natural environment as the overall merit of the agglomeration or city [8];
- Maintaining the energy security through the participation of urban transport in the realization of the growth of good economy in conditions limiting its demand for non-renewable energy e.g. from fossil fuels;
- Maintaining health, improve the personal safety of people using urban transport systems and the overall transport safety, in the terms of continuous conflicts in relationships: a man - a transport movement, man - means of transport;
- Improving the efficiency of passenger and freight transport in the cities and agglomerations of a given region;
- Improving the financial results of transport operations, which should be reflected in lower costs of service provision, access and mobility, and infrastructure projects.

Achieving these objectives requires programming and implementation of specific actions. They are included in given agglomeration transport policy, the city in the region and increasingly prepared by the local government plans for sustainable urban mobility. Among the major groups of activities, taking into account their objective aspect should be mentioned [3, 7, 20]:

- Reducing the need for travel of people and cargo transport;
- Creating an attractive offer of public transport;
- Offsetting the transport distance;
- Using of innovative transport technologies;
- Changes in the modal split aimed at reducing the level of car dependency.

On the other hand extension of objective aspect of the tool issues allows distinguishing the following areas of work which are included in Table 1. It shows the strongest relationship between actions/solutions and the challenges (problems to solve) of the mobility in cities and agglomerations[2].

The presented data show that in practice, the most commonly used technologies ITS relate to e-ticketing, management systems and traffic control and passenger and travel information. The strongest impact of these solutions is recorded in relation to such problems of urban transport like:

- Movement congestion negatively affecting the economic performance and availability of agglomeration or city;
- Widely understood transport safety;
- Global climate changes caused by emissions coming from urban transport.

In the next part of the present article stated assertions were confronted with two ITS projects: for Bielsko-Biala - the central city urban agglomeration, as well as for the

---

[2] Symbols: ++++ - the strongest relationships, + - the weakest relationships.

Table 1. Specification of undertaken actions (proposed solutions) and the challenges of urban mobility (problems to solve) with mutual relations and their [20]

Nature of undertaken action / proposed solutions	Challenges of urban mobility (problems to solve)					
	Health	Congestion	Safety and security	Public participation	Strategic planning	Global climate changes
Clean fuels and vehicles	++++	+	+	+	++	+++
Freight transport	++++	+++	+++	+++	+++	++
Demand management strategies	++++	++++	++	++++	+++	+++
Access restrictions, environmental zones	++++	++++	++	++++	+++	+++
Fees for entry	+++	++++	++	++++	+++	+++
Mobility management	++	++++	+	++++	+++	+++
Agency for Mobility	++	++++	+	++++	+++	+++
Ekopoints	++	++++	+	++++	+++	+++
Collective transport	++	+++	++++	+++	+++	++
New forms of public transport services	++	++++	+++	+++	+++	+++
Availability for the elderly / disabled passengers	+++	+	++++	+++	+++	++
The integration between transport modes	++	+++	+++	+++	+++	+++
Telematics in transport	+	++++	+++	+	+	++
eTicket	+	+++	+	+	+	++
Management and traffic control	++	++++	+++	+	+	+++
Passenger and travel information	++	++++	+++	+	+	++
Mobility options dependent on the car	+++	+++	++++	++++	+++	+++
Car-sharing	++++	+++	+++	++++	+++	+++
Carpooling	+	++++	+++	+++	++	+++
Pedestrian and bicycle traffic	++++	++++	++++	++++	++++	++++

municipalities participating in the Municipal Transport Union of the Upper Silesian Industrial District (KZK GOP) Katowice [11, 15, 17].

3   ITS Technology as an Instrument for Sustainable Urban Mobility

In the previous considerations of the basic tools of sustainable urban mobility have also been mentioned technologies of Intelligent Transport Systems. It is therefore necessary to put the question of what constitutes high usefulness of ITS solutions in the implementation of projects related to sustainable mobility in cities and agglomerations.

The answer to such questions requires consideration of the scope and objectives of the overall analyzed ITS projects and those of modules that most affect urban mobility, and its level of sustainability. Tables 2 and 3 shows listed issues for two projects in the

Table 2. Scope and general objectives established of the analyzed ITS projects in Silesia [own study based on: [4, 17]]

Project name	General scope of the project		General objectives of the project (expected results)
	Territorial	Technical (system components)	
1	2	3	4
1. The concept of Intelligent Transportation System for the city of Bielsko-Biała (2014)	area of Bielsko-Biała	−Road Traffic Management Centre −road traffic control −observance of regulations in road traffic supervising system −information system on operating conditions - analysis of the VMS −video monitoring system −meteorological station for the winter action −the priority system of public transport and emergency vehicles −information system for drivers and travellers	−improvement of runnability in the city of Bielsko-Biała −improvement of pedestrians and vehicles safety −improvement of vehicles movement, including buses in urban collective public transport −creation of information resources about transport functioning in the city of Bielsko-Biala
2. The concept of Intelligent Traffic Management System in the area of KZK GOP activities (2015)	−functional areas of the cities forming KZK GOP −28 municipalities participating in the KZK GOP	−system of area traffic light −information system for drivers −video monitoring system −traffic control system −public transport management system	−improvement of safety, comfort and speed of travel in the area of KZK GOP −increase the attractiveness of the KZK GOP transport system −increase the share of public transport of travels carried out within the area of KZK GOP −reduction of fuel consumption and air pollution and reduction of energy consumption and maintenance costs −increase the attractiveness of municipalities forming KZK GOP as a potential location for new investments by improving traveling conditions −reduction of the economic and social costs and environmental impacts associated with the accumulation of traffic on the Upper Silesian Metropolitan Area

Table 3. Specification of the objectives pursued by the selected modules analyzed ITS projects in Silesia [own study based on: [4, 17]]

ITS project name	The module name in the ITS project affecting the level of sustainable urban mobility	The objectives pursued by the model of analyzed ITS project									
		better access to key objectives	meet the quantitative and qualitative needs	supporting economic activity ensuring continuous improvement in quality of inhabitants life	increasing the economic attractiveness ensuring the continuous improvement of public transport	preserving the environment	reduction of emissions from transport	reducing the consumption of non-renewable energy	improving transport safety	improving transport efficiency	improvement of financial results of transport entities
1	2	3	4	5	6	7	8	9	10	11	12
1.The concept of Intelligent Transportation System for the city of Bielsko-Biała	road traffic control system	+	+	+	+	+	+	+	+	+	+
	information system on the movement conditions - VMS signs	+	+	+		+	+	+	+	+	+
	system for measuring the traffic congestion	+	+	+	+	+	+	+	+	+	+
	bus stops dynamic information	+	+		+	+	+	+		+	+
	priority system of public transport and emergency vehicles	+	+		+		+	+	+	+	+
2.Intelligent Traffic Management System in the area of KZK GOP activities	system of area traffic light control	+	+	+	+	+	+	+	+	+	+
	drivers information system	+	+	+		+	+	+	+		
	passenger information system	+	+		+	+	+	+		+	+
	public transport management system	+	+		+	+	+	+	+		

preparation phase to take their implementation. These are the previously mentioned concepts for IT Bielsko-Biala and municipalities in KZK GOP Katowice[3].

The data in Tables 2 and 3 clearly demonstrate a strong link between activities that increase the level of innovation of transport in cities and metropolitan regions related to intentions leading to an increase the level of sustainability of urban mobility. The activities using ITS technologies perform in fact an extensive range of objectives equilibration processes.

The analyzed concepts of ITS systems for selected areas of Silesia, due to its comprehensive solutions are a good example of properly formulated projects also in the field of urban mobility. This statement, however, is only the result of thoughtful analysis of ITS projects, and not a reflection of clearly defined objective in this regard. In the case of both projects should be noted that, despite supporting in the current financial perspective, by the European Commission concept of sustainable urban mobility and related appropriate changes in transport systems of cities and agglomerations, there are no references to them in this kind of action. This deforms the image of usability of the proposed solutions using ITS.

Changing the approach in this field could contribute to increase the number of stakeholders of ITS systems in cities and agglomerations, and thus to increase the expected effects of implemented solutions. The combination of innovative transport concepts and the concept of sustainable transport expands the scope of applying for funds from EU funding, thus increasing significantly the possibilities obtaining them.

4 Evaluation of Use the ITS Technology as a Tool for Sustainable Urban Mobility

Adoption of the approach that combine innovativeness of transport and sustainability urban mobility in solution projects based on ITS technologies requires the application of appropriate ways of evaluating such projects. In this regard are required two-stage evaluation activities:

- The first phase is to construct a suitable model of movement, so that it will be possible to identify all the sizes necessary for evaluating the actions using ITS technologies in the process of sustainability urban mobility;
- While the second phase includes specific methods for assessing applications for ITS systems to sustainability of mobility, using data obtained from the constructed model of movement.

The work on the model of traffic for the city of Bielsko-Biala, defined structural requirements for the four-phase model of the movement so that the representation of the actual model of urban transport system also takes into account solutions for Integrated Transport Management System in the city of Bielsko-Biala [10, 15, 22]. Synthetic approach of these demands construction is shown in Table 4. The information

---

[3] It is also considered a broader territorial scope of the project includes all the municipalities of the Central Subregion of Silesia.

Table 4. Selected structural requirements for the model of movement taking into account the use of ITS technologies on the example of the Integrated Transport Management System in the city of Bielsko-Biala [own study based on [15]]

ITS (modules) according to the concept of the Integrated Transport Management System in the city of Bielsko-Biala	Expected effects of the set ITS system implementation	Phenomena and processes in the ITS system associated with the expected results	Requirements relating to the construction of the model related to the expected results
1	2	3	4
1. The road traffic control system (individual transport) 2. System for measurement the traffic congestion	−increase the road traffic flow −shorten the driving time in the individual transport	−increase the traffic flow at crossings under the linear and territorial control −shorten the travel time for torsional relationship at crossings covered by the linear control	−modification time sanctions at crossings located in stretches of road covered by the linear control and in the area covered by the territorial control −change the functions of resistance of part of road −tCUR time change for passenger cars
3. The information system of the movement conditions - VMS signs; information system on available parking spaces	−reduction of travel time (transit journeys) in the city	−reducing the time to search for a parking space −guidance on parking spaces through a system of information −shorten the travel time for journey undertaken by individual transport especially in the areas of the paid parking zones and parking facilities for integrated interchanges and parking Park and Ride	−introduction to the model the concept of perceived travel time (PTT) for individual transport taking into account the time aspects of parking in the city −tCUR time change for passenger cars

(continued)

Table 4. (*continued*)

ITS (modules) according to the concept of the Integrated Transport Management System in the city of Bielsko-Biala	Expected effects of the set ITS system implementation	Phenomena and processes in the ITS system associated with the expected results	Requirements relating to the construction of the model related to the expected results
1	2	3	4
4. Dynamic passenger information	−reduce the waiting times being a component of the perceived travel time −increase the attractiveness of public transport in the city	−change the behavior of users of public transport related to travel beginning or the transfer −increase the attractiveness of public transport by changing the perceived travel time	−increase the attractiveness of waiting times in the perceived travel time and use that shaped parameters for the distribution of traffic flows on the transport network −increase the attractiveness of the perception of public transport through a specific indicator of perceived travel time and use of this phenomenon by distribution the traffic flows on the transport network
5. The priority system of public transport and emergency vehicles	−shorten or maintenance of an acceptable travel times on public urban transport −increasing the share of public transport	−shorten the times constituting the travel time included in the perceived travel time	−increase the attractiveness of driving time taken into account in the perceived travel time and the use of that shaped parameter at the stage of distribution of traffic flows on the network transport −increase the attractiveness of the perception of public transport and the use of modifying the perceived travel time in the distribution of traffic flows on the transport network (only for X PuT mode)

contained herein clearly indicate that model of movement, taking into account solutions using ITS technologies, reproduces them in the model activities. They illustrate using defined parameters, typical phenomena identified in urban transport systems.

Listed in Table 4 phenomenon confirm the general opinion that intelligent transport systems allow, among others, to [19]:

- Effective support for transport users;
- Reduce or eliminate the negative effects of transport on the environment;
- Further development of innovative transport, taking into account modern trends in the development of ICT.

Evaluation of ITS applications in the process of sustainability urban mobility will only apply to the first two areas of allocation effects. An example of this approach among others is to assess the effectiveness of activities promoted in the Sustainable Urban Mobility Plan (SUMP). Gauges and indicators the reduction in transport demand, increase the share of pedestrian traffic and changes in energy consumption and greenhouse gas emissions (pollution) are useful tools to evaluate the effectiveness of using ITS technologies in the plans SUMPs [18]. They can also be used in ITS projects implemented in the agglomerations of Silesia.

5 Conclusion

Analysis of selected area concepts of its systems allows drawing the following conclusions:

1. Comprehensive use of ITS technology allows achieving its targets for sustainable mobility in cities and agglomerations.
2. It is possible in fact an increase in the share of collective public transport while reducing the overall demand for public transport in the region.
3. Changes in demand will favorably affect the energy consumption and pollution emissions.
4. Joining in ITS systems projects issues of innovative transport and sustainable urban mobility is beneficial to transport users and local authorities. Increasing the number of stakeholders, and the same, the range of projects usability can be obtained the opportunity to achieve greater benefits in terms of improving living conditions of inhabitants.
5. The presented mode of action has currently only postulate character. However, it should be implemented at later stages of the implementation of ITS projects analyzed in Silesia.
6. Evaluation of implemented ITS projects should also consider the effects related to balancing urban mobility. In this regard, it is possible to use previously used instruments, such as method of evaluating the effectiveness of activities promoted in sumps or to develop a method taking into account the specificity of the agglomeration of Silesia.

References

1. Agenda 21 and the Rio Declaration (1992)
2. Atkinson, G., Dietz, S., Neumayer, E. (eds.): Handbook of Sustainable Development. Edward Elgar Publishing Ltd., Cheltenham (2008)
3. Banister, D.: The sustainable mobility paradigm. Transp. Policy 15(2), 78–80 (2008)
4. Cichoński, J., et al.: The concept of intelligent transportation system for the city of Bielsko-Biala. Internal publication, Gliwice (2014)
5. Czech, K.: Rio +20 Earth summit - what future of sustainable development. In: Sporek, T. (ed.): International Economic Relations; Selected Institutional Factors and Real Processes in a Global Mobility, No. 170, pp. 33–34. Scientific Papers Faculty of the University of Economics in Katowice. Economic Studies (2013)
6. Gudmundson, H.: Sustainable transport and performance indicators. In: Hester, R.E., Harrison, R.M. (eds.): Transport and the Environment. Issues in Environmental Science and Technology, No. 20, pp. 35–52. The Royal Society of Chemistry, Cambridge (2004)
7. Janecki, R.: A new culture of mobility as the direction of the development of urban and regional transport in Silesia region. In: Michałowska, M. (ed.): Contemporary conditions of transport development in the region, No. 143, pp. 142–144. Scientific Papers Faculty of the University of Economics in Katowice. Economic Studies (2013)
8. Janecki, R., Karoń, G.: Concept of smart cities and economic model of electric buses implementation. In: Mikulski, J. (ed.) TST 2014. CCIS, vol. 471, pp. 100–109. Springer, Heidelberg (2014). doi:10.1007/978-3-662-45317-9_11
9. Janecki, R., Karoń, G.: Stimulating sustainable mobility in urban agglomerations in industrial regions. In: The 5th World Sustainability Forum: Transitioning toward Sustainability. Basel, 7–9 September 2015
10. Karoń, G.: Travel demand and transportation supply modelling for agglomeration without transportation model. In: Mikulski, J. (ed.) TST 2013. CCIS, vol. 395, pp. 284–293. Springer, Heidelberg (2013). doi:10.1007/978-3-642-41647-7_35
11. Karoń, G., Mikulski, J.: Problems of systems engineering for its in large agglomeration – upper-silesian agglomeration in Poland. In: Mikulski, J. (ed.) TST 2014. CCIS, vol. 471, pp. 242–251. Springer, Heidelberg (2014). doi:10.1007/978-3-662-45317-9_26
12. Mobility for Development. Facts and Trends, p. 1. The World Business Council for Sustainable Development, Geneva (2007)
13. Molina, L.T., Molina, M.J.: The MCMA transportation system: mobility and air population. In: Molina, L.T., Molina, M.J. (eds.): Air Quality in the Mexico Megacity: An Integrated Assessment, p. 241. Kluwer Academic Publishers, Dordrecht (2002)
14. Rogers, P.P., Jalal, K.F., Boyd, J.A.: An Introduction to Sustainable Development. Earth-scan, London (2008)
15. Sobota, A., et al.: Integrated Transport Management in the city of Bielsko-Biala. Stage I - The execution of a traffic model for the city of Bielsko-Biala. Research work NB-148/RT5/2014. Silesian University of Technology Faculty of Transport, Katowice (2015)
16. Stockholm Declaration (1972). Report Brundtland, G.H. "Our Common Future"
17. The concept and architecture Intelligent Traffic Management System within the area of Communications Municipal Association of Upper Silesian Industrial District. SSM Silesian Metropolitan Network Sp. z o.o. Internal publication, Katowice (2015)
18. Suchorzewski, W.: The planning of sustainable urban mobility in the cities. http://docplayer.pl/15218739-Ch4llenge-planowanie-zrownowazonej-mobilności-w-miastach-wojciech-suchorzewski-politechnika-warszawska.html. Accessed 10 Nov 2015

19. Sussman, J.S.: Perspectives on Intelligent Transportation Systems (ITS). Springer, New York (2005)
20. Wefering, F., et al.: Guidelines. Developing and Implementing a Sustainable Urban Mobility Plan, pp. 7, 58, 63. Rupprecht Consult-Forschung and Beratung GmbH, European Commission Directorate – General for Mobility and Transport, Brussels (2014)
21. White Paper. Plan of Creating a European Transport Area - Towards a competitive and resource efficient transport system. The European Commission COM (2011) 0144, Brussels (2011)
22. Żochowska, R., Karoń, G.: ITS services packages as a tool for managing traffic congestion in cities. In: Sładkowski, A., Pamuła, W. (eds.). SSDC, vol. 32, pp. 81–103Springer, Heidelberg (2016). doi:10.1007/978-3-319-19150-8_3

Intelligent Solutions in Sustainable Transport Upper Silesia Agglomeration

Robert Tomanek[✉]

University of Economics in Katowice, 1 Maja 50 str., 40-287 Katowice, Poland
tomanek@ue.katowice.pl

Abstract. The metropolization processes in Poland are currently in progress. The tools for these changes are the following: implementation of EU urban policy (in particular by implementing Integrated Territorial Investments (ITI) strategy) and formation of metropolitan associations. The implementation of the ITI strategy requires formation of ITI associations. In the case of Upper Silesia, such association covers a territory extending beyond a highly urbanized area - the Association of Municipalities and Districts of the Central Subregion includes over 80 municipalities and districts. This situation causes the dispersion of the integration of the metropolitan area. The Metropolitan Association (MA) will be formed in accordance with the Act of 6 October 2015 on Metropolitan Associations, which became effective on 1 January 2016. Both ITI and MA will deal with sustainable transport and mobility, and the basic instrument for sustainable transport is the development of public transport. The sustainable mobility plan, which is already being developed for the ITI strategy of Upper Silesia, provides for the application of Intelligent Transport Systems (ITS). ITS enables the restriction of the fundamental barrier of metropolization - the activity of 5 public transport organizers characterized by different systems of organizing transport market, financing and transport tariffs. The purpose of the article is to present the conditions for use of selected types of ITS in order to provide sustainable transport for the Upper Silesia Agglomeration, which should lead to an increase in the importance of sustainable public transport.

Keywords: Sustainable transport · Sustainable mobility · Integrated Territorial Investments · Metropolization · Intelligent Transport Systems · Public transport

1   Introduction

The metropolization processes in Poland are accelerating. The tools for these changes are the following: implementation of EU urban policy (including in particular through implementation of Integrated Territorial Investments (ITI)) and formation of metropolitan associations. The implementation of ITI requires formation of ITI associations. In the case of Upper Silesia, such association covers a territory extending beyond a highly urbanized area - the Association of Municipalities and Districts of the Central Subregion includes over 80 municipalities and districts. The Metropolitan Association will be formed in accordance with the Act of 6 October 2015 on Metropolitan Associations,

© Springer International Publishing AG 2016
J. Mikulski (Ed.): TST 2016, CCIS 640, pp. 342–353, 2016.
DOI: 10.1007/978-3-319-49646-7_29

which became effective on 1 January 2016. Both ITI and MA will deal with sustainable transport and mobility, and the basic instrument for sustainable transport is the development of public transport. The sustainable mobility plan, which is already being developed for the ITI strategy of Upper Silesia, provides for the application of Intelligent Transport Systems (ITS). The implementation of MA tasks will also require the application of ITS, including the existing solutions already applied by transport organizers in the Upper Silesia Agglomeration. ITS enables the restriction of the fundamental barrier of metropolization – the activity of 5 public transport organizers characterized by different systems of organizing transport market, financing and transport tariffs. The purpose of the article is to present the conditions for use of selected types of ITS in order to provide sustainable transport for the Upper Silesia Agglomeration, taking into consideration the specific character of this urban area.

2 Metropolization Tools in Poland

In metropolitan areas, there are numerous problems concerning the cooperation between local government units, in particular related to network tasks - public transport, road management, water supply and waste management. At the same time, the attempts to create a uniform administrative structure have not fully succeeded anywhere - apart from Verband Region Stuttgart, where representatives of the metropolitan association are elected directly [7].

At the beginning of the political transformation in Poland, in 1991, municipalities were allowed to form voluntary intermunicipal associations and enter into agreements. For various reasons, these associations were formed slowly and many issues requiring cooperation did not function efficiently. Therefore, certain regulations intended to integrate the largest agglomerations were implemented in 2013 and 2016 respectively. The created tools include:

– Integrated Territorial Investments (ITI), which are implemented by ITI associations,
– metropolitan associations which are to be formed in the areas inhabited by over 500,000 citizens, pursuant to the Act on Metropolitan Associations, which became effective on 1 January 2016.

The adopted solutions are not free from defects, however, each of them provides financial resources directed towards municipalities and districts in metropolitan areas. According to the assumptions of ITI, they are intended to integrate metropolitan cities (province capitals) with their immediate surroundings. Therefore, it was emphasized in the Partnership Agreement that integrated activities for sustainable development of urban areas will be particularly supported - pursuant to Art. 7 of the Regulation of the European Parliament and Council (EC) 1301/2013 of 17 December 2013. It was specified as mandatory that "provincial" ITI shall be implemented in the territories of province capitals and related cities, indicated in the functional areas. Accordingly, 17 mandatory areas for implementing ITI were determined (Bydgoszcz and Toruń form one area in the Kujawy-Pomerania Province, whereas in the Lubusz Province there are two such areas). Provinces can also launch ITI in the functional areas of cities of regional and subregional significance. Currently, EUR 3.5 billion was allocated for this instrument (including domestic and regional funds), and the total number of ITI is 24 [16]. The concept of ITI implementation in the Silesia Province was based on allocating funds in subregions and related agglomeration centers. In particular, ITI financed at the national level was included in the central subregion (including Katowice). The allocation provided for ITI for the central subregion of the Silesia Province is the highest in Poland, with the value of EUR 793 million. The dominating activity within this ITI is "low emission urban transport", to which 50.2 % of funds were allocated. The majority of financial resources for ITI in the central subregion are intended for tasks located in the central area of Upper Silesia Agglomeration. Therefore, although its scope of activity is too broad, we can say that ITI instrument is a factor of metropolization, also in the case of Upper Silesia Agglomeration.

Table 1. Basic measures and indicators characterizing the population of the province of Silesia (as of 31.12.2014) [own study based on 12]

Specyfication	Silesian province	Poland
Total population in thousands. People as of 31.12.2014	4 585,9	38 478,6
Participation in [%] of the total Polish population	11,9	–
The age structure of the total population [%]:		
The population of pre-working age	16,8	18,0
The population of working age	63,2	63,0
The population of post-working age	20,0	19,0
The density of population per 1km2 in [persons/km2]	372	123
Natural increase per 1,000 people	–1,1	0,0
The feminization	107	107
Net migration per 1,000 population	–1,6	–0,4
The level of urbanization in [%]	77,3	60,3

Metropolitan associations can be formed pursuant to the Act of 6 October 2015 on Metropolitan Associations, which became effective on 1 January 2016. The Act had been drawn up for several years and its provisions currently raise many doubts. The formation of metropolitan associations is encouraged by statutory funding at the level of 5 % of tax on personal income of citizens of metropolitan municipalities. According to the Act, the only network services provided by metropolitan associations are transport-related tasks. In particular, this concerns public transport and cooperation for determining the course of national and regional roads. In this situation, the problems related to organization of public transport are becoming a key issue, especially that the Act specifies that an association should be responsible for organizing metropolitan transport and a source of integrated tariff within a metropolis. However, the conditions and procedures for formation of associations specified in the Act are relatively complicated and not free from discrepancies. There is a high risk that the Act will become a dead letter, since a metropolitan association can only be formed on the condition that the prime minister issues the relevant regulation until the end of April in the year preceding actions taken by an association. The very problem of delimiting metropolitan area raises doubts - according to different studies, metropolization processes in Poland are poorly advanced [7].

Metropolisation in the Silesian province covers the area of the Upper Silesian agglomeration. Province is the largest and most urbanized areas in Poland (indicators presented in Table 1). Silesian province is characterized by the high levels of industrial development, investment attractiveness, as well as a good location.

A particularly important factor for mobility is the growing number of cars per 1,000 inhabitants (data presented in Fig. 1), which in the metropolitan area increases congestion. The growing number of cars will increase in car journeys, and thereby causes that mobility is not sustainable.

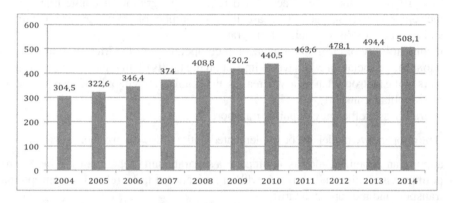

Fig. 1. The number of cars per 1,000 inhabitants in the province Silesian [own study]

3 Models of Organizing Metropolitan Transport in the Upper Silesia Agglomeration

Assuming the statutory formation of metropolitan association in the Katowice Agglomeration, certain model solutions can be adopted in the field of performing tasks related to public transport by the Metropolitan Association.

Depending on the scope of metropolitan transport system, two approaches to organization of public transport in a metropolitan area can be distinguished:

- cooperative model: metropolitan association is the organizer of metropolitan transport and coordinator of tariff system, whereas other transport (municipal, district, regional) in the metropolitan area is organized by the existing organizers (with the exception of tariff determination),
- centralized model: metropolitan association organizes the entire public transport in the metropolitan area (including coordination of the tariff and ticket system).

The cooperative model seems to be easiest for implementation, in particular for the following reasons:

– within the meaning of the Act, metropolitan transport is a small fracture of the currently provided public transport. In fact, it seems that in metropolitan areas it will be difficult to identify such transport (it will probably be transport between municipal operators and a part of intercommunal transport),
– required level of co-funding for such a small system will be significantly lower than statutory resources from personal income tax, therefore, it will not be necessary to provide additional funds for public transport to the association from municipalities and districts within the metropolitan area.

At the same time, it shall be noted that such narrow approach to transportation tasks of the Metropolitan Association will have the following negative effects:

– low progress in the field of actual and required integration of public transport in metropolitan areas (especially in case there is a number of organizers, actual barriers and instances of discontinuity of integration),
– incompatibility of tariff and organization competences of the Metropolitan Association (the Association will have an impact on the tariffs of all public transport organizers in the metropolitan area, whereas as an organizer, it will only be responsible for metropolitan transport),
– overall: lesser efficiency of public transport.

The cooperative model can be implemented in the following manner (alternatively):

– appoint an organizer, such as an office or a separate unit (department) of the Metropolitan Association and entrust it with organizational tasks related to metropolitan transport and metropolitan tariff,
– entrust the duties of organizer of metropolitan transport to another organizer (e.g. leading organizer of public transport in the metropolitan area) and provide supervision of transport and tariff determination performed by a separate small organizational unit of the Metropolitan Association (department).

The centralized model has its origin in the experiences of European metropolitan areas and it can be considered as a recommended and universally applied solution. However, in the conditions of the Polish Act on Metropolitan Associations, it will be hard to implement in the areas where the lack of integration causes the biggest problems (Tricity, Upper Silesian Industrial Region, Warsaw), in particular the following (shows in Fig. 2):

– appointment of one metropolitan transport organizer in the metropolitan area - due to the scope of activity, it would be a large unit with a significant budget. This would lead to the process of liquidating the existing structures of public transport organization,
– development of the metropolitan system of financing public transport - tax revenue provided for the Metropolitan Association will definitely not cover all expenses related to public transport in the Metropolitan Association, and various methods of financing public transport used by particular organizers are a further complication.

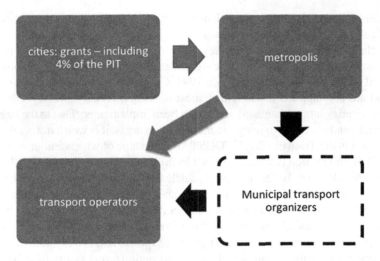

Fig. 2. Cash flow in agglomeration [own study]

The centralized model ensures full integration of public transport in a metropolis. Although it will result in issues related to transfer of competences, it will ultimately allow to:

- intensify transport integration in the metropolis,
- increase the efficiency of transport system,
- increase sustainable mobility.

The centralized model can be implemented by appointing one organizer of metropolitan transport, such as:

- separate Metropolitan Association unit (large department),
- metropolitan budget entity,
- external unit (however, Polish experiences and conditions allow to conclude that this solution is rather unlikely).

4   ITS in Sustainable Mobility in Metropolitan Areas in Poland

The development of transport is one of key factors behind the formulation of the smart city concept, and smart mobility is one of the key areas of smart cities [10]. At the same time, such shape of cities is becoming one of the key megatrends in the economic development of urban areas [15]. Projects undertaken in the field of smart mobility are implemented in accordance with the principles of sustainable development, which means focusing on minimization of the negative impact of transport on the environment, especially through the development of public transport, preferential traffic and access to services. Particularly significant in this field are the projects which use Intelligent Transport Systems (ITS), including projects related to: vehicle traffic management, congestion charges and smart cards [10]. In Poland, projects concerning urban traffic control system and tariff and ticket integration by means of smart cards are implemented.

Vehicle traffic management is the key problem solved by means of ITS - one of the pioneers in this field is USA, where the significance of this element was emphasized in the federal ITS programme as early as in 1991. Currently, urban traffic control systems in Poland are only applied in few cities, most notably in Tricity and Wrocław, where solutions covering large urban zone areas have been implemented. Due to the large scope of the system and necessity to integrate metropolitan areas, it is worth noting the system implemented in the Tricity in 2015 - TRISTAR, the value of which amounts to approximately PLN 160 million (including PLN 136 million co-financed by EU). The system takes into consideration the fact that both vehicle flow management and public transport management are dispersed in particular cities, and the integrated metropolitan transport management is assumed only as a target. The current regulations concerning metropolitan areas will not ensure integration of these systems. However, with reference to the impact on drivers in both individual and public transport, TRISTAR system offers access to advanced traffic information, which allows to optimize behaviors of drivers (apart from that, TRISTAR system also provides information for passengers in public transport, as well as parking information). By means of traffic detection, data on road incidents and meteorological information, TRISTAR system makes use of information boards to provide drivers with information concerning:

– traffic obstacles,
– journey times on main and alternative routes (e.g. journey time to Gdańsk via Tricity bypass and intercity road),
– detour recommendations [11].

In conjunction with adapting traffic signals to traffic conditions, TRISTAR system should support sustainable transport in the Tricity agglomeration, in particular reduce congestion and increase road transport security through more efficient use of transport infrastructure [8]. This is consistent with the expectations related to the benefits of implementing ITS. The literature on the subject widely presents relatively optimistic forecasts, which in practice seem to be unrealistic. It is sufficient to recall predictions from almost 20 years ago, when it was assumed that the implementation of ITS in road traffic management would lead to 50 % reduction in the number of car accidents, 25 % reduction in journey times and 50 % reduction of air pollution caused by transport in cities until 2017 [6]. In practice, ITS systems are not implemented according to the assumptions: even if the ITS architecture advocated by the specialists exists, the organization of transport systems for which ITS solutions are prepared does not correspond to it. Moreover, the costs and duration of project implementation are increasing, which is the effect of underestimating the risk resulting from the specific character of the decision-making processes in the public sector responsible for the functioning of transport infrastructure [2]. In particular, it is difficult to achieve the scale effects related to integration, which is the key factor limiting IT effects.

The barriers of integration in metropolitan areas, where it is frequently necessary to diversify tariff solutions applied in different parts of metropolis, have a similar limiting impact on the effects of applying ITS in tariff systems. Striving for tariff integration in metropolitan areas requires applying flexible and intelligent ticket systems. An example could be universal tickets, providing access to transport services and other urban services, especially when the ticket performs the function of the so-called electronic purse. It seems that progress in this field will be encouraged by dissemination of modern ticket systems. They do not necessarily have to be tickets in the form of electronic cards. The so-called virtual tickets, once rare, are developed thanks to the widespread use of smartphones. In Poland, these two directions of ticket system development are widely used, however, the biggest number of implementations are related to electronic cards applied mainly in public transport systems [14].

Cards allow to:

– integrate payments in transport systems (not only public) in metropolitan areas where many transport organizers operate,
– perform measurements of transport size, which is particularly significant for financial settlements of municipalities subsidizing public transport,
– integrate urban functions.

The main problems characterizing electronic card systems in Poland include:

– different technical standards of cards,
– adjustment of solutions to possibilities of obtaining support from EU funds - adapting programmes to conditions specified in operational programmes leads to creation of functionalities which will be characterized by low efficiency during the period of system utilization,
– closing the areas of card application within the existing organizational structures of public transport,
– investment expenditure and costs of system operation [5].

Implementation of electronic card systems faces smaller problems than in the case of urban traffic control systems, however, in the case of large projects and extended functions, the investment expenditure is increased and the implementation period is clearly extended. An example of such problems is Silesian Public Services Card (ŚKUP), which was put into use on 1 November 2015.

ITS projects applied in public transport undoubtedly raise the level of economy digitization and have a positive impact on the development of social capital, however, large projects are almost entirely financed from EU funds. Formal restrictions are a source of expensive project management during the so-called project durability period, moreover, they petrify the existing organizational structures due to the necessity of determining the so-called support beneficiary. In particular, this could be the reason for preserving ineffective solutions which block changes in the organization of metropolitan structures.

5   Directions for Application of ITS in Sustainable Transport in the Upper Silesia Agglomeration

The implementation of intelligent transport systems in Silesia is in the initial phase. In this respect, it is clearly noticeable delay in relation to the countries of Western Europe.

So far completed projects concerts in the area of public transport, among them can be replaced:

– dynamic passenger information system comprising a number of main lines of communication on the network KZK GOP,
– a network of electronic passenger information boards in the city of Rybnik including 167 stops of public transport,
– card systems enabling urban performing the function of an electronic ticket and e – wallet (SKUP, e-Bilet in Jastrzębie, Electronic City Card in Rybnik, Electronic City Card in Jaworzno, Electronic City Card in Częstochowa).

The problem of sustainable transport in the area of Upper Silesia Agglomeration is the subject of programme documents prepared within the framework of changes occurring in metropolitan area management. These issues were exposed in the "Strategy of Integrated Territorial Investments for the Central Subregion of Silesian Province for the years 2014-2020" [13], where sustainable mobility was distinguished among 9 strategic activities. Currently, the subregion prepared "Sustainable Mobility Plan" within the framework of ITI strategy, which should take into consideration key areas of activity for sustainable mobility, including in particular [3, 12]:

– in the field of public transport, e.g. application of attractive and flexible tariff solutions, electronic tickets and increased transport integration,
– in the field of vehicle traffic, especially dynamic systems of information for drivers are recommended.

The activities mentioned as elements of sustainable mobility plan will be implemented (not only by means of the ITI instrument) by the application of ITS, including also development of urban vehicle traffic control systems and electronic ticket systems. Apart from ITI Association (Subregion), the institutions which implement or may implement such projects include Komunikacyjny Związek Komunalny GOP (Communal Transport Association of Upper Silesia, associating 29 municipalities) and future Metropolitan Association (it is currently assumed that 24 municipalities will be included in the Association, including 10 cities of Upper Silesia Metropolitan Union). The organizational complexity of metropolitan management in the area of Upper Silesia Agglomeration will definitely increase the investment costs and risk in the field of ITS (shows in Table 2).

Until now, urban traffic control system in the Upper Silesia Agglomeration on a larger scale has only been implemented in Gliwice (for the cost of over PLN 30 million). KZK GOP has developed the concept of a system covering the Association area, whose functionalities are similar as in the case of TRISTAR system, although the costs of such project would undoubtedly be several times higher [9]. The uncertainty concerning the implementation of such project is increased due to the initiatives taken within the framework of ITI - including Gliwice and Tychy, which want to either expand (Gliwice) or create (Tychy) urban traffic control systems (at the total cost of approximately PLN 120 million). The projects related to urban traffic control are so little advanced and the number of stakeholders is so high that quick improvement of the situation concerning this system cannot be expected. It seems that such system should be designed and implemented in a completely different manner than resulting from the experiences of Tricity and Wrocław. The basis should be the approved system architecture and strategy of solution implementation in stages (per function and region).

Table 2. Result indicators SUMPs Central Subregion (2016) [12]

Strategic goal	Result indicator	Target value	Data source
1. Increase the competitiveness of sustainable transport	The number of units purchased rolling stock of public transport [szt.]	165	Operators of public transport
	The capacity of rolling stock purchased in the public transport [person]	8 250	Organizers and i operators of public transport
	Length built bicycle paths [km]	1 113	Local government units
	The length of newly built sections of the tram network [km]	23	Tramwaje Śląskie SA
	The length of the modernized sections of the tram network [km]	100	Tramwaje Śląskie SA
2. Integration of transport	Number of new integrated transport interchanges	53	Local government units
	The number of new buildings Park & Ride	53	Local government units
	Number of parking spaces in built facilities Park & Ride [pcs.]	3 225	Local government units

With reference to ŚKUP, KZK GOP is the unit which coordinates implementation and manages the system, therefore, the situation looks much better from the organizational perspective than in the case of urban traffic management system. The system was launched with two-year delay. It had been prepared since 2008 (design) and expected to be put into use in 2013, however, the system launch did not begin until the end of 2015. Within two months, only 20,000 cards out of the expected number of 700,000 cards were issued. It appears that the system has too many functionalities, which were expected to be ready and provided "on a turnkey basis" [4], and such ambitious assumptions increased the financial risk, as well as the risk related to the deadline for completing the project. This will probably also be the reason of errors in the operation of the system, which is public in its essence.

6   Conclusion

The regulatory changes, including especially changes related to implementation of ITI and formation of metropolitan associations, provide the opportunity to accelerate activities related to sustainable transport and mobility in agglomerations in Poland. The efficiency of ITS means that these systems are increasingly used for providing sustainable transport. In metropolitan areas, a significant problem is the integration of operations of many public administration units. ITS, and particularly urban vehicle traffic control systems and systems applied mainly in public transport of urban electronic cards are financed from public funds, especially EU funds. Therefore, the problem of cooperation between public administration and other stakeholders of metropolization processes is becoming a factor of success and risk related to implementation of large projects.

In the area of Upper Silesia Agglomeration, urban vehicle traffic control management is used to a limited extent, whereas e-ticketing in public transport, common in Poland, is implemented in the form of the largest domestic electronic card system, which has the most expanded functionality. The specific spatial and organizational character of the existing metropolitan area means that ITS system architecture should take into consideration the barriers of integration. Moreover, the system of ITS implementation has to be different than in other agglomerations - the functionality and scope of systems should be implemented in stages. This concerns not only the presented examples, but also other e-systems, which should function in the territory of a metropolis striving towards the concept of smart city.

References

1. Benson, B.: Implementing intelligent transportation systems. In: Button, K., Henser, D. (eds.) Handbook of Transport Strategy, Policy and Institutions, pp. 651–663. Elsevier, Oxford (2005)
2. Borkowski, P.: Metody obiektywizacji oceny ryzyka w inwestycjach infrastrukturalnych w transporcie. Uniwersytet Gdański, Gdańsk (2013)
3. Duportail, V., Meerschaert,V.: Final ADVANCE Audit Scheme and Guidelines (2013). http://eu-advance.eu/docs/file/d2_5_final_advance_audit_scheme_including_guidelines_en.pdf. Accessed 10 Dec 2015
4. Dydkowski, G.: Elektroniczne karty płatnicze a ŚKUP. Komunikacja Publiczna 3(36), 33–38 (2009)
5. Dydkowski, G.: Transformations in the ticket distribution network for public urban transport in the processes of implementation of electronic fare collection systems. In: Mikulski, J. (ed.) TST 2015. CCIS, vol. 531, pp. 198–209. Springer, Heidelberg (2015). doi:10.10 07/978-3-319-24577-5_20
6. Garret, A.: Intelligent transport systems – potential benefits and immediate issues. Road Transp. Res. 7, 61–69 (1998)
7. Grzelak, G., Jałowiecki, B., Smętkowski, M.: Obszary metropolitalne w Polsce: problemy rozowjowe i delimitacja. In: Raporty i Analizy EUROREG, Warszawa (2009). http://www.euroreg.uw.edu.pl/dane/web_euroreg_publications_files/602/obszary_metropolitalne_w_polsce_problemy_rozwojowe_i_delimitacja.pdf. Accessed 10 Dec 2015

8. Jamroz, K., Oskarbski, J.: TRISTAR- trójmiejski inteligentny system transport aglomeracyjnego. In: Transport Miejski i Regionalny, vol. 07–08, pp. 82–88 (2006)
9. Koncepcja i architektura Inteligentnego Systemu Zarządzania Ruchem na obszarze działania Komunikacyjnego Związku Komunalnego Górnośląskiego Okręgu Przemysłowego. Śląska Sieć Metropolitalna, Katowice (2015). http://bip.kzkgop.pl/pdf/uchwaly_zarzadu/2015/82_2015_zal1.pdf. Accessed 10 Dec 2015
10. Maping smart cities in the EU. European Parliament, Brussels (2014). http://www.smartcities.at/assets/Publikationen/Weitere-Publikationen-zum-Thema/mappingsmart cities.pdf. Accessed 10 Dec 2015
11. Oskarbski, J., Zawisza, M., Miszewski, M.: Information system for drivers within the integrated traffic management system - TRISTAR. In: Mikulski, J. (ed.) TST 2015. CCIS, vol. 531, pp. 131–140. Springer, Heidelberg (2015). doi:10.1007/978-3-319-24577-5_13
12. Plan zrównoważonej mobilności miejskiej Subregionu Centralnego Województwa Śląskiego. Uniwersytet Ekonomiczny w Katowicach, Katowice (2016). http://www.subregioncentral ny.pl/materialy/_upload/Adam/PZMM/PZMM_v.pdf. Accessed 10 Dec 2015
13. Strategia integrowanych Inwestycji Terytorialnych Subregionu Centralnego Województwa Śląskiego na lata 2014–2020. Związek Gmin i Powiatów Subregionu Centralnego Województwa Śląskiego, Gliwice (2015). http://www.subregioncentralny.pl/strategia-ITI.html. Accessed 10 Dec 2015
14. Urbanek, A.: Pricing policy after the implementation of electronic ticketing technology in public urban transport: an exploratory study in Poland. In: Mikulski, J. (ed.) TST 2015. CCIS, vol. 531, pp. 322–332. Springer, Heidelberg (2015). doi:10.1007/978-3-319-24577-5_32
15. Zawieska, J.: Smart cities – koncepcja i trendy rozwoju miast przyszłości. In: Gajewski, J., Paprocki, W., Pieregud, J. (eds.) Megatrendy i ich wpływ na rozwój sektorów infrastrukturalnych, pp. 26–55. IBNGR, Gdańsk (2015)
16. Zintegrowane Inwestycje Terytorialne. Nowe rozwiązania dla miast w polityce spójności na lata 2014–2020. Ministerstwo Infrastruktury i Rozwoju, Warszawa (2015). https://www.mr.gov.pl/media/9915/Broszura_PL.pdf. Accessed 10 Dec 2015

Geometry-Based Matching of Routes Between Dissimilar Maps

Rüdiger Ebendt[✉] and Louis Calvin Touko Tcheumadjeu

German Aerospace Center (DLR), Institute of Transportation Systems,
Rutherfordstr. 2, 12489 Berlin, Germany
{Ruediger.Ebendt,Louis.ToukoTcheumadjeu}@dlr.de

Abstract. Protocols for dynamic location referencing like e.g. OpenLR and AGORA-C tackle the problem of accurately matching locations between dissimilar digital maps. While this is done with the aim of limiting the amount of descriptive data to reduce bandwidth, in some projects bandwidth is not a problem (e.g. in the EC-funded project ROSATTE, and in the DLR projects MobiLind and KeepMoving), and without constraints regarding bandwidth it might be possible to learn from methods in similar areas like road network matching and map conflation to achieve a more reliable solution. Following this path, this paper presents an approach called Geometry InterMapMatching Extension (GIMME) which has been developed in the ongoing DLR project I.MoVe, and is currently capable of mapping short routes from a TeleAtlas map to a NAVTEQ map on-the-fly, with a success rate of 99.7%, and when compared to OpenLR, with up to 28.7% more correctly decoded routes.

Keywords: Dynamic location referencing · Road network matching · Map conflation · OpenLR

1 Introduction

There is a strong interest in protocols for dynamic location referencing like e.g. OpenLR and AGORA-C addressing the problem of accurately matching locations between dissimilar digital maps. The reason is that "conventional geo-referencing methods using coordinates or a pre-defined set of identifiers have structural limitations; they may fail to match the same location in different maps due to discrepancies with respect to level of detail, spatial and temporal accuracy, and semantic dissimilarities" [9]. For example, location references based on road names and house numbers could fail due to misspellings and notation differences, see e.g. [12]. Location references solely based on geographical coordinates (e.g., WGS84 coordinates) also have severe problems. Often there are significant offsets between different maps or topological differences due to varying accuracy of the digitalization from analogue map data, different manufacturing methods, and changes of the road network over time [12]. Therefore, a geographical coordinate located on a road in one map might be situated off-road in another. Moreover, different maps may contain roads which only exist in one of the maps (this holds even for different releases of maps from the same vendor). Finally, for structural and practical

© Springer International Publishing AG 2016
J. Mikulski (Ed.): TST 2016, CCIS 640, pp. 354–365, 2016.
DOI: 10.1007/978-3-319-49646-7_30

reasons, the use of the well-established RDS-TMC service is restricted to European major roads, such as the motorway network, the national main roads, and only some important parts of urban roads. This means that large portions of the road network are left uncovered, whereas every location in a map (including locations covering minor roads) can be transferred using a dynamic location referencing protocol like OpenLR or AGORA-C: such protocols describe how to match locations between dissimilar digital maps on-the-fly [9, 12].

On the other hand, all approaches to dynamic location referencing have the disadvantage of possibly failing to properly encode and decode a location in some cases. As a generally agreed industry goal, a dynamic location referencing method should perform at a hit rate of 95% and message size should be below 50 bytes on average [19].

OpenLR is an open standard for dynamic location referencing, i.e. for encoding, transmitting, and decoding location references in digital maps [13]. It was launched by TomTom International B.V. and developed for the use case of transferring traffic information from a centre to in-vehicle systems, built-in or used as an add-on (PND, Smart Phone). TomTom has provided test results, e.g. using a source map from Tele-Atlas and a destination map from NAVTEQ: success rates of 93% were achieved both for TMC paths and non-TMC paths [14]. Here, "success rate" is the percentage obtained by only counting the correctly decoded locations, i.e. those where encoder and decoder location were equal, and dividing this number by the number of all locations that could be decoded (for a discussion of the difference to the definition of "hit rate", see Sect. 4.1). In the AGORA project an attempt was made to combine three dynamic location referencing methods, namely an extension of the ILOC approach [2, 4], the Pivot Point approach [7], and the GOODLANE approach [6]. The resulting hit rate was good enough to meet the stated industry goal in terms of a 95% hit rate [8, 12], but the message size was unacceptably high [18, 19]. Therefore in the successor AGORA-C (the "C" stands for "compact", akin to a reduced message size, and also hints for "ALERT-C", the official name for the TMC standard [19]), complexity was reduced by focusing on two approaches only, the ILOC extension and the Pivot Point approach. The AGORA-C method is reported to have achieved the industry goal of an acceptably small message size and a sufficiently high hit rate [18, 19].

In contrast to other approaches to dynamic location referencing like AGORA-C which have licensing fees meaning extra cost, e.g. [15], OpenLR is a royalty-free, open standard under a creative commons license (CC BY-ND 3.0 [3]), and therefore fosters interoperability and promotes free choice between different vendors and technology solutions [5]. Moreover, OpenLR is published as an open-source framework, including a reference implementation of the proposed standard (in contrast, no reference implementation is publicly available for any of the aforementioned approaches).

Summarized, current methods for dynamic location methods either imply the extra cost of licensing fees, or do not fully meet the generally agreed industry goal of a 95% hit rate yet, or suffer from both these drawbacks. The starting point for the present paper then is the observation that the second part of the aforementioned industry goal, i.e. a message size below 50 bytes on average, is not relevant for a number of applications for which bandwidth is not a problem: e.g. in the EC-funded project ROSATTE, static data is sent from a public authority to a map provider, and in two DLR projects, MobiLind and the ongoing project I.MoVe, data from a traffic information provider is transformed from one well-known map to another, and often this transformation takes place on the same server. Without constraints regarding bandwidth it might be possible to learn from methods in similar areas like road network matching and map conflation to achieve a more reliable solution.

This paper presents an approach called "Geometry InterMapMatching Extension" (GIMME) which follows this path and has been developed in the ongoing DLR project I.MoVe. It is currently capable of mapping short routes from a TeleAtlas map to a NAVTEQ map on-the-fly, with a success rate of 99.7%. Moreover, when compared to OpenLR, up to 28.7% more routes have been decoded correctly.

The paper is structured as follows: in Sect. 2, necessary background from road network matching, map conflation, and a geometry-based approach to matching of single edges between maps are given. The latter approach is extended to a new algorithm matching complete routes in Sect. 3. Next, in Sect. 4, the accuracy of the new algorithm is compared to that of OpenLR. Finally, the work is concluded in Sect. 5.

2   Background

To keep the paper self-contained, necessary background from the three most-cited works in road network matching, usually in the context of combining two distinct maps into one new map (map conflation), and from a previous geometry-based approach to match single edges between two maps is given briefly.

2.1   Iterative Closest Point

Iterative Closest Point (ICP) was formulated in [1] as a generic algorithm to match two clouds of points by minimizing the difference between them.

Subsequently, it has been adopted in the field of road network matching [20]: first all nodes of the two networks are extracted and then correspondences of nodes between the two resulting datasets are established by combining initial thresholds for similarity measures like the Euclidian distance, the number of incident edges (i.e. the valence) and the angle differences between emanating edges. Next, if line segments are enclosed between two nodes of one dataset, they are matched to those enclosed by the corresponding nodes. The rationale is that if two roads from different datasets have corresponding starting and ending nodes, then there is a high likelihood that the roads themselves are corresponding counterparts [10]. This works well when matching whole roads since this only requires establishing the node correspondences between intersections. However, finding the respective correspondences for the remaining starting and ending nodes in two digital road maps from different vendors is more difficult. The reason is that the geometry of road segments (e.g., length and the number of segments per road) varies strongly among different maps. Therefore, the algorithm allows to re-compute the initial thresholds based on data of the first computed node correspondences, and to iterate the approach with stepwise relaxed constraints. Nonetheless, this algorithm is designed for the larger data sets involved when matching large portions of two road maps and therefore seems to be more appropriate for map conflation than for matching (short) routes.

2.2 Buffer Growing

Buffer Growing (BG) [16] is an efficient algorithm for the general task of line matching [10]. BG accounts for the aforementioned problem of possible strong differences between corresponding line segments with respect to their starting and ending points and their lengths by introducing the concept of a growing buffer. At start, 1:1 vs. 1: n correspondences between line segments are considered: a spatial buffer around the one source segment is defined, and for a valid correspondence all n target segments must be completely confined to this buffer. If no fitting target segment can be found in the current buffer, it is stepwise expanded until a respective number of target segments can be found [17]. This process also facilitates computing m : 1 or m:n correspondences (i.e., matching of complete routes): for this purpose, new logical integrities are built from previously found correspondences, which are then subject to new applications of BG in subsequent steps.

After the Buffer Growing process, a list of potential candidates for the matching reference is computed. The list may be ambiguous and typically contains a large number of matching candidates. By computing the geometric and topologic similarity between each matched pair, the best matching candidate can be confirmed as the final solution [22].

2.3 Delimited-Stroke-Oriented Approach

The basic idea of the Delimited-Stroke-Oriented (DSO) approach is to exploit contextual information as much as possible. For this purpose, a kind of pre-processing step identifies potential "[. . .] fundamental elements at more abstracted levels", i.e. a "series of conjoint road objects [chained together and acting] as the fundamental element in the matching process" [21]. For this purpose, the Delimited Strokes are progressively constructed at three different levels: on the first level, a Delimited Stroke represents a series of connected segments which have "good continuity" to each other (which means that one segment follows the other in almost the same direction), and are delimited by "efficient terminating nodes" (which are either prominent crossings with at least 4 incident nodes or dead-ends), whereas on the second level it is an arbitrary series (i.e., sharp turns are now allowed) which is delimited by arbitrary crossings or dead-ends. Only on the third level single road segments (edges) are considered.

Experimental results showed that the outlined contextual approach increases the accuracy of road network matching significantly [21]. Of course, this algorithm is designed for matching a whole map to another (i.e., for map conflation), and requires a significant amount of pre-processing. It is left to show that there can also be benefits for the on-the-fly matching of single routes between dissimilar maps (i.e., for the more focused problem addressed in this paper).

2.4 Geometry Matching

In [11], digital road networks have been benchmarked with respect to their fitness for route finding and traffic simulation. For this purpose, a geometry-based approach was used to match single edges (not complete routes) in a source map to one or more edges in a destination map (thereby establishing $1 : n$, but not $m:n$ relationships). In the following, the basic ideas of this algorithm called Geometry Matching (GM) are given briefly.

A geometry-based similarity measure is given which is mainly based on three criteria: (I) average distance, (II) angular difference, and (III) length of the projection of the candidate edge to the source edge, respectively. First, a proximity search is done in the destination map within a certain radius around the position of the start node of the source edge, and the similarity measure is computed for the found edges. All edges with a similarity measure above a certain threshold are included in the candidate list.

To determine a $1:n$ relationship for a source edge, the first best edge is determined as the candidate with the highest measure, and the result list is initialized as the singleton with this best edge. Then, a forwards and a subsequent backwards search are performed in an iterative manner, using the topology of the destination map:

Firstly, the candidate list is re-initialized with those outgoing edges of the current best edge, for which the similarity measure is good enough, and ordered with respect to the measure. Then, a new best edge is calculated and appended to the result list, and so on until an empty candidate list is encountered. Secondly, an analogous process starts with the incoming edges of the current best edge until termination. In the end, the result list is either empty if no candidates could be found, or is a sequence of one or more connected edges forming a stretch of road in the destination map corresponding to the source edge.

Summarized, the GM method resembles the idea of the BG algorithm (see Sect. 2.2), although no building of new logical integrities from previously found correspondences takes place here, and consequently no $m:n$ correspondences, i.e. no matches for complete routes are constructed. It is worth mentioning that GM uses both geometric and topological attributes of the edges. The use of a threshold for the average distance between source and candidate edges is an interesting novelty (see Sect. 4.1 for the details), since previous methods were using measures on top of the Hausdorff or the Fréchet distance.

3 Geometry Inter-Map-Matching Extension (GIMME)

3.1 Algorithm

The proposed new approach called Geometry InterMapMatching Extension (GIMME) builds upon the results of the GM algorithm (see Sect. 2.4): GM computes an ordered list of candidate edges (or "candidate list" for short) for every edge of the source route (or "source edge" for short), and returns an associative array relating every source edge with a corresponding candidate list. The order of each such candidate list is with respect to the similarity to the source edge, see Sect. 3.2. The aforementioned array is essentially a "pool" of eligible candidate edges, and the objective of GIMME is to construct a best match for the whole route by drawing (zero, one, or more than one) candidate edge(s) from the pool for every source edge, and by finding the best possible combination of them. The following key observations lay the ground for the new algorithm:

1. Firstly, in the context of this paper, source and destination routes must be finite directed paths, i.e. finite sequences of edges which connect a sequence of vertices which are all distinct from one another. Notice that cycles have been excluded here.

2. Secondly, there must be a non-empty candidate list for every source edge: if this is not the case, then there is at least one source edge which cannot be matched to any edge in the destination map, and consequently also the whole route cannot be matched.

3. There are m:n relations between source and destination edges. This means that for a single source edge 1:n relations must be established, where $n \geq 0$. For an example of the case $n = 0$, consider e.g. a 3:2 relation between source and destination edges: if all source edges are mapped to at least one destination edge, then there must be one destination edge occurring twice in the resulting destination route, in contradiction to the assumption that routes are valid if and only if they are finite directed paths. Notice that the fact that a source edge might not become mapped to any destination edge does not contradict Observation 2: also in this case eligible candidates for the respective edge do exist, but all these candidates are "consumed" for other (e.g. adjacent) source edges.

4. Another crucial point is that the order of a candidate list has nothing to do with that of the edge sequence of a directed path, because the candidates are ordered with respect to similarity to the source edge. Therefore, to build the final destination route by combining partial sequences drawn from the pool for every individual source edge, it is necessary to consider all permutations of subsets of the candidate list (including the empty set because of the case $n = 0$ in Observation 3). Since this seems to introduce a very complex step to the algorithm, the following point is needed to ensure efficiency of the approach:

5. Since the final route must be a (finite) directed path, it is not necessary to consider partial sequences for which it is already known that this property does not hold.

6. Observation 5 gives rise to the following *recursive* formulation:

    (a) The final destination route is constructed by trying to connect partial routes (i.e., partial solutions) for the head and tail of the list of source edges. Whenever a partial solution for the head can be connected topologically to one for the tail, then we have a match for the whole source route, and this solution is stored as a potential solution.

    (b) The partial solutions are computed by invoking the outlined algorithm recursively on the tail. Every recursive call returns a list of all (partial) solutions found (where a solution must respect Observation 5, or it is abandoned), and the calling code then checks each combination of the solutions for head and tail for topological connectedness. The boundary case of the recursion is reached when the tails have shrunk to an empty list (for which an empty list of solutions is returned).

    (c) Finally, from the list of solutions for the whole route returned at the topmost calling level, the *best* solution must be chosen. In the current version of GIMME, the *best* solution is defined as the longest *admissible* solution (for an explanation of the term *admissible* see Observation 7). In the rare case that there is more than one longest admissible solution, one is picked randomly.

7. A solution is *admissible*, if and only if (a) it respects Observation 2, i.e. for every source edge at least one corresponding candidate edge is used in it (but not necessarily for that respective source edge, see the case $n = 0$ in Observation 3), and (b) the length of the solution does neither exceed 110% nor fall below 90% of the length of the source route matched by this respective solution.

Notice that a solution is already a (finite) directed path by construction; hence this is not mentioned again in Observation 7. This also holds for partial solutions and that way partial solutions are excluded early if they cannot be completed to a solution for the whole route, and never must be considered again at a shallower calling level of the recursion. Therefore, the list of partial solutions returned by a recursive call is typically not very long.

Moreover, for real-world urban road networks also the candidate lists are usually rather short: in our experiments (see Sect. 4.2) they rarely exceeded a maximum length $C = 3$. GIMME has to consider a maximum of $\sum_{k=0}^{C} \frac{C!}{k!}$ permutations of subsets of a candidate list at a recursive call. Then it checks every combination of a solution for the head for topological connectedness with a returned solution for the tail (the number of which remains small as already stated by the previous observation) in quadratic run time. Since $\sum_{k=0}^{C} \frac{C!}{k!} = 16$ in most practical cases, the run time of GIMME usually stays within feasible limits for routes up to 15 segments, and during the experiments for short routes of 1 to 5 segments (see Sect. 4.2) it was not larger than that of the OpenLR encoder/decoder.

3.2 Similarity Threshold

While GM [11] computes a similarity measure and then orders the candidates in the candidate list with respect to this measure, GIMME currently does not use this order (see Sect. 3.1). Nonetheless, GIMME bases the decision whether to include a potential candidate into the candidate list or not on the same thresholds as GM, which are briefly reviewed in the following.

Figure 1(a) exemplifies a pair of a source edge and a corresponding candidate edge. Figure 1(b) shows the angle α between the two edges, and their mutual projection. The length of the projection of the candidate to the source (i.e. reference) edge is called the reference length, shown in Fig. 1(c) together with the intersection of the areas confined to the edges and the perpendiculars of their mutual projection, respectively. The average distance is defined as the ratio A/l of this area A and the reference length l.

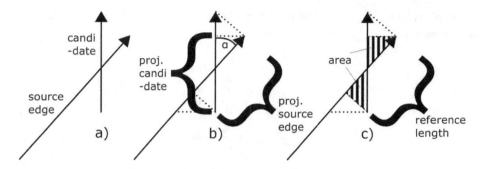

Fig. 1. Aspects of the geometry-based similarity thresholds [own study]

The used thresholds are:

- Maximum average distance: 20 m
- Maximum angle α: $40° \triangleq 0, 698$ rad
- Minimum projection length: 3 m

4 Evaluation

This section describes the results of an evaluation of the proposed algorithm GIMME. Firstly, evaluation measures are defined which correspond to well-known measures in e.g. binary classification in Machine Learning. Secondly, the results of experiments are given in terms of the introduced evaluation measures. For the experiments, the proposed algorithm and OpenLR have been used to match short routes consisting of up to 5 road segments between a TeleAtlas and a NAVTEQ map of the city of Potsdam, Germany.

4.1 Evaluation Measures

Let n be the total number of locations as sent in a test. For the scope of this test, let n_p be the number of positives, i.e. the number of matched routes, and let n_n be the number of negatives, i.e. the number of routes which could not be matched. Let n_{tp} denote the number of true positives, i.e. the number of correctly matched routes where source and destination route were equal. Let n_{fp} denote the number of false positives, i.e. the number of matched routes where source and destination route were not equal. Further, let n_{tn} be the number of true negatives, i.e. the number of routes truthfully identified not to be present in the receiver map, and let n_{fn} be the number of false negatives, i.e. the number of routes which could not be matched although they are present in the destination map. For a real-world evaluation, these numbers are subject to the following constraints:

$$n = n_p + n_n > 0$$
$$n_p \geq n_{tp} \geq 0$$
$$n_n \geq n_{tn} \geq 0$$

Further, let $n_t = n_{tp} + n_{tn}$. The percentage hit rate q_{hit}, success rate $q_{success}$, and error detection rate $q_{error\_detection}$ are defined by

$$q_{hit} = \frac{n_t}{n} \cdot 100\%$$

$$q_{success} = \frac{n_{tp}}{n_p} \cdot 100\%$$

$$q_{error\_detection} = \frac{n_{tn}}{n_n} \cdot 100\%$$

respectively, where $q_{hit} = 100\%$ if $n = 0$ and q_{hit} is undefined for $n < 0$, and analogously for $q_{success}$ and $q_{error\_detection}$.

Notice that the hit rate, success rate, and the error detection rate correspond to the well-known evaluation measures "accuracy", "precision", and "negative predictive value" for a confusion matrix in e.g. Machine Learning, respectively. The hit rate is combining the two aspects of good performance of a dynamic location referencing method, namely correct location decoding, and truthful absence detection. This however introduces a subtle dependency on the particular map, e.g. on the ratio of true positives to true negatives as found in it. Giving both success and error detection rates separately has also the advantage of providing more detailed information about the two different aspects of quality.

4.2 Experimental Results

For the evaluation, 1,000 short routes in a TeleAtlas map of Potsdam, Germany (TA 2012.06), consisting of 1 to 5 road segments have been created randomly. This set of test routes has been used in all experiments. For the first experiment, each route in the set was encoded with OpenLR 1.4.2 in the TeleAtlas map, and then decoded in a NAVTEQ map of Potsdam (NT 2012.Q2). The observed success and error detection rates were 91.9%, and 55.9%, respectively (for the definition of the two rates see Sect. 4.1). Both rates have been determined by manual inspection of the respective source and (potential) destination routes using TomTom's MapViewer tool (see Fig. 2).

Next, GIMME has been applied to the 149 routes which OpenLR had not been able to decode in the destination map (i.e., a hybrid approach of first using OpenLR and then GIMME has been applied). For this purpose, the routes were described by sequences of edge IDs in the respective native source and destination map format, and GIMME was given full access to both the source and destination map. In contrast, OpenLR as a map-agnostic dynamic location referencing protocol has only access to the source map at the sender side, and to the destination map on the receiver side. Moreover, in order to limit the amount of descriptive data, OpenLR only transfers intermediate location referencing points as map-agnostic WGS84 positions. As a result, GIMME decoded 149 more routes and all of them correctly (which has been validated by manual inspection). This is an increase in successfully decoded routes of 28.7%. As a consequence, the success and the error detection rate of the hybrid approach was 94.2% and 65.0%, respectively.

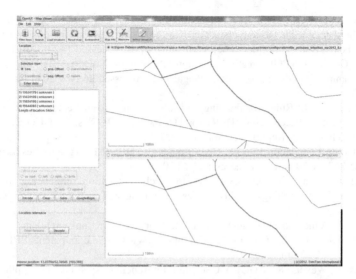

Fig. 2. Comparison of matched routes in a TA map (top) and an NT map (bottom) of Potsdam, Germany [own study]

Finally, GIMME was used as a standalone algorithm to find matches in the NAVTEQ map for the same set of 1,000 routes in the TeleAtlas map. This resulted in a success rate of 99.7%, and in an error detection rate of 69.0% (as determined by manual inspection). Compared to the previous run with OpenLR, 60 or 11.6% more routes could be decoded correctly by GIMME.

5 Conclusion

A new approach to match short routes between dissimilar maps has been presented. In contrast to known protocols for dynamic location referencing, it assumes that constraints regarding the bandwidth for the transfer of the descriptive data can be disregarded. E.g. this is the case whenever an omniscient matching centre with access to both the source and the destination map can be used.

The new approach exploits the availability of a more detailed geometrical description of the routes. It advances on the path of previous methods in the area of road network matching and map conflation like e.g. Buffer Growing and Geometry Matching. Experiments are given which clearly demonstrate the increased correctness and completeness of the proposed method.

References

1. Besl, P.J., McKay, N.D.: A method for registration of 3-D shapes. IEEE Trans. Pattern Anal. Mach. Intell. **14**(2), 239–256 (1992). IEEE Computer Society, Los Alamitos, CA, USA

2. Bofinger, J.M.: Analyse und Implementierung eines Verfahrens zur Referenzierung geographischer Objekte, Diplomarbeit, German (2001). http://www.ifp.uni-stuttgart.de/lehre/diplomarbeiten/Bofinger/Diplomarbeit%20Bofinger.pdf. Accessed 14 Dec 2015
3. Creative Commons (CC): creative commons Attribution-NoDerivs 3.0 Unported. http://creativecommons.org/licenses/by-nd/3.0/legalcode. Accessed 26 Nov 2014 [date of access: 14.12.2015]
4. Duckeck, R., et al.: Rules for defining and referencing an Intersection Location (ILOC); Detailed Location Referencing (DLR) for ITS based on ILOCs. ERTICO Committee on Location Referencing, Report, Version 1.0 (1997)
5. Free Software Foundation Europe: Open Standards. http://fsfe.org/activities/os/os.en.html. Accessed 26 Nov 2014 [date of access: 14.12.2015]
6. Hahlweg, C., et al.: GOODLANE - an approach to location referencing for telematic applications. In: Proceedings World Congress on Intelligent Transport Systems 2000, Torino, Italy (2000)
7. Hendriks, A.: A method and system for referencing locations in transport telematics. eP Patent App. EP20,010,101,123 (24 July 2002). http://www.google.com/patents/EP1225552A1?cl=en. Accessed 24 June 2002
8. Hendriks, T., Wevers, K.: AGORA-C location referencing - specification, applicability and testing results. In: Proceedings World Congress on Intelligent Transport Systems, Nagoya, Japan (2004)
9. Hiestermann, V.: Map-independent location matching certified by the AGORA-C standard. Trans. Res. C Emerg. Technol. **16**, 307–319 (2008)
10. Meng, Z.: Methods and Implementations of Road-Network Matching, Dissertation. Leibniz Universität Hannover, Hannover, Germany (2009)
11. Sämann R.: Bestimmung einer Bewertungsmetrik zum Vergleich digitaler Straßennetze für Verkehrsflusssimulation und Routing von Einsatzkräften, Master Thesis, German, Institut für Bauinformatik, Leibniz Universität Hannover in cooperation with Deutsches Zentrum für Luft- und Raumfahrt e.V. (DLR) (2014)
12. Schneebauer, C., Wartenberg, M.: On-the-fly location referencing—methods for establishing traffic information services. IEEE Aerosp. Electron. Syst. Mag. **22**(2), 14–21 (2007)
13. TomTom International B.V.: OpenLR website (2009–2016). http://www.openlr.org. Accessed 14 Nov 2014
14. TomTom International B.V, slides 42–43. http://openlr.org/data/docs/OpenLR-Introduction.pdf. Accessed 14 Nov 2014
15. Via Licensing Corp.: AGORA-C patent submission (2012). http://www.vialicensing.com/licensing/agorac-patentcall.aspx. Accessed 14 Nov 2014
16. Walter, V.: Zuordnung von raumbezogenen Daten - am Beispiel der Datenmodelle ATKIS und GDF, Dissertation, Deutsche Geodätische Kommission (DGK) Reihe C, Nummer 480 (1997)
17. Walter, V., Fritsch, D.: Matching spatial data sets: a statistical approach. Int. J. Geograph. Inf. Sci. **13**(5), 445–473 (1999)
18. Wevers, K., Hendriks, T.: AGORA-C on-the-fly location referencing. In: Proceedings World Congress on Intelligent Transport Systems 2005, San Francisco, California, USA (2005)
19. Wevers, K., Hendriks, T.: AGORA-C map-based location referencing. J. Transp. Res. Board **1972**(1), 115–122 (2006)
20. Volz, S.: An iterative approach for matching multiple representations of street data. In: Hampe, M., Sester, M., Harrie, L. (eds.) ISPRS, ISPRS Workshop - Multiple Representation and Interoperability of Spatial Data, Hannover, Germany, vol. XXXVI (2006)

21. Zhang, M.: Methods and Implementations of Road-Network Matching. Ph.D. thesis. Leibniz Universität Hannover, Hannover, Germany (2009)
22. Zhang, M., Meng, L.: An iterative road-matching approach for the integration of postal data. Comput. Environ. Urban Syst. **31**(5), 598–616 (2007)

Aircraft Taxi Route Choice in Case
of Conflict Points Existence

Przemysław Podgórski and Jacek Skorupski[✉]

Faculty of Transport, Warsaw University of Technology, Warsaw, Poland
przemek.podgr@gmail.com, jsk@wt.pw.edu.pl

Abstract. Aerodrome ground traffic management consists of many processes. Their complexity causes that telematic solutions intended to assist both the air traffic controller who coordinates taxiing and the pilot who executes them are introduced. An example of such a solution is A-SMGCS system. At the moment, intensive research on advanced system functions are carried out. They include smart taxi route choice taking into account the current traffic situation. The paper presents the model and the computer software tool implementing the dynamic taxi route choice module. It can be used in case of congestion in the aerodrome traffic described by the so-called conflict points. Presented taxi route choice system is integrated with the developed before system to identify conflict points. This allows intelligent traffic process management through the selection of an alternate taxi route. Such a solution could be implemented on the third level of A-SMGCS system.

Keywords: Aerodrome traffic management · A-SMGCS system · Aircraft taxi route choice

1 Introduction

Aircraft taxing before takeoff and after landing are standard procedures in aerodrome traffic. Their goal is to displace an aircraft between the runway in use and the designated parking place. In opposite to air traffic in airspace, taxing may seem to be quite simple. It takes place only in two dimensions and is much less dynamic at the same time. Unfortunately, this process is not so easy to organise and to supervise. At big airports a network of taxiways is usually complicated, many aircraft are taxing at the same time and their taxi routes intersect in many places. This is favourable for pilots mistakes. At the same time, the same factors, as well as air traffic services work technology based on visual observation of airfield, favour mistakes committed by an air traffic controller. All of that causes, that many incidents occur in aerodrome traffic, including serious incidents or even accidents with disastrous consequences [14, 15]. Common issues include: collisions with other aircraft, collisions with fixed elements of airport equipment, collisions with other motor vehicles, such as baggage trolleys, snow removal equipment, etc. The most serious incidents take place when it comes to accidental take-off from a taxiway or unauthorized runway crossing (Runway Incursion), or even taxiing on the runway. In these cases an accident with a large number of fatalities may even occur.

© Springer International Publishing AG 2016
J. Mikulski (Ed.): TST 2016, CCIS 640, pp. 366–377, 2016.
DOI: 10.1007/978-3-319-49646-7_31

This paper is a continuation of [1], which presented the method of determination of conflicts points in taxiways structure. Conflict points characterize areas of traffic congestion and more precisely the points where delays appear. Waiting to pass the conflict point takes place in sections adjacent to it. In case when these sections are short or many aircraft are waiting, this may cause creation of new conflict points. The method presented in the previous paper is based on a model developed by using Petri nets. In this paper we present the next step aiming at building the module of intelligent system which will support taxiing process. It allows finding an alternative taxi route, taking into account dynamic changes of time necessary for passing a conflict point. Therefore, algorithm for finding a dendrite of shortest routes was used [4]. Subsequently, it was computer implemented in CPN Tools environment as coloured, time Petri net.

2 Systems Supporting Aerodrome Traffic Control

Aerodrome traffic is supervised by air traffic control services, namely by the so-called ground controllers (GND). They must constantly observe traffic situation, not only aircraft but also movement of all technical service vehicles and people staying at the airfield. Ground controllers can use various supporting systems, especially in busier airports. One of the most advanced supporting system is A-SMGCS (Advanced Surface Movement Guidance and Control System). It provides control and guidance in manoeuvring area. It allows to improve safety of aerodrome operations, especially in low visibility conditions [5, 16]. An A-SMGCS should support following primary functions [2]: surveillance, routing, guidance and control.

Surveillance function provides pilots and controllers information about actual traffic situation. Each aircraft and ground service vehicle is identified and marked. Data about objects' positions are constantly updated for guidance and control requests.

Routing function is responsible for determining taxi route for every aircraft and vehicle in manoeuvring area. This function is crucial for aerodrome traffic safety and efficiency, because it helps to optimize routes to prevent conflict situations. Moreover, aircraft or vehicle moving correctly on planned route is easier to follow. Furthermore, routing function should allow to change designated route in case of occurrence of movement obstruction or as a result of changing the destination point. This may happen, for instance, when parking place assignment changes dynamically according to operational reasons. It is expected that routing should be implemented in an intelligent way, which means that system should anticipate difficulties caused by air traffic congestion. An algorithm presented in this paper, together with its implementation, will be an element of the module responsible for A-SMGCS system routing function. It will allow designation of the fastest taxi route for aircraft. Data continuously provided to this module will let it to determine alternative taxi routes and also dynamically change the route of already taxiing aircraft.

Guidance function provides tips for aircraft pilots and car drivers that allow them to move on designated track. Additionally, it allows them keeping awareness about traffic situation and monitoring operational status.

Control function, most of all, detects conflicts and provides problem solution. System alarms the controller about potential conflicts on runways and in manoeuvring area of the airport.

A-SMGCS has 4 functional levels of realisation [2]. Levels of traffic situation visualisation and monitoring, and warning about movement conflict as well as intrusion to the reserved area are already defined. However, levels of planning (routing) and automatic guidance are still tested. Algorithm described in this paper can support the research efforts on planning level regarding to aircraft movement.

The crucial problem is to supply the data to the A-SMGCS system. They will allow to locate aircraft and others vehicles. The most important sources of information are:

1. SMR (Surface Movement Radar). This radar allows detection of moving objects and showing the actual traffic situation in manoeuvring area via the suitable interface. Variable information is presented on the background of permanent elements, like runways, taxiways, buildings. SMR is especially useful in low visibility conditions. Thanks to the connection with ASDE (Airport Surface Detection Equipment), it is possible to correlate information included in flight plan with the actual position of an aircraft.

2. ADS-B (Automatic Dependant Surveilance - Broadcast). It is a system supplying data about positions of aircraft through the use of transponders (Mode-S).

3. MLAT (Multilateration). It is a technology of hyperbolic positioning, used to locate the aircraft. It is independent surveillance system, based on broadcasting ground stations, remote receiving ground stations and central ground station, calculating an aircraft position. The signal sent reaches several receivers and basing on difference in arrival time, the system is able to designate three-dimensional position of an aircraft [7].

4. Data Fusion. It is not a data acquisition system but rather a module which allows integration of data from different sources.

3 Method of Finding the Conflict Points

For the purpose of this study a model of the aircraft taxi process was developed in the form of coloured, hierarchical Petri net. It is presented in details in [1]. Petri nets are a powerful formalism for describing and modelling the dynamics of concurrent systems. In this section the concept and the basic elements of the model will be presented briefly. Also its implementation with the use of the CPN Tools 4.0 package will be discussed in brief.

3.1 Petri Net Model for Determining Conflict Points

The first phase of the model creation is the analysis of the structure of examined airport manoeuvring area. It aims at recognition of starting and ending points of aircraft movement (runways and terminal gates) and possible taxiways. The analysis of taxiways structure allows one to present them schematically in the form of a graph. The edges represent taxiways sections, and the nodes are the points of taxiways crossings. They are also the potential conflict points mentioned in this study.

Hierarchical, coloured, timed Petri net was used for mapping the dynamic taxiing processes occurring in the analyzed area. Processes were modelled in accordance with the actual rules applicable in air traffic. The Petri net was constructed using the general principles defined in [8, 9]. The following structure of the Petri net was adopted

$$S_{TX} = \{P, T, A, M_0, \tau, X, \Gamma, C, G, E, R, r_0, B\} \tag{1}$$

where:

P – set of places,

T – set of transitions $T \cap P = \emptyset$,

$A \subseteq (T \times P) \cup (P \times T)$ – set of arcs,

$M_0{:}P \to \mathbb{Z}_+ \times R$ – marking which defines the initial state of the system that is being modeled,

$\tau{:}T \times P \to \mathbb{R}_+$ – function determining the static delay that is connected with carrying out activity (event) t,

$X{:}T \times P \to \mathbb{R}_+$ – random time of carrying out an activity (event) t,

Γ – finite set of colors which correspond to the possible properties of tokens,

C – function determining what kinds of tokens can be stored in a given place: $C{:}P \to \Gamma$,

G – so-called "guard" function which determines the conditions that must be fulfilled for a given event to occur,

E – function describing so-called weights of arcs, i.e. the properties of tokens that are processed,

R – set of timestamps (also called time points) $R \subseteq \mathbb{R}$,

r_0 – initial time, $r_0 \in R$.

$B{:}T \to \mathbb{R}_+$ – function determining the priority of a given event, i.e. controlling the net's dynamics when there are several events that can occur simultaneously.

3.2 Software Tool CP-DET for Determining Taxiing Conflict Points

Calculation module CP-DET for determining conflict points was built based on the developed mathematical model. It was implemented with the use of CPN Tools 4.0 package [13], which allows to implement the hierarchical Petri nets and to divide the model by using the so-called page mechanism.

The main page of the model (*Main*) is shown in Fig. 1. It implements the general structure of the model together with the dynamics of the aircraft movement on individual taxiway sections in accordance with the rules adopted by airport traffic management. On this page there are two substitution transitions (marked as rectangles with double frame): *Permissions to use TWY* and *Permission to begin*. They model operations carried out by GND air traffic controller which is responsible for ensuring the safe taxi operation.

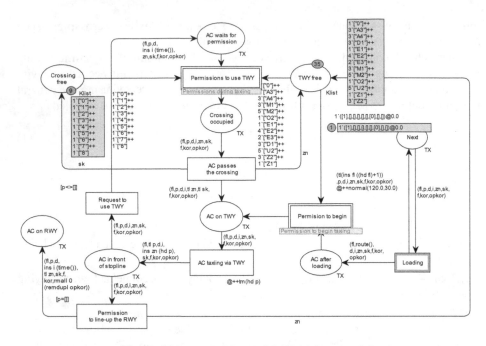

Fig. 1. Main page of the model (*Main*) [own study]

The set Γ consists of two colours in this model: $\Gamma = \{TX, Klist\}$. Tokens belonging to the *TX* colour represent taxiing aircraft and have the following structure

$$tx = 1'(\mathit{fl}, p, d, i, zn, sk, f, kor, opkor)@timestamp \tag{2}$$

where

fl – aircraft's number in the system,

p – planned taxi route (sequence of taxiway sections),

d – actual (realised) taxi route (sequence of taxiway sections),

i – moments in time when taxiing through individual taxiway sections finished,

zn – taxiway section currently occupied by the aircraft,

sk – taxiways crossing currently occupied by the aircraft,

f – total aircraft's delay [s],

kor – set of taxiway sections where the aircraft had to wait before entering,

$opkor$ – times of waiting before entering the sections from the *kor* set,

timestamp – represents the moment in time when the token becomes active.

The planned taxi route is defined by a series of taxiway sections and is generated as a result of execution of the *route*() function by the *Loading* transition. The output of this function is one of the predefined routes used at the airport.

The most crucial places endangered by a collision are taxiway intersections. Analysis of the situation and working out a decision allowing the aircraft to go through the intersection are included in the model on *Permissions during taxiing* page (Fig. 2). Due to its size, only a small portion of this page is presented in Fig. 2. It shows the analysis of only one taxiway section (designated as A3). The model assumes that only intersections endangered by a collision are sought, while during taxiing on a single taxiway section collisions do not occur.

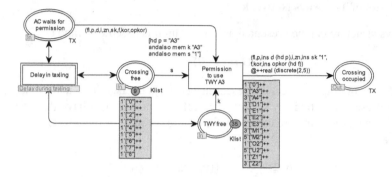

Fig. 2. Part of the page *Permissions during taxiing* [own study]

The decision allowing an aircraft to occupy the intersection is taken based on information in tokens stored in a place *Crossing free*. These tokens represent free intersections. Lack of token means that the intersection is currently occupied by an aircraft. The existence of free space on another part of the planned taxi route which permits an aircraft to easily leave the intersection is an additional condition required to occupy the intersection. Checking if this condition is met is based on the tokens contained in the *TWY free* place.

4   Supporting a Proccess of Finding a Taxi Route

Usually a taxi route between specified starting point and ending point (in case of taxing to takeoff - between parking place and runway in use) is statically predefined and used in most cases. On the one hand it is reasonable because it decreases the possibility of making mistakes in the taxi route designation or during the process of passing the taxiing instructions to the aircraft. On the other hand it is disadvantageous from a traffic management point of view, as it doesn't include actual (or expected) traffic situation.

The model, together with its implementation in form of CP-DET software presented in Sect. 3 allows designation of conflict points in the taxiways network. Conflict points may arise as result of traffic congestion, technical failures causing blockade of some taxiways sections, deterioration of weather conditions etc. The idea of the created system is to dynamically, intelligently designate the taxi route, taking into account the actual traffic situation represented by a set of conflict points. Thus, it will be possible to optimize a route with taxiing time taken as a criterion. Even better effects could be obtained if we use the forecasted traffic situation. An important element of the whole concept is a way of obtaining information about the actual taxiing time in particular sections. If we use the forecasted situation, there is a need for simulations, which include all future planned departure operations. If we use the current situation, the same information sources as for A-SMGCS system can be used. Installation of special sensors on the taxiways should be considered as well. This will make possible to track current traffic situation more efficiently and faster update all necessary taxiing times.

4.1 Model of Taxiways Network

Taxiways structure can be presented as a net, which is based on graph G

$$G = \langle V, B \rangle \tag{3}$$

where: V - set of graph vertices, representing taxiways intersections, and B – set of graph edges, representing a taxiway section. Typically, these sections have their own designations at the airport and are used to define the taxi route for the aircraft. Set B can be defined as:

$$B \subseteq V \times V = \{(u, v) : u \neq v\} \tag{4}$$

Graph G is a digraph, as taxiways have specified direction of movement. Finding the route with the shortest taxiing time requires determination of time characteristics on edges

$$t: B \rightarrow \mathbb{R}_+ \tag{5}$$

where $t(u, v)$ defines the taxiing time on the section connecting vertex u with vertex v.

In Sect. 5 an example of taxiways structure at London Heathrow airport, together with graph model of it, will be presented.

4.2 Algorithms for Finding Extremal Routes

There are a lot of algorithms for finding the shortest route in the network. These include Dijkstra's, Bellman-Ford's, Floyd-Warshall's algorithms. Because of the size of the paper, the comparative analysis of these algorithms will not be presented. As a result of this analysis the well known algorithm called maximum dendrite of shortest routes has been chosen to use [4]. In the literature also some specific algorithms suitable for tackling the ground movement problem can be found. One of the solutions is Taxi Route Planner (XRP) tool that aims at minimizing the holding time of aircraft that are maneuvering on airport taxiways, for both arriving and departing aircraft [11]. This is done in two consecutive steps, that is, a standalone, shortest path solution from runway to apron (or vice versa), neglecting the presence of other aircraft on the airport surface, followed by a conflict detection and resolution task that attempts to reduce and possibly nullify the number of conflicts generated in the first phase. Another interesting approach can be found in [6]. A sequential graph based algorithm to address the ground movement problem was introduced there. This algorithm aims to absorb as much waiting time for delay as possible at the stand (with engines off) rather than out on the taxiways (with engines running). The impact of successfully achieving this aim is to reduce the environmental pollution. Similar approach can be found in [3]. It presents Spot And Runway

Departure Advisor (SARDA) - an individual aircraft-based advisory concept for surface management. It utilizes the concept of the collaborative decision making with gate-holding as a measure to decrease aircraft fuel consumption.

4.3 Petri Net to Designate Alternative Taxi Route

Taxiway network defined by formulas (3)–(5) can be modelled as Petri net. This approach allows easy implementation of the algorithm for finding alternative taxi route, defined in Sect. 4.2. In general, the way of coloured timed Petri net creation proceeds according to the following scheme:

1. Graph G nodes become places in Petri net. They keep tokens of the colour T, which represents features of algorithms for finding the shortest routes dendrite. Colour T is defined as

    ```
    colset T = product INT*LP
    ```

    where INT represents an integer and LP is a list of integers.
    The token $md = 1'(t, lp)$ stored in the place modelling vertex v represents a route from the parking position (vertex 0) to the vertex v, where t is a total taxi time and lp is a list vertices in the shortest route.
2. Time characteristics $t(u, v)$ also become places in Petri net. However, they keep tokens of the colour which can be considered as integer numbers. They represent taxiing time on the section determined by (u, v) pair.
3. Edges (u, v) of graph G become transitions. The transition input places are: place representing vertex u and place representing characteristic on arc $t(u, v)$. Output place of the transition representing section (u, v) is the place modelling the vertex v.
4. Functions E, describing weights of arcs in Petri net (formula 1), realize steps of the algorithm for finding the shortest routes dendrite. They modify the special data structure containing the list of vertices in the shortest route and also the minimum taxiing time.

5   Example of Method Application

A part of taxiways system at London Heathrow airport is presented in Fig. 3. To make presentation clear, only few taxiways intersections were considered. They are marked by circles with figures. The taxiways not included, were marked by short lines crossing the taxiways.

The net representing part of taxiways system presented in Fig. 3 is shown in Fig. 4. In this case:

$$G = \langle \{0, 1, 2, 3, 4, 5, 6\},$$
$$\{(0, 1), (1, 2), (1, 3), (3, 2), (2, 6), (5, 6), (3, 4), (4, 5), (4, 6)\} \rangle \tag{6}$$

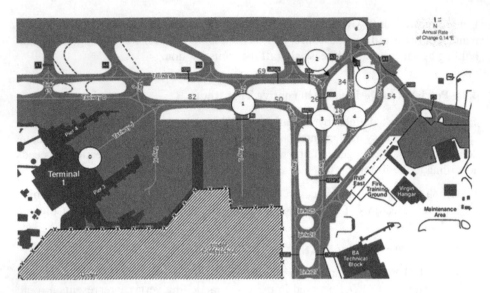

Fig. 3. London Heathrow airport with vertices and taxi times [own study]

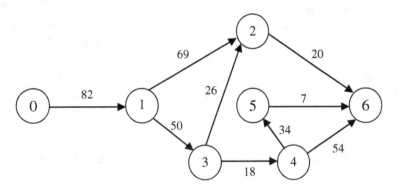

Fig. 4. Taxiways represented by a net [own study]

Values of function t were defined, based on distance between vertices and standard taxing speed, which includes aircraft movement on straight section and arcs. The taxiing speed was measured during real operation. It is usually about 10 knots, which means about 5 m/s. However, sometimes the taxiing speed at particular airport locations is a safety bottleneck, so it may be decided to implement speed restrictions [10].

Petri Net for this example, constructed according to principles described in Sect. 4.3 is presented in Fig. 5. Places named "0", "1", "2" etc. represent the nodes G of the graph shown in Fig. 4. Time characteristics $t(u, v)$ are stored in places named: "0–1", "1–2", "1–3" etc. Transitions are responsible for updating the total time necessary to taxi via a section (u, v). The list of vertices in the shortest route is represented by the lp variable, which is of colour LP defined as list of integers. An important part of this Petri net are the functions *ch2* and *ch3*, which choose the minimum taxi time. The final solution is stored in place named "D".

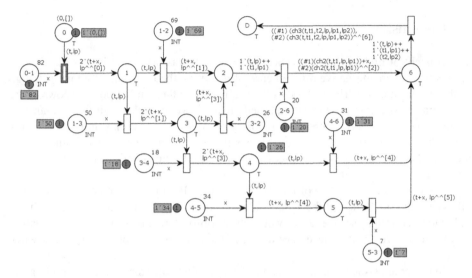

Fig. 5. Petri net for finding the fastest taxi route [own study]

Using the simulator which is a part of CPN Tools package allows finding the fastest taxi route. The sequence of vertices in this route is: $\langle 0, 1, 2, 6 \rangle$. The shortest taxi time which corresponds to it equals $t(0, 1) + t(1, 2) + t(2, 6) = 171\,\text{s}$.

The algorithm for determining conflict points which was discussed in Sect. 3 allows modification of transit times represented by the places keeping the tokens of colour INT. Therefore the integration of both applications, which is the next step of research, will let automatic generation of an alternative taxi route in case of the occurrence of traffic interferences which are represented by conflict points. To illustrate applicability of this solution, two simulation experiments will be presented.

Experiment 1. In the first experiment the following scenario will be examined. The aircraft is still at parking position (vertex 0) when, because of increased traffic, also the taxi times increase. The new values are (in seconds): $t(0, 1) = 100$, $t(1, 2) = 83$, $t(2, 6) = 40$, $t(1, 3) = 55$, $t(3, 4) = 25$, $t(3, 2) = 35$, $t(4, 6) = 40$, $t(4, 5) = 45$, $t(5, 6) = 20$. In this situation, there is a possibility to recalculate the taxi time again and look for alternate taxi route which is now more beneficial than route which was planned originally. After the simulation, it turns out that the alternative taxi route is described by a sequence of vertices $\langle 0, 1, 3, 4, 6 \rangle$ and the new shortest taxi time is 209 s.

Experiment 2. In the second experiment we assumed that the aircraft is already taxiing and it is approaching to the vertex 3. Taxi time to the vertex 3 was 132 s and the scenario assumes that in this moment, because of intensive precipitations, the taxi times in particular sections change in the following way (in seconds): $t(2, 6) = 30$, $t(3, 4) = 30$, $t(4, 5) = 25$, $t(4, 6) = 38$ and $t(5, 6) = 20$. There is a possibility to recalculate the taxi route and change the taxi instructions appropriately. The fastest route for the new traffic situation is $\langle 0, 1, 3, 2, 6 \rangle$ and the new shortest taxi time is 188 s.

6 Conclusion

The concept of finding the fastest taxi route, which was presented in this paper, also allows finding an alternative route in case of receiving the information about a change of the traffic situation. The method of determination of the conflict points allows at the same time predicting a new situation based on the movement plans within the manoeuvring area of the airport. There is a possibility of integration of both solutions (which is planned in the next stage of research). This will allow determination (by simulation) of the new, alternative routes if the prediction of future traffic situation suggests that it is necessary.

The hierarchical structure of the coloured, timed Petri net, presented in this paper, allows for the simulation of the actual and projected air traffic. Experiments carried out indicate full usefulness and efficiency of such an approach for determining the alternative taxi route. However, it is necessary to check the concept on more complex taxiways network.

An important issue which needs to be verified, is whether the concept of finding the best route for each aircraft individually (which was assumed here) gives a globally optimal solution. Knowing the typical properties of this kind of dynamical systems, it can be expected that the solution obtained here is better than standard, but it isn't globally optimal. Therefore it is necessary to modify the approach, by applying network planning mechanisms and simultaneous optimisation of taxi routes of all aircraft at the same time. Such research will be undertaken and for that purpose we plan to use a Dynamic Programming method.

Recently, the research on airport ground movement has started to take into account a speed profile optimisation problem so that not only time efficiency but also fuel saving and decrease in airport emissions can be achieved at the same time [12]. This problem is difficult to solve due to its computational load. However, using different objective functions (not just the time) is very promising research direction, which we plan to undertake in the future.

References

1. Czarnecki, M., Skorupski, J.: Method for identification of conflict points in the intelligent system of an aircraft taxi route choice. Arch. Transp. Syst. Telematics **8**(3), 9–14 (2015)
2. EUROCONTROL (European Organisation For The Safety Of Air Navigation): Definition of A-SMGCS Implementation Levels, edn. 1.2, Brussels (2010)
3. Jung, Y., et al.: Performance evaluation of SARDA: an individual aircraft-based advisory concept for surface management. Air Traffic Control Q. **22**, 195–221 (2015)
4. Korzan, B.: Elementy teorii grafów i sieci. Metody i zastosowania, Wydawnictwa Naukowo-Techniczne, Warsaw (1978). (in Polish)
5. Łabuś, M.: Wpływ A-SMGCS na bezpieczeństwo operacji lotniskowych. Prace Naukowe Politechniki Warszawskiej. Transport **103**, 147–155 (2014). (in Polish)

6. Ravizza, S., Atkin, J.A.D., Burke, E.K.: A more realistic approach for airport ground movement optimisation with stand holding. J. Sched. **17**, 507–520 (2013)
7. Siergiejczyk, M., Siłkowska, J.: Analiza możliwości wykorzystania techniki multilateracji w dozorowaniu przestrzeni powietrznej. Prace Naukowe Politechniki Warszawskiej. Transport **102**, 119–133 (2014)
8. Skorupski, J.: Airport traffic simulation using Petri nets. In: Mikulski, J. (ed.) TST 2013. CCIS, vol. 395, pp. 468–475. Springer, Heidelberg (2013). doi:10.1007/978-3-642-41647-7_57
9. Skorupski, J.: The risk of an air accident as a result of a serious incident of the hybrid type. Reliab. Eng. Syst. Saf. **140**, 37–52 (2015)
10. Stroeve, S.H., Blom, H.A.P., (Bert) Bakker, G.J.: Systemic accident risk assessment in air traffic by Monte Carlo simulation. Saf. Sci. **47**, 238–249 (2009)
11. Tancredi, U., Accardo, D., Fasano, G., Renga, A., Rufino, G., Maresca, G.: An algorithm for managing aircraft movement on an airport surface. Algorithms **6**, 494–511 (2013)
12. Weiszer, M., Chen, J., Stewart, P.: A real-time active routing approach via a database for airport surface movement. Transp. Res. Part C Emerg. Technol. **58**, 127–145 (2015)
13. Westergaard, M., Kristensen, L.M.: The access/CPN framework: a tool for interacting with the CPN tools simulator. In: Franceschinis, G., Wolf, K. (eds.) PETRI NETS 2009. LNCS, vol. 5606, pp. 313–322. Springer, Heidelberg (2009). doi:10.1007/978-3-642-02424-5_19
14. Wilke, S., Majumdar, A., Ochieng, W.Y.: Airport surface operations: a holistic framework for operations modeling and risk management. Saf. Sci. **63**, 18–33 (2014)
15. Wilke, S., Majumdar, A., Ochieng, W.Y.: The impact of airport characteristics on airport surface accidents and incidents. J. Saf. Res. **53**, 63–75 (2015)
16. Zhu, X., Tang, X., Han, S.: Aircraft intersection collision conflict detection and resolution under the control of A-SMGCS. In: Proceedings of 2012 International Conference on Modelling, Identification and Control, ICMIC 2012, pp. 120–125 (2012)

Effective Decision Support System
Based on Statistical Tools

Pavel Přibyl[✉] and Ondřej Přibyl

Faculty of Transportation Sciences, Czech Technical University,
Konviktská 20, 110 00 Prague, Czech Republic
{pribyl,pribylo}@fd.cvut.cz

Abstract. This paper focuses on an important topic – on evaluation of various transport structures. It is a typical situation that a decision maker must choose from various alternatives and with respect to different, often contradicting criteria. In this paper, a method adopted from the field of risk management, SAFMEA, was adopted. This is a novel approach that however introduces statistical evaluation of expert feedback to provide a robust decision support materials. Here, in a structured way, the expert know how is extracted and processed. In this approach, not only average aggregate values are evaluated. An important contribution of this approach is in looking into the variances in the answers from particular experts in order to capture the consensus or the lack of it. The method was validated on a case study from the Czech Republic, particularly on evaluating four different approaches to modernization of a highway D1. The method provided a recommendation that has been as a result actually implemented by the ministry of transport. The results confirm that the approach is worth consideration.

Keywords: Alternative selection · SAFMEA · Risk management · Transport structures

1 Introduction

Evaluation of several variants of transport structures is generally difficult problem, because of necessity to look on it with respect to the sustainability which is a complex task. It is necessary to address social impact, ecological aspects and economy during long term time horizon. Nevertheless, each of these three pillars is composed by a set of heterogeneous parameters.

One of the options that explicitly considers multiple criteria in decision-making environment is multiple-criteria decision analysis (MCDA). There are typically multiple (often also conflicting) criteria that need to be evaluated in making decisions. In making the decision of how to reconstruct a highway, there are not only very complex issues involving multiple criteria, but there are also multiple parties who are deeply affected from the consequences. One of the benefits of those methods is well structuring of complex problems and explicitly definition of the criteria. The disadvantage is heterogeneity of criteria and the possibility to express them as a unified set. Very sensitive is the weighting of criteria which could be influenced by subjective feeling.

© Springer International Publishing AG 2016
J. Mikulski (Ed.): TST 2016, CCIS 640, pp. 378–387, 2016.
DOI: 10.1007/978-3-319-49646-7_32

Ministry of Transport of the Czech Republic needed to assess the strategic plan for modernization of D1 highway. Highway D1 is the oldest one (first part was opened in 1971) and most loaded highway connecting capital Prague with the second large city Brno and Ostrava. The aim was to assess four variants of modernization of the D1, with respect to comfort and safety for users and minimizing operating costs in the strategic horizon of thirty years. The proposed variants were called as:

A. Modernization
B. Reconstruction
C. New
D. Zero

The term "Modernization" refers to a solution where both directions of the highway are extended for 75 cm, i.e. from the category "D26,5" to "D28". This solution allows ride in 4 lanes in one band during future maintenance works in the opposite direction. Also, a new concrete road surface will be provided. The term "Reconstruction" was proposed by an independent group and in short it concerns fragmentation of the existing concrete cover and covering by new asphalt carpet. The category D26,5 remains the same. The variant "New" was included in order to detect the expert opinion on complex rebuilding of highway on 2 × 3 lines in both directions. The "Zero" alternative expresses the present crisis status. This solution assumes a preservation of present quality of traffic only through maintenance works.

2 Basic Idea of Solution

At any comparative analysis can be a significant problem subjective view of the evaluator-expert. To reduce the influence of subjective evaluation, it was decided to use a modified method of qualitative risk assessment SAFMEA. The general methods of risk analysis are described in Refs. [1–10]. The detailed description of SAFMEA is in [12]. The value of risk is done as the product of possible damage resulting from hazard scenario and the probability of this scenario formation. The value of risk is possible to express as:

$$RS_i = Sv_i \times Lk_i, \tag{1}$$

where Sv_i is damage done by the scenario a Lk_i is the likelihood of this scenario. The classical SAFMEA assess separately severity Sv_i and separately the likelihood. Then unlikely severity with big consequences gives the same result as a small-scale damage that occurs with greater frequency. At the first sight it is obvious that the psychology of risk perception is different in a society. The fact that on roads die three people in average each week is socially perceived conciliatory than when a fire killed ten people in a tunnel at once, although once in decades. Following this situation, unlikely hazard situation with big damage provides the same value of RS as a small damage however more frequent. Apart from difficult comparison of social risk the disadvantage of risk evaluation according Eq. (1) is uncertainty given by multiplying of two estimated values. The estimation of "right" value depends on knowledge and experiences of expert and always is possible to speak about probability to catch a right value. Uncertainties as fundamental phenomena are related to each expert decision.

Incomplete knowledge dominate the problems of risk analysis is described in [11]. Uncertainties in decision making and risk analysis can be divided into two categories: those that steam from variability in known situations. It represents randomness in choosing of samples. Second one comes from basic lack of knowledge about fundamental phenomena. In our case is a lack of knowledge related to probability of risk situation because of rare occurrence similar situations.

The expert brings a uncertainty of damage estimation and uncertainty in estimation of probability in the Eq. (1). The resulting uncertainty is

$$u = \sqrt{\left(A_1^2 \cdot \left(\frac{z_{1max}}{k} \right)^2 + A_2^2 \cdot \left(\frac{z_{2max}}{k} \right)^2 \right)}, \tag{2}$$

where $z_{i\,max}$ is known deviation of i-th estimation, k is a coefficient based on the distribution law and A is a coefficient of sensitivity of inputs. The experience and the availability of expert knowledge from practice reduces the degree of uncertainty.

Therefore, to be limited to imprecision in estimating the two parameters, the proposed method takes the probability of an event pi as one. This consideration is relevant speaking about evaluation of road constructions because it says that the realistic severity is evaluated in a given time frame and not a severity which might be. Reducing uncertainty of the consequences D_i is achieved by the participation of more experts. As will be shown below, the four variants have been evaluated by 39 experts. In addition, they were also from different work areas.

3   The Risk Assessment by Modified SAFMEA

The modified SAFMEA denoted as SAFMEA-M lies in the cluster of the quantitative risk analysis methods. The whole process should be chaired by specialist on risk analysis ERA (Expert on Risk Analysis). The process of risk evaluation contains following stages:

A. Identification of possible hazard situations/scenarios (next RF "Risk Factors").
B. Quantification of possible severity of the scenario on non-linear scale
C. Calculation of the value of risk RPN and calculation of others statistical parameters
D. RPN arrangement in terms of severity
E. Interpretation of risk to the customer

Term RPN (Risk Priority Number) is adapted from FMEA (Failure Mode and Effects Analysis) method, [13], and reflect the relative risk as a number between zero and the maximal risk. The following text describes shortly the single stages of the proposed process:

3.1   Preparatory Stage

Risk expert (ERA) studied the project in detail and he selected segments of the project which will be investigated by risk analysis. This project evaluated the whole life cycle since the segment "Design", through "Building" until "Operation". The time frame is 30 years.

Consequently ERA formed basic group of experts BET (Basic Expert Team). In our case they were specialist in the area of road construction and operation. It was six persons in our case. It is not desirable to include some risk analysis specialist.

3.2   Identification of Risk Segments and Risk Factors (BET)

The BET experts optionally modified segments of project primarily chosen by ERA. The main task is to find out Risk factors (RF) and exactly define them. Risk factor could be, for example "Risk of increased financial costs" or "The risk of non-compliance with standards". The BET experts select a range of metric of scale of evaluation and they decide if a linear or non-linear scale will be applied.

The risk experts plug each Risk Factor with unique definition into database. Database has as output forms which are sent to the members of extended expert team EET.

The BET team prepares potential candidates for EET as well.

3.3   Expert Evaluation of Risk Factors (EET)

The forms are send with detailed explanation how to fill it and with description of each variant to all potential candidates. It was emphasized that it is not necessary to evaluate some RF, unless expert has not competence in this area. The questionnaires are anonymous and experts are not informed who is member of EET. Contact is mediated only through ERA. In our project 56 questionnaires have been sent and 33 came back (Fig. 1).

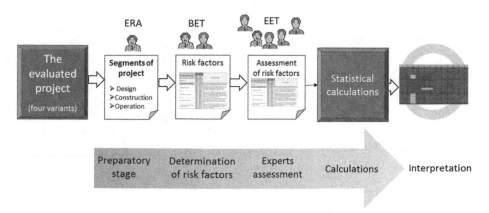

Fig. 1. Schematic description of modified method [own study]

3.4 Statistical Calculation and Interpretation of Results

Questionnaires returned from experts are statistically evaluated. The procedure of calculation is described in the following chapter. The most important stage is detailed analysis of results and convincing presentation to the customer.

4 Example of Evaluation for Highway

Risk specialist ERA decided to analyse the life cycle of highway in time horizon of thirty years and to evaluate three stages: design, building and operation. The segments of projects which are potentially risky for the whole project have been identified in the next step of the procedure. In this concrete case the following segments will be investigated:

A. Financial demand of the selected variant;
B. Technological, construction and project risk;
C. Demands on technical feasibility, construction period and lifetime
D. Operational aspects, quality of service and safety
E. Ecological and sociological aspects.

Each segment consist of several risk factors RF. Risk factor A.1 "The risk of failure of funding model", B.1 "The risk of delay in the realization of construction", D.1 "The safety for the users of highway" or E.2 "Noise pollution for inhabitants after construction". The BET team identified 20 risk factors at all. Each RF is completed by detailed explanation, see Table 1.

Table 1. Sheet of questioner "Financial demand of the selected variant" with risk factors A.1 and A.2 and four variant of solution [own study]

RISK SEGMENTS (I - IV)		Risk assesment of variants			
Risk Indicators		Modernization	Reconstruction	New	Zero
A. Financial demand of the selected variant					
A.1	The risk of failure of funding model				
A.2	Risk of financial increase				
A.3	Need to stop the renovations				
A.4	Impossiobility to meet business conditions				
B.1	The risk of delay in the realization of construction				
B.2	Technical difficulaties in the renovations				
B.3	Hidden failures				
B.4	Not meeting the technical requirements				

The scale for assessment of risk factors was chosen as non-liner (1; 2; 8; 16) to emphasise the importance of risk severity. This corresponds to verbal expression: small/medium/big and very big. Each of the experts therefore independently assessed the degree of risk by a numerical value 1, 2, 8 or 16.

The evaluation of risk is based on the statistical calculations. Due to the relatively large number of input variables (n-RF x E-experts) was necessarily to prepare specific software for the evaluation. The primary outcome is assessment of the risk factor as RPN value. The calculation proceeds in the following steps:

1. In each raw of the j form the expert E rated values of RPN_{jv}^{E} for v (v = variant 1 till 4) which directly corresponds to severity

$$RPN_{jv}^{E} = Sv_{jv}^{E} \tag{3}$$

For each raw is known $4xE$ experts estimations.

For each raw (j-th risk factor) and the v – variant is calculated the average value of the risk $mRPN_{jv}$

$$mRPN_{jv} = \frac{\sum_{k=1}^{E} RPN_{jk}}{E} \tag{4}$$

If there more than five experts answered and distribution of answers follows approximately Gaussian dispersion it is possible to calculate the standard deviation of the risk factor:

$$sRPN_{jv} = \sqrt{\frac{1}{E-1} \sum_{k} \left(RPN_{jk} - mRPN_{jv}\right)^{2}} \tag{5}$$

and further it is possible to estimate the quantile of distribution RPNE (Quantile is the approximate value of "average plus standard deviation" and it gives an idea of randomness of dispersion. In the case of normal distribution the value of qRPNjv is 0,84 quantile of the probability distribution)

$$qRPN_{jv} = mRPN_{jv} + sRPN_{jv} \tag{6}$$

5   Interpretation of Results

The final evaluation of each of the four variants and of all experts was lined up in a spreadsheet and processed graphically according average value of mRPNjv and standard deviation sRPNjv.

The sample of results for four randomly selected RF is in Table 2. The summary value of all risk factors since A.1 till E.2 for A-variant is 13,033; var. B = 27,218;

var. C = 22,799 a var. D = 22,819. Taking in account these values is possible to generally say that A-variant (Modernization) has the lowest summary risk.

Table 2. Resulting RF for different variants and selected RF [own study]

Risk factor	Value of $mRPN_{jv}$				Standard deviation $sRPN_{jv}$			
	var. A	var. B	var. C	var. D	var. A	var. B	var. C	var. D
A.1	3,622	8,784	6,459	10,054	3,990	5,963	4,897	6,896
B.1	3,667	4,821	13,641	2,231	2,950	3,641	4,252	3,602
D.1	2,872	8,974	1,333	11,026	3,062	5,081	1,155	5,561
E.1	2,861	4,639	1,389	8,500	2,356	3,595	0,494	5,715

It follows from the table that expert evaluated very positively safety (D.1 = 1,333) and ecology (E.2 = 1,389) for the new highway (var. C). On the other side this variant has the highest risk in delay of finalizing (B.1 = 13,641) and thus it is really excluded from the considerations. Similarly the experts evaluated the safety risk (D.1 = 11,026) if present status of the highway will be preserved and only maintenance will be provided. The resulting analysis and interpretation of results is beyond the scope of this article and it possible to find it in the report [14].

Figure 2 depicts average value of mRPN for a few of risk factors (A.1–B.5) and variant A "Modernization". The graphical presentation gives also overview how evaluated each expert. It is evident that expert Nr. 14 evaluated very differently from other experts. The risk factor A.1–A.4 as most risky as number 16. As mentioned before this deviation could not influence result because of big group of 39 experts.

	Risk Indicator	Average value of risk $mRPN_j^E$	Standard deviation $sRPN_j^E$	Nr. of responders n	1	2	3	4	5	6	7	8
A.1	The risk of failure of funding model	9,435	6,896	23		16	16	16	1	16	16	1
A.2	Výrazné zvýšení ceny stavebních prací	4,625	5,149	24		2	2	1	8	1	1	1
A.3	Zastavení stavby z jiných důvodů	1,920	2,777	25	1	1	1	1	2	1	1	1
A.4	Nemožnosti dodržení smluvních podmínek	3,636	4,391	22		2	1	1		1	1	1
B.1	The risk of delay in the realization of construction	2,560	3,602	25	1	1	2	1	1	1	1	1
B.2	Technické náročnosti při přípravě stavby	2,520	2,969	25	1	1	1	1	1	1	1	1
B.3	Skrytých vad	6,870	5,677	23			2	16	8	16	2	1
B.4	Nesoulad s rezortními předpisy	5,952	5,840	21	16	2	16	2	1	2	1	8

Fig. 2. The results of calculations of the mean risk for a variant of "Modernization." The figure shows how experts no. 4–16 assessed risk factors A.1 to B.4 (in Czech) [own study]

The final evaluation is based on comparison of the risk factors of all variants. The illustrative graphs as in Fig. 3 depict theirs interpretation. These comparative diagrams emphasize the risks factors that must be considered.

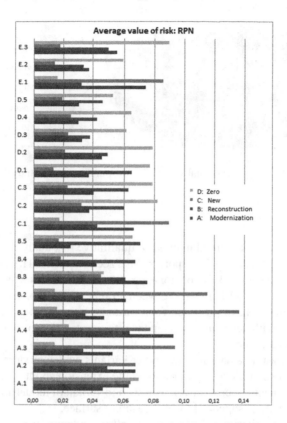

Fig. 3. Average value of RPN for risk factors A.1–E.3 and all four variants [own study]

6 Professional Structure of Experts

The first stage of the project and elaboration basic documents was prepared by team of experts from the Faculty of Transportation Sciences and Faculty of Civil Engineering – BET team. Thanks to the 57 letters which were sent to EET experts explaining the issue and its importance we got back 32 answers. As can be seen from the graph below, the same number of respondents was from academia field and from design and engineering organizations. It is from theoreticians and practitioners. Together it was 56% of questionnaires. Only a few smaller was representation of state administration and investor organizations (21%). Questionnaires were also sent to professional organizations and associations such as the Chamber of Commerce or ESMAD (association of transporters). Their responses represented approximately 14%. The last group with 8.7% of the responses covered independent experts from various fields (Fig. 4).

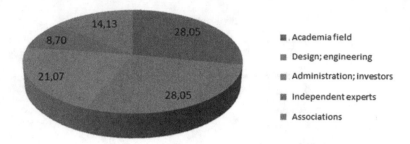

Fig. 4. Professional structure of respondents [own study]

7 Conclusion

The article describes the modified method of qualitative risk analysis, which was used in the project of the Ministry of Transport of the Czech Republic. The aim of this project was to emphasize the main risk factors for each of four alternatives which were proposed for rebuilding most loaded highway between Prague and Brno. Experts, as well as various interest groups have proposed four variants, from leaving the highway in its current form, only with necessary maintenance, to construction of new highways with three lanes in both directions.

Very sensitive was evaluation of two other variants, which modernized the existing highway. The Ministry of Transport promoted the extension about 75 cm in both directions and a new cement-concrete cover. The second variant of modernization pushed by another group of experts supposed to apply asphalt surface.

The use of multi-criteria analysis to assess all variants was too sensitive due to the subjective choice of parameters and weights of individual parameters. Therefore, the modified quantitative risk analysis has been applied which is in presented form less sensitive to subjective evaluation or even to some influencing of results.

The uncertainty is reduced because the evaluation of the probability of an event is neglected, since it is assumed that the event will always occur and probability is one. The proposed process starts with creating a core team that precisely defined a set of risk factors. The extended team of respondents should be as far as large for reducing of the uncertainties in the assessment of risk factors. It is possible to consider the evaluation as relevant in our case where 39 respondents completed the questionnaire.

The analysis of all segments of the project confirmed the basic hypothesis that the modernization according the concept presented by Ministry of Transport is the least risky in the time frame of thirty years. The construction has been divided into 21 segments. In the time of writing this article is finished or under construction 9 sections and modernization goes ahead successfully.

References

1. Přibyl, P., Janota, A., Spalek, J.: Bezpečnost v tunelech, BEN - technicka literatura, Praha (2008)

2. Steward, M.G., Melchers, R.E.: Probabilistic Risk Assessment of Engineering System. Chapman & Hall, London (1997)
3. Persson, M.: Quantitative risk analysis procedure for the fire evacuation of the road tunnel, Department of Fire Safety Engineering, Lund University, Report 5096, Lund (2002)
4. Sørensen, J.D.: BU2005 – Belastning og sikkerhed - Note 1: Introduction to Risk Analysis. Institute of Building Technology and Structural Engineering, p. 26. Aalborg University, Denmark (2005)
5. Carter, D.A., Hirst, I.L.: 'Worst case' methodology for the initial assessment of societal risk from proposed major accident installations. J. Hazard. Mater. **71**, 117–128 (2000)
6. Vrijling, J.K., van Gelder, P.H.A.J.M.: Societal risk and the concept of risk aversion. In: Advances in Safety and Reliability, vol. 1, pp. 45–52 (1997)
7. Smets, H.: Frequency distribution of industrial accidents caused by hazardous substances. Draft version, March 1993
8. Kroon, I.B., Hoej, N.P.: Application of risk aversion for engineering decision making. In: Safety, Risk and Reliability – Trends in Engineering, Malta (2001)
9. Vrijling, J.K., Wessels, J.F.M., van Hengel, W., Houben, R.J.: What is acceptable risk, Delft (1993)
10. FSV Austria Research Association Road-Rail-Traffic, Guideline RVS 09.03.11: Tunnel Risk Model – TuRisMo (2008)
11. Paté-Cornel, E.M.: Uncertainties in risk analysis: six levels of treatment. In: Reliability Engineering and System Safety, Northern Ireland, vol. 54 (1996)
12. Eltodo, EG. TP229: Bezpečnost v tunelech na pozemních komunikacích, pp. 45–53. Ministry of transport, Czech Republic (2010). ISBN 978-80-254-7953-7
13. Guide to FMEA and resources, Copyright 2006-Present by FMEA-FMECA.com. http://www.fmea-fmeca.com/fmea-rpn.html. Accessed Feb 2016
14. Přibyl, P.: Research report for project HADES. Submitted to the Highway directory of the Czech Republic (2010)

Communication Based Train Control and Management Systems Safety and Security Impact Reference Model

Marek Pawlik[✉]

Warsaw Railway Institute, Chłopickiego 50, 04-275 Warsaw, Poland
mpawlik@ikolej.pl

Abstract. Communication based train control and management systems in railway transport are only subdivided into imprecisely defined classes, which do not take into account, in a satisfactory way, their influence on railway safety and security. Paper defines innovative assessment method taking into account operational safety, technical safety and security. Assessment is based particularly on the scope of functions working in different operational situations, failure resistance both regarding systematic and random failures, protection against vandalism and terrorism as well as data being entered and being presented during system exploitation.

Keywords: Transport · Railway · Safety · Security · Communication · Train operation · Control command and signalling

1 Introduction

Railway transport from the freight point of view can be seen as: loading of wagons, trains' preparation, scheduling, trains' running from origin to shunting yard, shunting including creation of new rakes of wagons, trains' running from yard to yard, and finally from yard to destination and unloading of wagons. In case of passenger transport: trains, composed of locomotives and coaches or alternatively formed by one or more multiple units, are running according predefined schedule between passenger stations ensuring connections between trains for easy transfer for passengers.

As all transport modes, railway can be subdivided into transport means, transport infrastructure and transport operation. Transport means are: airplanes in case of air transport, vessels in case of water transport, trucks and cars in road transport, railway vehicles in case of railway transport. Transport infrastructure inherent to transport mode is formed respectively by: airports, inland water ports and seaports, roads, railway stations and railway lines. Transport infrastructure for intermodal and multimodal transportation additionally include different kinds of logistic centres. Transport operation is formed by operational rules and procedures, and includes also staff which is responsible for safety and security as well as different kinds of registers, which are supportive not only for improving transport services but also for different kinds of investigation bodies.

On the edge between transport means [1], transport infrastructure [2, 3] and transport operation [4, 5] there are transport control and management systems [6] ensuring proper

© Springer International Publishing AG 2016
J. Mikulski (Ed.): TST 2016, CCIS 640, pp. 388–400, 2016.
DOI: 10.1007/978-3-319-49646-7_33

use of transport means moving on transport infrastructure keeping operational rules and procedures and supporting exploitation management. In case of railway transport control and management is ensured by: railway signalling systems, railway control command systems, railway traffic control and management systems, and different miscellaneous automated supplementary systems.

Railway signalling systems are based on subdivision of tracks into sections, which can be used for setting routes for trains only if they are not occupied. Key role is therefore played by track occupancy checking devices. Information about track occupancy is used by station interlockings for safe route setting protecting trains against 'cornering' meaning collisions, when one train hits another train aside on the railway track switch. Information about track occupancy is also used by line-block systems for safe train spacing on route sections between stations protecting trains against 'end-on collisions' meaning collisions, when one train hits another train's end. Together station interlockings and line-block systems are protecting trains against 'head to head' collisions. Information about track occupancy is also used by level crossing protection systems, as barriers which are closing roads go down when trains are approaching and go up when they left crossing area.

An authority for the train driver to run the train on the defined route is given to the train driver by colour light signals. That is introducing risk associated with human factor to individual train operations. Removing human factor can only be achieved by introduction of a system verifying train move against authority. Such solutions have to be based on track-train data transmission and intervene by braking, when movement exceeds authority. Such systems are known as railway control command systems. Depending on technical solution, source information can be taken from signalling systems only or from signalling and management systems. Control command is also used as a source of information for traffic control and management especially in case of operational disturbances.

Additionally some functions, which are not safety critical can be provided by different miscellaneous automated supplementary systems, like for instance: ticketing, passenger information at stations, passenger information in trains, exchange of consignment letters for including railway freight transport in logistic networks, running gear diagnostic, security monitoring at stations.

Train control command together with:

- signalling, or interfaces to signalling,
- traffic control and management, or respective interfaces,
- miscellaneous automated supplementary systems, or interfaces to them,

form 'communication based train control' (CBTC) systems. The scale of implementation of the communication based train control and management systems in railway transport is presently growing significantly. High diversity of conceptual and technical solutions require tools, which will offer good way for choosing technical solution fitting needs. Assessment has to take into account not only scope of functions but also safety and security [7–9]. Unfortunately, at presence, CBTC systems in railway transport are only generally classified in a few groups. These groups, known as classes, are not defined precisely enough for assessing specific solutions' influence on safety and security.

Therefore it is necessary to define Communication Based Train Control and management systems Safety and Security Impact Reference Model – CBTC SSIRM model.

2   CBTC Classes

Within railway transport CBTC systems are subdivided into: Automatic Warning Systems (AWS), Automatic Train Protection (ATP) and Automatic Train Control (ATC). In rail transport additionally two complementary systems have been defined: Automatic Train Operation (ATO), which is a control command system, and Automatic Train Supervision (ATS), which is a traffic control system. All such systems are comprised of trackside devices (TRD) mounted on tracks or widely on railway lines and on-board devices (OBD) mounted in traction units – in locomotives (LOC), in electrical and diesel multiple units (EMU & DMU) and in special vehicles like e.g. on-track machines (OTM) and dual-mode vehicles (DMV) that can run on conventional road surfaces and on railway tracks used as a 'guide way'.

AWS class systems automatically warn drivers when trains are approaching danger points, where drivers have to be concentrated particularly. Depending on technical solution, as a danger point, systems consider signals which are capable to show danger aspect (STOP signal) or level crossings or other types of danger points, which are present on the tracks. Thanks to AWS TRD devices, which are mounted in front of such points AWS OBD devices know when they have to warn the drivers. To achieve such warning AWS systems use uplink transmission from TRD to OBD.

Track-train transmission (TTT) between TRD and OBD devices is used by all CBTC systems. TTT always ensure uplink - transmission from TRD to OBD devices, and depending on technical solution may require and ensure downlink – transmission from OBD to TRD devices. OBD devices always comprise on-board data receivers (ORx), and depending on technical solution, may comprise on-board vital computer (OVC), and may comprise interface to train – a train interface unit (TIU), and may comprise interface for driver – a driver machine interface (DMI). DMI always comprise displaying information for driver – DMI uplink, and depending on technical solution, may comprise DMI downlink, when some data is required to be entered by driver.

AWS systems do not comprise DMI downlink and may not comprise OVC and may not comprise TIU. They do comprise on-board logic, however such logic is not necessarily constructed in a way that respects technical safety rules. If it does, it forms OVC. If AWS system comprises TIU, then on-board logic is connected to train equipment. Minimum linking with braking system.

ATP class systems automatically protect train movements on the basis of information received from trackside, and reflecting changing operational situation. For instance in Poland in case of the line-block system colour light signal located just before station entrance signal and showing S3 aspect (one green flashing light) train is authorised to run with maximum speed 160 km/h up to the entrance signal and to pass station entrance signal with speed equal or lower than 100 km/h. ATP TRD sends stepwise profile taking into account real distances between signals authorising train to run x meters with maximum speed 160 km/h and following y meters with maximum speed 100 km/h up to the end of authority, which is in that case a station exit signal.

ATP TRD devices comprise signalling interface unit(s) (SIU) and trackside data transmitter(s) (TTx). ATP OBD devices comprise ORx, OVC, TIU and DMI. DMI may comprise also downlink. ATP OBD device may comprise on-board transmitter (OTx), which will ensure data downlink transmission if ATP TRD comprise trackside data receiver (TRx). Data downlink may be used for initiation of signalling functions e.g. automatic closing of the level crossings taking into account train speed in order to mini-mise level crossings close time. The TIU must ensure braking initiation as a minimum. Intervention is triggered in case of overspeeding in relation to stepwise speed profile and in case of passing end of authority. The TIU may comprise other interfaces e.g. to doors, to traction main switch, to pantograph accessories ensuring up and down panto-graph movement.

ATC class systems automatically control train movements on the basis of informa-tion received from trackside and reflecting changing operational situation. However ATC systems take into account dynamic character of the train movement. Therefore when the train is approaching location where the maximum allowed speed becomes lower, ATC OBD is verifying speed lowering against dynamic profile instead of step-wise profile. For instance in S3 aspect case described above ATC TRD, like ATP TRD, sends stepwise profile, which is, in case of ATC class system, used in significantly more advanced way, as it is used to calculate dynamic profile thanks to data obtained via DMI data downlink. Calculation is done by OVC equipped with inherent technical safety functionality. Dynamic profile is shown to the driver by DMI. The TIU must ensure braking initiation as a minimum. Intervention is triggered in case of overspeeding in relation to dynamic profile and in case of passing end of authority. The TIU may comprise other interfaces.

The ATO class systems ensure automatic train operation. This requires addi-tional data transmitted from TTx to ORx like exact location of platforms to ensure precise stopping positions and train-side on which doors are due to opened. ATO class systems are driver-less and therefore data for calculation of the dynamic speed profile must be pre-programmed and data presented to driver are useless. This means that DMI upload and download functionalities are not necessary. Moreover speed profile calculation has to cover not only speed decrease but also speed increase. Driving initiation, speed increase, cruise as a steady speed running, and braking must be initiated, performed and supervised. ATO OBD by definition comprise OTx and ORx as bidirectional transmission is a must. Necessary data is provided by ATS, which is a trackside system providing all data, which are necessary for several trains equipped with ATO and running on the same infrastructure. ATS TRD ensures TRx and TTx functionalities. ATS requires traffic controller or dispatcher responsible for traffic management in degraded circumstances. ATO and ATS systems together are usually used together with many safety and security related miscellaneous functions like: platform and train interior monitoring, passenger information systems, coordi-nation of opening train doors and platform doors, fire-fighting, security alarms together with coordination of fire brigades, medical emergency, railway police, police and even antiterrorist brigades for instance in case of driverless rail connections between terminals and parking places at airports. Emergency procedures and supporting technical systems and devices are crucial as such systems are used for downtown transport systems, for metro lines, for airport connections, where amount of passengers is high and frequent service is required. The ATO and ATS systems used together are sometimes called Automatic Peoples Mover (APM).

If a train is foreseen to run not only on ATS equipped lines but also on others equipped with ATP or ATC or even not equipped with control command system it has to be prepared to perform ATC OBD functionalities. This means for instance, that driver seat and DMI uplink and downlink have to be available in the train and dedicated functionality for automatic transitions between differently equipped infrastructure must be provided.

3   Safety and Security

The difference between safety and security is not always obvious. In many languages the same word is used for both expressions. Moreover both can be seen from different angles and provided and supported by different technical solutions. We can distinguish railway transport security, railway technical safety as well as railway operational safety. The question is how they should be understood, and what are the differences between them.

All technical, organizational and procedural means which are used to support passengers' health and protect railway against vandalism, against terrorism as well as against natural disasters support railway transport security. Also security staff like for instance railway police together with their officers, rules and procedures support railway transport security. All technical means, which are used to protect railway against incorrect preparation of the movement authority (MA), overpassing distance limits, overspeeding, and electrical hazard support railway technical safety. All organizational and procedural means, which are used to protect railway against incorrect preparation of the MA, overpassing distance limits, overspeeding, and electrical hazard support railway operational safety.

Some technical, organizational and procedural means support safety, some others support security, some support both although this is not usual. Use of single technical or procedural solution may ensure many protection functions ensuring protection against different kinds of hazards including those related to safety and those related to security.

Safety and security general hazards are summarised on Fig. 1 forming overall part of the CBTC SSIRM model, which is subsequently subdivided below into safety and security part.

Fig. 1. Railway transport general hazards [own study]

3.1 Protection Against Railway Safety Hazards Ensuring Safe Railway Operation

Railway safety hazards pointed above create danger for railway operation. Without any doubt operation must be safe - operational safety must be ensured. Following danger situations can be pointed as examples:

- 'cornering collisions' when one train hits another train aside on the switch,
- 'end-on collisions' when one train hits another train's end,
- 'head to head collisions' when two trains hit each other after running against each other on one track,
- 'switch movement derailments' when track switch is operated under the train causing redirection of a part of the train e.g. wagons go right while LOC already went left,
- 'overpassing distance limits',
- 'over-speeding derailments',
- 'level crossing accidents'.

It would be perfect if there will be no such dangers at all. To minimise occurrence of such situations railways, from the very beginning, introduced strict rules and procedures regulating duty of the trackside signalling staff (e.g. dispatchers, signalmen, lengthmen – persons operating manual level crossing protection systems) and duty of the on-board staff (e.g. train drivers, conductors, inspectors). Growing speed of trains together with enlargement of braking distances as well as growing need to enlarge railway lines capacity and therefore lower headways between trains, shortly put railway engineers in front of the question how to support safety related railway staff and minimise risk associated with human errors.

3.2   Protection Against Railway Safety Hazards Safety Related Railway Technical Means – CBTC and Signalling

Functions of the signalling systems and devices as well as CBTC systems are already described above. However one extremely important aspect needs to be pointed. All technical means, including the ones dedicated for ensuring railway safety, have limited reliability. Failures occur and will occur even in perfect systems and devices. The key questions are how to divide failures into safety-critical and non-safety-critical, and how to minimise occurrence of safety-critical failures, and how to minimise consequences which are caused when they occur.

The basic rule for railway engineers is to construct systems and devices in such a way, that failures lead to safe operational situations. On one side, when interlocking fails, colour light signal protecting non-occupied track section may display STOP aspect (red) instead of GO aspect (e.g. green). On the other side, when interlocking fails, colour light signal protecting occupied track section must not display GO aspect (e.g. green) instead of STOP (red). This is known as a fail-safe principle (FS principle). The correctness of fail-safe principle's application in case of mechanical, electro-mechanical and electrical systems, including relay systems, was and still is being verified during commissioning by simulating failures e.g. by blocking some elements and introducing short circuits.

In case of the electronic programmable technical means, like computerised interlockings, application of the fail-safe principle is possible only on the level of functional modules. It is possible to disconnect one of the computing modules and verify result, but it is not possible to verify consequences of incorrect functioning of such a module. As a result introduction of electronic programmable systems and devices took place after creation of a safety-integrity-level principle (SIL principle). This principle defines applicable rules separately for random failures and for systematic failures. Protection against random failures is based on reliability, availability and maintainability of modules and elements, while protection against systematic failures is based on rules applicable to design, production, installation, commissioning, exploitation, repair and utilisation, which must be obeyed by staff. For safety-critical solutions level SIL 4 is required, which means e.g. that probability of a random failure causing safety-critical-failure must be lower than 10–9 (10E−09) per function per hour, safety plan and reliability plan must be prepared, different programs must be used in different computing channels, etc. Finally safety proof as a document with predefined structure and required minimum content must be prepared and verified by independent safety auditor [8].

Due to complexity of the CBTC systems both principles - fail-safe and safety-integrity-level are applicable. However that does not mean, that both principles apply to each function, which is performed by defined communication based train control and management system. In each case details are given in a dedicated safety proof known as a 'safety case'.

CBTC SSIRM model in its safety part is shown on Fig. 2. It is taking into account safety principles described in this chapter and safety related signalling, train control and traffic management functions described above. Proposed model is created for assessment of technical systems and therefore takes into account technical means, which are used to protect railway against incorrect preparation of the movement authority (MA), over-passing distance limits, overspeeding, and electrical hazard. Railway system use also organizational and procedural means to protect railway against those hazards, however they have to be treated as a fall-back for technical means. Organizational and procedural means also have additional roles e.g. in maintaining infrastructure, maintaining vehicles and maintaining staff competences.

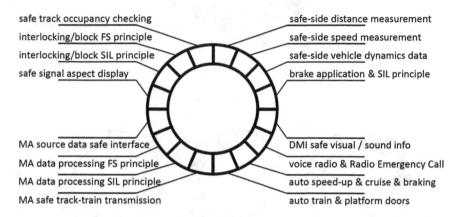

Fig. 2. Communication Based Train Control and management systems Safety and Security Impact Reference Model CBTC SSIRM - impact on railway safety [own study]

The safe-side expression needs to be well understood. As an example when on-board device know that distance to the end of authority is e.g. 1200 m, it needs to know how it is decreasing. Measuring is never 100% accurate. Measured distance may be under-estimated or over-estimated. Being on the safe-side requires on-board equipment to assume that measured distance is over-estimated, which means that maximum measurement error must be subtracted to achieve due stopping position. Safe-side principle means that train can be stopped some meters before the end of authority and must not pass end of authority due to incorrect distance measurement.

Figure 3 shows grouping of safety related functions, which allow everyone to understand character of system in question instantly. Fields associated with well-defined functionalities are marked – e.g. covered by defined colour – when respective functions are offered and left empty when they are not. As these functions can be grouped in five groups different colours are used. To make sure, that CBTC functions, which are based on safe TTT transmission are seen together although they are separated into track-side functions and on-board functions the same colour is used for these two groups.

As an example Fig. 4 shows: case A – station interlocking/line block, case B – TTT based ATC system (e.g. ERTMS/ETCS) as ATP will not include for instance safe-side vehicle dynamic data, case C – supporting train driver by voice radio and emergency stop (e.g. Polish 150 MHz radio system).

396   M. Pawlik

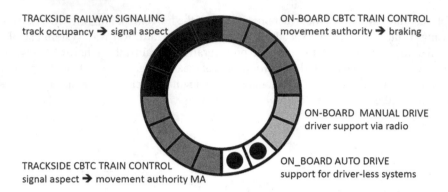

TRACKSIDE RAILWAY SIGNALING
track occupancy ➜ signal aspect

ON-BOARD CBTC TRAIN CONTROL
movement authority ➜ braking

ON-BOARD MANUAL DRIVE
driver support via radio

TRACKSIDE CBTC TRAIN CONTROL
signal aspect ➜ movement authority MA

ON_BOARD AUTO DRIVE
support for driver-less systems

Fig. 3. Grouping of functions impacting railway safety [own study]

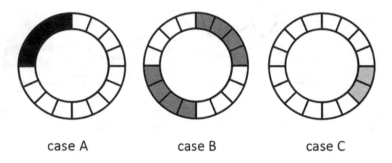

case A   case B   case C

Fig. 4. Grouping of functions impacting railway safety – cases A, B, C [own study]

Case A – shows no TTT transmission and no CBTS functionality both on-board and trackside. It shows that there is no on-board auto drive functionality which should be expected as TTT transmission is first of all used for CBTC functions and later for supporting ATO functionality as ATO requires additional data. Case A shows that in case of traffic disturbances and operational emergency caused e.g. by another train driver is not supported neither by visual or audio warning nor by voice radio.

Case B – shows, that TTT transmission and CBTC functions are provided. System has to be used in cooperation with case A type system as data for CBTC trackside functions have to be prepared in a safe way and taken is a safe way. In emergency however driver is supported.

Case C – shows mobile emergency stop, which can be initiated by track-side staff and on-board staff together with full radio coverage. Such systems are introducing traffic disturbances but ensure significant safety in case of emergency. Typically, when stop request is send by staff, train drivers in all affected trains are required to switch to radio emergency channel on which person that initiated stop request has to inform all concerned about situation.

3.3   Protection Against Railway Security Hazards Security Related Railway Technical Means – CBTC and Monitoring

CBTC SSIRM model in its security part is shown on Fig. 5 Protection against security hazards is sometimes provided by individual security management systems frequently requiring dedicated staff. Due to different kinds of interconnections between different security hazards such systems are frequently combined and centralised. This however creates a human attention barrier. Specialised technical means are therefore used to minimize required human attention keeping high security.

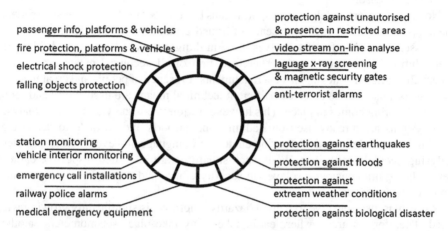

Fig. 5. Communication Based Train Control and management systems Safety and Security Impact Reference Model CBTC SSIRM - impact on railway security [own study]

Special attention is always required in relation to fire protection. In case of rolling stock recently after long and difficult standardisation works, national differentiated standards are being replaced by an European one defining in details requirements for materials, for fire barriers and many more. It is also required to remember about fire sensors, fire protection equipment e.g. fire extinguishers, etc. Types and quantities are usually defined e.g. by national construction law, but the way how evacuation and rescue is organized and what type of technical means are therefore required is within individual solutions.

Passengers need information about timetable. Traffic disturbances influencing timetable need to be reflected in passenger information as soon as possible, both on platforms and in vehicles. Lack of information e.g. due to lack of necessary technical means is dangerous as may lead to panic.

Passengers must also be protected against electrical shock Such risk is eliminated by defining restricted zones. Protection against electrical hazards have to be monitored continuously e.g. by continuous checking that traction power substations and power supply and signalling cabinets remain closed. In some cases also protecting staff and passengers against falling objects has to be taken into account and ensured.

Vandalism need to be punished. This requires continues video monitoring. The amount of data is a challenge from the point of view of data transmission, data storage, and length of video stream, which need to be observed. Platforms and other zones, which are open for public, need to be equipped with emergency call installations like yellow phones, red buttons, passenger emergency brakes. All of them should be connected with railway police headquarter. Additionally free access zones can be equipped with medical emergency like e.g. defibrillators or oxygen stands. Medical service or minimum railway police need to be immediately warned when someone is going to use them e.g. thanks to associated sensors.

Nowadays unfortunately all transport means have to be protected against terrorism. The simplest way is to ensure presence of armed paramilitary staff. However in such case passengers rather feel themselves to be in danger than to be protected. Morover such solution is extremely costly. Therefore, first of all, technical means are used to protect all restricted zones against encroachments. Presence of people and creatures like e.g. dogs or rats must be detected. Different technical means are used. More and more video monitoring comes in place. This however requires on-line video stream analyse. Such systems are already used on stadiums and airports and are able to detect e.g. crossing virtual safety borders, leaving unattended language and even picking prowlers analysing people's behaviour. When such situations are identified respective camera view is being immediately displayed in security centre and due staff is being warned acoustically and visually.

Railway is also sensitive to natural hazards. Such risk is significantly growing with speed. Therefore in areas, where earthquakes may take place seismometers or other devices are used to automatically impose speed restrictions when necessary. This is important not only along fault lines and near volcanos but also for instance in mining areas.

Figure 6 shows grouping of security related functions, which allow everyone to understand character of system in question instantly.

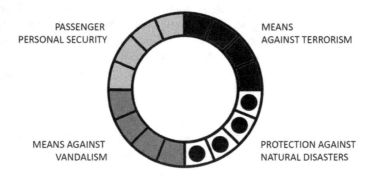

PASSENGER
PERSONAL SECURITY

MEANS
AGAINST TERRORISM

MEANS AGAINST
VANDALISM

PROTECTION AGAINST
NATURAL DISASTERS

Fig. 6. Grouping of functions impacting railway security [own study]

As an example Fig. 7 shows: case A – zones open for public equipped with passenger info, and basic passenger safety means, case B – open zones protected against vandalism and additionally open and restricted zones protected against terrorism (e.g. Warsaw metro lines), case C – all infrastructure protected against natural possible disasters (e.g. against earthquakes in case of French and Japan high speed railways).

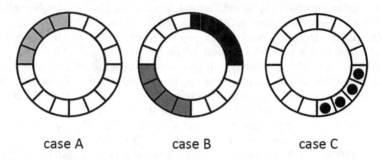

case A   case B   case C

Fig. 7. Grouping of functions impacting railway security – cases A, B, C [own study]

4   CBTC Safety and Security Impact Reference Model Sum-up

Safety and security are separate domains however they are complementary, especially in case of fully automated APM type functionally separated rail transport systems. Large variety of solutions used for ensuring safety and security need reference model which makes it possible to understand safety and security character of system in question instantly. The CBTC Safety and Security Impact Reference Model SSIRM allows decision makers to express their expectations. Additionally it allows experts preparing applicability analyses, e.g. for feasibility studies, to reject some solutions at the beginning of their work. Alternatively experts may use the CBTC SSIRM model to verify if common use of some systems covers safety and security well enough.

The CBTC SSIRM model does not exclude use of any existing system. It shows which safety and security functions are performed and allows comparison between needs and offered range of functions and their impact on safety and security as well as inherent safety. The CBTC SSIRM model does not take into account economical and financial consequences, respectively associated with society profits and with future expected business profits.

References

1. Rail Vehicles: German Railbuses, Locomotives, Multiple Units, Proposed British Rail Vehicles, Rail Vehicle Manufacturers, Rolling Stock, Trains, Rail. University-Press.Org, Joint Publication (2013). Joint publication. ISBN 123063228X, 9781230632285
2. R.C. Dorf (ed.): The Engineering Handbook, 2nd edn. Library of Congress Card Number 2003069766. CRC Press, LCC, New York, Joint Publication (2005). ISBN 0-8493-1586-7
3. Lichtberger, B.: Track Compedium, Track System – Substructure – Maintenance – Economics. Eurail Press, Hamburg (2011). ISBN 978-3-7771-0421-8

4. Convention concerning International Carriage by Rail (COTIF) Appendix C – Regulations concerning the International Carriage of Dangerous Goods by Rail (RID), Bern (2012). applicable from 1 January 2013 in Albania, Algeria, Armenia, Austria, Belgium, Bosnia and Herzegovina, Bulgaria, Croatia, Czech Republic, Denmark, Estonia, Finland, France, Georgia, Germany, Greece, Hungary, Iran, Iraq, Ireland, Italy, Latvia, Lebanon, Liechtenstein, Lithuania, Luxembourg, former Yugoslav Republic of Macedonia, Monaco, Montenegro, Morocco, Netherlands, Norway, Poland, Portugal, Romania, Serbia, Slovakia, Slovenia, Spain, Sweden, Switzerland, Syria, Tunisia, Turkey, United Kingdom, Ukraine
5. GE/RT8000 Rule Book, Railway Safety and Standards Board RSSB, London (2003). Applicable from 6 December 2003 in Great Britain
6. European Rail Traffic Management System, functions and technical solutions overview – from an idea to implementation and exploitation (Europejski System Zarządzania Ruchem Kolejowym, przegląd funkcji i rozwiązań technicznych - od idei do wdrożeń i eksploatacji), Marek Pawlik, KOW, Warsaw (2015). ISBN 978-83-943085-1-3
7. Nowakowski, T. (ed.): Joint work: Safety and Reliability: Methodology and Applications. Taylor & Francis Group, London (2015). ISBN 978-1-138-02681-0
8. EN 50129:2003/AC:2010: Railway applications - communication, signalling and processing systems - safety related electronic systems for signalling
9. Philpott, D.: Personal Security and Safety, A Comprehensive Handbook and Guide for Government & Corporate Managers & Leaders, Address All Hazards - Natural & Manmade. Government Training Inc., Longboat Key (2011). ISBN 978-1-937246-84-6

The Concept of the Integrated Global Maritime Transportation Space-Based Data Collection and Distribution Telematics System

Ryszard K. Miler[1][(✉)] and Andrzej Bujak[2]

[1] Faculty of Finance and Management, WSB Gdańsk University,
Aleja Grunwaldzka 238 A, 80-266 Gdańsk, Poland
rmiler@poczta.onet.pl
[2] Faculty of Finance and Management, Wroclaw Banking School,
Fabryczna 29-31, 53-609 Wrocław, Poland
andrzej.bujak@interia.pl

Abstract. The need for global and integrated maritime situational awareness capability is caused by unstable (in terms of security and safety) situation in global shipping represented by recently occurred issues (piracy, terrorism), having additionally an impact on economical conditions for modern shipping. As a response to growing demand for integration of all existing maritime data collection systems the concept of common platform for maritime data exchange has been developed. This telematics solution must be a cost-effective tool in gathering, transforming and dissemination of an integrated and comprehensive space-based maritime transportation data. This paper, based on existing systems research, represents the holistic approach to identification of important factors suitable for the concept of global integrated system. The working hypothesis stating that there are systematic reasons for implementation of a single, common and integrated telematics platform of maritime data exchange has been positively proved. As an example of efforts already taken the GLADIS solutions have been presented.

Keywords: Satellite AIS · Shipping monitoring and telematics systems · Maritime data collection · Ocean Data Telemetry Microsat Link (ODTML) · Global Awareness Data Extraction International Satellite constellation (GLADIS)

1 Introduction

Safety, security and freedom of navigation come as "constitutional" rights of all maritime states, and nowadays they have become one of the main determinants pertaining to economic efficiency of maritime transport processes [5, 11]. The problem of providing safety and security of maritime transport by efficient counteraction becomes crucial. It is no longer just a local or even a continental question; its character has become global and international [14], however without relevant data there is no possibility to fulfill this mission. Because much of the oceans and open seas surface is "unwired", and because illicit activity may be developed in areas where there is a lack of transparency almost immediately the need for data collecting from such as "unwired" places appears. And

© Springer International Publishing AG 2016
J. Mikulski (Ed.): TST 2016, CCIS 640, pp. 401–414, 2016.
DOI: 10.1007/978-3-319-49646-7_34

furthermore, due to technical limitations, it suggests solutions based on global space telematics systems focused on open access and financial and technical "affordability" [15].

Although a significant amount of national agencies, consortia and commercial institutions have already introduced space-based AIS (AIS-Satellite) systems with data extraction and handling capabilities, there is still an issue of such disparate systems sufficient level of interoperability, affordability and transparency and first of all cooperative possibilities to not exclude those nations with significant shortfalls in finances and technical infrastructure. Diversification of existing maritime transportation data collection and shipping monitoring systems is based on a different range of covered areas, direct benefits for end-users as well as different technical specifications of modern maritime sensors. In decent years global maritime transportation monitoring and awareness (including environmental aspects) has become more important that is why all supporting technical devices (such as sensors, acquisition subsystems and effectors) are becoming more ubiquitous as well.

In the light of the above mentioned reasons it becomes obvious that there is a need to establish a common, space-based and integrated telematics system in order to achieve internationally shared maritime data collection, analysis and distribution ability. In order to explore complex of issues linked to the technical and financial aspects as well as policy and international law this conceptual system should be managed by dedicated entity. This idea is possible to achieve by establishing "a public/private partnership that would create a low-cost, internationally shared data collection and distribution backbone in space with exceptionally low barriers to entry for any participating nations" [15].

2   Post 9/11 Maritime Security Backgrounds

The tragedy of 9/11 clearly indicated for all nations worldwide the need for global maritime monitoring and international collaboration and, in consequence, highlighted the potential of already existing systems for further consolidation, integration and free information sharing in order to better understanding situation in the maritime domain. "Most countries saw increased maritime domain awareness (MDA) as of first importance to the smooth functioning of commerce on the world's oceans, the crucial supporting frame of the world's economy, and crucial to their national interests. The potential unique contributions of current and planned space systems, owned by a wide range of nations and available to many others, to international global maritime awareness is a subject of growing interest to many" [16].

From definition Maritime Domain Awareness is an "effective understanding of maritime activity and its impact on safety, security, the environment and the economy" [17] and from definition it needs a contribution of maritime data. The fundamental step in order to achieve ship tracking ability and gather the maritime data in post 9/11 world was an introduction by the International Maritime Organization (IMO) in 2004 the Automatic Identification System (AIS), that became mandatory for all ships of 300 GT and above and immediately has opened ability for global monitoring of maritime transportation.

AIS is the system based on "self reporting" configuration which means each ship over 300 GT is obliged to broadcast over VHF channels its identity and position (among many other data fields such as heading, nature of cargo, next port of call etc.). Despite the original intent of AIS was to provide only a short-range collision-avoidance information (ship-to-ship, ship-to-shore stations) it has become a principal source of maritime transportation monitoring and data collection (with certain limitations) [18]. According to the technical demands all participating ships send their reports in two seconds interval with typical coverage of twenty nautical miles radius (with higher elevation of terrestrials/antennas the range is extended up to 50 Nm) [15].

As it was stated the AIS as short-range system has never been considerate as a long-range (space-based) system. However the range of this system has been significantly extended by the introduction of possibility to use a constellation of microsatellites that can communicate with "classic" AIS transponders via satellite connections. This way upgraded new system has been introduced as AIS – Satellite (AIS-S) with the bundle of totally new solutions and applications extremely useful for competent maritime authorities. What is more important, due to use of microsatellites the visibility scope is significantly enhanced up to 600-800 Nm from one satellite and thanks to this significant change the quality and range of the maritime situational awareness has been extended dramatically [3, 19].

Furthermore, a unique technology which is able to collect, process, distribute and what is also very important, to archive AIS messages received from ships all over the globe has been developed by utilizing advanced Satellite AIS detection capability as well as ability to capture thousands of vessels in a single satellite pass [15].

The use of AIS-S system has not raised any questions on its navigational (safety aspects) utility, however in the same time some important questions regarding collection (the type of sensors), processing and dissemination (the type of data format) have raised [20]. Entities involved in security and safety processes such as maritime, defense and environmental agencies (being also stakeholders in national/international maritime governance) are primarily users and may deeply benefit from AIS-S integrated data in their way to produce much more relevant and developed "operating picture". In addition, any other organizations involved in global maritime trade and transportation, such as maritime transport operators (MTO), terminal, and logistics operators may also have an interest in having access to the AIS-S date, since extended knowledge of their ship location, traffic condition, sea state etc. can contribute to better business processes understanding and optimization of all logistics processes. Apart of above mentioned near real-time data use there is an additional benefit for the entitled users from historical data (e.g. discovered patterns, anomalies, hidden capabilities) [15]. Taking into account systematic demands from such subsystems as Search and Rescue (SAR), Vessel Traffic and Monitoring Systems (VTMIS), Vessel Traffic Systems (VTS), Maritime Assistance Systems (MAS), SafeSeaNet, CleanSeaNet, Maritime Domain Awareness (MDA) and Maritime Situational Awareness (MSA) as well as ocean monitoring and weather monitoring (for disaster management and emergency response) it is strongly recommended to support all efforts in order to integrate all available maritime data from all existing sensors (systems) into one integrated system. What is more these systems (even treated separately) are more effective, flexible and near real-time with data collected from space and if integrated an effect of synergy will certainly occur [5–7, 9, 15].

3   A Survey on Current Space-Based Maritime Data Telematics Systems

Developed by D. Meldrum, the theory of mobile and small satellite systems of data collection recognized two trends in such systems development [21]:

- The economic effects of introduction of small satellites (lower costs);
- The technical effects of introduction of small satellites (power reduction, technical complexity of e.g. minisats, microsats, nanosats, picosats and even femtosats).

In all maritime countries on all continents (especially Europe, North America, Australia) a number of maritime data extraction and monitoring systems has been already introduced. Some of these systems utilize AIS-S capability with focus on micro-satellite technology.

Despite of all efforts most space-based AIS systems have been implemented at the national level and among regional organizations (e.g. European Maritime Safety Organ-ization) having limited possibilities for further integration. Due to the fact that there are more than 100 different systems already introduced worldwide authors decided to cover in their research selected shore-based and space-based systems (AIS, AIS-S) that can easily give an overview on existing systems capabilities and their potential for possible further integration.

3.1   International Worldwide Systems

In order to improve and maintain maritime database the Maritime Safety and Security Information System (MSSIS) has been introduced and made available to maritime nations worldwide. The MSSIS is built as a platform for maritime situational awareness exchange based on unclassified, near real-time data collection and distribution network. System is currently used by approximately 60 nations and maritime data is freely shared between all participants (data is shared without filtering, analysis or other "value added" options, so there is no demand for advanced technological and financial conditions/ prerequisites). It makes this system available for all countries regardless their econom-ical and technological status [15]. MSSIS supports bilateral collaboration between inter-national bodies involved in safety and security processes in maritime domain and thanks to maintaining an original (open to all) format is able to integrate data from single national sensors as well as a regional vessel-tracking network. Due to above mentioned ability MSSIS benefits its users with near real-time data that can be utilized by nations according to their individual demands and mission requirements having still possibility to deliver a "common picture". Figure 1 shows the current interface of MSSIS (with use of Transwiev - TV 32 software) [11].

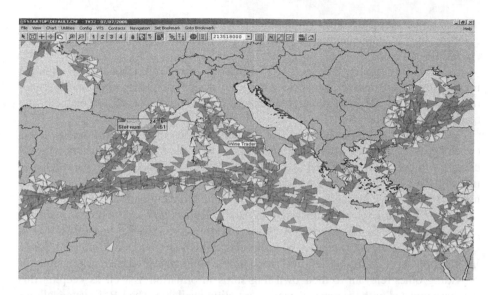

Fig. 1. Interface of MSSIS extracted from Maritime Operations Centre, Gdynia Poland systems [own study]

As a common system's interface and vessel tracking display serves the TV32 application, which offers a variety of additional features and also functions as a gateway for international users to access and contribute in building aggregated picture. The MSSIS has been upgraded permanently and amount of participating nations, organizations and other entitled bodies is growing. It allows stating that such a system is extremely useful in improving and maintaining MDA worldwide. However some regions (e.g. EU) has developed their own systems.

3.2   EU Systems and They Transition into Telematics Solutions

The European Maritime Safety Administration (EMSA) has developed the maritime data exchange system SafeSeaNet with ship monitoring and tracking module called SafeSeaNet Tracking Information and Exchange System (STIRES) [1, 14]. The information in SafeSeaNet is gathered from several message types (based on AIS data), including port and ship notification messages, Hazardous Materials notification messages (HAZMAT) and other reports (such as incident or weather reports). Thanks to the STIRES module activities linked to the surveillance, monitoring, tracking and shipping patterns discovery are more efficient, interoperable and time relevant [2, 15].

In order to improve the ability of EU maritime (coastal) countries in building an integrated satellite-based ship tracking information system the Long-Range Identification and Tracking of ships (LRIT) system has been introduced. Data gathered by this system is integrated into the STIRES display available to all authorized users. [11, 15]. The initial purpose of LRIT is to enhance safety and security by providing details of ships' identity and current position in order to evaluate and proper management of the existing risk posed by any single ship in configuration with other traffic details. It also helps in proper respond to already identified risk by the Governments of all participating countries (e.g. by implementing necessary measures, contingency and emergency plans etc.) [23].

Despite of its security utilities, LRIT has also potential safety benefits for example for maritime search and rescue (SAR) emergency actions delivering accurate and time-relevant picture of maritime situation on any given sea region that affects positively any rescue mission [23].

Technically the equipment of LRIT system consists of ship borne transponders (working as a information transmitting devices), satellite system and ashore national centers equipped with dedicated IT technology (severs, computers, software). Operationally, the LRIT consists of the national LRIT Data Centers (supported by any related national Vessel Monitoring System), the LRIT Data Distribution Plan and the International LRIT Data Exchange subsystem. System assumes that each national maritime administration (National Shipping Authority) is obliged to provide to the LRIT Data Centre "a list of the ships entitled to fly its flag, which are required to transmit LRIT information" [18]. Maritime administration on the national level is also responsible for providing other relevant to their own flag ships information as well as keeping this database updated with the most current information available. The LRIT system allows ships to transmit information only within its operational structure through National, Regional, Co-operative and International LRIT Data Centers by using the LRIT International Data Exchange subsystem [4, 23].

For the logistics purposes LRIT system should be treated as an additional to the "standard" AIS source of vessel positioning information, which can improve safety, timely and cost effective sea transportation when organized either by MTO or by selected carrier on his own risk.

In order to achieve and provide the European Maritime Authorities with more comprehensive and complex information adjusted to their requirements and to support the data exchange between particular system users, the project called Integrated Maritime Data Environment (IMDatE) has been launched [12, 14]. The scheme presenting the architecture of the IMDatE system is depicted in Fig. 2 [4].

IMDatE is assumed to integrate European maritime data and to establish platform for system integration for four applications (systems) important from European maritime safety and security perspective, already operated by the EMSA [9, 12]:

- SafeSeaNet (SSN) – the European coastal system with over 700 shore-based AIS stations responsible for automatically tacking all ships in the vicinity of 100 Nm from EU coastline and maintaining the database (cargo, voyage details) of all ships of interest;
- EU LRIT Data Centre (LRIT DC) – the European long-range system using satellite communication for mandatory tracking of all ships under EU country flag worldwide and any ship (irrespective of its flag) within maximum range of 1,000 Nm from the EU coastline [7];
- CleanSeaNet (CSN) – the European satellite-based system introduced for oil spills detection and evaluation with use of Satellite Aperture Radar (SAR) images;
- THETIS – the European ship inspection web-based application with aim on providing information to all Port State Control (PSC) officers.

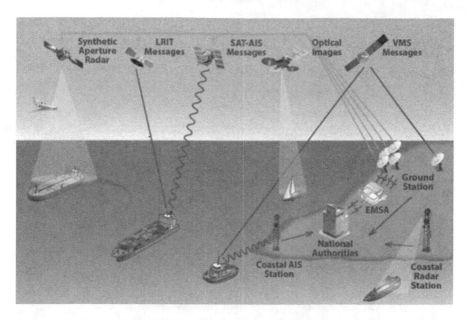

Fig. 2. Basic architecture of the integrated IMDatE system [4]

The IMDatE is being developed as a complex (still experimental) system integrating already existing capabilities into one common picture and database aiming to provide more complex, relevant, updated information for all EU member states utilizing basic telematics architecture. The IMDatE experimental solution is based on the latest state-of-the-art satellite technology as well as the latest principles on the field of Service Oriented Architecture (SOA) and Integrated Product Development (IPD) [13]. Apart of commercial efforts there are some military concepts and systems (e.g. NATO) supporting idea of space based maritime data integration.

3.3 NATO Systems

In partnership with the US Naval Forces Europe, the NATO Component Command Maritime (CCMAR) in Naples, Italy, has developed the MSSIS and both have been responsible for large dissemination of this system among European and North African countries (initially among the Mediterranean riparian states, later among all EU states and seas) [10]. CCMAR, until was closed, maintained also responsibility for the maritime safety and security on Mediterranean, using MSSIS as the primary tool enhancing near real-time shipping picture and in the same time opening window of opportunity to feed with data stream two more visualization and analysis tools: Baseline for Rapid Iterative Transformational Experimentation system (BRITE), and Fast Connectivity for Coalition Agents Program (FastC2AP) application. The interface of BRITE application is depicted in Fig. 3 [11].

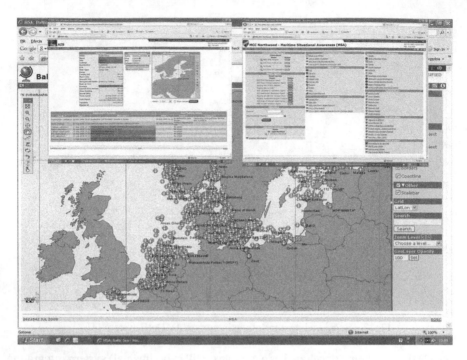

Fig. 3. Interface of BRITE application extracted from Maritime Operations Centre, Gdynia Poland systems [own study]

Initially, BRITE as an experimental system was responsible for fusion of entire available maritime AIS and AIS-S data with other relevant to shipping information (e.g. ownership, insurance, ship behavior patterns, anomalies, etc.) in order to achieve wider platform for analysis, comparison and anomalies discovery by using FastC2Cap functions [15]. FastC2AP, from the other hand, was among the newest intelligent IT solutions applied to the maritime awareness giving for the first time in this area a foundation for telematics system development. FastC2AP is based primarily on AIS (AIS-S) data and is fitted with program triggers based on smart agents; it enables to create alerts for numerous previously defined "suspicious" conditions or movements (e.g. two ships in drift caused by potential trans-shipment of illegal/excise goods, personnel or illegal migrants) without need for operator's analysis [24].

Virtual Regional Maritime Traffic Centre (VRMTC) established within the Italian Navy organizational structure is an additional noteworthy example of European military (navy) initiatives. VRMTC is being run as a network enabling connections with national maritime military operational centers located in the all EU/NATO countries altogether with selected African countries based on a specific "Operational Agreement" among 23 "Wider Mediterranean Community" navies. VRMTC is responsible for collecting, fusion and dissemination of unclassified maritime information that can improve MDA and MSA ability. Military related information exchanged between participating nations may be different while the nature of positional data is the same due to use of the same sensors (AIS, AIS-S based) [15].

Due to the necessity of further development of NATO ability to integrate maritime data a concept of Maritime Functional Services has been developed through Future Maritime Functional Area Services (FUMARSER) that is under development phase and is being created with focus on further telematics capabilities. FUMARSER has to be compliant with NATO Network Enabled Capability (NNEC) and therefore will take into account NATO C3 System Architecture Framework (NAF) [25]. The initial FUMARSER architecture is depicted in Fig. 4.

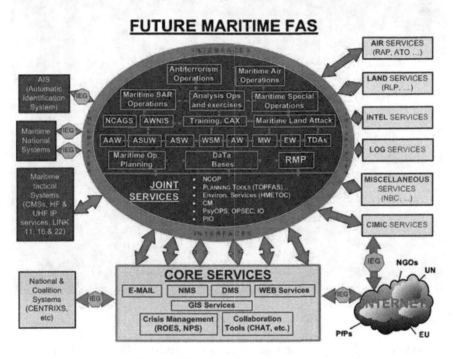

Fig. 4. FURMARSER initiative basic architecture [25]

Apart of the development of NATO integrated systems the US Maritime authorities supported by the US Coast Guard present national and different approach.

3.4 US Systems

The US Coast Guard (USCG) has implemented system known as Nationwide Automatic Identification System (NAIS). As all systems based on AIS transmission, NAIS is designed to collect AIS data from all vessels entering or maneuvering within the US exclusive economic zone – EEZ and territorial waters - TTW, primarily focused on traffic in 58 critical US ports, however system enables to collect data from adjacent sea areas out to approximately 2,000 Nm. The main goal of NAIS is to build a platform for maritime data exchange in order to achieve better MDA understanding through data dissemination and analysis. NAIS users are mainly USCG maritime traffic stations, SAR and environmental protection services [15].

The collected data is combined and fused with other government intelligence and surveillance to form an updated and holistic overview on maritime traffic within or near (USTTW) and is shared with other governmental entities and institutions for further use an analysis [15, 22]. The NAIS operational architecture is depicted in Fig. 5.

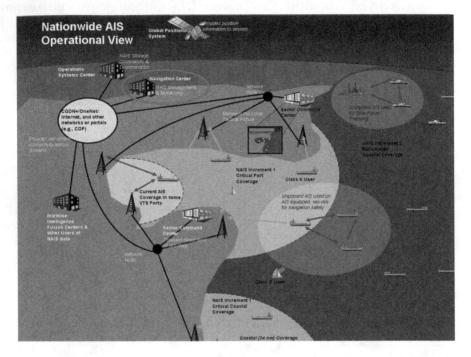

Fig. 5. NAIS Integrated System [22]

In addition, with cooperation with business entities, the USCG has introduced the program of migration into AIS-S platform and use of Long-range satellite systems. It determines increase of ability to provide better MDA solutions. There is, however, at least one restriction that can reduce utility of NAIS worldwide. There is still commercial ban preventing freely sharing US maritime data with international partners, which makes significant constrain for the idea of maritime systems global integration.

4   Space-Based AIS and Maritime Data Extraction Systems Integration Possibilities

Based on provided up to this point the holistic research, authors have assessed the current status quo of existing satellite systems and AIS, AIS-S based applications leading to the conclusion that there is a need for its further integration. The level of disintegration of existing space-based maritime data extraction and dissemination systems is relatively high, thus the idea of common space-based maritime data system (as a backbone for development of further concepts) has to be introduced. In further conceptual processes a high level described model of ideal system should be described. Such a model should present affordable and highly accessible space communication infrastructure as well as solutions for technical specification of deployed satellites having impact on political and operational aspects of proposed model [5–8].

A good candidate for such an "ideal" concept is the Global Awareness Data Extraction International Satellite constellation (GLADIS). GLADIS is the United States Naval Research Laboratory (USNRL) project, which ideally frames indicated on the beginning of this article initial systematic assumptions (system should be affordable, internationally shared long-range space based AIS with exceptionally low barriers to entry for possibly participating nations) [15].

The concept of GLADIS is based on the constellation of 30 low-cost nano-satellites arranged in five planes at altitude 550 km internationally owned by a consortium of nations. This solution enables sharing not only costs of ownership and maintenance but also resources, responsibilities and most of all benefits. GLADIS can provide a data link (via satellite connections) within 10 min for any location (including so far un-wired

Fig. 6. The space-based AIS and data extraction backbone [15]

areas of open sea and ocean). This includes ability to transmit routinely the maritime data of over 80 thousand of ships nearly continuously. Data from maritime sensors is packaged for secure transmission and immediately sent to the nearest GLADIS ground station for further processing and delivery via Internet to the sensor owners and entitled entities. Figure 6 illustrates the concept [15].

The satellites designed for the GLADIS constellation are intentionally not technical sophisticated in order to reduce initial costs, however they are capable of providing all necessary functionalities (such as certain data stream capability, data link payload, at least 3 years in orbit) in order to deliver maritime data from un-wired regions straight to anywhere located "wired" end users. The satellites used for GLADIS constellation are relatively small (size approximately 30 cm and 15 kg weight), which gives an additional opportunity to lower launch costs (they are small enough to be launched with other satellites) [15]. Thanks to specially dedicated application of Ocean Data Telemetry Microsat Link (ODTML), raw data from sensors is translated into standard RF message and transmitted to both facilities (GLADIS satellites or ground-located receivers). Data stream should be unclassified and easy to share (without political, economical and technical boundaries) and not to be sold on a e.g. subscription basis [15]. Above mentioned functionalities make GLADIS uniquely suited for further development of common integrated satellite based international maritime data collection system.

5 Conclusion

The GLADIS concept makes an excellent candidate for "ideal" concept and can be used as a basis for further works especially for further introduction of telematics capabilities (gathering the megadata, analysis and decision making, automatic and effective decision execution). Development and progress for defining and deploying a global space-based maritime data collection integrated system demand at least to accept such as assumptions:

- System must be capable to operate with low bit-rate data flows (raw data streams) and the constellation of satellites (as a systematic component) must be prepared for integration of all data formats by the use of specific application (e.g. ODTML), it must be possible to integrate and accommodate all data regardless of content, format and type of sensor;
- System must be intended for further introduction of telematics solutions (e.g. automatic standard procedures execution, intelligent data processing, evaluating and analysis);
- System has to cope with data "collisions" seem as a problems related to achieving information from various data sources and sensors (e.g. AIS-S and shore-based AIS stations), effective picture can be in such situation difficult to achieve due to signal and data duplication) and problems often occurred in very congested areas (where the signal of AIS transmitter/transponder may be disrupted);
- System must be open-architecture to enable standardization of any future formats of data from any future sensors allowing its integrating, formatting and protecting without any high additional costs;

- System may serve to improve international co-operation leading to increase of global safety and maritime security regardless of separate states current economic and technical conditions;
- The maritime data collected and stored (data base) should be shared between all participants (member nations) in the free basis restricted to the authorized end users only;
- The data (sensor) owner should be responsible for data accuracy and value;
- System should not allow any security (in terms of IT protection) breeches to occur and to be disseminated through the entire system network.

References

1. http://emsa.europa.eu/combined-maritime-data-menu/data-sources.html. Accessed 24 Jan 2015
2. http://emsa.europa.eu/operations/maritime-monitoring/86-maritime-monitoring/1520-integrated-maritime-data-environment-imdate.html. Accessed 5 Feb 2015
3. http://sevenseas-marine.com/services/. Accessed 5 Feb 2015
4. https://portal.emsa.europa.eu/web/imdate. Accessed 24 Jan 2015
5. Januszewski, J.: New satellite navigation systems and modernization of current systems, why and for whom? Zeszyty Naukowe Akademii Morskiej w Szczecinie **Nr 32-2**(14), 58–64 (2012). Szczecin
6. Januszewski, J.: Satellite navigation systems applications, the main utilization limits for maritime users. Zeszyty Naukowe Akademii Morskiej w Szczecinie **Nr 36**(108), 70–75 (2013). Szczecin
7. Januszewski, J.: Satellite navigation systems in the transport, today and in the future. Arch. Transport **22**(2), 175–187 (2010)
8. Januszewski, J.: Visibility and geometry of galileo satellites constellation. Annu. Navig. Part 1 **19**(201), 79–90 (2012)
9. Lechner, W., Baumann, S.: Global navigation satellite systems. Comput. Electron. Agr. **25**, 67–85 (2000)
10. Maritime Surveillance in Practice – Using Integrated Maritime Services, p. 7. EMSA Publication, Lisbon (2014)
11. Miler, R.K., Bujak, A.: exactEarthSatellite – AIS as one of the most advanced shipping monitoring systems. In: Mikulski, J. (ed.) TST 2013. CCIS, vol. 395, pp. 330–337. Springer, Heidelberg (2013). doi:10.1007/978-3-642-41647-7_40
12. Mylly, M.: EMSA's Integrated Maritime Environment – A Tool for Maritime Awareness, pp. 16–19. EMSA Publication, Lisbon (2014)
13. de Sousa, J.: EMSA's Integrated Maritime Environment – A Tool for Improved Maritime Domain Awareness, pp. 1–20. EMSA Publication, Lisbon (2012)
14. Szcześniak, J., Weintrit, A.: Europejskie systemy kontroli i śledzenia ruchu statków – geneza, zasady funkcjonowania oraz perspektywy rozwoju. Zeszyty Naukowe Akademii Morskiej w Gdyni **Nr 77**, 89–90 (2012). Gdynia
15. Earles, M.: International Space-Based AIS and Data Extraction Backbone: High Level Requirements. CANEUS International Publications, Montreal (2010)
16. Guy, T.: International Collaboration is the Silver Bullet, Global Space Partnership- Collaboration in Space for International Global Maritime Awareness (C-SIGMA), New York (2005)
17. http://www.gmsa.gov/twiki/bin/view/Main/MDAConOps. Accessed 14 May 2010
18. http://www.imo.org/conventions/contents.asp?topic_id=257&doc_id=647. Accessed 14 May 2010

19. http://www.exactearth.com/products/exactais/. Accessed 24 May 2010
20. http://www.thefederalregister.com/d.p/2010-01-15-2010-632. Accessed 12 May 2010
21. Meldrum, D.: Developments in Satellite Communication Systems, Useful Satellite Systems for Data Buoy Operators, Data Buoy Cooperation Panel, February 2008. www.jcommops.org/doc/satcom/satcom.pdf. Accessed 12 June 2015
22. http://www.navcen.uscg.gov/?pageName=NAISMain. Accessed 12 June 2015
23. MSC 81/25/Add.1, Annex 13 Resolution MSC.210(81): Performance standards and functional requirements for the Long-range identification and tracking of ships, pp. 12–16, Official Journal of the European Community, Luxembourg (2006). Accessed 19 May 2006
24. www.marina.difesa.it/vrmtc/2007/uk/vrmtcen.asp. Accessed 14 May 2010
25. www.tide.act.nato.int/mediawiki/index.php/FUMARSER. Accessed 14 May 2010

The Research of Visual Pollution of Road Infrastructure in Slovakia

Radovan Madleňák[✉] and Martin Hudák

The Faculty of Operation and Economics of Transport and Communications,
Department of Communications, University of Žilina, Žilina, Slovak Republic
{radovan.madlenak,martin.hudak}@fpedas.uniza.sk

Abstract. Visual smog becomes a social problem in recent decades. It constitutes contamination of public space by aggressive, size inappropriate ad, placed
often illegally. The object of the article is to study a visual smog and measure its
level on selected road communication. The number of ads and billboards near the
roads, distance between the billboard and road and also the density of billboards
will be taken into account. The analysis of traffic accidents on selected road
communication is also included in the article. The eye tracking technology is used
in research to collect data regarding the drivers' visual behaviour in real traffic
conditions.

Keywords: Road infrastructure · Road safety · Traffic accidents · Visual
pollution · Neuroscience

1 Introduction

Visual smog becomes a social problem in recent decades. It constitutes contamination
of public space by aggressive, size inappropriate ad, placed often illegally without
permission. The same situation concerns the advertisement placed next to roads. The
purpose of the article is to measure the level of visual smog in Slovak republic on selected
road [9].

The classic form of visual smog contained in public places, near the roads and highways, but also in the fields include putting up posters, light and non-light large advertising billboards, advertising on scaffolding, handrails or the means of public transport
[3, 5, 13].

Visual smog can take the driver's attention but also can influence driver's psyche.
Within the plenty of visual information can traffic signs merge between the visual smog.
In many cases, the visual smog negative influences the driver behaviour [6, 14, 17].

The term billboard refers to any external advertising sign that is permanently allocated along a roadway and carries any visual information. It is possible to define two
categories of billboards. The first one is "active" billboard, if it displays changes
frequently enough that a given driver could see more than one display on the billboard
during his approach. Active billboards can be multimedia (or electronic) billboards or
signs that change displays mechanically. A passive billboard does not meet this criterion.
There is also a specific category of active billboards ("visual billboards") and those

© Springer International Publishing AG 2016
J. Mikulski (Ed.): TST 2016, CCIS 640, pp. 415–425, 2016.
DOI: 10.1007/978-3-319-49646-7_35

billboards is displaying full motion. This kind of displays (billboards) may be used on multimedia billboards in some countries, but in Slovakia it is used rarely [1, 2, 4, 8, 10].

Regulation of placement of billboards and other advertising equipment in Slovakia is governed mainly by roads act. 135/1961 Coll. on the road communication, which partly regulates the placement of advertising equipment in road traffic. Placement of billboards and advertising equipment also regulates act. 50/1976 on territorial planning and building code (Building Act). More specific rules are municipalities enshrined in a generally binding regulation on the placing of advertising equipment. However, it is only a few cities and villages within such generally binding regulations are established [15]. There is no clear concept of placing advertising equipment and also of their appearance.

Visual smog becomes also a problem for local municipalities. Many of them are trying to mitigate its impact, but given the lack of legislation is the insufficient result. Advertising equipment is in many cases placed without the permission of the building office. Municipalities started to eliminate the illegal advertising, but the pace is not sufficient, as the new illegal advertising is emerging faster than the other removed. Changes in legislation in this regard is really needed.

The billboards threaten the safety of drivers by traffic experts. On the other side, the representatives of advertising agencies say that there is no study about the negative impact. However, the traffic experts have no doubts about the negative impact banners [7, 11].

František Kocúr from Transport Research Institute says: "*The billboards can distract the driver's attention at the expense of traffic signs. It depends mainly on the frequency, distance from the road and size of the billboards*". Milan Černák, the Vice President of Slovak Automobile Club notes: "*Many billboards are illuminated and the driver can fix them as fixed point*". He also states: "*The light reflection from the billboard in rear-view mirrors is very annoying*" [18].

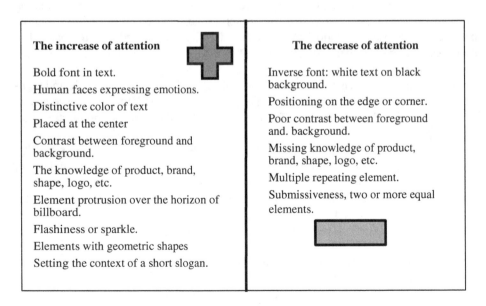

Fig. 1. The elements of billboard influencing the driver's behaviour [own study]

The number of vehicles on the roads is still increasing. Beside obvious benefits they also bring a big load growth of the road network. Transport requirements and its security is also still increasing. Transport safety is not only a serious traffic, social but also an economic problem. Traffic accidents are associated with major property damage, permanent bodily harm citizens and very often irretrievable loss of human life [12, 16].

The next figure (Fig. 1) describes the elements of billboards, which can more or less attract driver's attention what can cause a driver distraction.

2   Aim and Methodology

The first objective of this article is to *research the visual pollution* (visual smog) of selected road infrastructure. On the basis of this analysis we will try to achieve the second objective – to measure *how the visual pollution influencing the drivers* on the selected road.

To be able to solve the objectives of this article, we used the following methods, techniques and tools.

- The statistics about the traffic accidents were obtained from Traffic Inspectorate in Žilina.
- The data regarding the visual smog were collected using the GoPro camera. GoPro HERO features SuperView, a GoPro exclusive video mode that captures the world's most immersive wide-angle perspective.
- SMI Eye Tracking Glasses were used to collect data regarding the drivers' visual behaviour in real traffic conditions. This eye tracking glasses with the lightweight are designed to record a person's natural gaze behaviour in real-time in a broad range of applications, see Fig. 2. Glasses have on the bottom of the frame and also in the front frame placed high speed camera, which record the movement and position of the pupils in the eye and also the surroundings. Mobile phone is connected to glasses, which visualises the surroundings on the screen and shows the point of view by cursor.
- The SMI BeGaze software for eye tracking provides analysing and structuring information on experiments and displaying the eye tracking data as graphs, all in one sophisticated application. This application offers different types of analysis and results from experiments. One of the most useful for our purposes is Scan Path, or also called Gaze plot. It is a map which shows gaze fixations of driver and in which order they occurred, on the basis of numbered circles. The next interesting tool included in BeGaze software is AOI Editor (Areas of Interest). The AOIs can be defined for video or still images stimuli. In the function Move&Morph the AOIs change their position and size during the sequence of single video frames.

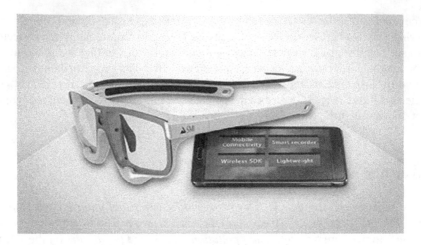

Fig. 2. SMI eyetracking glasses and mobile device [own study]

3   Analysis

The selected road communication is situated between the cities of Žilina and Martin, as it was mentioned above. It is a primary road number 18(I/18). This road was established in 1946 as one of the 16 original primary roads in Slovakia and its whole length is 304 km. The length of road stretch for research is 14 km. There is a more reasons for road stretch of primary road I/18. The first one is the high level of traffic density, the next one is the high number of traffic accidents. The third reason is the density of visual

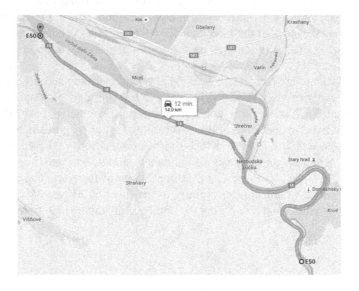

Fig. 3. Map of the selected road stretch [own study]

smog because there is a lot of billboards next to the road. The selected road stretch is shown in the following figure (Fig. 3).

Due to the absence of legislation in Slovakia is very frequent placement of illegal billboards. Specifically, in Žilina it was identified around 6,000 advertising spaces according to the portal www.zilinak.sk but only 200 of them were placed legally. The city of Žilina, together with other organizations dedicated to visual smog decided to remedy this situation. The steps that the city has taken yielded partial results of eliminating 100 illegally placed billboards, but this figure is still low compared with the total number of illegal billboards.

The problem is to eliminate the fact that almost all illegal billboards stand on private land. The owners do not have the permission from the building authority, police or authorization from the architect. To solve this unpleasant situation, the amendment to the Building Act came into force on 2nd January in 2015. The main goal is to remove all illegal advertising spaces by the end of July 2017. The main problem of the legislation was advertising equipment, which was not expressly fixed to the ground.

3.1 Research in Czech Technical University

A team of experts from the Czech Technical University have found that most drivers distracted ads with women. The aim of the that research was to research as billboards next to the road distract drivers. According to results of the research, billboard attract the driver's attention for 0,4 s. This time can be a crucial element in the meeting point of two cars, because it takes a half second, while a driver starts to brake. The car passes 10 m of roadway if the speed of car is 50 kilometres per hour in a half seconds.

A special simulator was used for this research, which simulates the driving in city but also outside the city and the researchers were measuring the eye movements of drivers. The volunteers passed the road on simulator several times. Before the last ride they were asked to notice and remember billboards. There were totally placed 23 billboards near the virtual road stretch of length 10 km. The researchers discovered several interesting results. When drivers drove for another car, then they looked at the billboard for 0,4 s in average. To remember the billboard, they needed 1,6 s. The longest fixation was at the billboard, which displayed a woman only in underwear. The average dwell time in this case was 0,6 s. One of the task for volunteers was to remember the billboards, but the research showed, that most of drivers did not remember that. The interesting finding was, that driver's fixation at the billboard was longer when driver was driving behind the another vehicle as there was no car in front of driver's car [18].

3.2 Swedish Study - Digital Billboards Distract Drivers

The most of digital billboards are created to draw attention using various types of pictures. To the year 2009, the Swedish administration didn't permit the electronic billboards and the traditional billboards next to the road was restricted. In the year 2009, the Swedish state administration gave provisional approval for the installation of several electronic billboards near the roads.

Driver's distraction is the main factor of many road accidents. The dynamic content and the position of the billboards The general results from the virtual studies showed that the multimedia content and the placement of the billboard have a major impact on driver's behaviour.

Eye-tracking research confirmed the attention grabbing nature of electronic billboards. One of the studies showed that digital billboards increased the number of driver's error (It doesn't matter on driver's experiences). Results from another study showed that roads with billboards caused more lane deviations of drivers and more dangerous situations on roads.

The main goal of Swedish study was to evaluate the impact of digital billboards on driver's performance and driving behaviour in a real traffic conditions. The number of participants was 41 and the data were collected on a motorway in Stockholm, in autumn of 2010. Twenty participants drove in daylight conditions and 21 drove in night-time conditions. The head mounted eye tracker was used to measure visual behaviour of participants [19].

Gaze analyses of the research were executed in SMI's BeGaze software. The four different performance indicators of driver's behaviour was analysed: visual time sharing, dwell time and number and duration of fixation. The results from this research are that The signs are less attractive for participants than the billboards. Regarding of driver's behaviour in daytime and night-time, there is no significant difference. There were more fixations at the billboards during the daytime than during the night-time (75 fixations during daytime and 61 during night-time totally). If we compare, for the other traffic signs were 23 fixations during daytime and 42 fixations during night-time. It means, that for digital billboards are higher the visual time sharing intensity, dwell times are longer and the number of fixations for digital billboards is greater.

According our measurement, the number of the advertisements next to the selected road stretch is 348. When the distance of analysed road was 28 km, it means that every 80 m of the road is placed 1 advertisement. The density of the advertisements is very high in certain stretches. Most of them are classical billboards and size of most billboards is 504*238 centimetres, see Fig. 4.

The average distance between the billboard and road is approximately one and half meter. There are no digital billboards on this road stretch. The following picture (Fig. 5) captures the image of classic billboards on selected road communication.

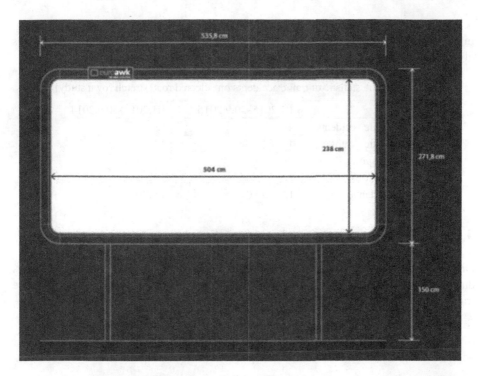

Fig. 4. Typical size of billboard near the road [own study]

Fig. 5. Example of visual smog next to the selected road [own study]

Transportation science researchers have long been aware of the negative impacts of driver's inattention on driving performance. The research over crash database has found that driver distraction is a key factor in approximately 30% of all actual crashes on roads.

Based on the data obtained from the Traffic Inspectorate in Žilina, on selected road stretch happened 21 traffic accidents for the period from 1 January to 30 September 2015. The most common cause of the accident was incorrect driving of driver, which occurred in 19 cases of accidents. Drivers inattention can cause a wrong way driving. In general, 90% of all traffic accidents caused a human error. A more detailed analysis of traffic accidents is shown in the table below (Table 1).

Table 1. The statistic of traffic accidents on selected road stretch [own study]

	1.1.2015–30.9.2015	1.1.2014–30.9.2014
Total traffic accidents	21	12
Road traffic deaths	4	3
Serious injuries	9	1
Minor injuries	7	3
Material damage	137 530 €	45 300 €

4   Results

Driving behaviour, especially drivers gaze of eyes was tested in good lighting conditions during the day. As it is mentioned above, the gaze of driver was captured by SMI eye tracking glasses. Before the experiment, it was necessary to do following steps:

– connect eye tracking glasses to mobile device,
– create and set up a new experiment for driver,
– calibrate glasses and record a driver's gaze.

When calibrating a device, there are three types of choice. The first one is zero calibration, it is automatic but least accurate. The second one is one-point calibration. The last type of calibration, used in our case is three-point calibration. To perform that, the man has to look at three different points and mark his gaze at mobile device. After

Fig. 6. The record of driver's gaze from eyetracking glasses [own study]

successful calibration, experiment regarding of driver's gaze may start. In Fig. 6 is shown the gaze of driver at billboard.

The total time to pass the selected road stretch and come back to starting point was 27 min (1633 s) by car approximately. The driver spent 5,6% of all his time (90,7 s) to looking at the visual smog.

During the ride, the driver totally looked at the 135 billboards or to other advertisements next to the road from all 348 advertisements (38,8% of near road advertisements was observed by driver). It means, he looked at more than every third billboard in average. According to the analyse in BeGaze software, the average time of one fixation at the billboard was 0,543 s and the longest fixation at the one billboard was 2 s. The average dwell time of one billboard (repeated fixation – gaze at one billboard more times) was 0,672 s and cumulative fixation at one billboard was 4,75 s! Detailed statistics about repeated driver's gaze fixation describes Table 2.

Table 2. The statistic of driver's gaze at the billboards (repeated fixations) [own study]

Number of driver's gazes at one billboard	Number of billboards
5	1
3	5
2	23
1	106

The advertisements near the roads can be placed at the both sides of the road – left and right. We analysed the fixation time of driver's gaze at the billboards on both sides of road. Number of fixations at billboards placed on the left side of the road was 40 and on the right side was 132. The total time of repeated fixation at the billboards on the left side was 22,14 s and the total time of repeated fixation at the billboards on the right side was 68,56 s. It means that driver looked at the advertisement on the left side of road approximately 25% of total time that he took care to near road advertisement. These findings are very dangerous, because the attention of the driver is focused out of the main roadway. The Fig. 7 captures the analysis in BeGaze software.

Another finding is very interesting too. The average dwell time of traffic signs was only 0,25 s. When driver passed a billboard, as compared to other traffic signs, the dwell time was longer, the number of fixations was greater and also the maximum fixation duration was longer.

There was also one billboard fallen on the ground and driver's fixation time at that billboard was 1 s. We can state that the billboards on the roads (or near the roads), in fact attract more sights than the traffic signs.

To conclude, visual pollution appears to have an effect on gaze behaviour as that they attract more and longer glances than regular traffic signs. This could be result of the fact that the drivers who look at the billboards becomes affected and interested in the billboard's message. Several driver's sights might follow to recognize the billboard's message entirely, which may lead to insufficient attention to traffic on the roads.

Fig. 7. Analysis in BeGaze software [own study]

5 Conclusion

The raw data were collected during real driving, so ensuring high external validity. The technology of eye track glasses used for collecting the data allowed detection of gaze target, which made the driver's sights evaluation reliable.

The statistics regarding of traffic accidents showed, that the most common cause of traffic accidents was incorrect driving of driver (19 traffic accidents from 21). Driver's inattention may be a one of the reason which may lead to incorrect driving. In case, the number of near road advertisements would be only half of total advertisements at the present, the number of driver's gaze at the billboard would be reduced (one billboard per 80 m reduced to one billboard per 160 m). It means, that there is not so many billboards and driver can pay attention more to roadway. The result is the decreasing number of time to spend of looking at the billboard, which may lead to safer road traffic and also can decrease the number of traffic accidents.

Finally, we can state (according our research) that the near road advertisements, in fact highly influence the drivers on the roads. This effect has an impact on inattention of the drivers, that may lead to decreasing safety on the roads.

Acknowledgements. This contribution was undertaken as part of the research project 1/0721/15 VEGA Research on the impact of postal services and telecommunication convergence on regulatory approaches in the postal sector.

References

1. Beijer, D., Smiley, A., Eizenman, M.: Observed driver glance behavior at roadside advertising signs. Transp. Res. Rec. **1899**, 96–103 (2004)
2. Chattington, M., et al.: Investigating driver distraction: the effects of video and static advertising (no. CPR208). Transport Research Laboratory, Crowthorne (2009)
3. Corejova, T., Imriskova, E.: Convergence at the postal market. Eksploatacja i niezawodnosc-Maintenance Reliab. **3**, 74–76 (2008)
4. Edquist, J., et al.: Effects of advertising billboards during simulated driving. Appl. Ergon. **42**(4), 619–626 (2011)
5. Kalasova, A., Faith, P., Mikulski, J.: Telematics applications, an important basis for improving the road safety. In: Mikulski, J. (ed.) Tools of Transport Telematics. CCIS, vol. 531, pp. 414–423. Springer, Heidelberg (2015)
6. Krizanova, A.: The current possition and perspecives of the integrated transport systems in Slovak Republic. Eksploatacja i niezawodnosc-Maintenance Reliab. **4**, 25–27 (2008)
7. Kubikova, S., Kalasova, A., Cernicky, L.: Microscopic simulation of optimal use of communication network. In: Mikulski, J. (ed.) Telematics - Support for Transport. CCIS, vol. 471, pp. 414–423. Springer, Heidelberg (2014)
8. Johansson, G., Rumar, K.: Drivers and road signs: a preliminary investigation of the capacity of car drivers to get information from road signs. Ergon **9**, 57–62 (1966)
9. Madlenak, R., Madlenakova, L.: Digital advertising system in urban transport system of Žilina town. Transp. Telecomm. **15**(3), 215–226 (2014)
10. Madlenak, R., et al.: Analysis of website traffic dependence on use of selected internet marketing tools. Procedia Econ. Finan. **23**, 123–128 (2015)
11. McMonagle, A.: Traffic accidents and roadside features. Highw. Res. Board Bull. **55**, 38–48 (1952)
12. Olson, R.L., et al.: Driver distraction in commercial vehicle operations (No. FMCSA-RRR-09-042) (2009)
13. Poliak, M., Konecny, V.: Factors determining the electronic tolling scope of road network. Ekonomicky Casopis **56**(7), 712–731 (2008)
14. Smiley, A., et al.: Traffic safety evaluation of video advertising signs. Transp. Res. Rec. **1937**, 105–112 (2005)
15. Vaculik, J., Michalek, I., Kolarovszki, P.: Principles of selection, implementation and utilization of RFID in supply chain management. Promet-Traffic Transp. **21**(1), 41–48 (2009)
16. Wallace, B.: Driver distraction by advertising: genuine risk or urban myth? Munic. Eng. **156**(3), 185–190 (2003)
17. Young, M.S., Mahfoud, J.M.: Driven to distraction: determining the effects of roadside advertising on driver attention (2007)
18. http://auto.sme.sk/c/5802498/vodicov-najviac-rozptyluju-billboardy-so-zenami.html. Accessed 29 Apr 2016
19. http://www.scenic.org/storage/PDFs/eebdd.pdf. Accessed 29 Apr 2016

A Method for Estimating the Occupancy Rates of Public Transport Vehicles Using Data from Weigh-In-Motion Systems

Wiktoria Loga[1], Krzysztof Brzozowski[2(✉)], and Artur Ryguła[2]

[1] University of Economics in Katowice, Katowice, Poland
wiktoria.loga@outlook.com
[2] Faculty of Management and Transport,
University of Bielsko-Biala, Bielsko-Biala, Poland
{kbrzozowski,arygula}@ath.eu

Abstract. The paper presents a method for estimating the occupancy rates of public transport vehicles directly on the base of data recorded by the Weigh-in-Motion system. Analysis of measurement data led to propose the analytical form of the function describing the occupancy rate of the vehicle, depending on the result of gross weight measurement. Moreover the correction factors were determined in order to achieve an acceptable accuracy of the method.

Keywords: Weigh-in-Motion systems · Public transport · Occupancy

1 Introduction

The dynamic development of the automotive industry has increased congestion of communication routes. This is particularly noticeable in urban areas. The density of traffic generators and absorbers and also a multitude of travel destinations especially in inner-city areas is forcing the restrictions on individual transport and at the same time attempt to popularize public transport. There is no doubt that this form of transport is the most economical, efficient and environment-friendly means of communication in urban areas. To make the idea of public transport promotion effective, it is necessary to improve the competitiveness of public transport vehicles. This is possible by increasing the quality of services, for which an essential tool is accurate information on the use and demand for this type of transportation. Data for these analyses provide research on occupancy rate in public transport vehicles. The most common way of obtaining this type of information are manual measurements, involving human resources in the form of observers. This involves significant investment in time spent on the preparation of research and development results, and thus also increases costs of analysis. In addition, research conducted by observers is also affected by measuring errors.

The researchers are constantly looking for alternative solutions to automate manual measurements performed so far. Automation of counting the number of passengers can have a positive impact not only on the accuracy by eliminating human error, but also to allow flexibility in the time period of conducted measurements. Support for counting

© Springer International Publishing AG 2016
J. Mikulski (Ed.): TST 2016, CCIS 640, pp. 426–435, 2016.
DOI: 10.1007/978-3-319-49646-7_36

passenger flows are systems of electronic tickets and city cards which allow to estimate the partial streams of passengers. More popular are also dedicated systems for calculating people installed in public transport vehicles. They are based on the video or infrared detectors [4].

The authors of the presented method propose the use of existing traffic infrastructure which are weight in motion (WIM) systems and the work is a continuation of earlier research [6]. The advantages of dynamic weighing, besides the continuous, automatic operation are also low cost of the measurements and the ability to carry out long-term statistical analysis. WIM stations make immediate classification of vehicles passing by sensors based on the number of axles and the distance between the axles. Additionally, the automatic number plate recognition (ANPR) cameras identify the vehicles and store information in the database. Knowing the design parameters and unique registration number of vehicle fleet, for which research is carried out, it is possible to identify the full desired means of transport. The method for estimating the occupancy rates of public transport vehicles, presented by the authors, is based on the corrected result of gross weight measurement.

2 Weight-In-Motion System

Preselection systems for weighing in motion use different technical solutions for the evaluation of individual parameters of the vehicle. The basic component of the WIM are loads sensors of vehicle wheels and axles. Nowadays, the following technologies are applied [3]:

- quartz sensors,
- piezoelectric polymer sensors,
- strain gauges,
- weighing plate.

Another important element of WIM stations are inductive loops which allow inter alia estimation of the magnetic vehicle length and speed measurement. Complement system are the ANPR cameras and sensors to evaluate the height of the vehicle.

From the standpoint of measuring the occupancy rate of public transport vehicles, the most important parameters recorded by the WIM station are: total weight of the vehicle (the sum of the load of individual axes) and the category of the vehicle. A widely used classification of vehicles which based on the COST 323 specification [2] is presented in Table 1.

An important parameter is a class of accuracy of the used measuring system. According to the COST 323 [2] a discrete classification accuracy of the system starting from class A (5) and ending with the class E is applied. In Poland, the most common class is B + (7) and B (10), which the tolerances are presented in Table 2.

Prior to the vehicle occupancy rate determination, the calibration for particular hour of the day needs to be calculated. The necessity of correcting the gross weight measurement is primarily a consequence of the influence of temperature and other fluctuating factors on the registered data.

Table 1. Vehicle classification according COST 323 [2]

Category	Description
1	Cars, cars + light trailers or caravans
2	Two axle rigid lorry
3	More than 2-axle rigid lorry
4	Tractor with semi-trailer supported by single or tandem axles
5	Tractor with semi-trailer supported by single or tridem axles
6	Lorry with trailer
7	Busses
8	Other vehicles

Table 2. Tolerances of accuracy class B + (7) and B(10) according COST 323 [2]

Criteria (type of measurements)	Domain of use	Accuracy classes: confidence interval width δ (%)	
		B + (7)	B(10)
Gross weight	> 3.5 · 10³ kg	7	10
Axle load Axle load > 10 kN			
Group of axles		**10**	13
Single axle		**11**	15
Axle of a group		**14**	20
Speed	> 30 km/h	**3**	4
Inter-axle distance		**3**	4
Total flow		**1**	1

3 Calibration of Measurement

The calibration of measurement was based on the method of auto-correction of weighing in motion stations presented in detail in [1]. As shown in the cited work lowest random variability has the first axle of selected category 5 vehicles. In the analyses the selection criteria for the reference vehicle include the following conditions:

- the speed of the vehicle must be in the range from 30 to 90 km/h,
- the gross weight of the vehicle (mc) must fulfil the condition:

$$\mu_{mc} - \sigma_{mc} \leq mc \leq \mu_{mc} + 2 \cdot \sigma_{mc} \tag{1}$$

where:

μ_{mc} – average gross weight of category 5 vehicles,

σ_{mc} – standard deviation of average gross weight of category 5 vehicles,

– the first axle load of the vehicle (nI) must fulfil the condition:

$$\mu_{nI} - \sigma_{nI} \le nI \le \mu_{nI} + 2 \cdot \sigma_{nI} \qquad (2)$$

where:

μ_{nI} – average first axle load of category 5 vehicles,

σ_{nI} – standard deviation of first axle load of category 5 vehicles,

The value of average weight of vehicle (μ_{mc}) has been set at level 35 ton, whereas the standard deviation $\sigma_{mc} = 0.17$. For first axle load criteria as average value was assumed $\mu_{nI} = 61.7\,\text{kN}$ and the standard deviation $\sigma_{nI} = 0.1$. Based on the above assumptions was defined calibration factor C_f:

$$C_f = \frac{nI_{ref}}{\frac{1}{n}\sum_{i=1}^{n} nI_i} \qquad (3)$$

where:

n – number of selected category 5 vehicles,

nI_{ref} – theoretical load of first axle of category 5 vehicle,

nI_i – first axle load of selected category 5 vehicles.

In estimation the calibration factor C_f was used the data set presented in Table 3.

Table 3. Dataset for estimating calibration factor

Date	Average first axle load [kN]	Number of vehicles
2015-10-12	54.61	276
2015-10-13	54.59	176
2015-10-14	53.60	175
2015-10-15	54.08	166
2015-10-16	55.41	169
2015-10-17	54.06	38
2015-10-18	55.98	43

An example of hourly variability for one week observation during the research is shown in Fig. 1. The value of calibration factor was calculated as the average one hour and three hours intervals.

In Fig. 2 is presented a summary of the gross weight of the vehicles included in the analysis as a function of the number of passengers before and after application of three-hour intervals calibration factor.

For registered data, the calibration process showed that vehicles passing through WIM station were underweight by about 10 %. After applying the C_f, factor of correlation R^2 decreased slightly.

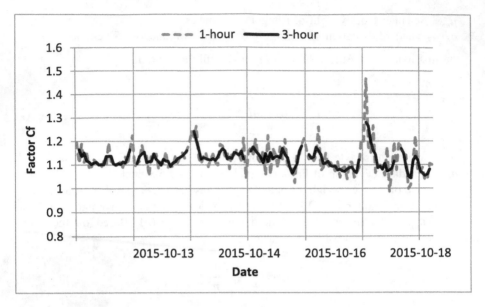

Fig. 1. The value of calibration factor C_f for one week [own study]

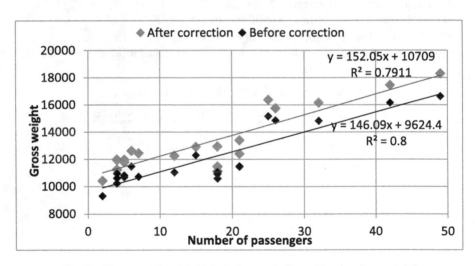

Fig. 2. Gross weight of vehicle before and after calibration [own study]

4   Occupancy Estimation

The occupancy estimation was conducted for selected public transport lines of administration of communications in Tychy (MZK Tychy). Number plate recognition using ANPR camera allowed to examine the four types of buses separately, as in Table 4.

Table 4. Vehicle of MZK Tychy for which the occupancy was estimated [own study]

Vehicle type	Number of Axels	Length [m]	Kerb weight [kg][a]
Solaris Urbino 12 CNG	2	12	10 900
Man NL 273 Lion's City CNG	2	12	10 400
Maz 103.465	2	12	9 500
SolarisUrbino 18	3	18	14 000

[a] *Estimated value*

The operator provides public vehicle services in the road DK 44, in which in Mikołów Śmiłowice, the WIM station is located (Fig. 3). Accuracy class of this station is B + (7), so that the measuring system would normally allow to record the gross weight with the accuracy to seven percent. The station consists of particular elements:

Fig. 3. WIM station in Mikołów Śmiłowice [own study]

– weight module with piezoelectric sensors and inductive loops,
– ANPR camera with IR illuminator,
– overview camera,
– height sensor.

In analysis, in process of vehicle recognition and verification, was taken into account data about: axle load, gross weight, vehicle length, vehicle speed and the distance between the axles.

Information on the occupancy rate of vehicle was obtained from the operator, who performed manual measurements in period from 07.10.2015 to 15.12.2015 on bus lines no. 33 and 82. The results of measurements are presented in Fig. 4.

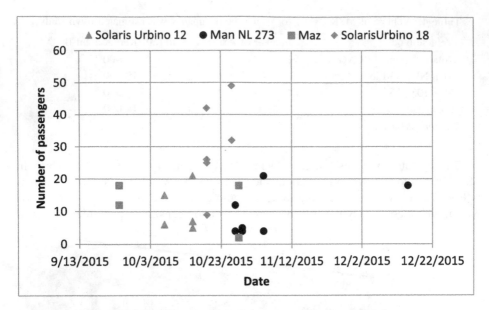

Fig. 4. Results of occupancy measurements [own study]

In order to generalize the method, the natural solution is the classification of the vehicle due to its length and taking in the calculation the average maximum number of passengers and kerb weight of the vehicle for each class. Using such an approach and the method of least squares were obtained a coefficients of linear function describing the occupancy rate of vehicles as in Table 5.

Table 5. Coefficients of a function for estimating the vehicle occupancy [own study]

Factor	Vehicle length	
	12 m	18 m
a	61.7	94.5
Δm	968	-486

For the estimation of the vehicles occupancy Eq. 4 is proposed, which is taking into account the weight of the vehicle measured by the WIM system W_{Wim} calibrated using the calibration factor C_f, the kerb weight of the vehicle W_{Veh}, the maximum capacity of the vehicle measured in passengers Q_{max} and coefficients from Table 5.

$$O_{Veh} = \left| \frac{W_{Wim} \cdot C_f - W_{Veh} - \Delta m}{a \cdot Q_{Max}} \cdot 100 \right| \; [\%] \tag{4}$$

where:
Δm – correction factor referred to gross weight,

a – correction factor referred to weight of a single public transport passenger.

On the basis of the Eq. (4) the estimated value of the vehicles occupancy for class 12-m and 18-m was determined and set together with the measured occupancy of vehicles (Figs. 5 and 6).

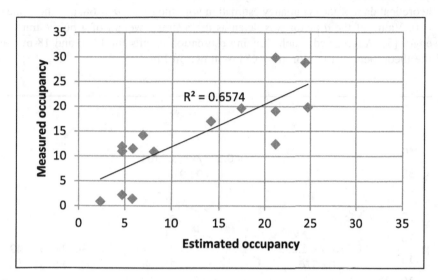

Fig. 5. Comparison of estimated and measured occupancy for 12-m class of vehicles [own study]

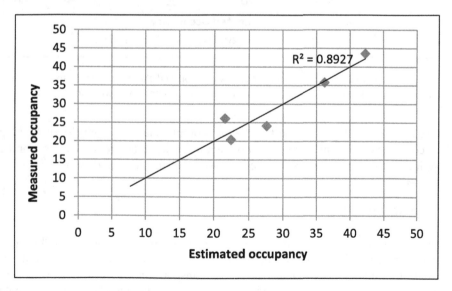

Fig. 6. Comparison of estimated and measured occupancy for 18-m class of vehicles [own study]

The shown dependence has a high degree of correlation between estimated and measured values especially for 18-m vehicles ($R^2 = 0.89$) and correlation at level $R^2 = 0.65$ for the 12-m class for which the applied model shows a slight tendency to overestimate.

Additionally, as the part of the analysis authors summarize the obtained results with the theoretical data of the occupancy estimation with the factor $a = 68$, and the value of $\Delta m = 0$. Value of the a factor was taken as the average weight of a public transport passenger [5]. Aggregated graph showing estimation results for 12-m and 18-m class and the theoretical distribution ($a = 68$) is shown in Fig. 7.

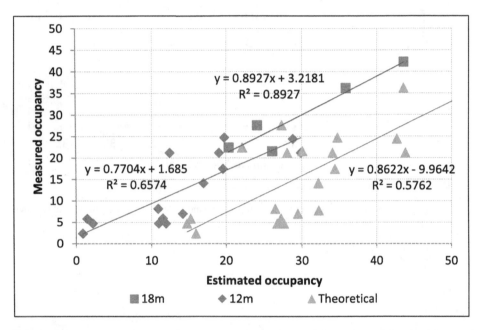

Fig. 7. Comparison of theoretical occupancy distribution and the estimated for the 12-m and 18-m classes of vehicles with regard to manual measured occupancy [own study]

Presented estimation results justify the division into classes of vehicles and taking into account the coefficient a and correction factor Δm separately for each class. For the theoretical distribution value of the coefficient of determination is $R^2 = 0.58$.

5 Conclusion

A method of public transport vehicles occupancy rate estimation, described in the paper and the obtained results of the proposed analytic function indicate the possibility of assessing the number of passengers in public transport vehicles using the weight in motion systems. The proposed method allows at the same time to obtain more accurate estimation than the use of theoretical approach. It should be recalled that the measurement of the gross weight is limited by the accuracy of WIM station and additional factors which take into account the dynamics of the vehicle motion. In addition, the accuracy of the proposed methods will be also affected by other factors related to the current exploitation of vehicles including, among others, the level of fuel and the kerb weight of the vehicle. The authors plan to continue research using extended data collection, and other WIM systems, including station equipped in the strain gauges load sensors.

References

1. Burnos, P.: Autokalibracja systemów ważących pojazdy samochodowe w ruchu oraz analiza i korekta wpływu temperatury na wynik ważenia. Rozprawa doktorska na Wydziale EAIiE, Akademii Górniczo-Hutniczej w Krakowie. http://winntbg.bg.agh.edu.pl/rozprawy2/10149/full10149.pdf. Accessed 12 Feb 2016
2. Jacob, B., O'Brien, E., Jehaes, S.: COST 3232 Weigh-in-Motion of Road Vehicles. Accessed 12 Feb 2016
3. Mitas, W.A., et al.: Elektroniczne narzędzia pomiarowe w transporcie – wagi preselekcyjne. Elektronika 52(12), 86–89 (2011)
4. Pixel System. http://www.pixel.pl/systemy/system-automatycznego-zliczania-pasazerow/. Accessed 12 Feb 2016
5. Rozporządzenie Ministra Infrastruktury z dnia 31 grudnia 2002 r. w sprawie warunków technicznych pojazdów oraz zakresu ich niezbędnego wyposażenia. Dz.U. 2003 nr 32 poz. 262
6. Ryguła, A., Loga, W., Brzozowski, K.: Estymacja napełnienia pojazdów komunikacji zbiorowej z wykorzystaniem preselekcyjnych systemów wazenia pojazdów. Technika Transportu Szynowego (12), 1341–1344 (2015)

Kamelot – Architecture of Unified Platform for Traffic Information Distribution

Petr Bures[1(✉)] and Jan Vlcinsky[2]

[1] Faculty of Transportation Sciences, Czech Technical University in Prague,
Konviktska 20, 110 00 Praha 1, Czech Republic
bures@fd.cvut.cz
[2] TamTam Research, Slunecnicova 338, Karvina, Czech Republic
jan.vlcinsky@tamtamresearch.com

Abstract. The paper focuses on architecting traffic information distribution system based on cloud services and open source applications. Firstly the requirement analysis is presented resulting later in description of core actors and set of high level use cases. Technology for those subsystems is reviewed listing available services and applications. Finally, architecture fulfilling given use cases and using appropriate technologies is proposed. At the end, changes in demand for such services, available technologies and similar solutions on the market which happened almost three years after initial project proposal are discussed.

Keywords: Traffic and travel information · Distribution · Requirements engineering · Use cases · Architecture · Subsystems · Providers · Location referencing transformation · OpenLR · DATEX II · TPEG · TPEG-Loc · Transcoding · Event description

1   Introduction

Equal and unobtrusive access to traffic information among the users from different member states of the EU is one of the goals of the European traffic policy, also mandated by the Directive No. 2010/40/EU and its Delegated Regulations.

Many countries have their traffic information systems already operational, but the problem is seamless Pan-European service, using same formats (proposed format is DATEX II) and same access conditions to access wide range of traffic data. To implement proposed functionality would be not only difficult, but also time and financially demanding. There are also many institutions/companies that create primary traffic data but do not have the distribution system set up yet.

To bridge the gap between the ability of an entity to create valuable primary content (traffic information) and a functional, efficient and quality provisioning of that content to a consumer, preferably in a standardized format is the aim of the project Kamelot. The paper presents main ideas behind the distribution system Kamelot and showcases the approach to design of an ITS system architecture [1–3].

© Springer International Publishing AG 2016
J. Mikulski (Ed.): TST 2016, CCIS 640, pp. 436–449, 2016.
DOI: 10.1007/978-3-319-49646-7_37

1.1 Core Concepts

This section presents core concepts of the application, depicted in Fig. 1.

1. Publisher sends primary data (in any documented format) to the broker.
2. Broker is central part of the chain
   (a) Receives the primary data, transforms them and distribute.
   (b) Data versions may be put into archive.
   (c) Data formats shall be documented.
   (d) System shall be monitored.
   (e) Archived content can be evaluated.
   (f) Publisher and consumers must have valid account in the system.
3. Consumer consumes data from the system.

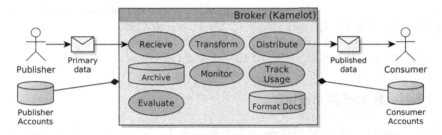

Fig. 1. Core concepts [own study]

1.2 Functional Requirements

Requirements are the foundation for all the project work that follows. Focus of the requirement engineering was at a general level, starting with identification of core concepts and elaborated into set of higher level use cases.

Requirements analysis took into account statements from the project Kamelot application and the Directive and other European documents analyzing its impact i.e. [4]. Following top level requirements for service were identified:

– Conversion of primary data to other formats
– Location reference conversion
– Distribute published data to consumers
– Provide framework for documenting published data
– Consumer account management
– Billing and invoicing
– Monitoring of services and content
– Archiving and access to archive

Expected subsystems and their features. Functional requirements lead to top level modules/subsystems. For each subsystem are listed features, provided by it.

- Publisher Management subsystem, that provides identity management, invoicing and configuration.
- Distribution subsystem, that provides primary data reception, metadata indexing, aggregation, publishing of simple feeds, aggregated data and filtered data (by push, pull methods) and usage tracing.
- Archiving subsystem that provides archiving of primary data, access to archived data and long term archiving (Amazon Glacier).
- Content Evaluation subsystem
- Consumer Management subsystem, that provides identity management, invoicing
- Documentation subsystem, that provides public and restricted access to the documentation
- Monitoring subsystem, that provides problem detection and alerting and parameter visualization
- Trans-Form subsystem (format to format transformation), that transforms plain data to DATEX II and TPEG and Transform DATEX II to TPEG and vice versa.
- Trans-Loc subsystem (location to location transformation), that transformation service between OpenLR, GDF, ALERT-C and TPEG-Loc.

2   Methods and Materials

Proposing such a system is challenging in regards to functionality, scalability and cost effectiveness. For this reason we had to carefully select innovative technologies (mostly cloud services). At the same time we had to take care to identify and fulfill requirements of our future users. We approached system architecture proposal in following steps:

1. Requirements engineering
2. Elaborate Use Cases
3. Research available building blocks
   (a) Services
   (b) Applications
4. Architecture proposal

Rest of this section summarizes the services and applications we were considering as building blocks of our architecture. Use cases and resulting architecture are described in following sections. There are two types of building blocks:

- Services: Virtual server, storage, archiving, API management and messaging
- Applications: Continuous integration server and monitoring application

2.1   Virtual Server Service

Virtual server provides virtual computer with defined processing power, local storage, memory and connectivity. It has pre-installed operating system and must allow remote access.

Examples are AWS EC2, Digital Ocean Droplets, Google Compute Engine, Active24 VPS, Active24 VMS.

AWS EC2 calls running instance of virtual server an "EC2 instance". EC2 servers are a bit more expensive than Digital Ocean Droplets, but in case there are huge amounts of data transferred between the server and other AWS services (typically AWS S3), EC2 could become more cost effective.

2.2   Storage Service

Storage service allows storing and accessing data in large volumes with on-line presence (data available at the time of request). It must provide secure access and data redundancy.

Examples are AWS S3, Google Cloud Storage, OpenStack storage Swift, Rackspace using Swift, Microsoft Azure Storage.

AWS S3 allows efficient and secure storing and retrieving of data by various APIs. Standard storage class stores data on 3 different places, reduced storage class on 2. AWS S3 provides notification on event "new object uploaded". Currently it seems to be the most effective solution for storing and accessing large number of objects on the market.

2.3   Archiving Service

Archiving service allows long term cost effective data storage. Data might be accessible with some delay after request. One example is Amazon Glacier.

Amazon Glacier allows long term cost effective archiving of data in the cloud. It complements with AWS S3 very well. Data archived in Amazon Glacier are not available immediately, they must be requested, and then they are within few hours restored to AWS S3. There are configurable policies, allowing removing restored data automatically to save storage costs. The costs are rather low, typically 2 to 3 times cheaper than AWS S3 using standard storage class. In case, the data are restored too often, Glacier becomes more expensive.

2.4   API Management Service

API Management Service allows managing many activities and services related to published API and its consumers. The services typically offers solutions for application identity management, API documentation, access management, tracking service usage, rate limiting, invoicing etc. Examples are 3scale (3scale.net), Apigee, Intel Mashery, Mulesoft Anypoint Platform.

3scale.net provides access control and security, developer's portal and documentation, analytics and reporting, billing and payments, service contracts and rate limiting. Comparing to other API management solutions, 3scale.net offers clearly defined price conditions incl. free tier plan and allows even long term free tier usage for smaller applications.

2.5 Messaging Service

Messaging service allows sending messages using queues (one message to one consumer) or topics (one message to multiple subscribers). Examples are Amazon Web Services SNS and SQS.

AWS SNS (Simple Notification Service) allows creation of so called topics, to which one can publish messages, others can subscribe to related topics and receive the messages published there. One of use cases is publishing to AWS SNS from AWS S3 as reaction to an object being stored to AWS S3. Notification to subscribers can be sent via e-mail, HTTP request, SMS etc.

AWS SQS (Simple Queue Service) allows creation of a queue, publishing to it and consuming the messages from the other side. Queue allows one message to be consumed only by one consumer.

2.6 Continuous Integration Application Jenkins CI

Jenkins CI (or Jenkins in short) is so called continuous integration server allowing management of large set of jobs being run on one or more servers. Typical use cases include automated builds and tests. Jenkins is very modular and it is rather easy to deploy test suites written in many different languages on it, run them and report the results out. Jenkins is Open Source MIT licensed product.

2.7 Monitoring Application Nagios

Nagios is computer system and network monitoring solution. It provides large set of pre-build checks and allows extending by writing custom plugins (in form of command line tools). The server runs on Linux, but it is capable of monitoring servers with different OS incl. MS Windows. Nagios is open source GPLv2 licensed product. It was used for example in the project Monitor [5].

3 Resulting Use Cases

First result is set of use cases (UC). This section identifies most important ones and describe them briefly. Use cases are grouped into packages as shown in Fig. 2 and are described in following subsections.

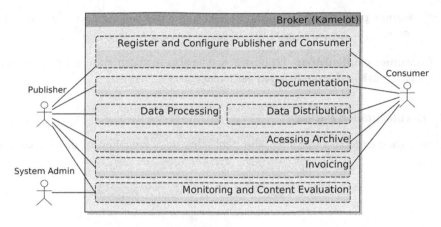

Fig. 2. Use Case packages [own study]

3.1 Register and Configure Publisher and Consumer Package

Figure 3 shows use cases dealing with registration of users and management of related user permissions.

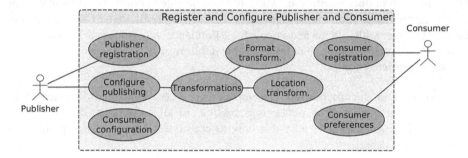

Fig. 3. Register and Configure Publisher and Consumer Use Cases [own study]

UC Publisher registration. Publisher will be able to create account in the system, allowing him to access it in role of publisher.

UC Configure publishing. Publisher will be able to configure all aspects of publishing his data. It is required to identify the primary data the publisher is planning to publish. The use case has few extension points dealing with:

- Configuring (optional) transformation of primary data into one or more secondary formats.
- Configuring (optional) transformation of location references used in the data.

UC Consumer registration. Consumer will be able to create account in the system, allowing him to access it in role of consumer.

UC Consumer preferences. Consumer will be able to set preferences for using the system and also ask for permission to consume some of provided data.

UC Consumer configuration. Publisher will be able to accept or reject requests sent by consumers asking for provision of particular data.

3.2 Documentation Package

Figure 4 shows use cases dealing with publishing documentation of distributed data formats.

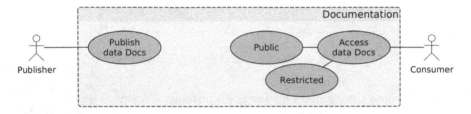

Fig. 4. Documentation Use Cases [own study]

UC Publish data documentation. Publisher will be able to publish documentation of data being provided. The documentation may consists of HTML, PDF or other document formats together with schema and sample files. Publisher shall set for each documented format if the documentation shall be accessible publicly or only for approved consumer accounts (restricted access).

UC Access data documentation. Consumer may access published documentation. Documentation published as public is accessible to all consumers, documentation published in restricted mode is visible only to consumers having related permission granted by publisher.

3.3 Data Processing Package

Figure 5 shows use cases dealing with reception of primary data and making the data ready for further distribution. Publisher may also opt for archiving.

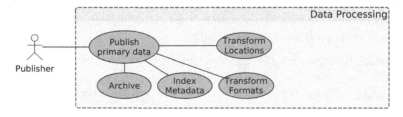

Fig. 5. Data Processing Use Cases [own study]

UC Publish primary data. Publisher sends primary data to the system which stores the data. According to feed configuration, other processing may take place:

– Archiving each published version.
– Transformation of data into one or more different formats.
– Transformation of location references.
– Indexing metadata.

Finally, the primary data becomes available for distribution in one or more formats.

3.4 Data Distribution Package

Figure 6 shows use cases dealing with distributing data to consumers.

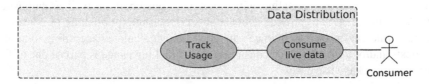

Fig. 6. Distribution Use Cases [own study]

UC Consumer live data. Consumer may consume live data. There are two patterns, PULL - when consumer repeatedly sends request for new data, and PUSH, when system sends the data at the moment new version becomes available.

Each request or attempt to send the data to consumer is tracked and may be used later on for invoicing.

3.5 Accessing Archive Package

Figure 7 shows use cases dealing with accessing archived data and with managing them. There are two forms of archives - direct access archive and long term archive. Moving data into long term archive is managed by predefined policies, moving the data after reaching defined age.

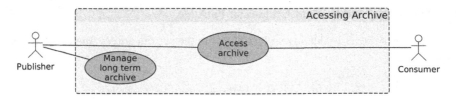

Fig. 7. Archiving Use Cases [own study]

Data can be retrieved only from direct access archive. Long term archive (which is supposed to be cheaper) stores data in remote storage. To read the data from long term archive, the data must be first copied into direct access archive, which process takes few hours.

UC Access archive. User can access data in direct access archive. The access allows listing available versions and retrieving them.

UC Manage long term archive. Publisher may list data in long term archive and request making some data to be copied into direct access archive, making them effectively available. Publisher can also modify rules for moving data from direct access archive to long term one.

3.6   Invoicing Package

Figure 8 shows use cases dealing with invoicing consumers as well as publishers. Minimal requirement is to provide data about service usage (publishing or consuming the data) sufficient to issue an invoice with costs being calculated according to number of requests and/or total amount of consumed data. Tracking usage is part of one of Data Distribution Package use cases.

Fig. 8. Invoicing use cases [own study]

UC Invoice Consumer. System collects data about usage of the service by particular consumer and sends them once a month to configured address. Optionally the system may create the invoice.

UC Invoice Publisher. System collects data about usage of the service by particular publisher and sends them once a month to configured address. Optionally the system may create the invoice.

3.7   Monitoring and Content Evaluation Package

Figure 9 shows use cases dealing with monitoring the system and evaluating published data.

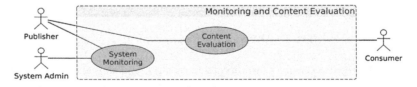

Fig. 9. Monitoring and content evaluation use cases [own study]

Monitoring is supposed to serve system administrators at publisher as well as by broker itself as a tool to alert in case there is a problem and to give overview about current runtime status.

Content evaluation is supposed to be quality management tool, running customized test suites and jobs on published data (available in archive) and evaluating various quality characteristics like conformance to agreed schema, number of published messages etc.

UC System Monitoring. System is continually monitored by monitoring tool. Users can view current runtime parameters and access historical values.

In case, there is some runtime parameter in non-standard status, system shall send notification to pre-configured parties.

UC Content Evaluation. User shall be able configuring jobs to be run regularly on archived data. Typical job can be e.g. validating all feed versions published last day against agreed schema, counting number of published messages etc. Jobs are defined by means of test suite or other job in form of source code repository.

The evaluation job can be configured, scheduled and results shall be visible by means of web application.

4 Resulting Architecture

This section describes proposal of architecture based on identified functional requirements and building on services and applications described in preceding sections. Following modules are proposed:

- Module Publisher Management
- Module Distribution
- Module Consumer Management
- Module Trans-Form
- Module Trans-Loc
- Module Archiving
- Module Content Evaluation
- Module Documentation
- Module Monitoring

4.1 Publisher Management and Consumer Management Modules

Both, publishers as well as consumers must have accounts and in both cases we need to allow tracking usage, issue invoices and maintain configuration.

For Identity management and Invoicing we are likely to use 3scale.net. Configuration must be maintained in our own configuration management application.

4.2 Distribution Module

The module shall receive primary data, do whatever internal processing is needed (indexing, aggregation, format or location transformation) and distribute it to consumers. It also must allow tracking how are consumers really consuming the data.

Figure 10 depicts the process. Publisher publishes the primary data into AWS S3 bucket what triggers S3 event starting processing. The event is available via AWS SNS messaging or by direct subscription from AWS Lambda function.

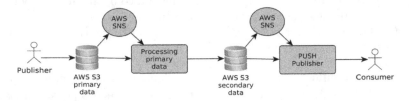

Fig. 10. Publishing primary data, processing & publishing secondary data via PUSH [own study]

The processing takes care of sending data to archive, triggering location and format transformations and possibly recording the publishing event for usage tracking.

Data to be published are always put into secondary bucket. For PUSH distribution we use again event triggered by AWS (secondary) bucket and post the data to the consumer. Consumer can also PULL the data from the secondary bucket accessing it via NGINX Proxy depicted on Fig. 11.

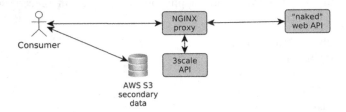

Fig. 11. NGINX proxy shielding API for PULL mode distribution

NGINX Proxy receives request from consumer. Before sending the request to "naked web API", it checks credentials of the user and tracks the request. If all is fine, the request is passed to "naked" web API, which creates temporary url for consuming requested data from secondary bucket and replies with redirect response. Consumer then requests actual data from provided temporary url. The temporary url is valid for short period (e.g. 10 min).

4.3 Trans-Form Module

Purpose of Trans-Form module is to allow various types of format transformations, triggered at the process of distribution. Basic library of transformation shall cover cases like conversion of plain data to DATEX II or TPEG and conversions between DATEX

II and TPEG. However, as each transformation can be specific, we assume, each will be run as separate process running either as AWS Lambda function. More complex transformation can be run as Docker based tasks on AWS EC2 instances.

4.4   Trans-Loc Module

The module allows transformation of location references between location referencing using OpenLR, GDF, Alert-C and TPEG-Loc.

Location transformation can be used as part of content transformation, or could be consumed independently. In all cases, it will be provided via API of custom application, running on virtual server. Usage of the API will be managed by 3scale.net.

4.5   Archiving Module

Process of archiving is depicted in Fig. 12. The module allows:

- Storing primary and secondary data in direct access archive.
- Accessing archived data via API.
- Packing older data into packages and moving them to remote archive.
- Requesting restoration of packages in remote archive to direct access one.

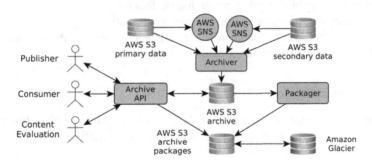

Fig. 12.   Archiving [own study]

Rules for what is to be archived and for how long are managed by Publisher Management module. Direct access archives are using AWS S3 and are populated by events described in Distribution module.

Web API allows accessing stored data (listing feeds, dates, versions and retrieving them).

Older archived data are regularly packed into larger packages and stored on "archive package" S3 bucket. Using policy files, the packages are moved to Amazon Glacier. We are taking advantage of the fact, that AWS S3 bucket provide handy interface for identifying archived packages and allowing requests for temporary restoring the packages back to AWS S3. If needed, policy files may define removal of package from Amazon Glacier.

4.6   Content Evaluation Module

The module manages jobs, evaluating content available in archive. The jobs are scheduled and run by Jenkins CI server, possibly running some of the jobs on Jenkins CI
workers.

Figure 13 depicts main content evaluation components. Jobs are defined by means
of source repository, being stored e.g. on Bitbucket or GitHub. The jobs contain test
suite or other work to be done. Jenkins CI updates local copy of the code and runs it.
Jenkins CI allows deploying multiple independent jobs. Results can be seen via Jenkins
CI web interface or be published to AWS S3 Quality reports bucket and seen from here.

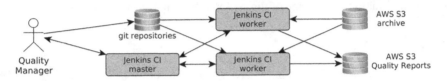

Fig. 13.   Content evaluation [own study]

4.7   Documentation Module

The module allows publishing documentation of data formats. Documentation shall be
managed by 3scale.net documentation module or alternatively by custom web application. In case of restricted documentation, permissions to see some documentation must
be set in Consumer Management module.

4.8   Monitoring Module

Monitoring will be based on Nagios, which allows status checks and notification and
which also provides web interface. Parameter visualization shall be managed by custom
application or by selected jobs in Jenkins CI server. Data will be stored in some database
and later can be visualized by means of dedicated web application. Alternative for
Parameter visualization is AWS CloudWatch.

5   Conclusion

Demand for traffic information distribution is growing, one of driving forces being
growing number of Delegated Regulations mandated by the Directive No. 2010/40/EU.
From this perspective seems the concept of Kamelot distribution system meeting real-
life requirements.

Comparing Kamelot concept to Mobility Data Marketplace (MDM)[1] developed in
Germany, we see significant overlap. While both systems are focusing on mediating
data distribution, MDM intentionally focuses only on transferring the data as they are,

---

[1] MDM: http://www.mdm-portal.de/en/.

Kamelot is on the other hand more focused on customizing the content, archiving and providing location reference transformations. For this reason one of deployment opportunities we see for Kamelot is serving as complementary component to content being provided by MDM (serving in role of Data Refinery service).

In course of the project we did not register any interest on using TPEG-Loc location referencing method as well as nobody seems to expect invoicing agenda being completely managed by third party application.

Building our solution on AWS cloud services proved to be very scalable and effective solution in some of other projects we have developed in last years, but we have also noticed signs, that MDM is also using this platform.

Technologies are evolving. We were for example planning to use AWS SNS as the only method to get quick notification of new feed version being published. Nowadays we can use AWS Lambda function, which allows running custom code written in Java, JavaScript or Python in reaction to such event.

Couple of services, originally planned to be solved by 3scale.net, have now alternative solutions, e.g. AWS CloudWatch provides alternative method for tracking usage of Kamelot services incl. visualization and Amazon Cognito provides base for authenticating users of web applications.

Our current implementation, which is to be completed by end of year 2016, proves proposed architecture for distribution, archiving, transformation, monitoring and content evaluation working very well. Remaining tasks are to make final decision about technologies to use for publisher and consumer management.

Acknowledgements.   The authors acknowledge the financial support provided by the Technology Agency of the Czech Republic through project Kamelot (TA04031524).

References

1. Bures, P.: The architecture of traffic and travel information system based on protocol TPEG. In: Proceedings of the 2009 Euro American Conference on Telematics and Information Systems: New Opportunities to increase Digital Citizenship, Prague, Czech Republic (2009)
2. Bures, P., Belinova, Z., Jesty, P.: Intelligent transport system architecture different approaches and future trends. In: Duh, J., et al. (eds.) Data and Mobility: Transforming Information into Intelligent Traffic and Transportation Services, Proceedings of the Lakeside Conference 2010, vol. 81, pp. 115–125. Springer, Berlin (2010)
3. Bures, P., Belinova, Z., Bures, P., Barta, D.: Evolving ITS Architecture - the Czech Experience. In: Mikulski, J., et al. (eds.) Modern Transport Telematics, vol. 239, pp. 94–101. Springer, Heidelberg (2010)
4. Algoe, R.: Trans: Study regarding guaranteed access to traffic and travel data and free provision of universal traffic information: D8 – Final Report, 1st edn., Brussels (2011) http://ec.europa.eu. Accessed 10 Dec 2015
5. Bures, P., Vlcinsky, J., Bures, P., Barta, D.: Monitoring of live traffic information in the Czech Republic. In: Mikulski, J., et al. (eds.) Modern Transport Telematics, vol. 239, pp. 9–16. Springer, Heidelberg (2010)

The Test of Traffic Control for Middle-Sized Towns Using Various Data Source

Tomáš Tichý[1]([⊠]), Zuzana Bělinová[2], Kristýna Cikhardtová[2], and Jiří Růžička[2]

[1] ELTODO a.s., Novodvorská 1010/14, 142 00 Praha 4, Czech Republic
tichyt@eltodo.cz
[2] Faculty of Transportation Sciences, Czech Technical University in Prague, Konviktská 292/20, 110 00 Praha 1, Czech Republic
belinova@k620.fd.cvut.cz,
{cikhakri,ruzicji4}@fd.cvut.cz

Abstract. The article presents urban traffic control method suitable and tested for a small urban areas enabling more fluent traffic and thus increasing the throughput of the controlled area. The model was implemented in the adaptive Traffic Dependent Control module for higher control of the area designed for the optimization of the intersection signal plans that uses various inputs (different detectors) and communication with special control unit for C2X and gives best results for intersections with irregular traffic. The whole system was tested in a small urban area proving its benefits in real traffic.

Keywords: ITS · Urban traffic control · Control unit · Pollutants

1 Introduction

In this paper we summarize results of several interesting applied research projects aiming to improve the traffic control and decrease the negative traffic impacts on the environment in cities. The basis for the improvements are inovative sensor networks used as input in the traffic information and control systems. Our goal was the development and testing of the sensorics network and further data integration in an unique complex enabling provision of useful information to end users helping to decrease the negative impacts on the environment and citizens' health in city areas.

2 Main Principles of Improved Traffic Control

The system is based on the standard principle of collecting, processing and using data, but it is unique with the wide spectre of data sources used as the system inputs.

The system is formed by a set of SW and HW subsystems enabling the processing of a wide spectre of input traffic, user and traffic engineering information in the output instruction to the end active elements capable of traffic control. The subsystems are designed to provide bound independent functionalities. Using them together with the support of correctly set interfaces the system is formed. This system can be easily and anywhere extended or modified (Fig. 1) [1, 2].

© Springer International Publishing AG 2016
J. Mikulski (Ed.): TST 2016, CCIS 640, pp. 450–461, 2016.
DOI: 10.1007/978-3-319-49646-7_38

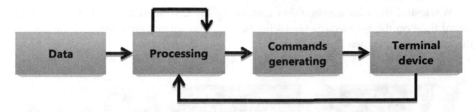

Fig. 1. Basic structure of a control system [own study]

During the control system design we have examined various data sources. The example are outputs from database RODOS where are implemented data from floating cars. For our test in area we have used these sources [1, 6]:

- Traffic point detectors from intersection
- Videodetectors
- Floating cars

These basic data sources we have for our system are extended with the data from:

- Bluetooth detectors
- Imission data
- Noise data
- C2X data

3   Starting Points for the New Traffic Dependent Control Implementation – the Data Sources

To integrate all the data sources, thorough analysis was needed at the beginning. To incorporate the environmental data, the analysis of the current usage of systems for measuring the traffic pollutants in relation with the methods for traffic control was done to ensure the positive impact on the environment. At the same time new traffic control was tested enabling also the cooperative C2X system and the floating car data usage.

As the testing area the city of Uherské Hradiště was chosen and a real test was conducted there [2].

3.1   C2X Detection

In the testing area, in the city Uherské Hradiště, we have also tested the communication units C2I for the traffic controller and the vehicle. In the traffic controller we installed the universal control unit, a modular system for a wide spectre of traffic applications. The universal control unit is able to cooperate with already existing subsystems and this makes the way to connect newly developed technologies with the old ones easier. For the test in the city Uherské Hradiště, there were constructed five pieces for each intersection based on prototype tested in laboratory conditions [2].

We tested the data transmission containing the information about the remaining signal time on the intersection signalling device. The special equipment of the car and the controller is shown in Fig. 2.

Fig. 2. The special equipment for C2I testing [own study]

The information was transmitted in the vehicle and presented as countdown on the monitor of the embedded car radio. For this test it was necessary to install both in the vehicle and in the traffic controller a special unit and establish the communication between the devices. That means there was the vehicle, traffic controller and the traffic control centre in the communication link. It is necessary to have in the vehicle an on-board unit (OBU) for the C2I and a mobile phone in a joint configuration with the dash-board monitor. In the controller the above mentioned universal control unit is installed. The principle is shown in Fig. 3 [2, 6].

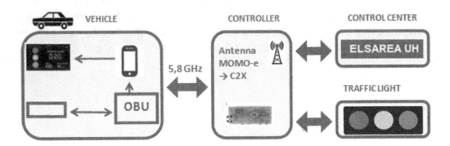

Fig. 3. Communication scheme [own study]

After establishing the communication it was possible to run the test drives. Information on the signal colour was presented in the according colour scheme, the countdown was numerical in the format m:ss, detail in the Fig. 4.

Fig. 4. Signal countdown in vehicle [own study]

The signal range is highly depending on the antenna used and its parameters. For this particular testing, antenna with a range 300 metres was used. The antenna devices have to be chosen with regard to the distances of the intersections and their particular traffic controllers.

In our specific case of testing in the Uherské Hradiště city, the unit sending the data in the vehicle was installed in just one traffic controller at the intersection. We chose the particular intersection to clearly demonstrate the benefits of such system, as the intersection is situated behind a curve causing the traffic lights can be seen only app. 100 m in front of the intersection. The antenna used thus sent the signal in the vehicle in the time the driver did not know at all in what state the intersection currently was [2, 5].

3.2 Pollutant Data Measurement Results

To measure the pollutants (NO, NO2, NOx, particle pollution, O3, SO2, CO, CO2 and VOC) we used the Airpointer ambient air quality monitoring system for airborne pollutants using internationally defined reference methods classified as relevant by the EU, the WHO and further responsible organizations all over the world. The other device was a sound meter. In the following graphs (Graph 1) there is the pollutant

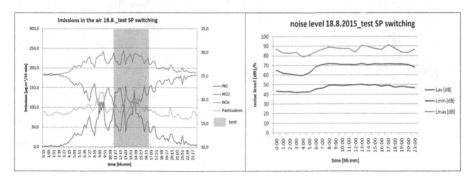

Graph 1. Air composition and noise level [own study]

amount in 15 min aggregations. The time of the traffic control testing is in the graph marked by the coloured bar. In the graphs there are the nitric oxide (NO), Nitrogen Dioxide (NO2), sum of them (NOx) on the main y axis and the particle pollution on the secondary y axis [2].

The noise level was measured using microphone mounted on a street lighting pole in the distance of approximately 10 m from the street centre line. The collected data contained only the dB noise levels, not the whole noise track, and that is why it was not possible to remove any irregular noise changes (e.g. transit of emergency service vehicle). The most relevant is therefore the mean value not containing the outstanding variations.

From the graphs it is seen, the level of the acoustic pressure is for the whole day from app. 6 to 22 h constant and then decreases slowly. The permitted basic noise level is for the day time from 6 to 22 h 55 dB and during night time 45 dB. The real level exceeds the permitted one practically all the time by app. 10 dB.

For better imagination there are shown complete one day data below in Table 1. The data are in one hour aggregations and the boxes are coloured according to the air quality and the share of each substance [1, 2, 6].

Table 1. The share of substances in the air [own study]

	$[\mu g.m^{-3}]$	$[\mu g.m^{-3}]$	$[\mu g.m^{-3}]$	$[\mu g.m^{-3}]$	$[\mu g.m^{-3}]$	$[\mu g.m^{-3}]$	$[\mu g.m^{-3}]$	$[\mu g.m^{-3}]$	$[\mu g.m^{-3}]$	$[\mu g.m^{-3}]$	$[\mu g.m^{-3}]$	$[\mu g.m^{-3}]$
	NO2	PM10	O3	SO2	NO2	PM10	O3	SO2	NO2	PM10	O3	SO2
time	Signal program switching				Dynamic algorythm testing				No test			
0:00	182,0	17,0	18,0	2,8	182,0	47,1	19,0	9,3	176,7	24,5	19,1	6,9
1:00	181,9	16,9	13,4	5,0	180,0	44,6	17,3	6,9	179,7	24,1	17,7	12,7
2:00	181,4	17,2	14,2	8,2	181,3	31,1	16,9	10,3	181,2	24,6	17,1	6,5
3:00	177,4	16,9	15,4	13,7	171,3	29,7	12,9	9,6	155,4	24,4	9,7	12,6
4:00	163,8	17,6	11,7	12,2	145,0	30,1	6,7	8,6	136,4	25,7	5,0	8,4
5:00	160,0	18,0	6,4	10,2	123,8	31,2	5,0	12,7	129,2	26,1	5,3	12,9
6:00	142,6	18,3	4,4	6,3	122,5	31,3	6,6	8,1	135,3	32,2	9,3	6,8
7:00	135,2	18,3	2,9	4,9	134,1	32,1	8,2	13,8	135,9	32,3	14,2	15,1
8:00	124,3	19,5	3,2	2,3	146,2	32,2	13,0	9,1	120,5	32,3	18,5	17,1
9:00	116,5	20,0	2,3	3,8	138,6	32,4	17,0	10,9	139,2	30,3	27,4	14,9
10:00	132,9	17,7	4,6	3,7	144,0	31,7	23,1	18,2	156,7	27,3	46,4	16,1
11:00	135,1	18,1	4,8	2,9	154,2	29,7	36,4	15,0	153,7	27,0	51,9	17,9
12:00	125,8	17,8	5,2	7,0	155,5	29,1	44,6	13,6	163,2	26,7	62,3	17,9
13:00	119,4	18,0	3,9	10,0	158,2	28,7	50,8	15,3	158,0	26,6	62,3	19,7
14:00	112,8	19,4	3,9	10,7	151,0	28,5	52,2	15,7	157,1	25,8	65,5	16,1
15:00	130,4	18,1	3,8	14,4	162,3	28,2	54,1	14,9	169,0	24,4	72,7	15,5
16:00	141,7	17,9	3,6	14,2	174,1	25,6	62,7	13,9	175,5	22,6	63,4	12,5
17:00	150,3	17,2	4,0	8,6	170,1	25,0	55,1	11,6	174,5	23,0	55,3	13,0
18:00	162,7	16,6	5,9	11,8	176,7	25,3	52,7	12,5	180,0	23,2	52,2	9,2
19:00	168,1	17,1	4,3	7,9	184,4	24,1	52,6	3,6	183,8	23,8	51,6	11,6
20:00	169,3	16,5	3,8	11,1	184,6	24,1	56,2	9,1	182,1	24,2	47,2	10,7
21:00	172,9	16,3	5,5	6,4	181,7	24,0	39,7	13,8	179,4	25,1	47,6	14,2
22:00	177,2	16,0	5,2	11,5	184,6	25,0	30,3	6,3	184,6	25,0	47,8	9,9
23:00	181,7	17,1	4,4	6,1	183,7	24,8	21,1	12,8	185,5	26,0	38,6	9,9

The explanation of the air quality evaluations, frames and corresponding colour is shown in the legend in Table 2. It is seen that the air quality of air is mostly good and getting worse after morning peak hour (about 10 AM). Unfortunately during the test the Czech Republic was hit by extreme warm weather without any rainfall. This generally influences the air quality in a bad way.

Table 2. Air quality levels [own study]

Index	Air quality	SO_2 1h pg/m^3	NO_2 1h pg/m^3	O_3 1h pg/m^3	PM_{10} 1h pg/m^3
1	very good	0 - 25	0 - 25	0 - 33	0 - 20
2	good	> 25 - 50	> 25 - 50	> 33 - 65	> 20 - 40
3	satisfying	> 50 - 120	> 50 - 100	> 65 - 120	> 40 - 70
4	suitable	> 120 - 350	> 100 - 200	> 120 - 180	> 70 - 90
5	bad	> 350 - 500	> 200 - 400	> 180 - 240	> 90 - 180
6	very bad	> 500	> 400	> 240	> 180

3.3 Bluetooth Device Detection

Bluetooth detectors are primarily used for the travel time or delay measurement, which was used for the evaluation of the traffic control systems impact on the traffic quality. Further the penetration in the traffic network was assessed, meaning how many vehicles from the traffic flow are the Bluetooth detectors able to detect (Graph 2). The Bluetooth detection has however the disadvantage of detecting duplications. If there is more than one on-line Bluetooth device, it is not possible to filter them in order not to create additional false unit, meaning in the traffic flow counts another vehicle [2, 4].

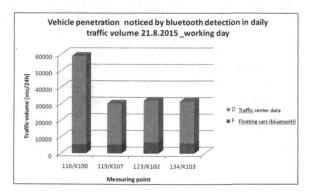

Graph 2. The percentage of vehicles equipped with Bluetooth on 21.8.2015 [own study]

On the following graph the penetration in the daily traffic flow is presented.

The portion of vehicles possible to detect by the Bluetooth detector was in the working days between 10% and 14%, during weekend the situation was worse and the percentage was scarcely 5% – less than a half compared to the working days. The difference between working days and weekends can be that company cars, that are very often equipped with Bluetooth hands free etc., are on the roads more during working days.

The aim was to find out if the floating cars are a relevant information source. So the data from the Bluetooth detection was verified and compared to the real traffic volume downloaded from the intersection controllers. There wasn't set any reliability limit, the point was to find out if we are able to quantify the penetration constantly. The detailed numbers including average values of working days and weekends are shown in Table 3.

Table 3. Detailed numbers of comparasion floating cars with controller data [own study]

	Flowting cars (bluetooth) [veh/day]				Controller data [veh/day]				Share	Penetration
	K100	K107	K102	K103	K100	K107	K102	K103	[veh/day]	Share [%]
Mean value during working days	6155	5947,5	7809,5	7103	52155	23941	24377,5	23048	46000	11,8466
Mean value during weekends	1753	1753	2081	2244	37281	11824	18142	17698	35528	4,70213

4 New Traffic Dependent Control Implementation

4.1 Principles of the Traffic Control

The overall goal of the research and development activities is to offer solution for smaller cities improving the traffic control, mainly in the cities using the traffic system ElsArea®. To achieve this, several phases were needed, starting from the development of an adaptive algorithm capable of using various data inputs, including the development of the Traffic Dependent Control (TDC) Module up to the testing of the traffic control together with various data sources. Most of the data it was possible to collect and send to a traffic centre (ElsArea®) through the universal communication unit (also used for cooperative systems).

As the Traffic Dependent Control Module is the main part contributing to the final traffic situation improvements, the TDC detailed description follows.

The TDC is a module for higher form of urban area traffic dependent control. The module is connected to a traffic centre and enables to control an intersection or group of intersections in a line or area. The advantage is the module not only enables the fully automatic traffic dependent control but also the implementation of traffic scenarios from traffic operator or engineer. The second version (advanced) of TDC is adaptive based. The algorithm changes thresholds and parameters setting of stages, cycle time, offset and green time of signal groups and calculates the new signal plans which are sent to controller intersection point [1, 6].

A user is able to observe the actual state of intersection through the ElsArea® web client (Fig. 5). The user can also intervene to the control operatively and switch a signal program or the control mode on each intersection connected to the traffic centre.

Fig. 5. The web client of ElsArea traffic center [own study]

The TDC function is launched by the user as one of the several levels of inter-section control. After its start the module checks the current situation and according to users preferences based on the mathematical rules or algorithm does the data evalua-tion. In case the preset limits are reached the intersection control is adjusted according the embedded equations [3].

The module has generally two parts – the configuration and the executive one. The configuration part is the graphical user interface and the executive part is conducting the calculation according the new control algorithm.

The graphical user interface (GUI) has to be well-arranged and present the infor-mation in an easily understandable way. The GUI reads the data on the intersections and detectors from the database and uses their parameters. The user is able to change most of the settings (the groups of intersections, find the devices and their parameters, add rules, change their priorities, etc.). The commands and the rules use the mathe-matical logic formulas.

The algorithm was based on simple logical conditions IF, THEN. One of the aims was easy configuration and setting of input variables to have good possibility for debug. Logical rules are based on detector data recording traffic volume. The basic principle of adaptive algorithm is seen in Fig. 6 [3].

4.2   Real World Testing in the City Area

As mentioned above, the testing has taken place together with the other measurements in the city Uherské Hradiště. This area was chosen due to its traffic parameters suitable for the desired testing. In this area the traffic volumes are relatively high and it was possible to access the real traffic data from the traffic detectors (inductive loops and video detection). Also in this area a test of adaptive algorithm took place previous year and the experiences that arise where useful for future and actual planed tests. In summary the tested area chosen for the TDC module testing consisted of five inter-sections with traffic signalization.

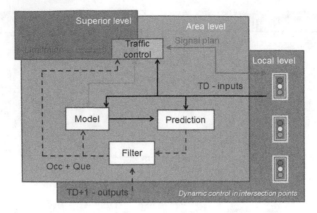

Fig. 6. Block schema of adaptive algorithm of TDC [own study]

Before the real world testing was conducted the whole area was modelled in the simulation environment and the TDC module was first verified in this environment. Both simulation software AIMSUN (Fig. 7) and VISSIM were used for the two developed types of the TDC system – the TDC-SP (Traffic demand control – signal plan) and TDC-A (Traffic demand control – adaptive) in the off-line mode [1, 6].

Fig. 7. Illustration of AIMSUN simulation [own study]

The simulation was formed especially to find out specific reliable parameters, that are going to lead to optimal result. These parameters are much easier to tune in simulation environment then in the real traffic flow. In the real traffic there is also the possibility to work with the parameters and algorithm, but the elementary test in laboratory conditions should ensure that the start of the test is not going to cause any traffic complications. In the simulation we created a model, that is the most similar to reality to reach the precise picture of the real traffic flow. The net was made based on satellite images and actual and new control system was created by floating chart diagrams or programmed in mathematical environment linkable with simulation.

For real control of traffic in the area Uherské Hradiště we tested both modes of TDC (signal plan selection and adaptive control) directly at intersection and compared it with model in VISSIM. VISSIM SW allows wide spread of evaluation. It is possible to define arbitrary amount of detectors and measured parameters. The system allows to generate not just traffic volume and occupancy, but also travel time or delay, that is in simulation environment perfectly precise. It is possible to monitor the simulation process, actual stages in time scale and protocol with variables from detectors, eventually from controllers – see Fig. 8.

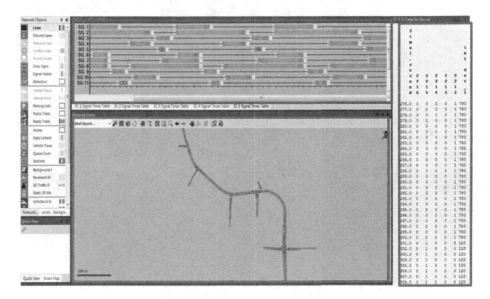

Fig. 8. Simulation process [own study]

The simulation outputs are exported in a text file. For a comparative purpose we used variables of delay and travel time. The most appropriate criterion was to evaluate either the whole area or the entrances from the secondary roads. The main flow cannot be used for evaluation as it is coordinated, so the transit duration is constant.

The traffic scheme of the testing is in the following Fig. 9. Basically every single module is connected to traffic centre: all intersections, adaptive control module and web client. These connections and information transfer works mutually [4].

As was mentioned above, all the intersections were equipped with universal communication unit. In case of demand, there were installed other detectors to cover needs of test. For example to add videodetectors to intersections, where inductive loops were missing. This kind of detection has also an advantage, that it is possible to detect the column length, so it is possible to use it like a strategic detector.

In the test also the ELS Area traffic centre was integrated. This traffic centre is able to display the driving data in the user interface, e.g. the current speed, depress of the

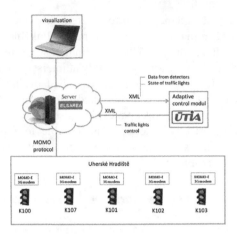

Fig. 9. Traffic schema pro testing of TDC and module C2X [own study]

brake/gas pedal, left/right indicator. In the traffic system ELS Area the virtual detectors were plotted with the possibility to display their data in the side bar. An example is in the Fig. 10.

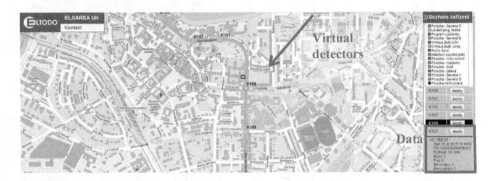

Fig. 10. Print screen of visualization in control centre with virtual detectors [own study]

The outputs from the simulation were in a text file, the main evaluated optional parameters were the travel time and delay. To monitor the traffic changes the situation on the side communications was monitored as in the time dependent control the main direction is the preferred one. Therefore the goal was to use spare time from the main stream to improve the situation on the side communications. That leads to improvement in whole network [3].

The application of the TDC module has proven the advantages primarily for the control of small urban areas resulting in the stabilization of the side directions and improving their traffic parameters significantly. The results concretely are global improvement of travel time and reduction of delay. Travel time in important entrances to the area decreased in average about 11 %, delay decreased about 10 %. The best result was approximately 40 % reduction of waiting time before the southernmost intersection on the main street in direction into the city centre. Another important result is reduction 18 % of waiting time before the northernmost intersection in the side road [3].

5 Conclusion

With today's wide spectre of available sensors and technologies, there are many new applications and improvements accessible for the traffic control. In our project we have developed the universal control unit enabling the interlacing of technological-telematic components in new applications in higher levels for city control with the system ElsArea®. In the testing we used the Traffic Dependent Control Module and the C2X systems. We have started also with the environment data monitoring that may be in future incorporated in these systems. The system ElsArea® is also ready for additional functions like guidance to parking places, informing the drivers, detection incidents, rerouting the traffic when additional information are used as the input, etc. We have tested and proved the benefits of the usage of the universal control unit in cooperation with a new traffic control method and with the C2X module enabling the communication among the vehicle, controller and traffic centre. In the same area the environmental pollutant data were collected to enable future incorporation of the traffic control results with these kind of data helping thus to decrease the traffic negative impacts [1, 6].

Acknowledgements. The research was supported by the Technology Agency of the Czech Republic under the research project Universal control unit UNIR (TA02031360) and the research project Possibilities of influencing the negative impact of transport on the urban environment through innovative sensor networks (TA04031418).

References

1. Tichý, T., et al: Závěrečná zpráva a zpráva o postupu prací a dosažených výsledcích za rok 2014. Číslo projektu: TA01030603. Název projektu: Nové metody pro řízení dopravy v kongescích v intravilánu. ELTODO, a.s (2015)
2. Tichý, T., Cikhardtová, K.: Testování řízení v oblasti prostřednictvím různých zdrojů dat. Konference Inteligentní doprava, Praha (2015)
3. Přikryl, J., et al.: Comparison by simulation of different approaches to urban traffic control. In: Mikulski J. (ed) Telematics in the Transport Environment, CCIS 329. Springer, Heidelberg (2012)
4. Joint work: Highway Capacity Manual 2000. Transportation Research Board, National Research Council, USA (2001)
5. Bělinová, Z., et al.: Smarter traffic control for middle-sized cities using adaptive algorithm, In: Smart Cities Symposium Prague 2015, pp. 1–4. ISBN 978-1-4673-6727-1. IEEE Catalog Number CFP15C83-ART
6. Tichý, T., et al.: Závěrečná zpráva a zpráva o postupu prací a dosažených výsledcích za rok 2015. Číslo a název projektu: TA02031360, Universální řídicí jednotka. ELTODO (2016)

Decision Support System in Marine Navigation

Zbigniew Pietrzykowski[✉] and Piotr Wołejsza

Faculty of Navigation, Maritime University of Szczecin, Szczecin, Poland
{z.pietrzykowski,p.wolejsza}@am.szczecin.pl

Abstract. The article presents basic functionalities of a navigational decision support system used in collision situations, illustrated with example performance of the NAVDEC system. Other similar systems have been characterized. The authors draw attention to the need to introduce legal regulations for the construction and operation of navigational decision support systems. The authors also attempt to outline directions of their development in the years to come in view of expected increase in number of unmanned and autonomous ships.

Keywords: Marine navigation · Decision support · Safety of navigation

1 Introduction

Increasing quantity and scope of information available on board a ship makes it easier for navigators to make a more complete situation analysis and assessment. At the same time, more data to be taken into account, with limited human perception, may hamper making decisions. Wrong decisions may result from fatigue or stress that lead to, inter alia, reduced mental toughness, lower personal safety, reduced self-esteem and situational awareness, disorders of leadership qualities, more time needed to make a decision and improper decisions [4].

Rapid progress in information and communication technologies facilitates finding new solutions. Decision support systems, functioning as an assistant/advisor, are a frequently proposed solution today. These systems analyze and assess a current situation and generate a solution recommended to the navigator. One should expect that the witnessed transformation of navigational information systems into decision support systems will allow us to reduce mistakes made by the humans, and consequently, reduce the number of accidents at sea.

On top of that, decision support systems may increase the efficiency of transport, as the proposed solutions are cost-effective, i.e. based on economic criteria. If we bear in mind sometimes huge losses caused by marine accidents, we can say these systems are important for assuring the safety of personnel, ship, cargo and environment.

This trend is also present in other modes of transport, including road transport. New solutions are dedicated to vehicle operators/drivers and traffic control centres. The scope and tools of decision support are obviously dependent on the specifics of transport mode, such as the characteristics of the operated vehicles and available infrastructure. The scale of applications is also essential, resulting from the purpose and availability of a given

© Springer International Publishing AG 2016
J. Mikulski (Ed.): TST 2016, CCIS 640, pp. 462–474, 2016.
DOI: 10.1007/978-3-319-49646-7_39

means of transport. It is particularly true for road transport, where solutions, so called assistants, are increasingly used for monitoring vehicle movement and, to some extent, taking over the control of the vehicle. The latter illustrates another trend, witnessed especially in air transport – development of autonomous vehicles. Unmanned remotely operated craft may be considered as a transitory stage. In both cases, for autonomous vehicles in particular, a decision support system seems necessary.

Traffic management systems also play a vital role [10]. Their number and areas of coverage are continually on the rise. In road transport, for instance, intelligent transport systems (ITS) have been implemented. Similar solutions have been used and systematically developed in shipping and aviation, known, respectively, as Vessel Traffic Service systems (VTS) and air traffic control (ATC) systems. It seems justified to employ decision support functions in these areas – generating solutions within traffic organization and management. It should be emphasized, however, that there are differences concerning the types and ranges of functionalities of decision support systems for sea-going vessels, airships or road vehicles as well as relevant traffic control centres.

The realization of decision support system requires various telematic solutions for the acquisition of information from different sources, data processing and transmission to traffic participants. Besides, these systems increasingly use methods and tools of knowledge engineering, including methods and tools of artificial intelligence: artificial neural networks, fuzzy logic, evolutionary algorithms, expert systems, approximate sets, knowledge bases etc. These tools allow us to provide technical systems, more specifically IT systems, with characteristics and abilities attributed to human intelligence, such as adaptation to substantial and unexpected changes, learning, autonomy and complexity [11].

As the indicated issues have a wide range, we will focus in this article on decision support systems for the navigator – operator of a sea-going vessel.

2 Areas of Decision Support at Sea

2.1 Standard and Emergency Situations at Sea

Making decisions in shipping can be considered from various perspectives: shipowner's, traffic service operators' controlling vessel traffic in approach channels and port waters, navigator's steering a ship and other traffic participants.

In the shipping context, decisions are related to the realization of specific transport tasks, including loading, voyage execution and unloading, with the vessel used for the carriage. Ship's operation, requiring constant maintenance activities, may be divided into standard (routine) and emergency situations. If we narrow down our considerations to voyage execution phase, decision making will cover various areas under the two categories of situations:

- standard situations:
  - open sea navigation – avoiding collisions and groundings, weather routing,
  - approach channel navigation – pilotage,
  - port manoeuvres – un/berthing, towing, turning;

- emergency situations [6]:
  - close quarters;
  - accidents involving personnel on board;
  - ship damage;
  - pollution;
  - assistance rendered to ships in distress;
  - other.

As far as information systems are concerned, decision support consists mainly in the automation of acquisition, processing and presentation of information. Decision support systems are expected to generate acceptable and optimal solutions, justify them and control process/es execution, using methods and tools for situation analysis and assessment, simulations and optimization. Such tasks are implemented by various methods and tools, including those for data integration, identification of dangerous situations and tools for communication with the system user.

2.2   Decision Support in Ship Conduct Process

The basic aim of navigation is the vessel's efficient and safe passage on an assumed trajectory. From this vantage point we can say that a navigational decision support system has to implement two basic tasks: conduct the vessel on an assumed trajectory and avoid collisions. The third phase to complete the voyage is to berth/unberth, which can include precise ship handling via dynamic positioning (DP), docking systems or pilot navigation systems.

The passage of the vessel from the port of departure to the port of destination requires acquisition, processing, analysis and use by the navigator of a large amount of information. The capacities of processing by a human are limited, which in turn affects the quality of decisions made. Wrong decisions made by navigators can result in collisions, particularly in areas with high traffic intensity, and consequently adversely affect the safety of people, vessel and the environment.

The decision support system carries out the task of steering a ship along a preset trajectory utilizing the data obtained from an Electronic Chart Display and Information System (ECDIS). Collision avoidance is based on information obtained from Automatic Radar Plotting Aid (ARPA) and Automatic Identification System (AIS) devices. The system, using the above information, works out alternative decisions for a given situation, which are in line with the specific character of an area and its restrictions on the one hand, and with the vessel traffic situation on the other hand. It may be said, therefore, that such a system should be capable of:

- assessing a navigational situation;
- indicating alternative decisions;
- proposing permissible and feasible decisions only;
- enabling the navigator to assess the decisions and to introduce additional assumptions.

Besides, such system should be a dialog system (question – answer – solution – decision), making it possible for the officer to update certain facts concerning the present situation of the ship. Therefore, problems could be solved in untypical situations (reduced visibility (fog signals) or in the night (ships' navigational lights). The major criterion for the system to be satisfied is that it should work in real time.

Berthing operation as well as DP require specific information. The main factor is precise positioning. But even a precise position is not enough if we do not know external factors which may affect our position, such as wind, waves, tide etc. Taking this into account, decision support system should:

- acquire, process, analyse and clearly present, on one display, information from DGNSS, anemometer etc.;
- present on the same display tools available for the operator, i.e. engine/engines, rudder/rudders, bow/stern thrusters etc.;
- "operate" between external factors and internal capabilities, i.e. provide the operator with solutions which enable him to perform tasks.

The major criterion for the system to be satisfied is that it should work in real time or even predict "behaviour" of own ship based on weather conditions and other factors.

3   Decision Support in Collision Situations

3.1   Existing System Functionalities

The navigational decision support system NAVDEC extends features of the ship navigational equipment [8]. It is an on-line system handled by the operator. The system records own ship parameters and the situation around the ship. The registered parameters are used for identification and assessment of the current navigational situation. The system works out and recommends solutions which assure the safe ship conduct. The system co-operates with standard equipment and systems installed on board, e.g. log, gyrocompass, Automatic Radar Plotting Aids (ARPA), Global Navigational Satellite System (GNSS), AIS (Automatic Identification System), Electronic Navigational Chart (ENC). The system performs information functions, similarly to the Electronic Chart Display and Information System (ECDIS). The information functions include presentation of bathymetric data from an electronic chart, an image of surface situation from a tracking radar, positional information from the AIS and GNSS receivers as well as calculation and presentation of targets movement parameters.

The correctness of situation identification and assessment as well as generated recommendations – recommended manoeuvers – depends on the accuracy of data used for these purposes. To improve the data quality used for calculations and presented to the operator. NAVDEC performs the fusion of own ship data and integration of data on targets from alternative sources. In the former case (fusion), measurements from a number of shipboard GNSS receivers are used to estimate more accurate own ship movement parameters, whereas in the latter the system integrates targets movement parameters delivered by alternative sources (ARPA, AIS).

A novel, important, functionality of NAVDEC is that it analyses and assesses the navigational situation in relation to all other or selected targets located within a distance pre-defined by the system operator. Normally this stage in the decision making process is performed by the navigator, because relevant regulations have to be taken into account. With NAVDEC on board, the operator is advised by the system on the identification of an encounter situation in compliance with the Collision Regulations. Such information is especially helpful in heavy traffic.

Apart from intense vessel traffic areas, one-to-one ships encounter situations in the open sea may end in a collision. Such events are known, for example the collision between m/v Gotland Carolina and m/v Conti Harmony in 2009 [13]. In situations qualified as a collision situation, the navigator decides which safe manoeuvre should be performed. He determines what to do (alter course and/or speed) and how to do it (manoeuvre parameters): moment to begin the manoeuvres and values of course and/or speed alteration to pass the target at a preset range considered as safe. NAVDEC is the only tool worldwide capable of performing this function because it incorporates the Collision Regulations, principles of good sea practice, and criteria used by expert navigators. The system calculates and offers on demand alternative solutions. Additionally, the system justifies the proposed manoeuvre. This function refers to all or selected targets.

3.2   New/Planned System Functionalities

NAVDEC system was installed on 12 ships of eight different shipowners. based on the feedback received from end users, the following functionalities are planned to be developed:

1. recommended save ship trajectory for an anticollision manoeuvre according to user defied criteria;
2. alternative ship trajectories calculated using additional (other) optimisation methods;
3. considering bathymetry when planning an anticollision manoeuvre;
4. automatic communication system for exchanging collision related messages and negotiation of anticollision manoeuvre;
5. taking into account wind direction when planning anticollision manoeuvre (version for sailors);
6. planning last minute manoeuvre (situation where a collision cannot be avoided by manoeuvre of one vessel only).

    These features greatly enhance the system capabilities.

4   NAVDEC - Examples of Use

4.1   Multi Ship Encounter Situation

The following figures present examples of NAVDEC use during a pilot test at sea on m/f Wolin, which belongs to Unity Line, the biggest Polish ferry operator. In Fig. 1 our vessel (5) is heading approx. NNW direction. There are 4 dangerous targets on her starboard side. According to the Collision Regulations [1] (1), our vessel is a give-way vessel. That is why the system suggests a new course (2), which enables it to pass all targets within a presumed distance, the closest point of approach (CPA) (7). Data of a selected target are displayed in (3), while the solution how to safely pass the target is in (4). The rosette (6) presents all solutions, which fulfils navigator requirements, for all targets within 8 Nm (in this particular case).

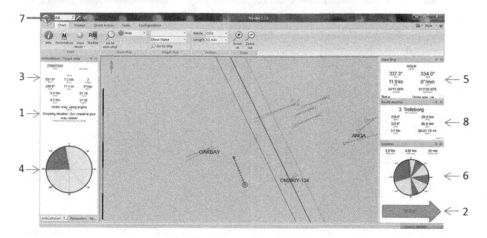

Fig. 1. Screenshot from the NAVDEC pilot test at sea on the m/f Wolin, showing main components [own study] (Color figure online)

NAVDEC uses navigation system data to provide fast and accurate options to the OOW (officer of the watch), including the most important variables to be considered:

1. Classification of encounter situation according to COLREGs ("**crossing situation**", "**head on situation**" or "**overtaking**") and which vessel is "*stand-on*" (has right of way) and which is "**give-way**" (must let the other vessel or vessels pass first),
2. Optimal course to avoid collision,
3. Target data,
4. Solutions for selected target,
5. Own vessel data: destination, distance to go, ETA, etc.,
6. Solutions how to pass all targets at predetermined distance from our vessel,
7. Assumed safe CPA,
8. Planned route.

When own vessel is a give-way vessel in relation to at least one target, NAVDEC displays a compass rosette with solutions (6), with red sectors indicating collision risk and yellow sectors indicating safe courses (respectively black and bright areas on figure) on which the vessel will pass other targets with predetermined CPA or larger distance. Displayed below the rosette is the optimal course requiring smallest deviation from current course: the green arrow (dark on figure) indicates starboard turn, a red arrow, not shown, a port turn.

Figure 2 presents another important functionality of NAVDEC. It is warning function connected with encounter parameters i.e. CPA and TCPA. Navigator can determine which targets will be considered as Dangerous Targets. In this particular situation, CPA limit was set up on 0.8 Nm and TCPA limit is 15 min and encounter parameters in relation to m/v Seagard are 0.6 Nm and 14 min. This is why tringle marking target "Seagard" changed colour to red (the oval on figure).

Fig. 2. Dangerous target [own study] (Color figure online)

In Fig. 3 another functionality of system is presented. After alteration course to starboard by m/f Wolin, present CPA to m/v Seagard is 0.8 Nm and is equal to CPA limit

Fig. 3. Keep course and speed in relation to selected vessel [own study] (Color figure online)

set up by navigator. System still classifies encounter situation according to COLREGs, but rosette with solution changed colour to green (dark on figure). Additionally information "Keep course and speed" appeared. It means that by keeping current parameters we will pass m/v Seagard at least on minimum required CPA.

There is no risk of collision for m/f Wolin in situation presented in Fig. 4. CPA in relation to all targets is bigger than minimum required by navigator. In case it is smaller, then TCPA is already negative, which means that vessels passed by.

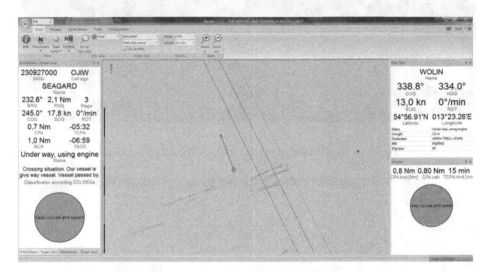

Fig. 4. Keep course and speed [own study] (Color figure online)

In situation presented above, both rosettes with solutions are green (dark on figure), which means that it is safe to keep course and speed until next encounter situation, which will be signalled by red-yellow rosette.

4.2 Multi Ship Encounters – Other Examples

The following two figures present m/f Wolin in three different encounter situations. In Fig. 5, southbound m/f Wolin met m/v Shetland Cement on opposite course. NAVDEC correctly classified that encounter as a head-on situation. In this case both vessels are obliged, according to the COLREGs, to alter course to starboard.

Figure 6 presents how the system utilises specific data from AIS, i.e. navigational status. Rule 18 of COLREGs "Responsibilities between vessels" is fully implemented in the NAVDEC system. It means that not only the positions, courses and speeds (like in ARPA) are taken into account in assessing encounter situation. In this specific case, the Arkona is a vessel engaged in fishing. The present CPA equals 5.7 Nm and is bigger than CPA limit preset by navigator. This is why the individual rosette (on the left) is green, which means that no action is required to pass the selected target at the preset CPA or farther.

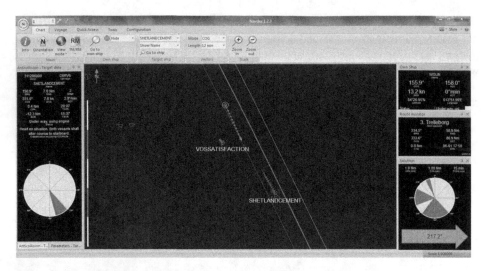

Fig. 5. A head-on situation [own study] (Color figure online)

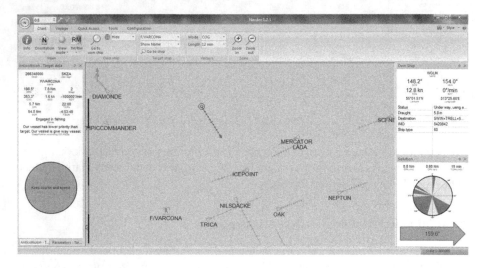

Fig. 6. Encounter situation with a fishing vessel [own study] (Color figure online)

5   Other NDSs on Ships

5.1   Totem Plus Decision Support Tool

The TOTEM DECISION SUPPORT TOOL [12] analyses target data and advises officers what action to take. Target CPA data are taken into account. If required, a "Course
To Steer" is displayed to the officer (Fig. 7). It is based on analysis of data from nearby
vessels, their CPA and TCPA, and the status based on the COLREGS. All the information is calculated automatically and is refreshed after a new message arrives [13].

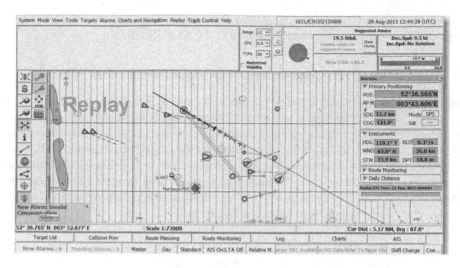

Fig. 7. Screenshot of Decision Support Tool [12] (Color figure online)

To present advice, AIS and ARPA targets within the required distance are constantly analysed. The Alert Radius, with a default value of 12 miles, (high seas) may be modified by the navigator. It also applies to the CPA, which is set by default to 0.4 mile, but can also be set by the operator according to the specific circumstances. Other parameters for instance Minimal Distance, have default values that only the Master can modify. Minimal values to be chosen are limited by specific vessel parameters such as ship's length and turning radius. Table 1 illustrates a system configuration [13].

Table 1. Example minimal values in the Decision Support Tool [11]

No	Parameter	Default value	Who can change
1	Minimum CPA allowed	0.4 mile	Watch officer
2	Alert radius	12 miles	Watch officer
3	Action time	20 min	Master
4	Vessel crossing from port - warning distance	20 min	Master
5	Vessel crossing from port - action distance	3 miles	Master
6	Vessel overtaking - warning distance	2.5 miles	Master
7	Vessel overtaking - action distance	1 mile	Master

In addition to the a/m information, the system displays an alert for "Approaching from Port" situations or "Overtaking" situations. In such cases, no advice is given as the ship is required by the COLREGS to keep her course and speed and the other vessel should take action, too. Once the approaching ship is closer than the defined distance, the system advises the officer to give such ships a warning signal. If the "give-way" ship is closer than the limit set for action - Action Distance - information on a new course to avoid collision will be given in accordance with the COLREGS [12].

5.2   Hyundai Intelligent Collision Avoidance Support System (HiCASS)

The system was developed to verify the model for use as a technological factor of e-navigation [3, 5, 9]. The system utilizes AIS information from vessels and establishes the risk of collision between own and another vessel as well as that between two other ships. The result of assessment, i.e. information on the degree of collision risk, is categorized by ship and by water area. The AIS Pilot plug is used to obtain AIS information from other vessels, which is transferred to the program using wireless communication. Where Wi-Fi is available, the navigator will be capable of using the program in a remote setting. The system display is presented in Fig. 8 [3].

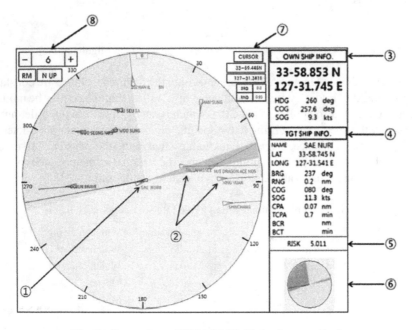

Fig. 8. Screenshot of HiCASS [3] (Color figure online)

The system display is similar to that of the radar and it can verify the degree of danger using a color code. Table 2 presents the functionalities of HiCASS [3].

The system marks vessels with a collision risk of 5.0 and above, which indicates a dangerous situation, with a line for clear identification. It also calculates the degree of risk of collision with any other vessel located around own ship. When risk is below 5.0, the degree of risk is displayed on the screen with a proper color code to indicate the level of risk [3].

Table 2. Functionalities of HiCASS [3]

No	Name	Description
1	Own vessel	Own vessel is always displayed on the center
2	Other vessel	Information of other vessels is collected from AIS
3	Own vessel data	Information of own ship including GPS position, heading, COG, SOG is displayed
4	Other vessel data	Information of other vessel including name, GPS position, relative heading/ distance, COG, SOG, CPA, TCPA, BCR, BCT is displayed
5	Degree of risk	The degree of risk of collision is displayed by the Model
6	Degree of risk around the area	Degree of risk around own ship is displayed through color code (1–4 green, 4–5 yellow, 5–7 red)
7	Cursor information	Displays relative heading/distance of the other vessel from the mouse cursor's position and own ship
8	Monitor calibration	Scale of monitor, direction of stern, rearrangement of other vessel's information

5.3 Challenges

The present technological development opens opportunities for building more advanced navigational systems and extending their applicability. These systems perform information functions and, increasingly, decision support functions. The test results concerning the navigational decision support system NAVDEC and analysis of similar existing or designed systems lead to a number of challenges that emerge before researchers and designers [2, 7].

The complexity and diversity of probable navigational situations impose a large number of factors to be considered before the generation of suggested solutions/decisions. This also refers to navigation in restricted waters, where navigator uses criteria of situation analysis and assessment.

To implement a system with such characteristics as adaptation, learning, autonomy and complexity it seems purposeful to continue and extend the use of artificial intelligence methods and tools.

Existing or designed navigational decision support systems assist the navigator mainly in standard situations. Decision support in emergency situations is often more difficult. This explains the lack or limited scope of use of decision support systems in emergency situations. Relevant solutions basically include procedures (scenarios) for previously defined cases of emergency.

Other challenges concern accuracy, validity, credibility of information, increased availability, reliability and security of the system. This entails the use of various navigational systems and equipment for assuring system redundancy and diversity. Hence, the implementation of such systems requires that relevant performance standards should be defined. The lack of standards concerning the requirements for services offered and failure to assure a specified quality level of the final product may even lead to the reduction of navigation safety.

The development of unmanned and autonomous ships observed particularly in air transport lets us conclude that also the shipping industry will witness a gradual growth of this type of craft. As a result, it may become necessary to build systems of traffic control and management and develop instruments for enforcing relevant regulations. A larger number of unmanned and autonomous vessels may in the future call for regulatory changes, which in the case of maritime transport will affect Collision Regulations, now in force for over 100 years.

References

1. COLREGs Convention on the International Regulations for Preventing Collisions at Sea, International Maritime Organization (1972)
2. Development of an E-Navigation Strategy Implementation Plan: Report on research project in the field of e-navigation, Submitted by Poland, Sub-Committee on Safety Of Navigation, 59th Session Nav 59/Inf.2 (2013)
3. E-Navigation Strategy Implementation Plan: A study on ship operator centred collision prevention and alarm system, Submitted by Korea, Sub-Committee on Navigation, Communications and Search and Rescue, NCSR 2/INF.10 (2015)
4. Gregory, D., Shanahan, P.: The Human Element, a guide to human behaviour in the shipping industry. The Stationery Office (TSO) (2010)
5. Hyundai heavy industries Developed HiCASS (2014)
6. International Convention for the Safety of Life at Sea (SOLAS), International Maritime Organization (1974)
7. Last, D., et al.: Navigation – Current Challenges, Coordinates, vol. XI, issue 3, pp. 8–12, March 2015
8. NAVDEC. www.navdec.com
9. Sangwon, P., et al.: PARK Model and Decision Support System based on Ship Operator's Consciousness, Marine Navigation and Safety of Sea Transportation, pp. 93–98. CRC Press (2015). Edited by Adam Weintrit and Tomasz Neumann, Print ISBN 978-1-138-02857-9
10. Pietrzykowski, Z.: Maritime intelligent transport systems. In: Mikulski, J. (ed.) TST 2010. CCIS, vol. 104, pp. 455–462. Springer, Heidelberg (2010). doi: 10.1007/978-3-642-16472-9_50
11. Pietrzykowski, Z., Borkowski, P., Wołejsza, P.: Marine integrated navigational decision support system. In: Mikulski, J. (ed.) TST 2012. CCIS, vol. 329, pp. 284–292. Springer, Heidelberg (2012). doi:10.1007/978-3-642-34050-5_32
12. Totem decision support tool. www.totemplus.com/DST.html. Accessed 14 Dec 2015
13. Wołejsza, P., Magaj, J.: Analysis of possible avoidance of the collision between m/v Gotland Carolina and m/v Conti Harmony, Annual of Navigation No. 16, Gdynia, pp. 165–172 (2010)

The Shortest Path Problem with Uncertain Information in Transport Networks

Tomasz Neumann[✉]

Faculty of Navigation, Gdynia Maritime University,
Al. Jana Pawła II 3, 31-345 Gdynia, Poland
t.neumann@wn.am.gdynia.pl

Abstract. The purpose of this paper is to find a solution for route planning in a transport networks, where the costs of tracks, factor of safety and travel time are ambiguous. This approach is based on the Dempster-Shafer theory and well known Dijkstra's algorithm. In this approach important are the influencing factors of the mentioned coefficients using uncertain possibilities presented by probability intervals. Based on these intervals the quality intervals of each route can be determined. Applied decision rules can be described by the end user.

Keywords: Transport networks · Dempster–Shafer theory · Dijkstra's algorithm

1 Introduction

Many papers meet a challenge with a route selection problem, navigation in the transport sector, some few examples take uncertainty into account evidence (or lack of knowledge). It is easy to find a few papers in transport study, which take account of the uncertainty by Dempster-Shafer theory. The primary application of citied above theory is the decision making, geographic information systems and statistics. Intelligent Driving recognition with Expert System used Dempster-Shafer theory for recognizing executed by the driver manoeuvres the evidence from the transducer fixed in a carriage. Uncertainty evidence models can been used only for the neighbourhood problem, but coverage of this problem is worldwide. The aim of the study is based information to find a solution of routing problems on uncertainty. Dempster-Shafer theory is a hypothesis of uncertainty that can appraise a specific statement the extent to which support a few sources of different data. In fact, there is an substitute to conventional probability theory, so that the accurate representation of ignorance and combination of evidence. Presented above approach was formerly introduced by Dempster [4] and then continued by Shafer in his publication in 1976, A Mathematical Theory of Evidence [12].

Routing problems in networks are the problem in the context of sequencing and in recent times, they have to receive progressive note. Congruous issues usually take places in the zones of transportation and communications. A schedule problem engages identifying a route from the one point to the other because there are many of optional tracks in miscellaneous halting place of the passage. The cost, time, safety or cost of travel are different for each routes. Theoretically, the method comprises determining the

© Springer International Publishing AG 2016
J. Mikulski (Ed.): TST 2016, CCIS 640, pp. 475–486, 2016.
DOI: 10.1007/978-3-319-49646-7_40

cost of all prospective tracks and the find with minimal expense. In fact, however, the amount of such options are too large to be tested one after another. A traveling salesman problem is a routing problem associated with preferably strong restrictions. Different routing problem emerges when it can to go from one point to another point or a few points, and choose the best track with the at the lowest estimate length, period or cost of many options to reach the desired point. Such acyclic route network problem easily can be solved by job sequencing. A network is defined as a series of points or nodes that are interconnected by links. One way to go from one node to another is called a path. The problem of sequencing may have put some restrictions on it, such as time for each job on each machine, the availability of resources (people, equipment, materials and space), etc. in sequencing problem, the efficiency with respect to a minimum be measured costs, maximize profits, and the elapsed time is minimized. The graph image and the example of costs of borders are given in the Fig. 1. In this hypothetical idea the tract network is illustrated by a graph. Presented graph is given with an ordered pair G: = (V, E) comprising a set V of vertices or nodes together with a set E of edges (paths), which connect two nodes. The task is to reach the N1 node from N3 node in the graph at smallest cost [9].

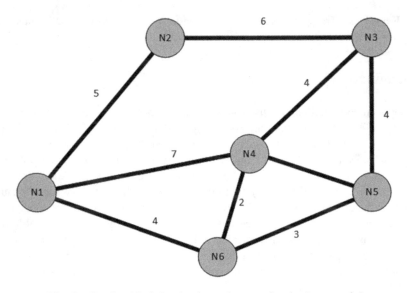

Fig. 1. Graph with defined values of costs of paths [own study]

Conventional Dijkstra's [3] should provide the shortest route in the graph with non-negative edge path costs, but described example include uncertainty. Therefore Dempster-Shafer theory is required, which concern with uncertainty by belief functions. In Dempster-Shafer theory the set $\Omega = A_1, A_2, \ldots, A_n$ of all the eventual conditions of the structure. It can be presented by $P(\Omega)$ the powerset 2^{Ω}.

$$P(\Omega) = 2^{\Omega} = \{\{\}, \{A_1\}, \{A_2\}, \ldots, \{A_1, A_2\}, \ldots, \Omega\} \tag{1}$$

Dempster-Shafer theory designate functions (m) called Basic Belief Assignment on the $P(\Omega)$.

$$m : 2^{\Omega} \to [0, 1] \tag{2}$$

It permits not commonly work out the portions of evidence presented by power set $P(\Omega)$. A basic belief assignment (m) appeases:

$$m(\phi) = 0 \tag{3}$$

$$\sum_{A \in P(\Omega)} m(A) = 1 \tag{4}$$

Belief function $Bel(A)$ for a set A is defined as the sum of all Basic Belief Assignment of subsets of A defined and said that a part of faith B are assigned, must be assigned to other hypothesis, that it means:

$$Bel(A) = \sum_{B | B \subseteq A} m(B) \tag{5}$$

The Dempster-Shafer theory also defines the plausibility $Pl(A)$ as the sum of all the Basic Belief Assignment of sets B that intersects the set of A:

$$Pl(A) = \sum_{B | B \cap A \neq 0} m(B) \tag{6}$$

2 The Problem

Since the entire linear programming model of an abridged interpretation of the problem of the routes of the ship is an unacceptable solution times for a typical daily planning process, a heuristic approach to decide on the hand. Author determined on this approach for its implementation comparatively straight computation, as well as its record of good results with congruous issues to the inherent.

There are a few other algorism such as Dijkstra's algorithm, which is an individual source-single goal shortest path algorithm, the Bellman-Ford algorithm to calculate the shortest path algorithm with a free hand, A* algorithm solves the single pair shortest path problems using a heuristic algorithm and Floyd Warshall algorithm to find all pairs of Johnson-perturbation and the shortest path algorithm to find the shortest path locally. Genetic algorithms are also used to finding shortest path [1]. In this article to calculating author decided to use algorithm presented by Dijkstra.

2.1 Model Input

Contribution to the example record vessel properties, motion report data and digital climate prognosis data: the example will join consumption curves, velocity diminution curves, vessel class, ship wind and weather sea borders, motion statement velocity, maximal permitted speed, motion statement trace data to contain waypoints, their latitude and longitude. On top of it to data related to the motion of the ship it is indispensable to the specification of the surroundings. In particular significant is the specification of the practicable routes between the first point and the last one.

2.2 Dijkstra's Algorithm

For a published source apex (node) in the graph, the algorism discovers the way with smallest cost (i.e. the shortest path) among that vertex and every other ones. It could also be used for discovering the smallest cost way from one vertex to a goal vertex by stoppage the algorism is intended by the smallest way to the goal vertex. For instance, if the apexes of the graph describe the cities and there are given costs of flowing ways distances among pairs of points combined immediately to the road, Dijkstra's algorism can be used to discover the briefest route among one city and all other cities. Consequently, the briefest path algorism is highly used in routing protocols in a web network, in particular the IS-IS and Open Shortest Path First.

3 Route Planning with Uncertain Information Model in Transport Networks

Mainly in the road transportation area the most important data for routing is the traffic on the roads. It can be easily calculate into two states: congestion or not congestion. Thus in this model the investigated two hypotheses in Dempster-Shafer theory are the *Congestion* (CO) and *No Congestion* (NC), so $\Omega = \{CO, NC\}$ and $P(\Omega) = \{\{\}, \{CO\}, \{NC\}, \{CO, NC\}\}$ characterized by the Basic Belief Assignment values of the focal elements $m(CO), m(NC), m(\Omega)$, where $m(\Omega)$ expresses the uncertainty.

On the case of congestion on a particular road costs twice considered by the predetermined cost. The evolving traffic congestion can be caused by several factors. These factors can be calculated as follows: weather, traffic density and closed track. So the basic belief assignment functions are the following: m_1: bad weather, m_2: high vehicle density, m_3: closed lane.

The mathematical theory of evidence deals with function combining information contained in two sets of assignments, subjective expert ratings. This process may be interpreted as a knowledge update. Combining sets results in forming of new subsets of possible hypotheses with new values characterising probability of specific options occurrence. The aforementioned process may continue as long as provided with new propositions. This function is known as Dempster's rule of combination. If more than one factor appears on an edge, then it is possible to cumulate them based on the following formula, where A is the investigated set, B, C are elements of $P(\Omega)$.

This equation is proposed by Dempster:

$$m(C) = \frac{\sum\limits_{A \cap B = C} m_1(A)m_2(B)}{1 - \sum\limits_{A \cap B = \phi} m_1(A)m_2(B)} \tag{7}$$

Combination rules specify how two mass functions, say m_1 and m_2, are fused into one combined belief measure $m_{12} = m_1 \times m_2$ (we here let the binary operator $\times$ denote any rule for mass function combination). Many combination rules have been suggested (several are presented in [2]), and below we briefly discuss the ones we use in this study.

For a given source vertex (node) in the graph, the algorithm finds the path with lowest cost (i.e. the shortest path) between that vertex and every other vertex. It can also be used for finding the shortest cost path from one vertex to a destination vertex by stopping the algorithm is determined by the shortest path to the destination node. For example, if the vertices of the graph represent the city and are the costs of running paths edge distances between pairs of cities connected directly to the road, Dijkstra's algorithm can be used to find the shortest route between one city and all other cities. As a result, the shortest path algorithm is widely used routing protocols in a network.

Short characteristic of Dijsktra algorithm [10] presented is in Fig. 2.

01. The input of the algorithm consists of a weighted directed graph G and a source
   vertex s in G
02. Denote V as the set of all vertices in the graph G.
03. Each edge of the graph is an ordered pair of vertices (u,v)
04. This representing a connection from vertex u to vertex v
05. The set of all edges is denoted E
06. Weights of edges are given by a weight function w: E → [0, ∞)
07. Therefore w(u,v) is the cost of moving directly from vertex u to vertex v
08. The cost of an edge can be thought of as (a generalization of) the distance between
   those two vertices
09. The cost of a path between two vertices is the sum of costs of the edges in that path
10. For a given pair of vertices s and t in V, the algorithm finds the path from s to t with
   lowest cost (i.e. the shortest path)
11. It can also be used for finding costs of shortest paths from a single vertex s to all
   other vertices in the graph.

Fig. 2. Dijkstra algorithm [own study]

4   Route Planning Algorithm with Uncertain Information

In the previous chapter, the decision of the relative interval was simply because the lowest one was worse than the highest value of the other. If there is an overlap between the intervals, then the decision is not easy. When both endpoints of the interval than the end points of the other less: In this case, the routing method may choose fewer values. However, when an interval is the inner part of another interval, the decision is not clear. A possible choice is the comparison of the center points of the intervals. The election rules depend on the human decision: The end user can develop worst-case design, or at best, design or other design choices [6].

So far it looked at the transportation planning problem as a static problem. Of course this is in fact not the case. Uncertainty can through events such as errors in the communication between automated guided vehicles and the system maintained stay reservations, break-down of a mobile unit (engine failure) or failures are caused (for example due to traffic accidents) in the transport network. Uncertainty can also be caused by a change in the transport requests. For example, does the arrival of a new transport request a current plan unworkable.

Uncertainty and especially incidents can be dealt with proactive or active. Proactive methods try to create robust plans, while reactive methods of incidents actually recover they occur. A typical proactive approach is to insert limp in plans, so that, for example, delays have no consequences and new demands can be easily inserted. If nothing unexpected happens these plans take much longer than necessary [13].

For a given source vertex in the graph, the algorithm finds the path with lowest cost between that vertex and every other vertex. It can also be used for finding the shortest cost path from one vertex to a destination vertex by stopping the algorithm is determined by the shortest path to the destination node. For example, if the vertices of the graph represent the city and are the costs of running paths edge distances between pairs of cities connected directly to the road, Dijkstra's algorithm can be used to find the shortest route between one city and all other cities. As a result, the shortest path algorithm is widely used routing protocols in a network.

The proposed scheme is shown in Fig. 3. Provides an overview of the optimal path, the removal of all further edges [11].

In Fig. 4 is presented an algorithm for route planning with uncertain information. The algorithm calculates the least cost intervals for each node of the source node. The main original calculating method on values with uncertainty is generated with Dempster-Shafer theory.

Because of the uncertainty of the road network after disasters, it is possible that the chosen path is blocked, although its reliability is very high and the journey time is short. In such cases, one can do nothing but set the path is blocked. So another option for the adjustability of the path is proposed. This ensures that if the chosen path is blocked, we do not have to circle. This problem is not solved in presented procedure.

When the assessment of the situation undergoes solely a subjective expert rating, the results are only to be obtained in form of linguistic variables. Theories presented show [14] possibility of transforming such values into figures with use of the fuzzy sets theory, a concept created by L.A. Zadeh in the sixties of the 20th century and

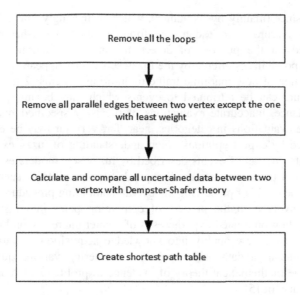

Fig. 3. Diagram of searching new paths [own study]

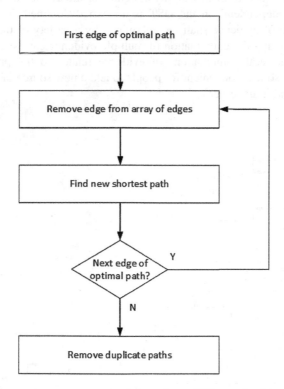

Fig. 4. First steps in proposed extended version to Dijkstra algorithm [own study]

developed ever since (mainly by its author), which increasingly intercedes in various economic issues. According to Zadeh, the aforementioned theory has not been suffi- ciently employed for the purpose of detection analysis of marine units. A more extensive use of possibilities offered by the fuzzy sets theory appears as a necessity for rational construction of new maritime traffic monitoring systems [7, 8].

A fuzzy nature can be attributed to events which may be interpreted in fuzzy manner, for instance, inaccurate evaluations of precisely specified distances to any point. Subjective evaluations in categories: near, far, very far may be expressed with fuzzy sets defined by expert opinions. Such understanding of fuzzy events is natural and common. Introduction of events described by fuzzy sets moderates the manner in which the results of processing are used, expands the versatility of such approach, as well as changes the mode of perceiving the overall combining procedure. Deduction of specific events involved in the process of combining pales into insignificance, as obtaining information on related hypotheses is of greater interest. Combining evidence of fuzzy values brings new quality into knowledge acquisition due to the usage of combination results as a data base capable of answering various questions. Other possibilities of the mathematical theory of evidence in problems of transport in navi- gation can be found in [5].

The principle of connection DST allows people to connect two independent sources of evidence in one or two basic probability assignments are defined on the same frame. Here, the term "independent" in the DST is not strictly defined. The word simply means that a range of evidence shall be determined by a variety of means.

Figure 5 shows that the combination of multiple evidence can be converted into a double recursion several combinations of evidence relates to the properties of the combination rules, so it is convenient for people to add a new source of evidence to the old system in an arbitrary order.

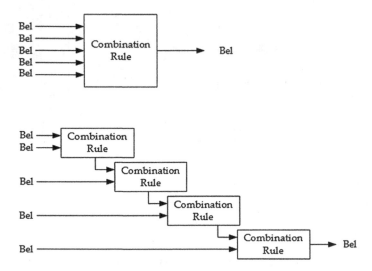

Fig. 5. Transformation of multiple combined evidence [own study]

5   Numerical Example

The Fig. 6 shows an example of the unreal maritime restricted area. It consists of eight obstacles (in the form of islands), 21 turning points and 29 edges. Each edge is described by value of the distance between two vertices.

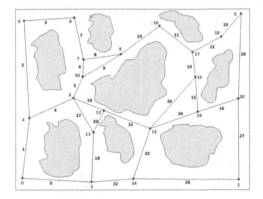

Fig. 6. Scheme of traffic among islands [own study]

All connections are shown in the Table 1. Number of edge corresponds to the edge shown in Fig. 1. Each connection is between initial node and final node. The distance between nodes describes a dimensionless measurement of the distance between the nodes.

For the purposes of demonstration and calculations developed computer application [8] which graphically created schema presented in the article. For middle-class computer all the calculations were done in less than 150 ms. Such a short calculation time can be a prerequisite for further research into the search for alternative paths.

Result of the algorithm is a path with a length equal 1365. The algorithm indicated that the shortest distance between the vertex labelled 0 and 3 leading by edges: 0, 22, 23, 25, 14, 12, 13. The shortest path is presented in Table 2 and in Fig. 7.

Selecting one of several paths is a multi-criteria problem. Conventional multi-criteria decision making (MCDM) techniques were largely non-spatial. Use medium or cumulative effects that are considered appropriate for the entire area into account. In this case, subjective approach is proposed to solve the problem. Based on expert opinion, it is possible to submit the results and select the appropriate transition path. It seems that an appropriate tool for this may be Dempster-Shafer theory. Dempster-Shafer theory (DST) is a promising method to deal with certain problems in data fusion and combination of evidence. Based on statistical techniques for data classification, it is used when the evidence is not sufficient to assign a probability of individual events and declares that are mutually exclusive. Also, both input and output may not be accurate and defined by sets. DST concept is relatively simple, and the technique is easily extensible. In the case of maritime transport, as an international business with a high risk, new evidence will appear and become available once the war,

Table 1. Table of all edges [own study]

Number of edges	Initial vertex	Final vertex	Distance between vertices
0	0	1	296
1	0	4	258
2	4	5	416
3	5	6	216
4	4	9	208
5	9	10	98
6	10	7	73
7	7	6	183
8	7	8	160
9	10	8	187
10	8	16	203
11	16	17	181
12	17	18	139
13	18	3	130
14	17	15	107
15	15	19	148
16	19	20	183
17	9	11	168
18	9	12	137
19	1	11	210
20	11	12	102
21	12	13	212
22	1	14	176
23	14	13	225
24	13	19	216
25	13	15	293
26	14	2	452
27	2	20	343

Table 2. Description of shortest path [own study]

Number of edges	Initial vertex	Final vertex	Distance between vertices
0	0	1	295
22	1	14	176
23	14	13	225
25	13	15	293
14	17	15	107
12	17	18	139
13	18	3	130
			Total 1365

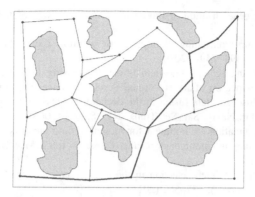

Fig. 7. The shortest path [own study]

diplomatic events or other hazards. DST-based model, which allows incremental addition of knowledge, can satisfy the needs of those conditions. Compared with Bayesian probability theory time zone avoids the necessity of assigning prior probability, and provides intuitive tools to manage uncertain knowledge.

6 Conclusion

The Dijkstra algorithm is well known. It was first published half a century ago. To this day, finding connections between vertices is used. But not always the shortest path is the best. It is to consider various criteria. This paper is an introduction to further research.

Shortest path problems widely exist in real world applications. The paper presents a model to be considered and an algorithm for routing in road network of uncertainty of status information of roads, cost factors and their uncertainty. In presented model uncertainty have the probability values using defined probability of at least and maximum values using Dempster-Shafer theory. Decision rules can be defined for nodes by the end user. The calculations are based on basic belief assignment values. Results of presented paper can be used for travel decisions, in which the decision is a binary, crisp values, intervals and fuzzy numbers.

References

1. Bagheri, H., Ghassemi, H., Dehghanian, A.: Optimizing the seakeeping performance of ship hull forms using genetic algorithm. TransNav Int. J. Mar. Navig. Saf. Sea Trans. **8**(1), 4957 (2014)
2. Bostrom, H.: On evidential combination rules for ensemble classifiers. In: 11th International Conference on Information Fusion. IEEE (2008)
3. Dijkstra, E.W.: A note on two problems in connection with graphs. Numer. Math. **1**, 269–271 (1959)

4. Dempster, A.P.: A generalization of Bayesian inference. J. Roy. Stat. Soc. B **30**, 205–247 (1968)
5. Filipowicz, W.: Fuzzy reasoning algorithms for position fixing. Pomiary Au-tomatyka Kontrola **2010**(12), 1491–1494 (2010)
6. Szucs G.: Route planning with uncertain information using Dempster-Shafer theory. In: International Conference on Management and Service Science, September (2009)
7. Neumann, T.: Multisensor data fusion in the decision process on the bridge of the vessel. TransNav Int. J. Mar. Navig. Saf. Sea Trans. **2**(1), 85–89 (2008)
8. Neumann, T.: A simulation environment for modelling and analysis of the distribution of shore observatory stations - preliminary results. TransNav Int. J. Mar. Navig. Saf. Sea Trans. **5**(4), 555–560 (2011)
9. Neumann, T.: Method of path selection in the graph - case study. TransNav Int. J. Mar. Navig. Saf. Sea Trans. **8**(4), 557–562 (2014)
10. Neumann, T.: Good choice of transit vessel route using Dempster-Shafer theory. In: International Siberian Conference on Control and Communications (SIBCON). IEEE (2015)
11. Neumann, T.: Parameters in the softwares model to choose a better route in the marine traffic. Int. J. Mach. Intell. 454–457 (2015)
12. Shafer, G.: A Mathematical Theory of Evidence. Princeton University Press, Princeton (1976)
13. Zutt, J., et al.: Dealing with uncertainty in operational transport planning. Intell. Infrastruct. Intell. Syst. Control Autom. Sci. Eng. **42**, 349–375 (2010)
14. Zadeh, L.A.: The concept of a linguistic variable and its application to approximate reasoning. Inf. Sci. **8**, 199–249 (1975)

Relationships Between e-Navigation, e-Maritime, e-Shipping and ITS

Adam Weintrit[✉]

The Faculty of Navigation, Gdynia Maritime University,
Al. Jana Pawla II 3, 81-345 Gdynia, Poland
weintrit@am.gdynia.pl

Abstract. In the paper the Author try to explain relationships and connections Intelligent Transport Systems (ITS). Thanks to advances in information technology, free communication between ocean and land is now available and all maritime society carry forward the e-Navigation for maritime accident prevention, transport efficiency, energy conservation and marine environment protection purpose. The e-Navigation is an International Maritime Organization (IMO) led concept based on the harmonization of marine navigation systems and supporting shore services driven by user needs and is establishing the free communication between land, offshore and vessel for traffic control, survey activity, maritime traffic facility management, vessel monitoring and navigation without any limit of time and space.

Keywords: e-Navigation · e-Maritime · e-Shipping · Marine navigation · ITS · Safety of sea transportation · Transport telematics

1 The IMO e-Navigation

The IMO e-Navigation is defined as "the harmonized collection, integration, exchange, presentation and analysis of marine information on board and ashore by electronic means to enhance berth to berth navigation and related services for safety and security at sea and protection of the marine environment" [4, 11][1].

It should be noted that the term e-Navigation is often used in a generic sense by equipment manufacturers and service providers. This claim should be seen as an aspiration, rather than an indication of compliance.

The e-Navigation is a concept to support and improve decision-making through maritime information management and it aims to [4]:

- facilitate the safe and secure navigation of vessels by improved traffic management, and through the promotion of better standards for safe navigation;
- improve the protection of the marine and coastal environment from pollution;
- enable higher efficiency and reduced costs in transport and logistics;
- improve contingency response, and search and rescue services;

---

[1] Since 2006 the Author is a member of IMO's Expert Group on e-Navigation.

© Springer International Publishing AG 2016
J. Mikulski (Ed.): TST 2016, CCIS 640, pp. 487–498, 2016.
DOI: 10.1007/978-3-319-49646-7_41

– enhance management and usability of information onboard and ashore to support effective decision making, and to optimize the level of administrative workload for the mariner.

The e-Navigation aims to provide digital information for the benefit of maritime safety, security and protection of the environment, reducing the administrative burden and increasing the efficiency of maritime trade and transport [3].

The work conducted by the IMO during the last years lead to the identification of specific user needs and potential e-Navigation solutions. The e-Navigation Strategy Implementation Plan (SIP), which was approved in 2014, contains a list of tasks required to be conducted in order to address 5 prioritized e-Navigation solutions, namely [14]:

– improved, harmonized and user-friendly bridge design;
– means for standardized and automated reporting;
– improved reliability, resilience and integrity of bridge equipment and navigation information;
– integration and presentation of available information in graphical displays received via communication equipment; and
– improved Communication of VTS Service Portfolio (not limited to VTS stations).

It is expected that these tasks, when completed during the period 2015–2020, should provide the industry with harmonized information in order to start designing products and services to meet the e-Navigation solutions.

The ultimate goal of e-Navigation is to integrate ship borne and land based technology on a so far unseen level. e-Navigation is meant to integrate existing and new electronic navigational tools (ship and shore based) into one comprehensive system that will contribute to enhanced navigational safety and security while reducing the workload of the mariner (navigator) [1, 12]. The bridge between those two domains will be broadband communication technology which is about to arrive in regular commercial shipping within the next years to come. The constituting element of this integration is a common maritime data model. The existing concept of the Geospatial Information Registry can be adapted to the enhanced scope of a future Marine Information Registry covering additional maritime domains by expansion, amendment and moderate rearrangement. Though the basic philosophy of the IHO S-100 Registry prevails, virtual barriers for maritime stakeholders to associate with the Registry concept must be lowered by all means. This includes options to adopt existing register structures including identifier systems and stewardship for selected areas and elements of additional maritime domains in contrast to the possibly daunting overall third party ownership for a wide scientific field by potential contributors. Besides the recognized international organizations like, IALA, IMO and IHO who are currently discussing the further steps in e-Navigation, a grass root movement may take place with several stakeholders involved populating the Marine Information Registry. Such a grass root movement would truly demonstrate that e-Navigation has been understood and accepted. To allow for the orderly development of that stage of e-Navigation in accordance with the IMO defined goals and aspirations of e-Navigation, it would be required to activate the appropriate IMO instruments already in place to define elementary principles and structure of the Marine Information Registry, to assign roles and responsibilities amongst international organizations and stakeholders, and thereby facilitate the main pillar of e-Navigation, its "grout", namely the Common Maritime Data Structure (CDMS) [5] (Fig. 1).

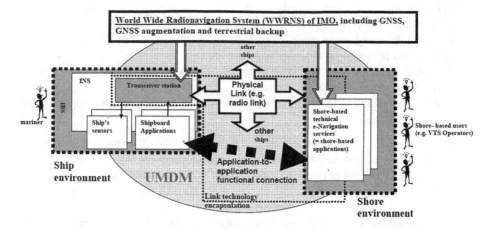

Fig. 1. Conceptual, e-Navigation compliant architecture overview [4, 5, 13]

In e-Navigation it has been decided to develop the Common Maritime Data Structure (CMDS) in the IHO S-100 format. IHO S-100 is based on the ISO 19100 series of geographic information standards. While ISO 19100 is mainly intended for geospatial data, it is believed that S-100 will also be able to handle, e.g., data integration standards such as the IEC 61162-1. The use of GIS (Geographic Information System) has seen unprecedented growth in the last twenty years. With more and more sophisticated, powerful technology becoming cheaper and system memories expanding, which means that we can handle much bigger volumes of data. We can say that GIS is in a golden age. It was once the preserve of the cartographer or surveyor - recently GIS has become a core part of modern sciences and technologies.

2 The EU e-Maritime

Maritime transport is a major economical contributor in the EU as well as a necessary component for the facilitation of international and interregional trade on which the European economy is strongly dependent. The EU e-Maritime initiative [9, 11], is seen as a cornerstone for the achievement of the strategic goals of the EU Maritime Transport Strategy 2018 and related policies, recognising the critical role of ICT for productivity and innovation, and anticipating a new era of e-Business solutions, based on integrated ICT systems and tools.

Whereas "e-Maritime" stands for internet based interactions between all the different stakeholders in the maritime sector, the EU e-Maritime initiative is aimed at supporting the development of European capabilities, strategies and policies facilitating the adoption of upgraded e-Maritime solutions in support of an efficient and sustainable waterborne transport system fully integrated in the overall European transport system.

Upgraded e-Maritime solutions should facilitate decision making and information exchange between different stakeholder groups involved in [9, 11]:

1. Improving the safety and security of maritime transport services and assets and environmental protection;
2. Increasing the competitiveness of the EU maritime transport industry and strengthening the EU presence on the international scene;
3. Integrating sustainable waterborne transport services into efficient door-to-door transport services in Europe and beyond;
4. Reinforcing the human factor particularly supporting competence development and welfare for seafarers.

The ultimate goal for the EU e-Maritime initiative is to make maritime transport safer, more secure, more environmentally friendly and more competitive by improving knowledge, facilitating business networking, and dealing with externalities. A specific focus for e-Maritime is increasing the efficiency of compliance to various reporting regulations.

The scope of the EU e-Maritime initiative illustrated in Fig. 2 encompasses the following domains [6, 9, 11]:

1. Improved Administration Domain Applications:
   a. Common Reporting Interface including dynamic integration with Single Windows;
   b. Integrated Maritime Surveillance for cargo and ship movements facilitating EU and national administrations to collaborate in safety, security and environmental risk management;
2. Improved Business Domain Applications:
   a. Improved Shipping Operations;
   b. Improved Port Operations;
   c. Integration into Logistic chains;
   d. Promotion of seafaring profession.

The e-Maritime is aimed at supporting the development of maritime transport in Europe through the development of a framework that will be based on the latest information, communication, and surveillance technologies. In line with the EU transport policy objectives, e-Maritime solutions must offer a complete approach that extends beyond pure transport services addressing logistics, customs, border control, environmental and fishing control operations. The e-Maritime must be considered in its broadest sense as it promises to provide interoperability between all maritime administrative functions, with important applications in commercial activities [6].

Europe's e-Maritime focuses primarily on the shore-based facilitation and on the development of electronic technology, processes and services to facilitate the flow of goods over sea – and consequently the ships that carry these goods – to, from and around Europe. The European Commission intends to develop applications for administrations, ship operations, harbours/ports/terminals, transport logistics and improving life at sea and promoting seagoing [3].

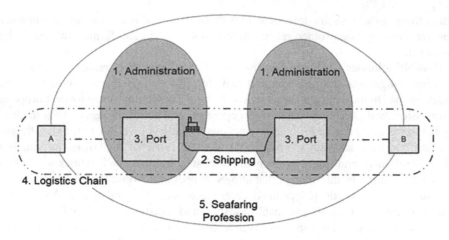

Fig. 2. Integrated view of e-Maritime domains [6]

If the main aim of e-Navigation is to enhance the navigation capabilities of a ship without compromising its efficiency, e-Maritime aims to increase its profitability without compromising its safety. Due to the cooperation of the European partners involved, close coordination had been established between the two initiatives. The EU e-Maritime initiative supported the deployment of e-Navigation services in Europe, while e-Navigation provided a global perspective for the EU initiative. The EU's e-Maritime and IMO's e-Navigation both make use of the same electronic technology, processes and service, and the European Commission wants to make use of those being developed by IMO for e-Navigation wherever possible in the e-Maritime concept development [3].

In summary the e-Maritime initiative aims at optimizing maritime related processes and reducing the administrative work. This will be done by identifying existing practices and regulations and by proposing improvements and simplifications deriving from use of electronic systems and information.

2.1 e-Maritime and ICT

The e-Maritime concept aims at promoting the competitiveness of the European maritime transport sector and a more efficient use of resources through better use of Information and Communication Technology (ICT) tools [2].

The comprehensive use of electronic information and the Internet are changing the world. In maritime transport and transport in general, notifications, declarations, certifications, requests and service orders are increasingly submitted, managed and stored in electronic rather than paper format. Modern ICT systems provide undeniable benefits that are not allowed by paper based information as automated information verification and analysis, processing of data and optimisation routines, easy sharing of information already submitted or stored and so on. However, many of the current processes and regulations, even if electronic, are still based on procedures established for paper transactions decades ago. For example, notifications are still required to be submitted at certain times, as when using telexes and telefaxes, even if this information has already been received by some other related authorities in a digital format and could be shared [2].

Possible solutions to allow re-use of data already submitted, remove unnecessary reporting obligations and optimise port and ship processes will be discussed in the context of the Digital Transport and Logistic Forum. The Forum, set up by the European Commission, will gather experts, business operators and policy makers in order to identify the needs and prepare for common actions at the EU level for improving freight transport and logistics with more efficient use and reuse of digitalised information currently produced and stored by many different stakeholders. Such actions would aim to improve sharing of information that allows shippers to choose the transport service most suited to their needs, reduce the time and resources absorbed by compliance with administrative requirements and enable transport and logistic service providers to optimise the management of transport assets in real-time, thus facilitating the establishment of environmentally efficient transport and logistic services for all users [2].

2.2   Challenges to Be Addressed by e-Maritime

The following challenges are to be addressed by e-Maritime [6]:

1. The simplification and automation of message exchanges between Maritime Administrations and maritime operators to achieve quantum improvements in maritime safety, security, customs control and environmental protection;
2. The facilitation of commercial transactions in the maritime industry, including the transformation of intermodal networks into efficient open networks with risks distributed amongst operational participants.

The e-Maritime aims to achieve standardisation, security and interoperability of information exchanges between Administrations and maritime operators in Europe. The freeing up of information exchanges arising from e-Maritime will result in a simplification and automation of messages, resulting in real-time digital information becoming available to Maritime Administrations – enabling them to improve their safety, security, customs controls and environmental protection functions. It will also enable them to convey helpful information readily and selectively to maritime operators. Similarly, e-Maritime will reduce the workload on ships' personnel through extensive automation of message exchanges between themselves and administrations.

It is expected that the ICT infrastructure that will be developed to facilitate information exchanges between Maritime Administrations and maritime operators will be available for use by operators to facilitate commercial transactions. The scope for such transactions is virtually limitless. One application would the transformation of intermodal networks into efficient open networks with risks distributed amongst operational participants and without having to incur the major costs of Information and Communication Technology (ICT) infrastructure for each network.

2.3   Objectives of EU Initiative

The EU e-Maritime aims to promote coherent, transparent, efficient and simplified solutions based on advanced information technologies. This would allow reaching the following three policy objectives [2]:

– Improving the safety and security of maritime transport services and assets and environmental protection: Port and ship security and safety increasingly require integrated surveillance, monitoring and control systems, incorporating adequate 'intelligence' for proactive, remedial and cross-border operations;
– Increasing the competitiveness of the EU maritime transport and logistics industry: Improved utilisation of advanced ICT will lead to innovation regarding the quality of shipping services and will facilitate reduction of operational costs and increased competitiveness of the sector. At the same time, the performance of the whole EU transport system can be improved by better integrating waterborne transport into efficient door-to-door transport services in Europe and beyond;
– Reinforcing the human factor: the EU seafaring and maritime professions experience a serious shortage of qualified people. Young people do not go to the sea as they used to. An important factor is the lack of continuing professional education offered to the mariners in a flexible manner at sea and ashore, as well as difficult reconciliation of family life and working life. The e-Maritime solutions can support competence development (improved long-distance training) and improve welfare for seafarers (access to long-distance health services; connectivity with families; etc.).

3   Commonalities

Based on the above mentioned similarities and differences between e-Navigation and e-Maritime concepts, there are some important questions that could be raised by the experts [3]:

– What could be the common synergies and potential benefits for e-Navigation and e-Maritime?
– Should the EU consider adopting e-Navigation solutions at an early stage?
– What support could be given from EU to facilitate the implementation of e-Navigation:
  – using IHO S-100 as the common Data Standard for information exchange?
  – leading the development on automated ship reporting (further develop Safe-SeaNet/Single Window)?
  – further development of e-Navigation shore based Maritime Service Portfolio (digital information services to ships such as update of chart, weather, port, safety etc.)?
  – initializing EU Research and Development (R&D) projects on developing viable and functional e-Navigation/e-Maritime models?
  – As the two projects are of similar nature, it is important that close cooperation is maintained between IMO and EU. Both bodies are very important for the benefit

for safety at sea, optimizing maritime related processes and reducing the administrative burden.

4   e-Shipping

The term shipping originally referred to transport by sea, but is extended in American English to refer to transport by land or air (International English: "carriage") as well. "Logistics", a term borrowed from the military environment, is also fashionably used in the same sense.

Nowadays discussion examined some of the factors that influence the regulatory decisions in international shipping and addressed some of the key challenges the shipping industry is facing today in relation to the use of cyber-physical systems on board ships.

We should look at the impact of cyber-physical systems on board ships and how these systems affect navigation, seafarers, safety and security of the vessels. The e-Shipping should be 'user needs led' rather than led by technologists or regulators. Furthermore, it should be added that the success of "e-Navigation" will rely heavily upon the proper involvement of all parties concerned and in particular the seafarers throughout its development and implementation. Additionally, aim of "e-Shipping" is not to replace the seafarers on board vessels, but to assist seafarers in taking more informative decisions thus making the ships safer and more efficient.

Concluding, despite the advanced technological developments, "e-Shipping" has not yet matured and further studies are required especially with regards to maintaining the cyber security which is vital for the ship and the port facilities.

4.1   On-Line Shipping

In the other hand there are quite other meaning for "e-Shipping". The e-Shipping, which is also called "online shipping", is the ordering process of a transport service (usually parcel) made entirely on the Internet. The packing slip is generated instantly online and the collection order of the shipment is sent directly to the shipping department of the carrier. The first companies that have developed this service appeared in the late 1990s in the United States. Since then, the number of e-Shipping offers increased strongly in Anglo-Saxon countries in particular through the development of electronic commerce.

Most major integrators (UPS, DHL, FedEx, TNT) have developed their own e-Shipping tool reserved to their business customers.

The e-Shipping has made possible a new profession, that of the comparison and of intermediation in online delivery of benefits. Several e-Shipping companies (e-Shipping sites) facing small businesses or individuals, and developed rapidly in the late 1990s.

The e-Shipping may apply to all freight services (folds, parcels, heavy or bulky items, pallets, containers) and covers all transport stakeholders (express integrators, postal operators, companies urban race, relay point networks, courier, road freight, air or water etc.).

4.2   e-Freight and Intelligent Transport Systems (ITS)

The European Commission in line with one of the main measures of the 2007 Freight Transport Logistics Action Plan, established a roadmap for the development of an integrated electronic application that is capable of following the movement of goods into, out-of and around the European Union.

This concept 'e-Freight' and will operate within and across all freight transport modes. Through e-Freight there will be a paper-free, electronic flow of information associated with the physical flow of goods. The system will allow tracking of freight along its journey across transport modes and automate the exchange of content-related data for regulatory and commercial purposes. A necessary condition for this is that standard interfaces within the various transport modes are in place and their interoperability across modes is assured.

The implementation of a system for the maritime exchange of information from ship to shore, shore to ship and between all stakeholders, using services such as SafeSeaNet, LRIT (Long-range Identification and Tracking) and AIS (Automatic Identification System), will facilitate safer and more expedient navigation and logistics operations, thereby improving maritime transport's integration with other transport modes.

4.3   e-Customs

While all EU Member States have electronic customs systems, they are not interconnected. The European Commission considers that, if customs legislation were simplified, customs processes and procedures streamlined and IT systems converged, traders would save money and time in their business transactions with customs. In addition to improving safety and security checks, this would contribute to the competitiveness of European business.

The European Commission has adopted two proposals to modernise the EU Customs Code and to introduce an electronic, paper-free customs environment in the EU. The result should be to increase the competitiveness of companies doing business in Europe, reduce compliance costs and improve EU safety and security.

The proposal for a Regulation to modernise the Customs Code would simplify legislation and administration procedures both from the point of view of customs authorities and traders. It would:

- simplify the structure and provide for more coherent terminology, with fewer provisions and simpler rules;
- provide for radical reform of customs import and export procedures to reduce their number and make it easier to keep track of goods;
- rationalise the customs guarantee system; and
- extend the use of single authorisations (whereby an authorisation for a procedure issued by one EU Member State would be valid throughout the Community).

5   Intelligent Transportation Systems (ITS)

Developing an European transport system that is sustainable requires best use to be made of existing infrastructure. Intelligent transport systems (ITS) are vital for this. By integrating technologies for information, communications and control, they enable authorities, operators and individual travellers to make better informed and co-ordinated decisions. For example, ITS can enable more effective planning, help travellers and freight distributors to avoid delays and congestion, and increase the productivity of transport operations. In addition, ITS applications can reduce energy use, accidents and environmental damage.

ITS applications cover all modes of transport and provide a vast range of services:

- in the management of road, rail, air, waterborne and urban traffic, including: advanced information for users; traffic control; incident management; navigation, surveillance and guidance; and vehicle safety and control systems;
- in electronic payment and the enforcement of regulations;
- in the management of public transport, freight movements and other fleet applications;
- in planning and policy-making activities.

Intelligent transport systems are also a key enabler of the integration of different transport modes to provide door-to-door transport services.

Intelligent transportation systems (ITS) are advanced applications which, without embodying intelligence as such, aim to provide innovative services relating to different modes of transport and traffic management and enable various users to be better informed and make safer, more coordinated, and 'smarter' use of transport networks.

The concept of intelligent transportation systems, developed since 80s of the last century includes all modes of transport. Relevant development in road transport is the most advanced. Recently waterborne transport, especially maritime transport (and inland shipping as well) has gained more attention in this regard, in connection with the construction and development of maritime intelligent transport systems [8].

5.1   Maritime Information Technology Standard

This section is dedicated to the distribution of information and resources related to ICT standardization within the area of maritime systems and operations. This is mainly related to ships, ports and authorities' operated infrastructure [7].

e – NAVIGATION, e – MARITIME AND MARITIME ITS = MITS

The e-Navigation is an initiative by IMO to develop a strategic vision for e-Navigation, to integrate existing and new navigational tools, in particular electronic tools, in an all-embracing system that will contribute to enhanced navigational safety (with all the positive repercussions this will have on maritime safety overall and environmental protection) while simultaneously reducing the burden on the navigator.

The e-Maritime was coined by the European Commission to describe a more extensive initiative. The EU e-Maritime initiative aims to foster the use of advanced information technologies for working and doing business in the maritime transport sector. This will in principle cover all aspects of maritime transport and trade.

The Fig. 3 illustrates the relationship between the two initiatives and it is quite clear that the one is completely enclosed by the other. Also, developments within the ICT work on e-Navigation has further restricted this work to SOLAS related tasks, i.e., only navigational safety and security. This means that, e.g., Single Window technology mostly is out of scope for e-Navigation.

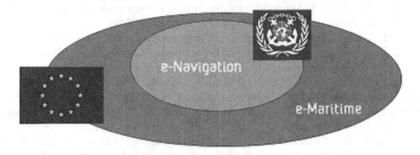

Fig. 3. The relationship between the two initiatives: the IMO's e-Navigation and the EU e-Maritime [7]

Maritime Intelligent Transport Systems [7, 10] is not so well defined, but will for most people be more or less the same as e-Maritime, although on a world-wide level. The MiTS pages will cover what basically is the e-Maritime or Maritime ITS domain.

6   Conclusion

Maritime transport can be said to be the original Intelligent Transport System and developments in this sector should be of interest for ITS research in other modes. A driving force behind this architecture is the international nature of sea transportation and the need to establish standards for information exchanges to further increase efficiency, reduce exhaust gas emissions and fuel consumption and improve the security in the sector. Obviously, reduction of greenhouse gas GHG emissions through operational measures can also reduce costs. This has been realized by the shipping community and the e-Navigation (by International Maritime Organization - IMO) and e-Maritime (by European Union - EU) initiatives testify to this. Both initiatives have identified the information architecture as critical for the future development of the ship transport area. The development of a maritime ITS architecture needs to consider legacy systems, the international nature of shipping, international legislation and standards as well as highly varying quality of service on available communication channels.

Regardless of the body that takes the initiative (IMO, IALA, EU, etc.), all such initiatives like e-Navigation and e-Maritime, that can improve safety at sea are beneficial and should be considered as a desirable.

References

1. Bibik, L., et al.: Vision of the decision support model on board of the vessel with use of the shore based IT tools. TransNav. Int. J. Mar. Navig. Saf. Sea Transp. **2**(3), 255–258 (2008)
2. EC. European Commission – Transport - Transport Modes - Maritime - Mobility and Transport - National Single Windows and e-Maritime. European Commission, Brussels (2015). http://ec.europa.eu/transport/modes/maritime/e-maritime_en.htm. Accessed 15 Dec 2015
3. Hagen, J.R.: eMar project – facilitating information exchange. Decision Dynamics Ltd on behalf of the eMAR Consortium (2013). www.emarproject.eu. Accessed 15 Dec 2015
4. IMO MSC 85/26/Add.1, Annex 20. Strategy for The Development and Implementation of e-Navigation. International Maritime Organization, London (2008)
5. Jonas, M., Oltmann, J.-H.: IMO e-navigation implementation strategy – challenge for data modelling. TransNav Int. J. Mar. Navig. Saf. Sea Transp. **7**(1), 45–49 (2013)
6. Lynch, C.: e-Maritime overview. Sustainable Knowledge Platform for the European Maritime and Logistics Industry (SKEMA) Periodic Study: e-Maritime Task 1 report, 5th Feb. 2010
7. MITS. Web site provided by the Department of Maritime Transport Systems at the Norwegian Marine Technology Research Institute (MARINTEK) (2015). http://www.mits-forum.org/. Accessed 15 Dec 2015
8. Pietrzykowski, Z.: Maritime intelligent transport systems. In: Mikulski, J. (ed.) TST 2010. CCIS, vol. 104, pp. 455–462. Springer, Heidelberg (2010). doi:10.1007/978-3-642-16472-9_50
9. Pipitsoulis, C.: e-Maritime: Concept and Objectives. European Commission, DG Energy and Transport, 26 March 2009
10. Rødseth, O.J.: A maritime ITS architecture for e-Navigation and e-Maritime: supporting environment friendly ship transport. In: IEEE Conference on Intelligent Transportation Systems, October 2011
11. Theologitis, D.: Deployment of e-Maritime systems, maritime transport and ports policy; maritime security. In: Joint Meeting at Short Sea Shipping and Motorways of the Sea, Brussels, 8 July 2009
12. Weintrit, A.: Development of e-Navigation strategy. Advances in Transport Systems Telematics 2. Chapter 9 of Section III: Systems in Maritime Transport. Monograph edited by J. Mikulski. Faculty of Transport, Silesian University of Technology, Katowice (2007)
13. Weintrit, A.: Telematic approach to e-Navigation architecture. In: Mikulski, J. (ed.) TST 2010. CCIS, vol. 104, pp. 1–10. Springer, Heidelberg (2010). doi:10.1007/978-3-642-16472-9_1
14. Weintrit, A.: Prioritized main potential solutions for the e-Navigation concept. TransNav Int. J. Mar. Navig. Saf. Sea Transp. **7**(1), 27–38 (2013)

A Concept of Weather Information System
for City Road Network

Krzysztof Brzozowski[1(✉)], Andrzej Maczyński[1], Artur Ryguła[1],
and Paweł Piwowarczyk[2]

[1] Faculty of Management and Transport, University of Bielsko-Biala, Bielsko-Biała, Poland
{kbrzozowski,amaczynski,arygula}@ath.eu
[2] APM PRO sp. z o.o., Bielsko-Biała, Poland
pawel.piwowarczyk@apm.pl

Abstract. In the article, the authors presented preliminary concept for an information system on road conditions in the city's transport network, which also
makes it possible to carry out short-term weather prediction to provide users with
early warning of impending deterioration of traffic conditions. The system
consists of three layers: the measuring, computing and information. In this paper,
the authors are especially focused on the first layer, which is a fundamental part
of the system.

Keywords: Traffic safety · Smart city · Mobile sensors

1 Introduction

A well-functioning city's transportation network largely determines the quality of life
of its inhabitants. It affects not only the speed and convenience of movement, but it
shapes the image of the city in the eyes of potential investors from various industries.
The proper functioning of the communications network depends not only on available
infrastructure, but also on implemented maintenance and management systems. The
growing importance in this field have different types of ITS solutions related to the area-
controlled traffic lights, giving priority to emergency vehicles and public transport,
informing about congestion, indicating alternative routes and parking occupancy or
cooperating with the passenger information system etc. [3, 6]. It appears that in this
multiplicity of ITS applications in urban areas, there is no dynamic information system
about the road surface condition. In particular, it is about acquiring data which are related
to the variability, in longitudinal road section, of the parameters such as surface temperature, its condition (dry, wet, icy, etc.) and etc. This information would be useful not
only for drivers but also for traffic maintenance services and transportation companies.
Such system would be particularly useful in the autumn and winter, when black ice may
occur locally, which dramatically decreases the safety of road users. Thanks to the
system, traffic maintenance service would gain information about the current road state
in their operating transportation network area. In addition, thanks to the supplementary
weather prediction module, the system could generate warnings related to the expected

© Springer International Publishing AG 2016
J. Mikulski (Ed.): TST 2016, CCIS 640, pp. 499–508, 2016.
DOI: 10.1007/978-3-319-49646-7_42

deterioration of road conditions. Such information would be also useful for companies engaged in transportation tasks in the city, including primarily municipal and taxi companies.

In this paper, the authors present a preliminary concept of dynamic information system on the road surface state in the city. The system would ensure the online information to the road users through websites and mobile applications. In particularly important points of transportation networks, such information would be also displayed on variable message signs. All information together with relevant messages would be also sent to interested institutions and companies i.e. public transport authority and municipal transport company. Weather prediction algorithm, as an elementary part of the proposed system, would also provide messages containing warnings about the expected deterioration of road conditions. Thanks to that, road maintenance services would be able to react on time.

2   The Overall Concept of the System

In the proposed system, information on road conditions in the city's transport network can be divided into three main layers with clearly separated functions (Fig. 1).

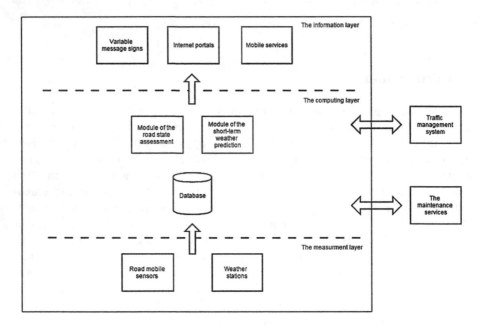

Fig. 1. Layer of the road conditions information system [own study]

The basic flow of data is to be held from the base layer, which in proposed system consists of measurement devices. The main task of the layer is in fact gathering data about the current pavement state in certain critical points of the city's transportation network and along chosen routes. The principal measurements would be executed using mobile road sensors, which could provide information about the pavement condition along the road. In order not to increase the cost of system maintenance, the installation of the measuring system could be carried out on public transport vehicles. In addition, the measurement layer would be responsible for obtaining information from stationary weather stations about air temperature, wind speed and direction, humidity and precipitation intensity recorded at the specific measuring points.

The computing layer would consist of three main modules. The first is a database, in which all measurement data would be collected, together with the coordinates defining the location of their acquisition. The second module, on the basis of the appropriate calculation algorithms, would be responsible for forecasting condition of roads in urban areas not covered by direct mobile measurements, which is possible as far as the weather conditions and traffic volume are known. Therefore, the computing layer of the system must also have direct access to information from the traffic management and winter maintenance services system. The possible solution would be installing mobile road sensors on salt spreader vehicles. The third module of the system would be related to carrying out short-term weather prediction. According to PN-EN 15518 [4], the time period of short-term prediction could be 4 h. Processing data collected within the computing layer will require the use of spatial information system. In particular, such solution will simplify the analysis of time series of the road condition data and the subsequent presentation of the results for the entire transportation network system - from the information layer. A hardware implementation requires a central server, which will be a destination node for data sent from mobile sensors as well as weather stations and at the same time will provide the appropriate computing resources for implemented prediction algorithms.

The communication with the final user in road conditions information system would be held via the information layer. In this layer, depending on the needs, different methods of communication could be implemented. One of the communication channels could be variable message signs placed in the most important nodes of transportation network. The processed data and predictions could be shared with users online through the web server as well as be available in a variety of formats for mobile devices. After the large-scale implementation of I2 V (Infrastructure to the vehicle) communication, information about the road state could also be sent directly to the vehicle. Properly prepared data would be also transmitted directly to the municipal services and transport companies.

3   Measurement Layer

From the point of view of ensuring adequate efficiency and accuracy of the results, the most significant layer of the system is the measuring module. The quality of information collected in the database depends on the configuration and proper selection of the sensors, as well as the routes and trips frequency of vehicles equipped with mobile measuring units. Consequently, the accuracy of short-term weather prediction will be determined by correctness and representativeness of the obtained data. It is assumed that the measurement layer will consist of three basic types of devices:

- mobile road sensors,
- stationary flush-mounted road sensors,
- weather stations.

The basis for the proposed system would be mobile road sensors. Such sensors are mounted on the vehicle and they evaluate the road condition online (Fig. 2). They are currently manufactured by several vendors, inter alia, the German company Lufft [1], US High Sierra Electronic [2] or Finnish Vaisala [7]. Generally, they all qualify the road condition to one of the following classes: dry, wet, snow/ice, black ice, the presence of a deicing substance. In addition, the sensors measure the road temperature, dew point temperature, relative humidity in the layer above the pavement and the water film height. They also define the grip in dimensionless scale. Table 1 summarizes the basic parameters of the MARWIS sensor from the Lufft company. As already mentioned, the optimal solution would be to place sensors on public transport vehicles, especially on buses or trolleybuses. This solution provides data from strictly defined routes and known intervals. The choice of the number of necessary sensors and routes as well as determination of the frequency of trips, (would sensor be used in each course or just some) requires of course carrying out accurate, separate study for each specific city.

Fig. 2. An example of a MARWIS mobile road sensor from Lufft company [1]

The measurement system should be supplemented by flush-mounted road sensors placed in critical point of the transportation network. Firstly these would be the places at which, due to landform or other factors, there is a local occurrence of icing or long-lasting "wet" or "moist" road state. Therefore this would be primarily viaducts, bridges, runs through a forest area, depressions, etc. The indication of city's areas particularly sensitive to weather conditions should not cause any major problems. In the process of their designation, it is sufficient to refer to the experience of institutions responsible for winter maintenance and the road incidents statistics. Stationary surface condition sensors could be also located at selected, most representative points of the network. Data obtained from these sensors, similarly as information from mobile sensors, would serve as input data to algorithms responsible for assessing the roads condition in city's areas not covered by the direct measurements, including short-term road state prediction. Due to the fact, that this sensor opposite to the mobile units, provides continuous information about road state at a given point in the network [5], it can be additionally used as data for verifying the accuracy of weather prediction algorithms.

Table 1. Basic technical parameters of the MARWIS mobile road sensor from Lufft company [1]

Basic technical parameters	
Weight	1.7 kg
Surface conditions	Dry, moist, wet, snow/ice, critical/chemical wet
Friction (grip)	Measuring range: 0...1 (smooth...dry) Sampling rate: 100 Hz
Dew point temperature	
Measuring range	−50...60 °C
Accuracy	±1.5 °C (temperature 0...35 °C)
Road surface temperature	
Measuring range	−40...70 °C
Accuracy	±0.8 °C @ 0 °C
Resolution	0.1 °C
Relative humidity above the road surface	
Measuring range	0...100 %
Water film height	
Measuring range	0...6000 μm
Resolution	0.1 μm

As stationary road sensors, classic units installed in the road such as passive sensor IRS31Pro could be used [1]. According to the authors, it's worth considering the use of optical sensors mounted over the road. Their application would be particularly useful at the stage of study, during which the locations of the aforementioned representative points of the transportation network would be performed. The supplier of optical road sensor may be a number of companies. Such devices are offered for example by Lufft (non-invasive road sensor NIRS31) and Vaisala (DSC111 sensor).

These sensors are easy to install, moving them to another location is not associated with significant costs and most of all they are non-invasive for the road surface. According to the manufacturers, using measuring method allows early detection of the presence of ice crystals, long before road surface becomes slippery. For example, the phenomenon of ice formation through cooling surface at night, is detected by the sensor up to two hours earlier than with conventional road sensor [8]. NIRS31 sensor is shown in Fig. 3 and its main parameters in Table 2.

Fig. 3. Non invasive road sensor NIRS31 from Lufft company [1]

Table 2. Basic technical parameters of the Non Invasive Road Sensor NIRS31 from Lufft company [1]

Basic technical parameters	
Weight	10 kg
Surface conditions	Dry, Damp, Wet, Snow, Ice
Friction (grip)	Measurement range 0…1 (critical…dry)
Layer thickness (water, snow, ice)	
Measuring range	0…2 mm (snow 0…10 mm)
Resolution	0.01 mm
Surface temperature	
Measuring range	−40…70 °C
Accuracy	±0.8 °C
Resolution	0.1 °C

Therefore, the basis for measuring layer would be mobile sensors, which would measure the state of the surface along selected routes of public transport vehicles with additional stationary sensors. In order to provide the necessary data for computing layer, the measurement module must also include meteorological data from the local road weather stations. The primary importance will have parameters such as air temperature, wind speed and direction, humidity and precipitation intensity. Properly located weather stations will help to determine in the city, the field of air temperature and precipitation, which is an essential element in weather prediction and assessment of the roads condition in areas not covered by direct measurements.

4   Test Measurements Using Mobile Road Sensors

As mentioned above, the most important element of measurement layer will be mobile road sensors. These devices have been introduced onto the market relatively recently and therefore their capabilities, advantages and disadvantages are not fully recognized, particularly with regard to continuous working conditions - as in the proposed system. For this reason, the authors decided to carry out a preliminary test of the devices.

The experiments were to confirm or deny the usefulness of sensors for building the measurement layer of the proposed system. Test study comprised sample drives on the express road S69 (bypass of the city of Bielsko-Biala) using a vehicle equipped with a sensor MARWIS (Fig. 4).

Fig. 4. Mobile road sensor mounted on the vehicle [own study]

The test was performed in winter conditions, on 26/01/2015 from 12:30 to 13:45. During the experiment, 20 km long sections of a road were taken into account and test drives were done for each lane in the direction of Żywiec and Cieszyn – see Fig. 5.

Fig. 5. Test road section – express road S69 [own study]

During the test drive, data on surface temperature, surface condition, the friction and the water film height were registered. In addition, GPS sensor was used, in order to determine the vehicle's position and timestamp. The profile of the road is shown in Fig. 6.

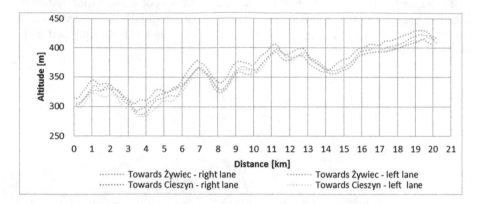

Fig. 6. Profile of the test road section [own study]

Analysis of the surface temperature showed visible value variability in the road longitudinal section (Fig. 7). This variability is particularly important in the case of surface temperatures oscillating around 0 °C and may result in very different road conditions. Test rides also allowed to observe similar road conditions at particular lanes in cross-section of the road. There was a discrepancy of data in the first part of the road section, only on the right lane in the direction of Żywiec. The temperature differed by an average of 2 °C. The time difference between the measurement for the left and right lane was about 30 min.

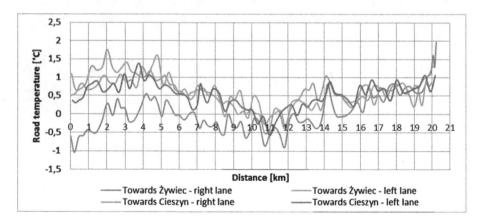

Fig. 7. Road temperature distribution on the test road section [own study]

In addition, as a part of the analysis, the recorded value of the surface temperature and water film height were also combined (Fig. 8). The presence of water film within

certain surface temperature is especially important information for both winter maintenance services as well as traffic management system and the transport network users.

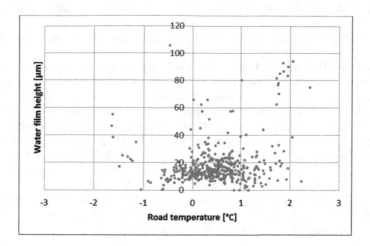

Fig. 8. Road temperature and water film height on the test road section [own study]

5 Conclusion

The concept of the road conditions information system in the city's transport network
presented in this work, is a preliminary idea. Its possible implementation should provide
drivers with access to current data on the roads condition in the area covered by the
system. This access would be possible mainly at the level of web applications but the
most important messages would be also displayed on variable message signs. The information generated by the system enables maintenance services to take appropriate predictive actions related to the deterioration of weather and allows directing services vehicles
in particular areas of the city, where the most difficult conditions prevailed. In winter
time, municipal transport companies will be able to optimize fleet management and also
use information from the system in their own passenger information systems.

The primary measurement tool provided for use in the described system are mobile
road sensors. Hence, it was decided to conduct a test study aimed to confirm the usefulness of these sensors for presented system. Experimental study confirmed that these
devices can be implemented at the level of measurement layer of the system. The use
of this class of devices will ensure the supply of reliable information about road conditions and variation in the road cross and longitudinal section.

The authors of this concept are aware of the need to perform a wide variety of
research and analysis, with varying degrees of detail, which can fully determine the
appropriateness of the system in case of particular topographic conditions, as well as
estimate the costs of its implementation and maintenance. Currently a search for institutions or companies that would be interested in developing the proposed concept and
conducting pilot studies is going on.

References

1. G. Lufft Mess- und Regeltechnik GmbH (2015). http://www.lufft.com/pl/. Accessed 25 Aug 2015
2. High Sierra Electronics, Inc. (2015). http://www.highsierraelectronics.com. Accessed 25 Aug 2015
3. Oskarbski, J., Jamroz, K., Litwin, M.: Intelligent transportation system – advanced management traffic systems (2006). http://www.pkd.org.pl/pliki/referaty/oskarbski,_jamroz,_litwin.pdf. Accessed 25 Aug 2015
4. PN-EN 15518:2011 Winter maintenance equipment. Road weather information systems
5. Ryguła, A., Brzozowski, K., Konior, A.: Utility of information from road weather stations in intelligent transport systems application. In: Mikulski, J. (ed.) TST 2015. CCIS, vol. 531, pp. 57–66. Springer, Heidelberg (2015). doi:10.1007/978-3-319-24577-5_6
6. Ryguła, A., Konior, A.: The integrated driver's warning system for weather hazards. Logistic No. 5/2014. The Institute of Logistics and Warehousing, Poznań (2014)
7. Vaisala (2015). http://www.vaisala.com/en/Pages/default.aspx. Accessed 25 Aug 2015
8. Ząbczyk, K., Pierzchała, K.: Technika laserowa w meteorologii drogowej, Kraków (2007). http://www.telway.pl/download.php?id=9. Accessed 25 Aug 2015

Author Index

Printed in the United States
By Bookmasters